Annem Emine Parlar'a ve rahmetli babam Sait Parlar'ın anısına

To my mother and to the memory of my father

Mahmut Parlar

Interactive Operations Research with Maple

Methods and Models

Birkhäuser
Boston • Basel • Berlin

Mahmut Parlar
DeGroote School of Business
McMaster University
Hamilton, Ontario, L8S 4M4
Canada

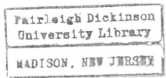
Library of Congress Cataloging-in-Publication Data

Parlar, Mahmut, 1952-
 Interactive operations research with Maple : methods and models/ Mahmut Parlar.
  p. cm.
 Includes bibliographical references and index.
 ISBN 0-8176-4165-3 (alk. paper) – ISBN 3-7643-4165-3 (alk. paper)
  1. Operations research–Data processing. 2. Maple (Computer file) I. Title.

T57.6.P38 2000
658.4'034'0285–dc21   00-031232
                                                          CIP

AMS Subject Classifications: 90C05, 90C15, 90C30, 90C39, 90C40, 90A09

Printed on acid-free paper
©2000 Birkhäuser Boston   *Birkhäuser* ®

ISBN 0-8176-4165-3   SPIN 10754229
ISBN 3-7643-4165-3

Typeset by the author in LaTeX.
Cover design by Joseph Sherman, New Haven, CT.
Printed and bound by Hamilton Printing, Rensselaer, NY.
Printed in the United States of America.

9 8 7 6 5 4 3 2 1

Contents

Preface

Interactive Operations Research with Maple: Methods and Models has two objectives: to provide an accelerated introduction to the computer algebra system Maple and, more importantly, to demonstrate Maple's usefulness in modeling and solving a wide range of operations research (OR) problems.

This book is written in a format that makes it suitable for a one-semester course in operations research, management science, or quantitative methods. A number of students in the departments of operations research, management science, operations management, industrial and systems engineering, applied mathematics and advanced MBA students who are specializing in quantitative methods or operations management will find this text useful. Experienced researchers and practitioners of operations research who wish to acquire a quick overview of how Maple can be useful in solving OR problems will find this an excellent reference.

Maple's mathematical knowledge base now includes calculus, linear algebra, ordinary and partial differential equations, number theory, logic, graph theory, combinatorics, statistics and transform methods. Although Maple's main strength lies in its ability to perform symbolic manipulations, it also has a substantial knowledge of a large number of numerical methods and can plot many different types of attractive-looking two-dimensional and three-dimensional graphs. After almost two decades of continuous improvement of its mathematical capabilities, Maple can now boast a user base of more than 300,000 academics, researchers and students in different areas of mathematics, science and engineering.

My goal in writing this book has been to show that Maple can be immensely useful in the type of modeling work that is common in operations research, management science and related disciplines of operations management and industrial and systems engineering. There are several excellent texts on OR *and* on Maple, but none of them describe the use of Maple in solving OR problems. This book fills the gap and provides the missing link between Maple and its successful use in OR.

Chapter 1 starts with an introduction to operations research. Chapter 2 is an accelerated introduction to Maple that covers the most important commands that are relevant in OR applications. This is followed in Chapter 3 by a treatment of some of the mathematical techniques useful in OR modeling: algebra and calculus, ordinary and partial differential equations, linear algebra, transform methods and probability theory.

Chapter 4 discusses linear programming, one of the oldest and perhaps the most frequently used OR technique. In this chapter Maple's symbolic manipulation capabilities are not used since linear programming problems must be solved numerically. After describing the graphical method to solve linear programming problems (which is simplified by using Maple's capability to plot regions of inequalities) we proceed to a discussion of the simplex method, which is algebraic in nature. Maple's own implementation of the simplex method is also demonstrated and integer linear programming is mentioned briefly.

Chapter 5 presents an exposition of nonlinear programming and many problems in unconstrained and constrained optimization are solved explicitly. Chapter 6 covers dynamic programming, in which Maple can be useful in finding optimal policies for dynamic optimization problems in closed form. After describing a simple numerical example (the "stagecoach problem"), several models with a linear system and quadratic costs in both discrete and continuous time are examined. An exact solution of a workforce planning model is followed by the analysis of some stochastic dynamic programming models, including a gambling problem that gives rise to a myopic policy. The chapter ends with a discussion of stopping time problems. Stochastic processes is presented in Chapter 7 where exponential/Poisson duality, renewal theory and discrete- and continuous-time Markov chains are described in detail. These chapters amply demonstrate the success of Maple in obtaining analytic solutions to many OR models.

Chapters 8 and 9 are concerned with the application of OR methods to inventory management and queueing models, respectively. Here, Maple's symbolic manipulation capabilities are used extensively to obtain optimal solutions in closed-form. For example, in the chapter on inventory models, implied backorder costs and optimal base-stock policy for a dynamic stochastic inventory problem are computed explicitly. In the chapter on queueing models, the operating characteristics of Markovian bulk arrival systems and the transient solutions for some queues are found in closed form. Where symbolic analysis becomes impossible to use, we employ Maple's extensive knowledge of numerical analysis. For example, solution of the exact model for the continuous-review inventory problem with backorders requires intensive numerical computations. Similarly, dynamic optimization of a Markovian queue involves solving differential equations numerically. These computations are performed easily by stringing together a few lines of Maple code.

In Chapter 10, Simulation, Maple's ability to generate random variates from nearly 20 probability distributions is used to find solutions to different problems. For example, simulation is used to compute a definite integral that could not be evaluated in closed form. In this chapter we also describe the famous Monty Hall "car and goats" problem and present its counterintuitive solution by simulation. Simulation is frequently used to analyze the properties of dynamic systems that evolve probabilistically over time. We use Maple's random variate generators to simulate an inventory system with random yield and find the distribution of inventory level. We also simulate a non-Markovian ($G/G/1$) queueing system and compute the average waiting time in the queue.

Each chapter (except Chapter 1) ends with a summary and numerous exercises. A Solutions Manual for the exercises will be provided to instructors who adopt the book for their course.

Internet Resources

Every Maple worksheet displayed in the book has been tested with Maple R5.1 and R6. These worksheets can be downloaded from my personal Web page at the address www.business.mcmaster.ca/msis/profs/parlar/.

Waterloo Maple's Web site www.maplesoft.com/ is always worth a look for announcements of new products, demos and Maple support information.

The following yahoo.com page has links to interesting sites concerned with Maple: dir.yahoo.com/Science/Mathematics/Software/MAPLE.

Readers interested in following the lively Internet discussions on symbolic algebra packages such as Maple and Mathematica may wish to peruse the sci.math.symbolic newsgroup. A recently established newsgroup that carries Maple-specific postings is comp.soft-sys.math.maple.

Acknowledgments

During the last three years of writing this book, I received feedback from many friends and colleagues. First and foremost, I would like to thank Ray Vickson (University of Waterloo) who encouraged me to undertake this project and provided useful comments on early drafts. Martin von Mohrenschildt (McMaster University) saved me many hours of work by quickly directing me to the correct Maple commands to solve some of the problems presented in the book. Elkafi Hassini (University of Waterloo) read many chapters and sent comments. George Wesolowsky (McMaster University), Paul Stephenson (Acadia University) and Mustafa Karakul (University of Toronto) are friends and colleagues who took the time to read parts of the manuscript and comment on its style.

I consider myself fortunate in having acquired my early training in OR under the guidance of two eminent operations researchers: I took many OR courses from Halim Doğrusöz and İzzet Şahin[1] who, in the mid-1970s, were professors in the graduate program in Operations Research and Statistics at the Middle East Technical University, Ankara. I learned from Dr. Doğrusöz that there is more to OR than just high-powered mathematics, and I learned from Dr. Şahin (who directed my master's thesis on renewal theory) that there is more to modeling than just deterministic analysis. I am grateful to both.

My editors Ann Kostant and Tom Grasso at Birkhäuser have always been very encouraging and positive about this project and I thank them for that.

Finally, a big thank you to my wife and children for their understanding during the long process of writing this book.

I would be delighted to hear from readers who have found the Maple worksheets contained in this book useful in their OR-type work.

MAHMUT PARLAR
parlar@mcmaster.ca
www.business.mcmaster.ca/msis/profs/parlar/

May 2000
Hamilton, Canada

---

[1] Deceased. (*Nur içinde yatsın.*)

Interactive Operations Research with Maple

Methods and Models

1

Introduction to Operations Research

1.1   A Brief History

The activity called *operations research* (OR) was developed during World War II when leading British scientists were consulted to improve military strategy against the enemy. A team of physicists, mathematicians, statisticians, engineers and sometimes even psychologists and chemists conducted *research* on military *operations* under the leadership of P. M. S. Blackett.[1] These scientists worked on important military problems such as the deployment of radar, management of bombing and antisubmarine operations and military logistics. Many of these problems were solved by applying the scientific method and using mathematical techniques such as algebra, calculus and probability theory.[2]

The news of the successful application of scientific method to military operations spread quickly, and the American military began establishing its own OR teams, which were instrumental in the successful supply and deployment of U.S. forces in the Pacific.

After the war ended, many of the scientists who worked on military OR were discharged from war duty, and they began applying the scientific method and OR

---

[1] This team was nicknamed "Blackett's Circus."

[2] For example, one interesting problem involved the maximization of the probability of hitting an enemy submarine. It was found that if a depth setting of a charge was made at 35 feet below the surface of water (rather than the usual 100 feet), then the probability of hitting the submarine would be maximized. It was reported that implementing this solution increased the casualty of enemy submarines by about 100%. For a more detailed description of these military-related problems and the history of OR, see Beer [20] and Singh [176].

techniques to the solution of industrial problems. Textbooks began appearing in the 1950s[3], and OR was recognized as a formal academic subject. At the same time many universities started offering courses leading to graduate degrees in OR. Societies such as the Operations Research Society of America, the Institute of Management Sciences (USA),[4] the Canadian Operational Research Society, The Operational Research Society (England), Société Française de Recherche Opérationelle (France) and Centro per la Ricerca Operativa (Italy) were formed. These societies also began publishing their own journals describing the theory and applications of OR to a wide variety of problems.

Since the publication of the early books on OR almost 50 years ago, operations researchers have written thousands of papers describing the new theoretical developments and practical applications of OR in different areas. These articles have appeared in journals such as *Operations Research, Management Science, Interfaces, Mathematics of Operations Research, Naval Research Logistics, IIE Transactions, Manufacturing and Service Operations, European Journal of Operational Research, INFOR* and *Decision Sciences. Interfaces* regularly publishes articles on the real applications of OR. For example, the November-December 1998 issue (Volume 28, Number 6) of *Interfaces* has articles on optimizing army base realignment (using integer linear programming), optimal decisions and the fourth-and-goal conferences at the end of a close football game (using dynamic programming) and trash-flow allocation (using linear programming under uncertainty). In the May-June 1998 issue of the same journal one can read about the application of OR techniques to modeling AIDS, and in the March-April 1998 issue one can learn about the impact of the 1990 Clean Air Act Amendments and an efficient technique for selecting an all-star team.

1.2   The Method of OR

Operations research can be defined as the scientific study of operations in an organization. Its object is to *understand* and *predict* the behavior of an organization (i.e., a system) by observation, experimentation and analysis in order to *improve* the behavior of the organization. This is achieved by first *formulating the problem* facing the organization. For example, a retailer may realize that the average level of inventory for an item is consistently high resulting in high inventory carrying

---

[3]The first formal OR textbook was written by Churchman, Ackoff and Arnoff [45] and appeared in 1958. This was followed by Sasieni, Yaspan and Friedman's book [164], published in 1959. A collection of notes on OR topics (such as probability, search theory, Markov processes, queueing, sequential decision processes and production scheduling) assembled by the OR Center at MIT [139] based on a summer program of OR lectures also appeared in 1959.

[4]In the 1990s, The Operations Research Society of America and The Institute of Management Sciences merged to form INFORMS (Institute for Operations Research and the Management Sciences), which has a total of approximately 12,000 members including some Nobel prize winners; see www.informs.org.

costs. In this case, an OR analyst would formulate the problem while interacting with the decision maker (i.e., the retailer). This would involve specifying the objectives (e.g., cost minimization) and identifying those aspects of the inventory system that can be controlled by the retailer (e.g., magnitude and frequency of orders placed). Problem formulation is perhaps similar to a medical diagnosis since in this stage of an OR study, one usually observes the symptoms (such as high inventory) and then attempts to make a diagnosis of the problem (such as very large order quantities).

The next step involves the *construction of a model* that represents the behavior of the system. In OR, mathematical models usually assume the form $V = f(\mathbf{x}, \mathbf{y})$ where V is the performance measure (value function) of the organization, system or decision maker under study (e.g., average cost, profit per month, etc.). The set $\mathbf{x} = \{x_1, \ldots, x_m\}$ is the set of variables that can be controlled by the decision maker, $\mathbf{y} = \{y_1, \ldots, y_n\}$ is the set of parameters that normally cannot be controlled by the decision maker and f is the relationship between V and $(\mathbf{x}, \mathbf{y})$. Additionally, there may be equality and/or inequality constraints limiting the values that can be assumed by the decision variables.

In the inventory example, the OR analyst attempting to solve the retailer's problem constructs a mathematical model of the inventory system and the ordering policy used by the retailer. As a very simple example, if the annual demand λ for the item is deterministic and uniform, a reasonable model of the average annual cost C may be expressed as $C = K\lambda/Q + \frac{1}{2}hQ$ where K is the (fixed) cost of ordering, Q is the order quantity (the decision variable) and h is the cost of holding inventory per unit per year. Once such a mathematical model is constructed, it becomes relatively easy to use calculus to find the optimal order quantity that would minimize the cost function C. In this example, Q is the decision variable, K, λ, and h are the uncontrollable parameters in the problem and C is the performance measure of the system.

Assuming that the model is properly validated (i.e., that it does represent the system under study reasonably accurately), the OR analyst must *derive a solution* from the model. In the simple case discussed here, in order to optimize (i.e., minimize) the average cost, one can use calculus and find the optimal order quantity. This involves differentiating the cost function, equating the result to zero and solving for the unknown decision variable Q. As we will show in Chapter 8, Inventory Models, the optimal solution Q^* for this model is

$$Q^* = \sqrt{\frac{2\lambda K}{h}}$$

and the minimum average annual cost under this decision is $C^* = \sqrt{2\lambda K h}$.

In the next stage of an OR analysis, an *evaluation of the solution* is performed and the solution is compared to the policy or decision it is expected to replace. At this stage, the analysts can highlight the differences between the current practice and the recommended optimal solution. In the inventory example, the analysis may reveal that under the current policy, the retailer was ordering unnecessarily large amounts (say, $2Q^*$ units instead of the optimal Q^* units) that resulted in high inventories and carrying costs.

In the final stage of an OR study, the solution is *implemented* in order to improve the performance of the system. For example, in the inventory problem, the order decision is revised to Q^* (say, down from the nonoptimal $2Q^*$) to reflect the optimality of the solution found. This assumes that the decision makers have accepted the results of the study and are willing to revise their current policy.

Normally, the five stages of an OR study are initiated in the order given. But very often each stage interacts with others before the termination of the study. For example, the analyst solving the inventory problem may discover that the demand is not quite deterministic but exhibits a fair degree of randomness. In that case, the model construction stage must be revisited and a more accurate model reflecting the randomness of demand must be constructed.

These nontechnical and philosophical—but important—issues were discussed and debated in great detail in the 1950s and 1960s by OR pioneers such as Ackoff and Churchman. For their views on problem formulation, model construction, systems approach and so on, see [3], [4] and [44].

1.3   About This Book

This book is about the computer algebra system Maple and its use in solving OR problems using mathematical models. There are no prerequisites for using the Maple component of the book other than the ability to start the Maple program installed in a computer. All necessary commands for Maple symbolics, numerics and graphics are summarized with examples in Chapter 2. However, in order to understand the OR techniques and applications such as optimization, stochastic processes, inventory models and queueing processes, the reader will need some familiarity with calculus, probability and linear algebra. Thus, it is assumed that the reader has such a background, but in Chapter 3 we do provide a quick review of these important topics that will be relevant in the subsequent chapters.

2
A Quick Tour of Maple

2.1   Introduction

Maple is a mathematical problem-solving environment that combines symbolic, numerical and graphical methods used in different areas of mathematics. This chapter provides an introduction to Maple and presents those aspects of it that can be useful in operations research applications.

Maple has more than 2700 commands and nearly 40 mathematical packages (such as linear algebra, integral transforms, Boolean logic, numerical approximations, financial mathematics and algebraic curves) that can be used to solve problems from basic arithmetic to general relativity theory. As we will see throughout this book, the solution of many challenging operations research problems becomes extremely easy by loading a package and simply combining a few Maple commands.

Maple has an extensive on-line help facility. To obtain help about a command, it is sufficient to type a question mark and then the name of the command. For example, typing `? int` (and pressing the <ENTER> key) provides information on definite and indefinite `int`egration, including a large number of examples where the `int` command is used. Readers who are new to Maple can take a "tour" by clicking on Help and then New User's Tour on the Maple menu bar.

We start this chapter by discussing some aspects of basic symbolic computations and expression manipulations such as simplifications. This is followed by numerical computations such as finding roots, computing definite integrals, solving differential equations. In many cases important insights about a problem can be obtained by a graphical representation of, for example, the implicit curves arising from the necessary conditions of an optimization problem or the plot of a three-dimensional surface of the objective function. We describe Maple's very impressive array of two- and three-dimensional graphics tools that facilitate such graphical analyses.

There are several excellent books on Maple dealing with different aspects of the software. One of the most lucid among these is the official Maple manual (*Maple V Learning Guide*) by Heal, Hansen and Rickard [85]. An advanced but still very readable book is Heck's *Introduction to Maple* [86]. Israel [97] is mainly concerned with the use of Maple in calculus and Corless [51] is written for scientific programmers who already have experience in other computer languages. Redfern and Chandler [154] discuss Maple's use in solving ordinary differential equations and Klimek and Klimek [111] present a comprehensive treatment of Maple's graphics command and structures used in plotting curves and surfaces. Currently there are more than 300 books on Maple written in different languages including French, German, Spanish, Russian and Japanese. An up-to-date list of these books can be obtained from Waterloo Maple's Web site at `www.maplesoft.com/publications/books/maplebooks.html`.

2.2 Symbolics

2.2.1 Expanding, Factoring and Simplifications

Maple's main strength is in its ability to manipulate symbolic expressions. For example, it can `expand` an expression such as $(x+y)^{10}$. When asked to `factor` the result, Maple gives the original expression.

As a simple example, consider the expansion of the expression $(x + y)^{10}$. Note that, in Maple, all characters which follow a sharp character (#) on a line are considered to be part of a comment. Thus, in the following example, the file name of the worksheet (i.e., `Expand.mws`) is understood by Maple to be a comment. Note that each Maple command must end with a semicolon (;) or a colon (:). Termination by a semicolon informs Maple to display the results. When the results are expected to be very long, the colon can be used to terminate the command thus suppressing the output.

```
>    restart: # Expand.mws
>    expand((x+y)^10);
```

$$x^{10} + 10\,x^9\,y + 45\,x^8\,y^2 + 120\,x^7\,y^3 + 210\,x^6\,y^4 + 252\,x^5\,y^5 + 210\,x^4\,y^6$$
$$+ 120\,x^3\,y^7 + 45\,x^2\,y^8 + 10\,x\,y^9 + y^{10}$$

```
>    factor(%);
```

$$(x + y)^{10}$$

Note that the ditto operator, %, is a shorthand notation for the result of the previous, i.e., the last, command.

A Maple expression can be named using the syntax `name:=expression`. Several examples using the simplification commands `simplify()`, `normal()` and `combine()` are given below.

>   `restart: # Simplification.mws`

The `simplify()` command is used to apply simplification rules to an expression:

>   `x:=a^3/((a-b)*(a-c))+b^3/((b-c)*(b-a))`
>   `+c^3/((c-a)*(c-b));`

$$x := \frac{a^3}{(a-b)(a-c)} + \frac{b^3}{(b-c)(b-a)} + \frac{c^3}{(c-a)(c-b)}$$

>   `simplify(x);`

$$b + a + c$$

The `normal()` command is used to simplify a rational function f/g where f and $g \neq 0$ are polynomials.

>   `y:=(w^2+3*w+2)/(w^2+5*w+6);`

$$y := \frac{w^2 + 3w + 2}{w^2 + 5w + 6}$$

>   `normal(y);`

$$\frac{w + 1}{w + 3}$$

Note that x, y and z were the names assigned to the preceding two expressions. If we wish Maple to "forget" these assignments, we can "unassign" them by placing the name inside single forward quotes:

>   `x:=' x' : y:=' y' : z:=' z' :`

The `combine()` command attempts to combine terms into a single term. In some sense, this is—like the `factor()` command—the opposite of the `expand()` command.

>   `A:=x^y*x^z;`

$$A := x^y x^z$$

>   `combine(A);`

$$x^y x^z$$

In this case `combine()` did nothing since very often one needs to specify an option such as `power`, `ln`, or `symbolic` to help `combine()` recognize the expressions and do the simplification.

>   `combine(A,power);`

$$x^{(y+z)}$$

>   `B:=log(x)/2+log(y)-log(z);`

$$B := \frac{1}{2}\ln(x) + \ln(y) - \ln(z)$$

> `combine(B,ln);`

$$\frac{1}{2}\ln(x) + \ln(y) - \ln(z)$$

> `combine(B,ln,symbolic);`

$$\ln(\frac{\sqrt{x}\,y}{z})$$

> `C:=sqrt(x)*sqrt(y);`

$$C := \sqrt{x}\,\sqrt{y}$$

In the following example, the combination of $\sqrt{x}\sqrt{y}$ into $\sqrt{xy}$ becomes possible only after informing Maple—using `assume`—that both x and y are positive quantities since $\sqrt{x}\sqrt{y}$ is *not* equal to $\sqrt{xy}$ for all values of x and y, e.g., $\sqrt{-1}\sqrt{-1} \neq \sqrt{(-1)(-1)}$.

> `combine(C);`

$$\sqrt{x}\,\sqrt{y}$$

> `assume(x>0,y>0);`
> `combine(C);`

$$\sqrt{x\,y}$$

2.2.2   Function Definitions Using "->" and unapply()

The assignment operation (`:=`) allows the user to name a Maple object as in `f:=1/x+x;`. Once this is done, one can even evaluate the expression at different values using `subs()` or `eval()` as in the following examples.

> `restart: # Function.mws`
> `f:=1/x+x;`

$$f := \frac{1}{x} + x$$

> `subs(x=1,f); eval(f,x=1); subs(x=(a+b)^2,f);`

$$2$$

$$2$$

$$\frac{1}{(a+b)^2} + (a+b)^2$$

If the expression is intended to define a function, then it is easier (and preferable) to define the function formally using either the arrow (`->`) notation or the `unapply()` command.

> `f:=x->1/x+x;`

$$f := x \to \frac{1}{x} + x$$

```
>   f(1);
```

$$2$$

```
>   f:=unapply(1/x+x,x);
```

$$f := x \to \frac{1}{x} + x$$

Now the function $f(x)$ can be evaluated at different values of x.

```
>   f(1); f(a); f((a+b)^2);
```

$$2$$

$$\frac{1}{a} + a$$

$$\frac{1}{(a+b)^2} + (a+b)^2$$

The unapply() command can be useful if, for example, in the middle of a Maple session, we find an expression that we want to define as a function:

```
>   restart: # Unapply.mws
>   Result:=int(int(lambda* exp(-lambda*u),u=0..t),
    lambda);
```

$$Result := \frac{e^{(-\lambda t)}}{t} + \lambda$$

Here, we create a function out of the Result that depends on two variables lambda and t.

```
>   F:=unapply(Result,lambda,t);
```

$$F := (\lambda, t) \to \frac{e^{(-\lambda t)}}{t} + \lambda$$

```
>   F(1,5); F(a,b);
```

$$\frac{1}{5} e^{(-5)} + 1$$

$$\frac{e^{(-a b)}}{b} + a$$

2.2.3   Lists, Sets and Arrays

Lists, sets and arrays are basic types of Maple objects and they play an important role in the effective use of Maple.

Lists

A Maple *list* can be created by simply enclosing any number of expressions or
other Maple objects in square brackets.

```
>   restart: # Lists.mws
>   L:=[ a,b,sin(x),int(x^2,x)];
```

$$L := [a, b, \sin(x), \frac{1}{3}x^3]$$

The number of elements of the list can be counted using `nops()` and the
membership of an element can be checked with `member()`.

```
>   nops(L);
```

$$4$$

```
>   member(b,L); member(cos(x),L);
```

true

false

To access, say, the first and the fourth elements of the list, we enter `L[1]` and
`L[4]`.

```
>   L[1]; L[4];
```

$$a$$

$$\frac{1}{3}x^3$$

The *ordered* elements of the list L can be converted to an expression sequence
by `op(L)`.

```
>   op(L);
```

$$a, b, \sin(x), \frac{1}{3}x^3$$

Sets

A mathematical *set* is an unordered collection of elements. In Maple a set is created by enclosing the elements in curly brackets.

```
>   restart: # Sets.mws
>   A:={ a,b,2,Q^2,1/x} ;
```

$$A := \{2, \frac{1}{x}, a, b, Q^2\}$$

```
>   B:={ c,f,10!} ;
```

$$B := \{c, f, 3628800\}$$

Operations on sets can be performed with `union` and `intersect`.

```
>   A union B;
```

$$\{2, \frac{1}{x}, a, b, c, f, Q^2, 3628800\}$$

Since A and B do not have any elements in common, their intersection $A \cap B$ is the empty set $\varnothing$, which Maple denotes by {}.

```
>    A intersect B;
```

$$\{\}$$

Solving a system of two equations in two unknowns gives the solution using the set notation. Before the expressions involving x and y can be manipulated, the solution found here must be assigned.

```
>    Sol:=solve({ a*x+b*y=c,d*x+e*y=f} ,{ x,y} );
```

$$Sol := \{y = \frac{af - cd}{ae - bd}, \; x = -\frac{-ec + fb}{ae - bd}\}$$

```
>    x; y;
```

$$x$$

$$y$$

```
>    assign(Sol); x; y;
```

$$-\frac{-ec + fb}{ae - bd}$$

$$\frac{af - cd}{ae - bd}$$

It is possible to convert() a list to a set and perform set operations on it.

```
>    L:=[ 9,6,t,4,2] ;
```

$$L := [9, 6, t, 4, 2]$$

```
>    LSet:=convert(L,set);
```

$$LSet := \{2, 4, 6, 9, t\}$$

```
>    LSet intersect A;
```

$$\{2\}$$

Arrays

As a data structure, an array is an extension of a list that is indexed by positive integers ranging from 1 to nops(list). In an array, one can have many dimensions and the indices can be negative or zero. Arrays are also related to objects such as vectors and matrices.

```
>    restart: # Arrays.mws
```

Here A is a 2 × 2 array with negative and zero indices.

```
>    A:=array(-1..0,0..1);
```

$$A := \text{array}(-1..0, \; 0..1, \; [])$$

Let us make assignments to the first three entries but leave the last one unassigned. (The presence of the comment symbol # before A[0,1] :=d indicates that this is treated as a comment, not as an assignment.)

```
>    A[ -1,0] :=a; A[ -1,1] :=b; A[ 0,0] :=c; #A[ 0,1] :=d;
```

$$A_{-1,0} := a$$

$$A_{-1,1} := b$$

$$A_{0,0} := c$$

```
>    print(A);
```

$$\text{array}(-1..0, 0..1, [$$
$$(-1, 0) = a$$
$$(-1, 1) = b$$
$$(0, 0) = c$$
$$(0, 1) = A_{0,1}$$
$$])$$

Since A has nonpositive indices, it is not treated as a matrix. But the following 2×2 array B acts like a matrix and thus one can manipulate it using matrix operations with the command evalm().

```
>    B:=array(1..2,1..2,[[ 3,4] ,[ 2,k]] );
```

$$B := \begin{bmatrix} 3 & 4 \\ 2 & k \end{bmatrix}$$

```
>    with(linalg):
```

```
Warning, new definition for norm
```

```
Warning, new definition for trace
```

```
>    evalm(A &* A); evalm(B &* B);
```

$$A^2$$

$$\begin{bmatrix} 17 & 12+4k \\ 6+2k & 8+k^2 \end{bmatrix}$$

2.2.4   Additional Commands to Manipulate Expressions

The map() Command

The map() command applies a procedure to each operand of an expression that can be a list, set, array. This is a very useful command as it can substantially simplify the evaluation of a large expression such as a list or a matrix in a single operation.

```
>    restart: # Map.mws
```

> `theta:=[ 0,Pi/4,Pi/2,3*Pi/4,Pi] ;`

$$\theta := [0, \frac{1}{4}\pi, \frac{1}{2}\pi, \frac{3}{4}\pi, \pi]$$

We can use `map()` to simplify the computation of the `sin` of each number in the list θ.

> `map(sin,theta);`

$$[0, \frac{1}{2}\sqrt{2}, 1, \frac{1}{2}\sqrt{2}, 0]$$

Differentiating each element of the following array A one by one would be very cumbersome. But by using `map()`, we can apply the differentiation operator `diff` to each element of A and automatically obtain the complete result.

> `A:=array(1..2,1..2,[[ sin(x),x^2* cos(x)] ,[ log(x),-x]] );`

$$A := \begin{bmatrix} \sin(x) & x^2\cos(x) \\ \ln(x) & -x \end{bmatrix}$$

> `map(diff,A,x);`

$$\begin{bmatrix} \cos(x) & 2x\cos(x) - x^2\sin(x) \\ \dfrac{1}{x} & -1 \end{bmatrix}$$

The `collect()` command

The `collect()` command is used to collect coefficients of similar powers.

> `restart: # Collect.mws`

Here, it can simplify expressions by regrouping terms involving `exp(x)`.

> `f:=int(x* (exp(-x)+exp(x)),x);`

$$f := -\frac{x}{e^x} - \frac{1}{e^x} + x e^x - e^x$$

> `collect(f,exp(x));`

$$(x-1)e^x + \frac{-x-1}{e^x}$$

Here, it collects the coefficients of `x^2`.

> `g:=a^4* x^2+x^2-a^2* x^2;`

$$g := a^4 x^2 + x^2 - a^2 x^2$$

> `collect(g,x);`

$$(a^4 + 1 - a^2)x^2$$

2.2.5 Solving Equations Analytically

As we will see in later chapters, Maple can find analytic (i.e., symbolic) solutions for a large class of algebraic and differential equations, hence making it very useful in operations research modeling and applications. Here are a few examples:

Example 1 *A function arising in deterministic inventory models.* The function $f(x) = a/x + bx$ arises frequently in operations research and it represents the average cost of operating an inventory system for particular values of a and b. The minimum-cost order quantity can be found by differentiating the function, equating the result to zero and solving for the positive root.

```
>   restart: # Solve.mws
>   f:=x->a/x+b*x;
```

$$f := x \rightarrow \frac{a}{x} + bx$$

```
>   fx:=diff(f(x),x);
```

$$fx := -\frac{a}{x^2} + b$$

```
>   solve(fx,x);
```

$$\frac{\sqrt{b\,a}}{b}, \; -\frac{\sqrt{b\,a}}{b}$$

To compute the second derivative of f, we use the shorthand notation `diff(f(x),x$2)` where the dollar sign $ between x and 2 is an operator that signifies differentiating the function with respect to x *twice*.

```
>   fxx:=diff(f(x),x$2);
```

$$fxx := 2\frac{a}{x^3}$$

Since the function is convex for positive x, (i.e., $f_{xx} > 0$), the minimum of f occurs at $\sqrt{ab}/b$.

Example 2 *A function arising in stochastic inventory models.* The problem of solving $1 - F(Q) = \beta$ where $F(t) = \int_0^t \lambda e^{-\lambda u}\, du$ arises in single-period stochastic inventory modeling where the demand is assumed to be exponentially distributed with parameter λ. The analytic solution for the optimal order quantity can again be found by solving the equation.

```
>   F:=unapply(int(lambda*exp(-lambda*u),u=0..t),t);
```

$$F := t \rightarrow -e^{(-\lambda t)} + 1$$

```
>   eq:=1-F(Q)=beta;  solve(eq,Q);
```

$$eq := e^{(-\lambda Q)} = \beta$$

$$-\frac{\ln(\beta)}{\lambda}$$

Thus, the optimal order quantity is obtained as $Q = -\ln(\beta)/\lambda > 0$ since $0 < \beta < 1$.

Example 3 *A differential equation arising in stochastic processes.* Maple can also solve differential equations analytically. The next example is encountered in the description of a Poisson process where λ is the arrival rate and $p_0(t)$ is the probability that there will be zero arrivals between time 0 and time t.

```
>    DE:=diff(p0(t),t)=-lambda*p0(t);
```

$$DE := \frac{\partial}{\partial t} p0(t) = -\lambda \, p0(t)$$

```
>    Sol:=dsolve({ DE,p0(0)=1} ,p0(t));
```

$$Sol := p0(t) = e^{(-\lambda t)}$$

```
>    assign(Sol); p0Sol:=unapply(p0(t),t);
```

$$p0Sol := t \rightarrow e^{(-\lambda t)}$$

Thus, $p_0(t) = e^{-\lambda t}$.

2.3   Numerics

In many practical problems the user may be interested in obtaining only a numerical solution to a problem. For example, finding the sample mean or sample variance of a data set or computing the probability of a specific event may require numerical (not symbolic) computations. In those cases, we can make use of Maple's abilities to perform the necessary numerics and obtain the results.

Although Maple's symbolic manipulation and solution capabilities can be very impressive, there are certain instances where it becomes impossible to find solutions analytically. In those cases, we again resort to numerics to find the solution.

Maple can also be used as an interactive calculator that can add and subtract real numbers.

```
>    restart: # BasicNumerics.mws
>    2+3; evalf(3452.51-sqrt(Pi));
```

$$5$$

$$3450.737546$$

It knows about complex numbers and their manipulation.

```
>    sqrt(-1); (1+6*I)*(3-5*I)*7*I;
```

$$I$$

$$-91 + 231\,I$$

Maple can deal with very large numbers and manipulate them using (almost) infinite precision. It can also do integer factorizations with ifactor() and perform floating-point evaluations to (almost) any precision with evalf().

```
>   Big:=100!;
```

$Big := 93326215443944152681699238856266700490715968264\backslash$
$81621468592963895217599993229915608941463976156\backslash$
$51828625369792082722375825118521091686400000000\backslash$
0000000000000000

```
>   ifactor(Big);
```

$(2)^{97}$ $(3)^{48}$ $(5)^{24}$ $(7)^{16}$ $(11)^9$ $(13)^7$ $(17)^5$ $(19)^5$ $(23)^4$ $(29)^3$ $(31)^3$ $(37)^2$
$(41)^2$ $(43)^2$ $(47)^2$ (53) (59) (61) (67) (71) (73) (79) (83) (89)
(97)

```
>   evalf(log(Big),35);
```

$$363.73937555556349014407999336965564$$

In the following examples we present additional Maple commands that can aid in numerical analysis.

2.3.1 Solving Equations Numerically

Example 4 *A function arising in stochastic inventory models (continued).* Consider again the problem of solving $1 - F(Q) = \beta$ for the inventory problem where $F(t) = \int_0^t f(u)\,du$ is the cumulative distribution function of the random demand. Let us now assume that demand is normal with mean μ and variance σ^2 and that the probability of negative demand (for given μ and σ^2) is negligibly small.

```
>   restart: # NumericSolve.mws
>   f:=x->1/(sqrt(2*Pi*sigma))*exp(-(x-mu)^2/
    (2*sigma^2));
```

$$f := x \to \frac{e^{(-1/2\,\frac{(x-\mu)^2}{\sigma^2})}}{\sqrt{2\,\pi}\,\sigma}$$

```
>   F:=unapply(int(f(u),u=0..t),t);
```

$$F := t \to -\frac{1}{2}\frac{\sqrt{\pi}\,\sigma\,(\mathrm{erf}(\frac{1}{2}\frac{\sqrt{2}\,(-t+\mu)}{\sigma}) - \mathrm{erf}(\frac{1}{2}\frac{\mu\,\sqrt{2}}{\sigma}))}{\sqrt{\pi}\,\sigma}$$

In this problem, Maple attempts to solve $1 - F(Q) = \beta$ symbolically using solve(), but it can only find a result in terms of the error function $\mathrm{erf}(x)$ defined as

$$\mathrm{erf}(x) = \frac{2}{\sqrt{\pi}}\int_0^x e^{-t^2}\,dt.$$

But as soon as we specify the values of the parameters (i.e., $\mu = 100$, $\sigma = 10$ and $\beta = .7$), Maple solves the equation and finds $Q = 86.88$ as the optimal order quantity.

```
>   solve(1-F(Q)=beta,Q);
```

$$\text{RootOf}(-\text{erf}(\frac{1}{2}\frac{\sqrt{2}\,(-\_Z+\mu)}{\sigma})\,\pi\,\sigma + \text{erf}(\frac{1}{2}\frac{\mu\sqrt{2}}{\sigma})\,\pi\,\sigma - 2\sqrt{\pi\,\sigma}\,\sqrt{\pi}$$
$$+2\sqrt{\pi\,\sigma}\,\sqrt{\pi}\,\beta)$$

```
>   mu:=100; sigma:=10; beta:=.7;
```

$$\mu := 100$$
$$\sigma := 10$$
$$\beta := .7$$

```
>   solve(1-F(Q)=beta,Q);
```

$$86.88641461$$

Note that the same solution can also be obtained by using the `fsolve()` command that solves equations only numerically. Here, the argument $0..150$ informs Maple that it should search for the numerical solution in the interval $[0, 150]$.

```
>   fsolve(1-F(Q)=beta,Q,0..150);
```

$$86.88641461$$

2.3.2   Statistical Computations Using the `stats` Package

Maple's `stats` package contains procedures for performing some of the basic statistical computations.

Example 5 *Some statistical computations.* Suppose the following numbers are the closing prices of a stock in the last 10 business days.

```
>   restart: # Stats.mws
```

```
>   with(stats);
```

[anova, describe, fit, importdata, random, statevalf, statplots, transform]

```
>   Prices:=[ 97.6,99.4,101.3,102.1,96,107,96.5,92,
    100.4,99.8] ;
```

Prices := [97.6, 99.4, 101.3, 102.1, 96, 107, 96.5, 92, 100.4, 99.8]

After loading the `describe` subpackage, we can numerically compute the mean, variance and the standard deviation of these prices.

```
>   with(describe);
```

> [*coefficientofvariation, count, countmissing, covariance, decile,
> geometricmean, harmonicmean, kurtosis, linearcorrelation, mean,
> meandeviation, median, mode, moment, percentile, quadraticmean,
> quantile, quartile, range, skewness, standarddeviation, sumdata,
> variance*]

```
> mean(Prices);
```

$$99.21000000$$

```
> variance(Prices);
```

$$14.74290000$$

```
> sqrt(variance(Prices)); standarddeviation(Prices);
```

$$3.839648421$$

$$3.839648421$$

It is also possible to group these prices into ranges of half-open intervals of our choice such as [85,90), [90,95) and so on. After loading the `transform` subpackage, we use the `tallyinto()` command to count the number of observations in each range.

```
> Ranges:=[ 85..90,90..95,95..100,100..105,105..110] ;
```

Ranges := [85..90, 90..95, 95..100, 100..105, 105..110]

```
> with(transform);
```

[*apply, classmark, cumulativefrequency, deletemissing, divideby,
frequency, moving, multiapply, scaleweight, split, standardscore,
statsort, statvalue, subtractfrom, tally, tallyinto*]

```
> PriceList:=tallyinto(Prices,Ranges);
```

PriceList := [Weight(85..90, 0), 90..95, Weight(95..100, 5),
Weight(100..105, 3), 105..110]

Thus, there are no observations in the range [85,90), one observation in the range [90,95), five observations in the range [95,100) and so on.

The result can be plotted as a histogram (not displayed here) after loading the `statsplots` subpackage.

```
> with(statplots);
```

[*boxplot, histogram, scatterplot, xscale, xshift, xyexchange, xzexchange,
yscale, yshift, yzexchange, zscale, zshift*]

```
> #histogram(PriceList);
```

And finally, we can even generate random prices distributed normally with a mean of 100 and standard deviation of 5.

```
> RandomPrices:=[ random[ normald[ 100,5]] (10)] ;
```

RandomPrices := [105.8791978, 97.18317935, 101.1769700,
92.78724855, 94.60401883, 99.88992677, 87.07360818,
97.78364360, 94.98359260, 99.86065130]

Maple's random number generation facilities will be useful in Chapter 10, Simulation.

2.3.3 The simplex *Package*

Linear programming (LP) problems are always solved numerically since it is impossible to find the solution to such problems analytically in terms of the problem parameters. Maple has an implementation of the simplex algorithm that finds the optimal solution to LP problems numerically. We will discuss different aspects of the simplex method and linear optimization in Chapter 4, Linear Programming.

Example 6 *A linear programming problem.* As a simple example, consider the problem of maximizing the objective function $z = 2x_1 + 3x_2$ subject to the three constraints (i) $x_1 + 1.75x_2 \leq 650$, (ii) $1.5x_2 \leq 200$, and (iii) $x_1 + x_2 \leq 600$. Maple finds the optimal solution as $x_1 = 533.33$, $x_2 = 66.66$ and $z = 1266.66$.

```
> restart: # Simplex.mws
> with(simplex);
```

Warning, new definition for maximize

Warning, new definition for minimize

[*basis, convexhull, cterm, define_zero, display, dual, feasible, maximize, minimize, pivot, pivoteqn, pivotvar, ratio, setup, standardize*]

```
> z:=2*x[ 1] +3*x[ 2] ;
```

$$z := 2x_1 + 3x_2$$

```
> C:={ x[ 1] +1.75*x[ 2] <=650,  1.5*x[ 2] <=200,
  x[ 1] +x[ 2] <=600} ;
```

$$C := \{x_1 + 1.75x_2 \leq 650, \ x_1 + x_2 \leq 600, \ 1.5x_2 \leq 200\}$$

```
> Sol:=maximize(z,C,NONNEGATIVE); assign(Sol);
```

$$Sol := \{x_1 = 533.3333333, \ x_2 = 66.66666667\}$$

```
> z;
```

$$1266.666667$$

2.4 Graphics

The third important aspect of Maple is its capability to plot impressive-looking two- and three-dimensional graphs. The dictum "one picture is worth one thousand words" also applies in operations research modeling and graphing curves on the plane or surfaces in space may reveal important information about a problem's characteristics.

In this section we will consider several two- and three-dimensional Maple graphs that can be useful in modeling and analysis.

2.4.1 Two-dimensional Plots

Example 7 *Plot of the Weibull density.* The Weibull random variable arises frequently in reliability and queueing theory where it is used to represent the lifetime of devices or the time to complete a specific task; see Law and Kelton [118, p. 161]. Figure 2.1 displays the Weibull density $f(x) = \alpha\beta^{-\alpha}x^{a-1}e^{-(x/\beta)^a}$, $x > 0$, for a specific value of the shape parameter $\alpha = 2.5$ and shift parameter $\beta = 1.7$. Naturally, these parameters can be easily changed to obtain and plot Weibull densities with different shapes.

```
>    restart: # WeibullPlot.mws
>    alpha:=2.5; beta:=1.7;
```

$$\alpha := 2.5$$

$$\beta := 1.7$$

```
>    W:=alpha*beta^(-alpha)*x^(alpha-1)
     *exp(-(x/beta)^alpha);
```

$$W := .6634645233\, x^{1.5}\, e^{(-.2653858093\, x^{2.5})}$$

```
>    plot(W,x=0..4);
```

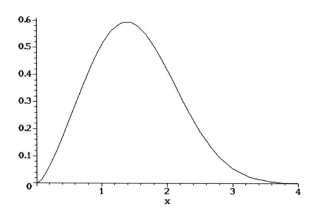

FIGURE 2.1. Plot of the Weibull density with shape parameter $\alpha = 2.5$ and scale parameter $\beta = 1.7$.

Example 8 *Implicit plot of necessary conditions in an optimization problem.* Suppose we wish to find the stationary points of the function $z = -x^3 - 2y^3 +$

$3xy-x+y$, i.e., the set of points that satisfy the necessary conditions $z_x = z_y = 0$. Before analyzing the problem mathematically, it may be useful to plot the necessary conditions as implicitly defined curves in the xy-plane.

```
>    restart: # ImplicitPlot.mws
>    z:=-x^3-2*y^3+3*x*y-x+y;
```

$$z := -x^3 - 2y^3 + 3xy - x + y$$

```
>    zx:=normal(diff(z,x));
```

$$zx := -3x^2 + 3y - 1$$

```
>    zy:=normal(diff(z,y));
```

$$zy := -6y^2 + 3x + 1$$

```
>    with(plots):
>    implicitplot({zx,zy},x=-1..1,y=-1..1,
        numpoints=1000,color=black);
```

We see in Figure 2.2 that the two curves for $z_x = 0$ and $z_y = 0$ intersect at two points.

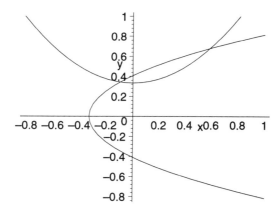

FIGURE 2.2. Implicit plots of the necessary conditions $z_x = 0$ and $z_y = 0$ for $z = -x^3 - 2y^3 + 3xy - x + y$.

Now solving $z_x = 0$ and $z_y = 0$ for (x, y) we obtain the two real-valued solutions.

```
>    Sol:=solve({zx,zy},{x,y});
```

$$Sol := \{x = \text{RootOf}(18\_Z^4 + 12\_Z^2 - 1 - 9\_Z),$$
$$y = \text{RootOf}(18\_Z^4 + 12\_Z^2 - 1 - 9\_Z)^2 + \frac{1}{3}\}$$

```
>    evalf(allvalues(Sol));
```

$$\{x = .5875626256, \ y = .6785631723\},$$
$$\{y = .3429560780, \ x = -.0980955898\}, \{$$
$$y = -.5107596261 - .4653767156 \ I,$$
$$x = -.2447335179 + .9507825483 \ I\}, \{$$
$$x = -.2447335179 - .9507825483 \ I,$$
$$y = -.5107596261 + .4653767156 \ I\}$$

Example 9 *Contour plot and implicit plot displayed jointly.* Consider the problem of minimizing the quadratic function $f(x, y) = (x - 2)^2 + (y - 3)^2$ subject to the linear constraint $g(x, y) = x + y \leq 4$ for $x, y \geq 0$. In order to obtain insight into the behavior of the objective function and its relation to the constraint, we can plot the contours of the function $f(x, y)$ using `contourplot()` along with the constraint boundary $g(x, y) = 0$ using `implicitplot()`. The graph in Figure 2.3 reveals that since the unconstrained optimal solution is outside the feasible region, the constrained optimal solution for this problem will be on the constraint boundary $x + y = 4$.

```
>    restart: # ContourPlot.mws
>    with(plots):
>    f:=(x-2)^2+(y-3)^2;
```

$$f := (x - 2)^2 + (y - 3)^2$$

```
>    g:=x+y=4;
```

$$g := x + y = 4$$

```
>    fCP:=contourplot(f,x=0..5,y=0..5,color=black,
     scaling=constrained,contours=[ 1,2,3,4,5] ,
     numpoints=1000):
>    gIP:=implicitplot(g,x=0..5,y=0..5,color=black,
     scaling=constrained):
>    display({ fCP,gIP} );
```

Example 10 *Graph of the numerical solution of an ordinary differential equation.* The differential equation encountered in Example 3 had a closed-form solution as the arrival rate parameter λ was a constant. Now suppose that this parameter is time dependent, i.e., it is given by $\lambda(t) = \log(t^2 + e^{-t^2})$. This time it becomes impossible to solve the resulting differential equation $dx(t)/dt = -\lambda(t)x(t)$ analytically since $\lambda(t)$ is now a very complicated function. We can, however, use Maple's knowledge of numerical solutions of DEs and even plot the resulting solution trajectory.

```
>    restart: # DEPlot.mws
>    lambda:=t->log(t^2+exp(-t^2));
     #plot(lambda(t),t=0..1);
```

$$\lambda := t \rightarrow \log(t^2 + e^{(-t^2)})$$

```
>    DE:=diff(x(t),t)=-lambda(t)*x(t);
```

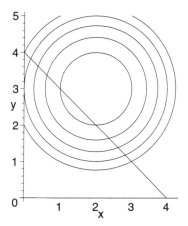

FIGURE 2.3. The contours of the objective function $f(x, y)$ displayed with the constraint boundary $g(x, y) = 0$.

$$DE := \frac{\partial}{\partial t} x(t) = -\ln(t^2 + e^{(-t^2)}) x(t)$$

Maple makes a good attempt to solve this DE, but it can give the solution only in terms of an integral.

```
>   dsolve({ DE, x(0)=1} ,x(t));
```

$$x(t) = e^{\left(-\int_0^t \ln(u^2 + e^{(-u^2)}) \, du\right)}$$

But using the type=numeric option of dsolve(), we can ask Maple to compute numerically the solution trajectory.

```
>   Sol:=dsolve({ DE, x(0)=1} ,x(t),type=numeric);
```

$$Sol := \mathbf{proc}(rkf45\_x) \ldots \mathbf{end}$$

Once the solution is found, it can be evaluated at different time points and eventually plotted using odeplot(), shown in Figure 2.4.

```
>   Sol(0); Sol(.5); Sol(1);
```

$$[t = 0, \; x(t) = 1.]$$

$$[t = .5, \; x(t) = .9970800089348753]$$

$$[t = 1, \; x(t) = .9296351615552543]$$

```
>   with(plots):
```

```
>   odeplot(Sol,[ t,x(t)] ,0..1,color=black,
    labels=[ t,x] );
```

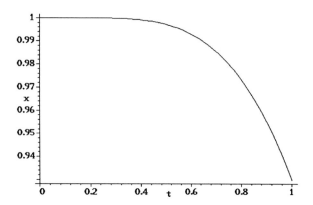

FIGURE 2.4. Graph of the solution trajectory for $dx(t)/dt = -\lambda(t)x(t)$, $x(0) = 1$ where $\lambda(t) = \log(t^2 + e^{-t^2})$.

2.4.2   Three-Dimensional Plots

Maple's plotting abilities also include plotting three-dimensional surfaces, which arise frequently in operations research modeling. In this section we consider two examples.

Example 11 *Plot of the objective function surface in an inventory problem.* In [81, p. 44] Hadley and Whitin consider a deterministic inventory problem where backorders are permitted. In this example we plot the objective function surface of this problem and calculate the optimal values of the decision variables.

```
>    restart: # BackorderPlot.mws
```

Here, λ is the annual demand rate, K is the order cost, h is the holding cost per unit per year and π_1 and π_2 are the fixed and variable (in time) costs, respectively, of backordering demands.

```
>    C:=lambda* K/Q+h* (Q-s)^2/(2*Q)
       +(1/Q)* (pi[ 1] * lambda* s+pi[ 2] * s^2);
```

$$C := \frac{\lambda K}{Q} + \frac{1}{2}\frac{h(Q-s)^2}{Q} + \frac{\pi_1 \lambda s + \pi_2 s^2}{Q}$$

```
>    lambda:=200; K:=8; h:=14; pi[ 1] :=.8; pi[ 2] :=25;
```

$$\lambda := 200$$

$$K := 8$$

$$h := 14$$

$$\pi_1 := .8$$

$$\pi_2 := 25$$

```
>    C;
```

$$1600 \frac{1}{Q} + 7 \frac{(Q-s)^2}{Q} + \frac{160.0\,s + 25\,s^2}{Q}$$

When we assign specific values to the parameters and plot the objective function $C(Q, s)$, we observe that the function is strictly convex with a unique minimum. The optimal solution is found as $Q = 16$ and $s = 1$ with $C(16, 1) = 210$ by solving the necessary conditions.

```
> plot3d(C,Q=10..30,s=0..5,axes=boxed,shading=none,
  orientation=[ 107,67] );
> CQ:=normal(diff(C,Q)); Cs:=normal(diff(C,s));
```

$$CQ := \frac{-1600 + 7\,Q^2 - 32\,s^2 - 160.0\,s}{Q^2}$$

$$Cs := -\frac{14\,Q - 64\,s - 160.0}{Q}$$

```
> Sol:=fsolve({ numer(CQ),numer(Cs)} ,{ Q,s} );
```

$$Sol := \{Q = 16.00000000,\ s = 1.000000000\}$$

```
> assign(Sol);
> C;
```

$$210.0000000$$

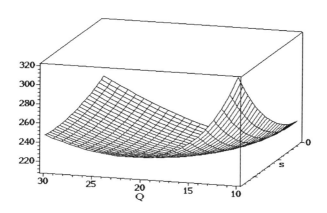

FIGURE 2.5. Objective function of the inventory problem where backorders are permitted.

Example 12 *Bivariate normal density.* Now we consider the surface plot of the density $f(x, y)$ of the bivariate normal random vector (X, Y). We will plot the density using the plot3d() command.

```
> restart: # BivariateNormalPlot.mws
```

```
>    q:=(x-mu[ 1] ) ^2/sigma[ 1] ^2
     -2* rho* (x-mu[ 1] )* (y-mu[ 2] )/(sigma[ 1] * sigma[ 2] )
     +(y-mu[ 2] ) ^2/sigma[ 2] ^2;
```

$$q := \frac{(x - \mu_1)^2}{\sigma_1{}^2} - 2\frac{\rho\,(x - \mu_1)\,(y - \mu_2)}{\sigma_1\,\sigma_2} + \frac{(y - \mu_2)^2}{\sigma_2{}^2}$$

```
>    c:=1/(2* Pi* sigma[ 1] * sigma[ 2] * sqrt (1-rho^2));
```

$$c := \frac{1}{2}\frac{1}{\pi\,\sigma_1\,\sigma_2\,\sqrt{1 - \rho^2}}$$

```
>    f:=c* exp (-q/(2*' (1-rho^2)' ));
```

$$f := \frac{1}{2}\frac{e^{\left(-1/2\,\frac{\frac{(x-\mu_1)^2}{\sigma_1{}^2} - 2\frac{\rho(x-\mu_1)(y-\mu_2)}{\sigma_1\sigma_2} + \frac{(y-\mu_2)^2}{\sigma_2{}^2}}{1-\rho^2}\right)}}{\pi\,\sigma_1\,\sigma_2\,\sqrt{1 - \rho^2}}$$

Note that in the previous expression we enclosed $1 - \rho^2$ in single quotes (' ')
so that Maple would not immediately evaluate $2(1 - \rho^2)$ to $2 - 2\rho^2$. The single
quotes can be useful in checking the accuracy of symbolic data entry.

```
>    mu[ 1] :=10; mu[ 2] :=5; sigma[ 1] :=7; sigma[ 2] :=12;
     rho:=.7;
```

$$\mu_1 := 10$$

$$\mu_2 := 5$$

$$\sigma_1 := 7$$

$$\sigma_2 := 12$$

$$\rho := .7$$

We check by double integration that the joint density integrates to 1, i.e., that
$\int_{-\infty}^{\infty}\int_{-\infty}^{\infty} f(x, y)\,dy\,dx = 1.$

```
>    int (int (f, x=-infinity..infinity),
     y=-infinity..infinity);
```

$$.9999999998$$

We may assume, for example, that the return X from stock A and the return
Y from stock B have a bivariate normal distribution with parameters $\mu_1 = 10\%$,
$\mu_2 = 5\%$, $\sigma_1 = 7\%$, $\sigma_2 = 12\%$ and $\rho = 0.7$. With these mean values, standard deviations and correlation coefficient, we plot the surface of the probability
density in Figure 2.6. The orientation argument is used in order to view the
surface from a position that is nearest to the lowest values assumed by x and y,
i.e., -5 and -15, respectively.

```
>    plot3d (f, x=-5..25, y=-15..25, shading=none,
     axes=boxed, orientation=[ -112, 54] );
```

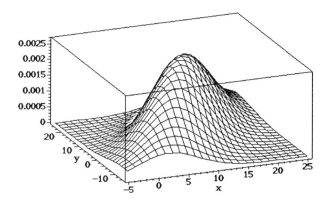

FIGURE 2.6. The surface plot of the bivariate normal density $f(x, y)$ with $\mu_1 = 10$, $\mu_2 = 5, \sigma_1 = 7, \sigma_2 = 12$ and $\rho = .7$.

2.5   Other Useful Commands and Packages

2.5.1   The `piecewise()` Command

In many important applications some functions are given in a "piecewise" form. For example, in inventory problems with quantity discounts the unit price $C(Q)$ is usually a step (i.e., piecewise) function of the order quantity Q. In probabilistic modeling, some random variables—such as the triangular—have densities that are specified as piecewise functions. Maple's `piecewise()` is a very powerful command that can manipulate expressions involving piecewise functions.

We discuss two examples.

Example 13 *Inventory problem with quantity discounts.* In many inventory problems, the unit price $C(Q)$ is a decreasing function of the order quantity Q given as a step function. For specified quantities (break levels) $q_0 = 0, q_1, \ldots, q_N$ with $q_i < q_{i+1}$ and $q_{N+1} = \infty$, if Q units are ordered such that $q_j \leq Q < q_{i+1}$, then the unit price is $C(Q) = C_i$ dollars per unit and the total purchase cost is $C_i Q$ dollars.

Consider the following case.

```
>   restart: # piecewise.mws
```

The break-points are given as 1000, 2000 and 3000.

```
>   q[ 0] :=0; q[ 1] :=1000; q[ 2] :=2000; q[ 3] :=3000;
```

$$q_0 := 0$$

$$q_1 := 1000$$

$$q_2 := 2000$$

$$q_3 := 3000$$

The piecewise unit cost function $C(Q)$ is defined in terms of the break-points.
```
>    C:=piecewise(Q<q[ 1] ,5,  Q<q[ 2] ,4,
     Q<q[ 3] ,3,  2);
```

$$C := \begin{cases} 5 & Q < 1000 \\ 4 & Q < 2000 \\ 3 & Q < 3000 \\ 2 & otherwise \end{cases}$$

```
>    eval(C,Q=1234);
```

$$4$$

The total cost $QC(Q)$ is also a piecewise function as plotted in Figure 2.7.

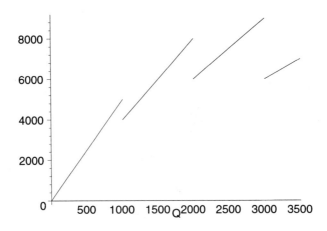

FIGURE 2.7. Plot of the total purchase cost function $QC(Q)$ in an inventory model with quantity discounts.

```
>    plot(Q*C,Q=0..3500,discont=true,color=black);
```

Example 14 *The triangular random variable.* An even more impressive use of `piecewise()` is demonstrated in the following example. We consider the triangular random variable X whose density $f(x)$ is given as a piecewise function [12, p. 218].

$$f(x) = \begin{cases} \frac{2(x-a)}{(b-a)(c-a)}, & if\ a \leq x \leq b \\ \frac{2(c-x)}{(c-b)(c-a)}, & if\ b < x \leq c. \end{cases}$$

```
>    assume(a<b,b<c);
>    f:=x->piecewise(a<=x and x<=b,
     2* (x-a)/((b-a)* (c-a)),
     b<=x and x<=c,
     2* (c-x)/((c-b)* (c-a))):
>    f(x);
```

$$\begin{cases} 2\,\dfrac{x-a}{(b-a)\,(c-a)} & a-x \le 0 \text{ and } x-b \le 0 \\[2ex] 2\,\dfrac{c-x}{(c-b)\,(c-a)} & b-x \le 0 \text{ and } x-c \le 0 \end{cases}$$

with assumptions on a, b and c

We check and see that the density indeed integrates to 1, i.e., $\int_a^c f(x)\,dx = 1$.

>   int(f(x),x=a..c);

$$1$$

Evaluating the integral $\int_a^c x f(x)\,dx$ to compute the mean of X gives $E(X) = \frac{1}{3}(a+b+c)$.

>   EX:=int(x* f(x),x=a..c);

$$EX := \frac{1}{3}b + \frac{1}{3}a + \frac{1}{3}c$$

with assumptions on b, a and c

We can easily compute the cumulative distribution function $F(x) = \int_a^x f(u)\,du$, $a \le x \le c$ and display the result as another piecewise function.

>   F:=normal(convert(int(f(u),u=a..x),
>   piecewise,x));

$$F := \begin{cases} 0 & x \le a \\[2ex] \dfrac{x^2 - 2ax + a^2}{(-b+a)\,(a-c)} & x \le b \\[2ex] \dfrac{-ac + ab - x^2 + 2cx - cb}{(-c+b)\,(a-c)} & x \le c \\[1ex] 1 & c < x \end{cases}$$

with assumptions on a, b and c

2.5.2   The interface() Command

Sometimes it may be useful to see the body of a procedure in order to examine the written code. This is made possible by first entering interface(ver boseproc=3) followed by readlib('procedurename').[1]

For example, to see the code for the futurevalue() command that is a part of the finance package, we do the following.

>   restart: # VerboseProcedure.mws

>   interface(verboseproc=3);

>   readlib('finance/futurevalue');

---

[1] The readlib() command became obsolete in Maple Release 6.

proc(*Amount, Rate, Nperiods*)
option
'*Copyright (c) 1994 by Je'ro^me M. Lang. All rights reserved.*';
description '*future value at Time of Amount invested at Rate*';
   Amount × (1 + *Rate*)^{Nperiods}
end

Thus, we see that `futurevalue()` takes three inputs (`Amount`, `Rate` and `Nperiods`) and produces one output computed using the formula `Amount*(1+ Rate)^Nperiods`.

To see the code for commands that are not part of a specific package, one should use `op(command)` as in `op('sqrt')`.

```
>    op('sqrt'):
```

2.5.3 The `infolevel[]` Command

In Maple it is possible to obtain very general information on how a problem is being solved with the `infolevel[]:=Level` command. For example, to see how an integral is being evaluated, we enter `infolevel[int]:=5;` before performing the integration. At each stage of solution process, Maple displays the technique it uses to attack the problem.

Example 15 *Definite integration.* Consider the problem of evaluating the definite integral of the exponential density, i.e., $\int_0^t \lambda e^{-\lambda u}\, du$.

```
>    restart: # infolevel.mws
>    infolevel[ int] :=5;
```

$$infolevel_{int} := 5$$

```
>    int(lambda* exp(-lambda* u), u=0..t);
```

```
int/elliptic:    trying elliptic integration

int/ellalg/elltype:    Checking for an elliptic integral
lambda* exp(-lambda* u)    freeof(u)    u

int/indef1:    first-stage indefinite integration

int/indef1:    first-stage indefinite integration

int/indef2:    second-stage indefinite integration

int/indef2:    applying derivative-divides

int/indef1:    first-stage indefinite integration
```

$$-e^{(-\lambda t)} + 1$$

Example 16 *Solution of a differential equation.* To follow Maple's steps in solving a differential equation, we enter `infolevel[ dsolve] :=5;`. Consider the problem of solving $dx/dt = ax(t), x(0) = x_0$.

```
>    infolevel[ dsolve] :=5;
```

$$infolevel_{dsolve} := 5$$

```
>    dsolve ({ diff (x(t),t)=a*x(t),x(0)=x[ 0]} ,x(t));

Methods for first order ODEs:

trying to isolate the derivative dy/dx

successful isolation of dy/dx

 -> trying classification methods

trying a quadrature

trying 1st order linear

1st order linear successful
```

$$x(t) = x_0 \, e^{(a t)}$$

2.5.4   Exporting to C, FORTRAN and LaTeX

Maple has facilities that allow exporting its output to some programming languages and typesetting processors. As an example, let us assume that we wish to export the expression $f := \pi \sin(\theta x) \ln(\sqrt{x^3})$ to C, FORTRAN and LaTeX. This is done as follows:

```
>    restart: # Export.mws
>    readlib(C): readlib(fortran): readlib(latex):
>    f := Pi*sin(theta*x)*ln(sqrt(x^3));
```

$$f := \pi \, \sin(\theta \, x) \ln(\sqrt{x^3})$$

```
>    C(f);

        t0 = 0.3141592653589793E1* sin (theta*x)* log (sqrt
        (x*x*x));

>    fortran(f);

        t0 = 0.3141593E1* sin (theta*x)* alog (sqrt (x**3))
```

Output from C and fortran can also be written to an external file for further processing; see ? C and ? fortran.

If a complete Maple worksheet is to be exported to LaTeX, it is preferable to do it by choosing File/Export As/LaTeX from the Maple menu bar. To format and print documents exported and saved in this manner, the LaTeX style package maple2e.sty is needed. In Windows installations, this file is located in the Etc folder where Maple is installed.[2]

```
>   latex(f);

\pi \,\sin(\theta\,x)\ln (\sqrt {{ x} ^{ 3}})
```

The C and FORTRAN output can be optimized resulting in a sequence of assignment statements in which temporary values are stored in local variables beginning with the letter t.

```
>   C(f,optimized);

      t2 = sin(theta*x);

      t4 = x*x;

      t6 = sqrt(t4*x);

      t7 = log(t6);

      t8 = 0.3141592653589793E1*t2*t7;

>   fortran(f,optimized);

      t2 = sin(theta*x)

      t4 = x**2

      t6 = sqrt(t4*x)

      t7 = alog(t6)

      t8 = 0.3141593E1*t2*t7
```

2.5.5   Programming in Maple

In this book we will be using Maple interactively to solve operations research problems. But Maple also has a very sophisticated programming language that can be used to develop programs (procedures) automating the computations performed.

---

[2]Maple worksheets used in this book were produced with maple2e.sty, mapleenv.def, mapleenv.sty, mapleplots.sty, maplestyle.sty, mapletab.sty and mapleutil.sty. These style files are written by Stan Dewitt of Waterloo Maple, Inc.

For example, suppose we wish to write a program to compute the factorial $n!$. This is done by creating a procedure, naming it (say, f) and then entering the statements that define the factorial function.

```
> restart: # Program.mws
> f:=proc(n) if n=0 then 1 else n*f(n-1) fi end;
```

$$f := \mathbf{proc}(n) \, \mathbf{if} \, n = 0 \, \mathbf{then} \, 1 \, \mathbf{else} \, n \times f(n - 1) \, \mathbf{fi} \, \mathbf{end}$$

Once the program is written, we can run it by simply evaluating the function $f(n)$ for different values of n.

```
> seq(f(n),n=0..5);
```

$$1, \, 1, \, 2, \, 6, \, 24, \, 120$$

As another example, the next procedure called `DiffInt` computes the derivative and the indefinite integral of a given function $y(x)$.

```
> DiffInt:=proc(y,x)
> local yDiff, yInt;
> yDiff:=diff(y,x);
> yInt:=int(y,x);
> [y,yDiff,yInt]
> end;
```

$$\mathit{DiffInt} := \mathbf{proc}(y, \, x)$$
$$\mathbf{local} \, \mathit{yDiff}, \, \mathit{yInt};$$
$$\mathit{yDiff} := \mathrm{diff}(y, \, x); \, \mathit{yInt} := \mathrm{int}(y, \, x); \, [y, \, \mathit{yDiff}, \, \mathit{yInt}]$$
$$\mathbf{end}$$

The input function x^2, its derivative $2x$ and integral $\frac{1}{3}x^3$ follow.

```
> DiffInt(x^2,x);
```

$$[x^2, \, 2x, \, \frac{1}{3}x^3]$$

For an excellent coverage of Maple's programming facilities, we refer the reader to Monagan, Geddes, Labahn and Vorkoetter [132].

2.5.6 The finance Package

In addition to its sophisticated mathematical packages such as integral transforms and linear algebra and others, Maple also has a somewhat simple finance package that performs basic financial calculations. Using this package, one can calculate present and future values, compute annuities and option prices and so on.

Basic Finance

```
> restart: # Finance.mws
> with(finance);
```

[amortization, annuity, blackscholes, cashflows, effectiverate, futurevalue, growingannuity, growingperpetuity, levelcoupon, perpetuity, presentvalue, yieldtomaturity]

Suppose that in three years we have promised to pay someone $100. If the interest rate is 6% per year, then $83.96 deposited now (i.e., the present value) will grow to $100 at the end of year 3.

```
>   presentvalue(100,0.06,3);
```

$$83.96192830$$

Suppose we deposit $100 in our bank account, which grows at a rate of 6% per year. The initial deposit will grow to $119.10 in 3 years.

```
>   futurevalue(100,0.06,3);
```

$$119.101600$$

If the interest rate is 6% per year, then the present value of an annuity paying $100 at the end of each year for the next 15 years is $971.22.

```
>   annuity(100,.06,15);
```

$$971.2248989$$

Option Pricing

Maple's blackscholes() command computes the arbitrage-free value of a European-style call option. If the spot price of a stock is $20, the strike price of the option is $18, risk-free interest rate is 7%, time to maturity of the option is 90 days and the volatility of the stock is .09, then the correct value of the option *now* using the Black/Scholes formula [34] is $2.63. Maple also finds that the option seller should hold 0.81 units of the stock now in order to *hedge* and eliminate the risk in his portfolio (consisting of the option and the stock).

```
>   BS:=blackscholes(20,18,0.07,90/365,sqrt(0.09),
    hedge);
```

$$BS := 10\,\text{erf}(.09259259260\,(\ln(\frac{10}{9}) + .02835616438)\,\sqrt{18}\,\sqrt{73}\,\sqrt{2})$$

$$+ 1.154009520 - 8.845990480\text{erf}(\frac{1}{2}($$

$$.1851851852\,(\ln(\frac{10}{9}) + .02835616438)\,\sqrt{18}\,\sqrt{73}$$

$$- .004109589041\,\sqrt{18}\,\sqrt{73})\sqrt{2})$$

```
>   evalf(BS); evalf(hedge);
```

$$2.630815160$$

$$.8153042532$$

For a discussion of Black/Scholes formula, financial derivatives and financial engineering, see Baxter and Rennie [17], Luenberger [128], Neftci [136] and Wilmott [194].

2.6 Summary

Maple's ability to perform symbolic and numeric computations and its two- and three-dimensional plotting features can be helpful in modeling and solving OR problems. In this chapter we provided a brief introduction to those Maple commands that will be used in subsequent chapters. In particular, we used examples to describe the following features of Maple: function definitions, lists, sets and arrays, algebraic simplifications and expression manipulations, solution of equations symbolically or numerically, performing statistical and financial computations, plotting graphs, obtaining information about Maple's internal operations and exporting the Maple output to other programming languages such as C and FORTRAN. Although Maple's individual commands are powerful enough to solve a wide range of OR problems by just pasting together a few commands, we also described briefly Maple's powerful programming language. With this basic Maple background, the reader should now be able to follow the solution of OR problems that will be covered throughout the book.

2.7 Exercises

1. Simplify
$$\frac{x^4 - 1}{x - 1}$$
using `normal()`. Is it always a good idea to use `normal()` to simplify expressions? Would you use `normal()` to simplify $(x^{50} - 1)/(x - 1)$? Why, or why not?

2. Let $f(t) = \lambda e^{-\lambda t}$ be the density of the exponential random variable T defined using the "->" notation. First compute the cumulative distribution function $F(t) = \int_0^t f(u)\,du$ by using the "->" notation. Now compute $F(t)$ by using the `unapply()` command. What is the difference between the two results? Which method evaluates its input?

3. Let L be a list defined as $[e^a, 1, e^{e^b}]$ where a and b are real numbers. Find the logarithm of the elements of this list using the `map` command.

4. Consider the nonlinear equations $x^2 + y^2 = 9$ and $x^2 - y^2 = 4$. First, use the `implicitplot()` command to plot and graphically locate the roots of this system. Next, use the `solve()` command to compute all the roots exactly. If the solution is obtained in terms of the RootOf function,

use `allvalues()` to generate all the roots. Find the floating point approximation for the root in the first quadrant using `evalf()`. For additional information on these functions, use Maple's help facility by entering a question mark followed by the command name, e.g., `? solve`.

5. Generate 50 random variates distributed uniformly between 0 and 100 and find their mean. Is the computed sample mean close to the true mean? Now plot these numbers in a histogram with five intervals [0,20), [20,40),. . .,[80, 100).

6. Maximize the linear objective function $z = x_1 + x_2 + x_3$ subject to the linear constraints $2x_1 + 3x_2 \leq 8$ and $2x_2 + 5x_3 \leq 10$ where the decision variables x_i, $i = 1, 2, 3$ are required to take nonnegative values. Find the optimal x_i, $i = 1, 2, 3$ and z.

7. Consider the differential equation $\dot{N}(t) = [b_1 N(t) - m] N(t)$, with $N(0) = N_0$ a given constant. Let $b_1 = 0.4$, $m = 50$ and $N = 100$. Solve this differential equation and plot its solution. Now increase $b_1 = 10$ and resolve the problem. What happens to the trajectory of $N(t)$ as b_1 increases? (This is a simplified model of a population growth—or decline—where the birth rate is proportional to the population size and m is the death rate; see e.g., Bellman and Cooke [25, p. 13].)

8. Consider the differential equation $\dot{x}(t) = u(t)x(t)$ with $x(0) = 1$ where $u(t)$ is a piecewise defined function given by

$$u(t) = \begin{cases} .2, & \text{if } 0 \leq t < 1 \\ .2, & \text{if } 2 \leq t < 3 \\ .1, & \text{otherwise.} \end{cases}$$

Solve this DE and plot the solution trajectory for $t \in [0, 5]$. Can you assign an economic interpretation to this problem?

3

Maple and Mathematical Foundations of Operations Research

3.1   Introduction

Although there have been some isolated cases[1] of successful applications of operations research that did not involve any mathematics, operations research modeling generally requires a good understanding of basic university-level mathematics such as algebra, calculus, linear algebra, differential equations, transform methods and probability theory.

In this chapter we review these important mathematical topics and describe the Maple statements that can be used to perform the required operations in order to solve equations and inequalities, evaluate integrals and derivatives, find the solution of linear systems, solve ordinary and partial differential equations and difference (recurrence) equations, invert Laplace transforms and generating functions and find the joint probability distribution of functions of random variables.

---

[1] When the human element is involved in a decision environment, it may be impractical or impossible to construct a mathematical model of the decision problem. A case in point is the widely circulated "elevator problem" (Taha, [182, pp. 1–2]). This problem involved elevators in an office building with slow service and waiting times that were perceived to be long by the users who complained loudly. A mathematical solution based on queueing theory was found unsatisfactory. But after a careful examination of the problem, it was discovered that the users' complaints were due to boredom since the actual waiting times were not that long. A nonmathematical solution for this problem was proposed that involved installing full-length mirrors at the entrances to the elevators. Once the mirrors were installed, the users' complaints disappeared as they kept occupied by watching the others (and themselves) while waiting for the elevators.

We will see that Maple's ability to perform symbolic manipulations will permit us to easily solve difficult mathematical problems that are relevant in OR. For example, evaluating manually an integral such as $\int (k \log(x) - 2x^3 + 3x^2 + b)^4 \, dx$ is a nontrivial task since expanding the integrand results in 35 terms such as $-144k \log(x)x^5$ and $12k^3[\log(x)]^3x^2$. While integrating each of these terms may not be very difficult, the sheer task of mathematical bookkeeping that involves collecting the terms, simplifying them makes this an error-prone operation if performed manually. We will see that Maple can evaluate this integral (and many others) symbolically and produce the correct result in a fraction of a second. As always, it is important to check the results obtained, but that too can be achieved relatively easily with Maple's help.

3.2 Algebra

In this section we describe the use of the `solve()` function which can successfully solve a large class of equations and inequalities.

3.2.1 Solution of a Single Equation or Inequality

First consider the quadratic equation $ax^2 + bx + c = 0$. Maple solves this easily in terms of the coefficients and gives the well-known result.

```
>    restart: # QuadraticCubic.mws
>    Quadratic:=a* x^2+b* x+c;
```

$$Quadratic := a x^2 + b x + c$$

```
>    SolQ:=solve(Quadratic,x);
```

$$SolQ := \frac{1}{2} \frac{-b + \sqrt{b^2 - 4ac}}{a}, \frac{1}{2} \frac{-b - \sqrt{b^2 - 4ac}}{a}$$

```
>    SolQ[ 1] ; SolQ[ 2] ;
```

$$\frac{1}{2} \frac{-b + \sqrt{b^2 - 4ac}}{a}$$

$$\frac{1}{2} \frac{-b - \sqrt{b^2 - 4ac}}{a}$$

Next, we attempt to solve a cubic equation given by $x^3 - 2x^2 + x + 21 = 0$. The exact solution of this cubic involves complex numbers (denoted by I in Maple). The solution is simplified using the `evalf()` command. The correctness of the solution is also checked by substituting it back into the original equation.

```
>    Cubic:=x^3-2* x^2+x+21;
```

$$Cubic := x^3 - 2x^2 + x + 21$$

```
>    SolC:=solve(Cubic,x);
```

$$SolC := -\frac{1}{6}\%1^{(1/3)} - \frac{2}{3}\frac{1}{\%1^{(1/3)}} + \frac{2}{3}, \frac{1}{12}\%1^{(1/3)} + \frac{1}{3}\frac{1}{\%1^{(1/3)}}$$

$$+\frac{2}{3} + \frac{1}{2}I\sqrt{3}(-\frac{1}{6}\%1^{(1/3)} + \frac{2}{3}\frac{1}{\%1^{(1/3)}}), \frac{1}{12}\%1^{(1/3)}$$

$$+\frac{1}{3}\frac{1}{\%1^{(1/3)}} + \frac{2}{3} - \frac{1}{2}I\sqrt{3}(-\frac{1}{6}\%1^{(1/3)} + \frac{2}{3}\frac{1}{\%1^{(1/3)}})$$

$$\%1 := 2276 + 36\sqrt{3997}$$

```
>   evalf(SolC);
```

$$-2.135720892, 2.067860447 - 2.357265088\ I,$$
$$2.067860447 + 2.357265088\ I$$

```
>   seq(simplify(subs(x=SolC[i],Cubic)),i=1..3);
```

$$0, 0, 0$$

Maple also knows how to solve inequalities such as $|x + 3| < 2$ and $(x - 3 + a)(x - 1 + a)(x - 2 + a) < 0$ for the unknown x. The solution of these inequalities is found as follows.

```
>   restart: # Inequality.mws
>   solve(abs(x+3)<2,x);
```

$$\text{RealRange(Open}(-5),\ \text{Open}(-1))$$

```
>   solve((x-3+a)*(x-1+a)*(x-2+a) < 0, { x } );
```

$$\{x < 1 - a\},\ \{2 - a < x,\ x < 3 - a\}$$

Example 17 *Aircraft engines.* Suppose that an airplane makes a successful flight if at least 50% of its engines are operating. If $1 - p$ is the probability that an engine will fail independently of the other engines, we want to find a range of values for p such that a 8-engine plane is preferable to a 4-engine plane.

Using simple probabilistic reasoning it is easy to see that the number of engines operating is binomial. Thus, it follows that an 8-engine plane will make a successful flight with probability

$$P_8 = \sum_{n=4}^{8} \binom{8}{n} p^n (1 - p)^{8-n}$$

and a 4-engine plane will make a successful flight with probability

$$P_4 = \sum_{n=2}^{4} \binom{4}{n} p^n (1 - p)^{4-n}.$$

Solving the inequality $P_8 > P_4$ manually would clearly be a challenging task. But using Maple's solve() command and then evaluating the result with evalf(), we find that $p > .586$.

```
>    restart: # Airplanes.mws
>    Digits:=6;
```

$$Digits := 6$$

```
>    P[ 8] :=sum(binomial(8,n)*p^n* (1-p)^(8-n),n=4..8);
```

$$P_8 := 70\,p^4\,(1-p)^4 + 56\,p^5\,(1-p)^3 + 28\,p^6\,(1-p)^2 + 8\,p^7\,(1-p) + p^8$$

```
>    P[ 4] :=sum(binomial(4,n)*p^n* (1-p)^(4-n),n=2..4);
```

$$P_4 := 6\,p^2\,(1-p)^2 + 4\,p^3\,(1-p) + p^4$$

```
>    Sol:=solve(P[ 8]>P[ 4] ,{ p} ):
>    evalf(Sol);
```

$$\{p < -.238776\},\ \{p < 1.,\ .586886 < p\},\ \{1.22332 < p\}$$

The graphs of both P_8 and P_4 as a function of $p \in (0, 1)$ can be plotted after removing the # sign before the plot command.

```
>    #plot([ P[ 8] , P[ 4]] ,p=0..1);
```

Of course, sometimes it may be impossible for Maple to find the solution of an equation using the solve() command. But in some cases where there are no symbolic parameters, using the floating point version of the solver, i.e., fsolve(), may result in a solution.

Example 18 *Solution of a single-period stochastic inventory problem.* As an example, consider the equation $1 - \int_0^Q f(x)\,dx = \beta$ that arises in stochastic inventory theory. Here, Q is the order quantity decision variable, $f(x)$ is the demand density and β is a constant between 0 and 1 that is computed using parameters such as the order cost, unit revenue, shortage cost and salvage value. Supposing that $\beta = 0.65$ and that demand is normal with mean $\mu = 1000$ and standard deviation $\sigma = 100$, we first attempt to solve this equation with solve(), which fails. Next, trying fsolve() but with a specific interval over which the search should be conducted, i.e., from 500 to 1500, we find the correct solution.

```
>    restart: # NewsboyEquation.mws
>    with(stats);
```

$$[anova,\ describe,\ fit,\ importdata,\ random,\ statevalf,\ statplots,\ transform]$$

```
>    Eq:=1-statevalf[ cdf,normald[ 1000,100]] (Q)=0.65;
```

$$Eq := 1 - statevalf_{cdf,\ normald_{1000,\ 100}}(Q) = .65$$

```
>    solve(Eq,Q);
```

$$\mathrm{RootOf}(20\,statevalf_{cdf,\ normald_{1000,\ 100}}(\_Z) - 7)$$

```
>    fsolve(Eq,Q=500..1500);
```

$$961.4679534$$

Thus, the optimal order quantity should be $Q^* = 961.46$ units.

3.2.2   Solution of a System of Nonlinear Equations

Maple's powerful `solve()` function can also solve systems of equations. First consider a simple example with two equations: a circle defined as $x^2 + y^2 = r^2$ and a line passing through the origin defined as $y = ax$. We know from the geometry of these equations that the system has two solutions. Maple finds them as follows.

```
>   restart: # CircleLine.mws
>   circle:=x^2+y^2=r^2;
```

$$circle := x^2 + y^2 = r^2$$

```
>   line:=y=a*x;
```

$$line := y = ax$$

```
>   Sol:=solve({ circle,line} ,{ x,y} );
```

$$Sol := \{x = \text{RootOf}((1 + a^2)\_Z^2 - 1)\,r,$$
$$y = a\,\text{RootOf}((1 + a^2)\_Z^2 - 1)\,r\}$$

Since the solution is not in a form that is very convenient to interpret, we use the `allvalues()` command to obtain the set of two solutions.

```
>   allvalues(Sol);
```

$$\{x = \frac{r}{\sqrt{1 + a^2}},\ y = \frac{ar}{\sqrt{1 + a^2}}\},\ \{x = -\frac{r}{\sqrt{1 + a^2}},\ y = -\frac{ar}{\sqrt{1 + a^2}}\}$$

Example 19 *A constrained multi-item inventory problem.* As another example, consider the problem of finding the solution to the system

$$\frac{1}{2}h_1 - \eta\frac{\lambda_1}{Q_1^2} = 0$$

$$\frac{1}{2}h_2 - \eta\frac{\lambda_2}{Q_2^2} = 0$$

$$\frac{\lambda_1}{Q_1} + \frac{\lambda_2}{Q_2} = B$$

with unknowns Q_1, Q_2 and η. This problem arises in the optimization of a constrained multi-item inventory system where there is a constraint on the total number of orders that can be placed. The parameters h_i correspond to the holding costs, λ_i are the annual demands and B is the maximum number of orders that can be placed per year; see Hadley and Whitin [81, pp. 56–57]. We assume $h_1 = 10$, $h_2 = 5$, $\lambda_1 = 100$, $\lambda_2 = 150$ and $B = 50$ and solve the system with Maple.

```
>   restart: # ThreeEqns.mws
>   h[ 1] :=10; h[ 2] :=5; lambda[ 1] :=100;
    lambda[ 2] :=150; B:=50;
```

$$h_1 := 10$$

$$h_2 := 5$$

$$\lambda_1 := 100$$

$$\lambda_2 := 150$$

$$B := 50$$

> eq[1] := (1/2)*' h[1] ' -eta*' lambda[1] ' /Q[1] ^2=0;

$$eq_1 := \frac{1}{2} h_1 - \frac{\eta \lambda_1}{Q_1^2} = 0$$

> eq[2] := (1/2)*' h[2] ' -eta*' lambda[2] ' /Q[2] ^2=0;

$$eq_2 := \frac{1}{2} h_2 - \frac{\eta \lambda_2}{Q_2^2} = 0$$

> eq[3] :=' lambda[1] ' /Q[1] +' lambda[2] ' /Q[2] =' B' ;

$$eq_3 := \frac{\lambda_1}{Q_1} + \frac{\lambda_2}{Q_2} = B$$

> solve({ eq[1] ,eq[2] ,eq[3]} ,{ Q[1] ,Q[2] ,eta});

$$\{Q_2 = \text{RootOf}(-3 - 6\_Z + \_Z^2), \ Q_1 = \frac{1}{2} \text{RootOf}(-3 - 6\_Z + \_Z^2) + \frac{1}{2},$$
$$\eta = \frac{1}{80} (\text{RootOf}(-3 - 6\_Z + \_Z^2) + 1)^2\}$$

> Sol:=evalf(allvalues(%));

$$Sol := \{\eta = .6964101616, \ Q_2 = 6.464101616, \ Q_1 = 3.732050808\},$$
$$\{Q_2 = -.464101616, \ \eta = .003589838475, \ Q_1 = .267949192\}$$

Thus the correct solution with positive order quantities is found as $Q_1 = 3.73$, $Q_2 = 6.46$, and $\eta = 0.69$.

Example 20 *A marketing game.* As a final example consider a problem adapted from Ackoff and Sasieni [4, p. 345]. In a duopolistic environment, there are two companies that spend x_i dollars ($i = 1, 2$) on advertising. Each company's advertising effectiveness factor is a_i. Total market size is S and the market is split in proportion to the money spent on advertising. Thus, if company 1 spends x_1 and company 2 spends x_2 and if the factors are a_1 and a_2, then their sales are

$$\frac{S a_i x_i}{a_1 x_1 + a_2 x_2}, \quad i = 1, 2.$$

Suppose that the gross profit of each company is g_i, $i = 1, 2$, so the profit P_i of company i is obtained as

$$P_i = \frac{g_i S a_i x_i}{a_1 x_1 + a_2 x_2} - x_i, \quad i = 1, 2.$$

The solution to this competitive problem can be obtained by computing each company's "reaction curve"

$$R_i = \frac{\partial P_i}{\partial x_i}$$

and solving for x_1 and x_2 from $R_1 = R_2 = 0$.

```
>    restart: # Marketing.mws
```

```
>    P[ 1] :=g[ 1] * S* alpha[ 1] * x[ 1] /(alpha[ 1] * x[ 1]
     +alpha[ 2] * x[ 2] )-x[ 1] ;
```

$$P_1 := \frac{g_1\, S\, a_1\, x_1}{a_1\, x_1 + a_2\, x_2} - x_1$$

```
>    P[ 2] :=g[ 2] * S* alpha[ 2] * x[ 2] /(alpha[ 2] * x[ 2]
     +alpha[ 1] * x[ 1] )-x[ 2] ;
```

$$P_2 := \frac{g_2\, S\, a_2\, x_2}{a_1\, x_1 + a_2\, x_2} - x_2$$

```
>    R[ 1] :=diff(P[ 1] , x[ 1] )=0;  R[ 2] :=diff(P[ 2] , x[ 2] )=0;
```

$$R_1 := \frac{g_1\, S\, a_1}{a_1\, x_1 + a_2\, x_2} - \frac{g_1\, S\, a_1^{2}\, x_1}{(a_1\, x_1 + a_2\, x_2)^2} - 1 = 0$$

$$R_2 := \frac{g_2\, S\, a_2}{a_1\, x_1 + a_2\, x_2} - \frac{g_2\, S\, a_2^{2}\, x_2}{(a_1\, x_1 + a_2\, x_2)^2} - 1 = 0$$

The optimal advertising amounts x_1, x_2 and the profits P_1, P_2 are found as follows.

```
>    sol:=solve({ R[ 1] , R[ 2]} ,{ x[ 1] , x[ 2]} );
```

$$sol := \left\{ x_1 = \frac{g_1^{2}\, g_2\, S\, a_1\, a_2}{g_1^{2}\, a_1^{2} + 2\, a_1\, a_2\, g_1\, g_2 + a_2^{2}\, g_2^{2}}, \right.$$

$$\left. x_2 = \frac{g_2^{2}\, g_1\, S\, a_1\, a_2}{g_1^{2}\, a_1^{2} + 2\, a_1\, a_2\, g_1\, g_2 + a_2^{2}\, g_2^{2}} \right\}$$

```
>    assign(sol);
```

```
>    simplify(P[ 1] );  simplify(P[ 2] );
```

$$\frac{a_1^{2}\, S\, g_1^{3}}{(g_1\, a_1 + g_2\, a_2)^2}$$

$$\frac{a_2^{2}\, S\, g_2^{3}}{(g_1\, a_1 + g_2\, a_2)^2}$$

3.3   Calculus

Calculus plays an important role in mathematical modeling of operations research problems. In this section we describe some examples of calculus techniques whose implementation can be substantially simplified using Maple.

3.3.1   Limits

Computing the limits using limit() is easy with Maple. Consider the following examples.

```
>   restart: # Limits.mws
```

Maple knows that one dollar invested now that compounds continuously at a rate of r% per annum will grow to e^{rt} dollars in t years.

```
>   f[ 1] :=(1+r/n)^(n*t);
```

$$f_1 := (1 + \frac{r}{n})^{(nt)}$$

```
>   limit(f[ 1] ,n=infinity);
```

$$e^{(rt)}$$

The next limit operation needs l'Hôpital's rule, which Maple supplies.

```
>   f[ 2] :=(a^x-1)/x;
```

$$f_2 := \frac{a^x - 1}{x}$$

```
>   limit(f[ 2] ,x=0);
```

$$\ln(a)$$

Maple recognizes that the next limit does not exist. However, if the direction is specified as right or left, then Maple finds the limits.

```
>   f[ 3] :=abs(x)/x;
```

$$f_3 := \frac{|x|}{x}$$

```
>   limit(f[ 3] ,x=0);
```

$$\textit{undefined}$$

```
>   limit(f[ 3] ,x=0,left);
```

$$-1$$

```
>   limit(f[ 3] ,x=0,right);
```

$$1$$

Maple also knows how to take multidimensional limits.

```
>   limit((x+y)/(x*y),{ x=Pi/4,y=Pi/4} );
```

$$8\,\frac{1}{\pi}$$

In the next example Maple appears to be unable to find the result since its initial attempt fails to produce a limit. However, we realize that Maple has done the right thing since it does not know the sign of the parameter b. As soon as we inform Maple that $b > 0$, the limit is correctly found as 0.

```
>    f[ 4] :=sin(a*x)*exp(-b*x);
```

$$f_4 := \sin(a\,x)\,e^{(-b\,x)}$$

```
>    limit(f[ 4] ,x=infinity);
```

$$\lim_{x\to\infty}\,\sin(a\,x)\,e^{(-b\,x)}$$

```
>    assume(b>0);
>    limit(f[ 4] ,x=infinity);
```

$$0$$

3.3.2   Derivatives

Partial Derivatives

Maple has two means of performing ordinary and partial differentiation: either it uses the diff() function or the D() operator. The main (and subtle) difference between the two is that the D operator computes derivatives of *mappings* (i.e., functions), while diff() computes derivatives of *expressions*.

Here are some examples.

We first define an expression in two variables f:=a*x^2*y+b*x*y^2; and differentiate it with respect to x and y using diff(). We also find the mixed partial derivative diff(f,x,y); of f.

```
>    restart: # Derivatives.mws
>    f:=a*x^2*y+b*x*y^2;
```

$$f := a\,x^2\,y + b\,x\,y^2$$

```
>    diff(f,x); diff(f,y);
```

$$2\,a\,x\,y + b\,y^2$$

$$a\,x^2 + 2\,b\,x\,y$$

```
>    diff(f,x,y);
```

$$2\,a\,x + 2\,b\,y$$

Next, we define a mapping g (i.e., a function) involving the same expression as in f and perform differentiation using the D operator. Note that the output is also obtained as a function.

```
>    g:=(x,y)->a*x^2*y+b*x*y^2;
```

$$g := (x, y) \rightarrow a x^2 y + b x y^2$$

>   D[1] (g) ; D[2] (g) ;

$$(x, y) \rightarrow 2 a x y + b y^2$$

$$(x, y) \rightarrow a x^2 + 2 b x y$$

>   D[1, 2] (g) ;

$$(x, y) \rightarrow 2 a x + 2 b y$$

Finally, we see that using the D operator, one can differentiate composite functions.

>   f:=' f' : g:=' g' :

>   f:=g->cos (g^2);

$$f := g \rightarrow \cos(g^2)$$

>   u:=v->cos (v^2);

$$u := v \rightarrow \cos(v^2)$$

>   v:=x->x^3;

$$v := x \rightarrow x^3$$

>   D(u@v) (x);

$$-6 \sin(x^6) x^5$$

Example 21 *Inventory model with lost sales.* As another example, consider the objective function of a deterministic inventory problem where lost sales are allowed. Defining Q as the order quantity and T as the length of time during which sales are lost, Hadley and Whitin [81, p. 48] show that the average annual cost C is obtained as a function of Q and T. This cost function and its partial derivatives are easily found as follows.

>   restart: # LostSales.mws

>   C:=lambda* K/ (Q+lambda* T) + (1/2)* h* Q^2/ (Q+lambda* T)
     +pi* lambda^2* T/ (Q+lambda* T) ;

$$C := \frac{\lambda K}{Q + \lambda T} + \frac{1}{2} \frac{h Q^2}{Q + \lambda T} + \frac{\pi \lambda^2 T}{Q + \lambda T}$$

>   CQ:=normal (diff (C, Q)) ;

$$CQ := -\frac{1}{2} \frac{2 \lambda K - h Q^2 - 2 h Q \lambda T + 2 \pi \lambda^2 T}{(Q + \lambda T)^2}$$

>   CT:=normal (diff (C, T)) ;

$$CT := -\frac{1}{2} \frac{\lambda (2 \lambda K + h Q^2 - 2 \pi \lambda Q)}{(Q + \lambda T)^2}$$

We can even solve for the (Q, T) pair satisfying the necessary conditions. It is clear that the sign of $(\pi \lambda)^2 - 2\lambda K h$ plays an important role in the determination of the optimal solution. For details, see Hadley and Whitin [81, p. 49].

```
>    solve({ numer(CQ),numer(CT)} ,{ Q,T} );
```

$$\{Q = \%1, \ T = -\frac{-2\,K + \%1\,\pi}{\%1\,h - \pi\,\lambda}\}$$

$$\%1 := \text{RootOf}(2\,\lambda\,K + h\_Z^2 - 2\,\pi\,\lambda\_Z)$$

```
>    allvalues(%);
```

$$\left\{Q = \frac{1}{2}\frac{2\,\pi\,\lambda + 2\,\sqrt{\%1}}{h}, \ T = -\frac{-2\,K + \frac{1}{2}\frac{(2\,\pi\,\lambda + 2\,\sqrt{\%1})\,\pi}{h}}{\sqrt{\%1}}\right\},$$

$$\left\{Q = \frac{1}{2}\frac{2\,\pi\,\lambda - 2\,\sqrt{\%1}}{h}, \ T = \frac{-2\,K + \frac{1}{2}\frac{(2\,\pi\,\lambda - 2\,\sqrt{\%1})\,\pi}{h}}{\sqrt{\%1}}\right\}$$

$$\%1 := \pi^2\,\lambda^2 - 2\,h\,\lambda\,K$$

Leibniz's Rule

In some operations research problems (especially those arising in inventory theory) the objective function involves an integral whose upper and/or lower limits may be a function of the decision variable. Such integrals must be differentiated using Leibniz's rule of differentiation under the integral sign. Fortunately, Maple is aware of this rule as the following statements demonstrate.

```
>    restart: # Leibniz.mws
>    C:=Int(g(Q,t),t=a(Q)..b(Q));
```

$$C := \int_{a(Q)}^{b(Q)} g(Q, t)\,dt$$

```
>    diff(C,Q);
```

$$\int_{a(Q)}^{b(Q)} \tfrac{\partial}{\partial Q} g(Q, t)\,dt + (\tfrac{\partial}{\partial Q}\,b(Q))\,g(Q, b(Q))$$
$$- (\tfrac{\partial}{\partial Q}\,a(Q))\,g(Q, a(Q))$$

```
>    C:=' C' :
```

As an example, consider the following cost function that appears in the analysis of a continuous-review inventory model where lead-time randomness can be reduced by investment; see Gerchak and Parlar [70]. The decision variables are Q (order quantity), r (reorder point) and α (investment amount). The partial derivatives of the cost function w.r.t. the three variables are easily computed using the `diff()` function.

```
>   C:=K* lambda/Q+h* (Q/2+r-mu) +b* lambda* alpha
    * Int (FBar(x),x=u(r,alpha)..infinity)/Q
    +G(Q,alpha);
```

$$C := \frac{K\,\lambda}{Q} + h\left(\frac{1}{2}\,Q + r - \mu\right) + \frac{b\,\lambda\,a\,\int_{u(r,\,a)}^{\infty} \mathrm{FBar}(x)\,dx}{Q}$$
$$+ G(Q,\,a)$$

```
>   CQ:=diff(C,Q);
```

$$CQ := -\frac{K\,\lambda}{Q^2} + \frac{1}{2}\,h - \frac{b\,\lambda\,a\,\int_{u(r,\,a)}^{\infty} \mathrm{FBar}(x)\,dx}{Q^2}$$
$$+ \left(\frac{\partial}{\partial Q}\,G(Q,\,a)\right)$$

```
>   Cr:=diff(C,r);
```

$$Cr := h - \frac{b\,\lambda\,a\,\left(\frac{\partial}{\partial r}\,u(r,\,a)\right)\mathrm{FBar}(u(r,\,a))}{Q}$$

```
>   Calpha:=diff(C,alpha);
```

$$Calpha := \frac{b\,\lambda\,\int_{u(r,\,a)}^{\infty} \mathrm{FBar}(x)\,dx}{Q}$$
$$- \frac{b\,\lambda\,a\,\left(\frac{\partial}{\partial a}\,u(r,\,a)\right)\mathrm{FBar}(u(r,\,a))}{Q} + \left(\frac{\partial}{\partial a}\,G(Q,\,a)\right)$$

As discussed in Gerchak and Parlar [70], these equations are solved numerically to find the optimal decisions.

Implicit Differentiation

In some problems it may be impossible to express uniquely the dependent variable in terms of the independent variable. For example, the expression (i.e., the relation) $x^2 + y^2 = r^2$ represents a circle of radius r, but solving for the dependent variable y one obtains $y = \pm\sqrt{r^2 - x^2}$, indicating that y is not a function of x. In those cases, to compute the derivative of y w.r.t. x, we use implicit differentiation that has been implemented by the Maple function `implicitdiff(f,y,x)`, which computes dy/dx, the (implicit partial) derivative of y with respect to x.

In the first example, we implicitly differentiate $x^2 + y^2 = r^2$ and obtain $dy/dx = -x/y$.

```
>   restart: # ImplicitDiff.mws
>   implicitdiff(x^2+y^2=r^2,y,x);
```

$$-\frac{x}{y}$$

Example 22 *Continuous-review inventory problem.* A more difficult case involves (yet another) inventory example where the partial derivatives of a particular cost function $C(Q,r)$ are found as

$$
\begin{aligned}
C_Q &= -\lambda[A + \pi\,B(r)] + \frac{1}{2}hQ^2 = 0\\
C_r &= hQ - \pi\lambda K(r) = 0
\end{aligned}
$$

where $B(r)$ and $K(r)$ are functions of r. Implicitly differentiating C_Q and C_r to compute dr/dQ, we obtain the following.

> `CQ:=-lambda*(A+pi*B(r))+h*Q^2/2;`

$$CQ := -\lambda\,(A + \pi\,B(r)) + \frac{1}{2}h\,Q^2$$

> `Cr:=h*Q+pi*lambda*(-K(r));`

$$Cr := h\,Q - \pi\,\lambda\,K(r)$$

> `implicitdiff(CQ,r,Q);  # r' (Q) in CQ`

$$\frac{h\,Q}{\lambda\,\pi\,D(B)(r)}$$

> `implicitdiff(Cr,r,Q);  # r' (Q) in Cr`

$$\frac{h}{\lambda\,\pi\,D(K)(r)}$$

These results can be used to investigate the shape of the curves represented by $C_Q = 0$ and $C_r = 0$ in the Qr-plane to see if there is a unique solution to the problem; see Hadley and Whitin [81, p. 170].

Maple can also compute the second-order implicit derivatives d^2r/dQ^2 for each relation and obtain the following results.

> `implicitdiff(CQ,r,Q$2);  # r" (Q) in CQ`

$$-\frac{h\,((D^{(2)})(B)(r)\,h\,Q^2 - \lambda\,\pi\,D(B)(r)^2)}{\lambda^2\,\pi^2\,D(B)(r)^3}$$

> `implicitdiff(Cr,r,Q$2);  # r" (Q) in Cr`

$$-\frac{(D^{(2)})(K)(r)\,h^2}{\lambda^2\,\pi^2\,D(K)(r)^3}$$

3.3.3 Integrals

Maple possesses a truly impressive knowledge of integration. A symbolic integral such as $\int_a^b x^2\,dx$ can easily be evaluated to obtain $(b^3 - a^3)/3$:

> `restart:  # Integrals.mws`

```
>   Int(x^2,x=a..b);
```

$$\int_a^b x^2\,dx$$

```
>   value(%);
```

$$\frac{1}{3}b^3 - \frac{1}{3}a^3$$

This is, of course, a very simple problem that could have been quickly solved by anyone who has some passing knowledge of integrals. But consider the following problem that arose in aerospace engineering in the study of turbulence and boundary layers.[2]

Example 23 *An indefinite integral arising in aerospace engineering.* The function to be integrated is given by $f(x) = (k \ln(x) - 2x^3 + 3x^2 + b)^4$.

```
>   f:=(k*log(x)-2*x^3+3*x^2+b)^4;
```

$$f := (k \ln(x) - 2x^3 + 3x^2 + b)^4$$

When expanded, the expression whose integral is sought involves 35 terms. Evaluating this integral manually may be possible if one is willing to spend several hours to check and recheck every term in the integration process. Even then, there still is a possibility that human errors may creep into the computations and one may be unsure of manually obtained results.[3] However, by using Maple, we quickly find the result.

```
>   expand(f); nops(%);
```

$$16x^{12} - 96x^{11} + 216x^{10} + b^4 + 108\,k\ln(x)\,x^6 + 4\,k^3\ln(x)^3\,b$$
$$+\,24\,k^2\ln(x)^2\,x^6 - 72\,k^2\ln(x)^2\,x^5 + 54\,k^2\ln(x)^2\,x^4$$
$$+\,6\,k^2\ln(x)^2\,b^2 + 12\,k^3\ln(x)^3\,x^2 - 8\,k^3\ln(x)^3\,x^3$$
$$+\,108\,x^6\,b + 144\,x^8\,b + k^4\ln(x)^4 + 24\,x^6\,b^2$$
$$-\,216\,x^7\,b - 72\,x^5\,b^2 - 8\,x^3\,b^3 + 54\,x^4\,b^2 + 12\,x^2\,b^3$$
$$-\,216\,x^9 + 81\,x^8 + 48\,k\ln(x)\,x^6\,b - 144\,k\ln(x)\,x^5\,b$$
$$-\,24\,k\ln(x)\,x^3\,b^2 + 108\,k\ln(x)\,x^4\,b$$
$$+\,36\,k\ln(x)\,x^2\,b^2 - 24\,k^2\ln(x)^2\,x^3\,b$$
$$+\,36\,k^2\ln(x)^2\,x^2\,b - 32\,k\ln(x)\,x^9 + 144\,k\ln(x)\,x^8$$
$$-\,216\,k\ln(x)\,x^7 + 4\,k\ln(x)\,b^3 - 32\,x^9\,b$$

$$35$$

---

[2]This problem appeared in a Macsyma advertisement in the September 1984 issue of *Scientific American.*

[3]Legend has it that the engineer who was working on this problem tried to evaluate the integral for more than three weeks with pencil and paper, always arriving at a different result. With Maple the correct solution is found in a fraction of a second.

```
>    Result:=int(f,x);
```

$$Result := k^4\,x\,\ln(x)^4 - 4\,k^4\,x\,\ln(x)^3 + 12\,k^4\,x\,\ln(x)^2$$
$$- 24\,k^4\,x\,\ln(x) + 3\,k^2\,b\,x^4\,\ln(x) + 4\,k\,b\,x^6$$
$$- \frac{108}{25}\,k\,b\,x^5 - \frac{108}{25}\,k^2\,x^5\,\ln(x) - 2\,k^3\,x^4\,\ln(x)^3$$
$$+ \frac{3}{2}\,k^3\,x^4\,\ln(x)^2 - \frac{3}{4}\,k^3\,x^4\,\ln(x) + \frac{24}{7}\,k^2\,x^7\,\ln(x)^2$$
$$- \frac{48}{49}\,k^2\,x^7\,\ln(x) + 4\,k^2\,x^6\,\ln(x) - 4\,k\,b^2\,x^3 - 24\,k^3\,b\,x$$
$$+ 12\,k^2\,b^2\,x - 4\,k^3\,x^3\,\ln(x)^2 + \frac{8}{3}\,k^3\,x^3\,\ln(x) - 4\,k\,b^3\,x$$
$$- 6\,k\,b^2\,x^4\,\ln(x) + 4\,k\,b^3\,x\,\ln(x) + 4\,k^3\,b\,x\,\ln(x)^3$$
$$- 12\,k^3\,b\,x\,\ln(x)^2 + 24\,k^3\,b\,x\,\ln(x) - 6\,k^2\,b\,x^4\,\ln(x)^2$$
$$+ 6\,k^2\,b^2\,x\,\ln(x)^2 - 12\,k^2\,b^2\,x\,\ln(x) - 8\,k^2\,b\,x^3\,\ln(x)$$
$$+ \frac{8}{3}\,k^2\,x^3\,b + \frac{16}{13}\,x^{13} + \frac{8}{25}\,k\,x^{10} - \frac{16}{9}\,k\,x^9 - \frac{16}{5}\,b\,x^{10}$$
$$+ b^4\,x + \frac{24}{7}\,b^2\,x^7 + \frac{48}{343}\,k^2\,x^7 - \frac{108}{49}\,k\,x^7 + \frac{108}{125}\,k^2\,x^5$$
$$- \frac{8}{9}\,k^3\,x^3 - \frac{2}{3}\,k^2\,x^6 - 2\,b^3\,x^4 + \frac{27}{8}\,k\,x^8 - \frac{3}{4}\,k^2\,b\,x^4$$
$$- \frac{16}{5}\,k\,x^{10}\,\ln(x) - \frac{48}{49}\,k\,b\,x^7 + \frac{3}{2}\,k\,b^2\,x^4$$
$$+ \frac{48}{7}\,k\,b\,x^7\,\ln(x) + 24\,k^4\,x - 8\,x^{12} + \frac{216}{11}\,x^{11}$$
$$- \frac{108}{5}\,x^{10} - 12\,k^2\,\ln(x)^2\,x^6 + \frac{54}{5}\,k^2\,\ln(x)^2\,x^5$$
$$+ 4\,k^3\,\ln(x)^3\,x^3 - 27\,x^8\,b - 12\,x^6\,b^2 + \frac{108}{7}\,x^7\,b$$
$$+ \frac{54}{5}\,x^5\,b^2 + 4\,x^3\,b^3 + 9\,x^9 - 24\,k\,\ln(x)\,x^6\,b$$
$$+ \frac{108}{5}\,k\,\ln(x)\,x^5\,b + 12\,k\,\ln(x)\,x^3\,b^2$$
$$+ 12\,k^2\,\ln(x)^2\,x^3\,b + 16\,k\,\ln(x)\,x^9 - 27\,k\,\ln(x)\,x^8$$
$$+ \frac{108}{7}\,k\,\ln(x)\,x^7 + 16\,x^9\,b + \frac{3}{16}\,k^3\,x^4$$

```
>    simplify(diff(Result,x)-f);
```

$$0$$

Differentiating the (horrendous looking) result and subtracting it from the original expression gives 0, indicating that Maple was able to find the correct solution.

Example 24 *Definite integral of the unit normal density.* Maple's knowledge of integration also includes numerical evaluation of integrals for which no closed-form solution is available. For example, to evaluate the probability that the unit normal random variable $Z \sim N(0, 1)$ takes a value between, say, -0.5 and 0.35, we must compute

$$\Pr(-0.5 \leq Z \leq 0.35) = \int_{-0.5}^{0.35} \phi \, dz$$

where

$$\phi = \frac{1}{\sqrt{2\pi}} e^{-\frac{1}{2}z^2}.$$

Since functions of the form e^{-z^2} cannot be evaluated analytically, this integral has to be computed numerically.

```
>    restart: # NormalDensity.mws
>    phi:=(1/('sqrt(2*Pi)'))*exp(-z^2/2);
```

$$\phi := \frac{e^{(-1/2z^2)}}{\sqrt{2\pi}}$$

```
>    Int('phi',z=-0.5 .. .35); evalf(%);
```

$$\int_{-.5}^{.35} \phi \, dz$$

$$.3282931124$$

Thus, $\Pr(-0.5 \leq Z \leq 0.35) = .3282931124$. The fact that this is the correct result can be seen by finding the probability by an alternative method that uses Maple's knowledge of the distribution of the normal density. Defining $\Phi(z) = \int_{-\infty}^{z} \phi(u) \, du$, we find the same result from $\Phi(0.35) - \Phi(-0.5) = .3282931124$.

```
>    with(stats):
>    Phi:=z->statevalf[cdf,normald[0,1]](z);
```

$$\Phi := statevalf_{cdf, normald_{0,1}}$$

```
>    Phi(.35)-Phi(-0.50);
```

$$.3282931124$$

Example 25 *Integration of the exponential density.* As another example, let us consider the integration of the exponential density $f(t) = \lambda e^{-\lambda t}$, $t \geq 0$. Since $f(t)$ is a density, we should have $\int_0^\infty \lambda e^{-\lambda t} \, dt = 1$, provided that $\lambda > 0$.

```
>    restart: # ExponentialDensity.mws
>    f:=lambda*exp(-lambda*t);
```

$$f := \lambda e^{(-\lambda t)}$$

```
>   int(f,t=0..infinity);
```

Definite integration: Can't determine if the integral is convergent.

Need to know the sign of --> lambda

Will now try indefinite integration and then take limits.

$$\lim_{t \to \infty} -e^{(-\lambda t)} + 1$$

Naturally, Maple is unable to evaluate the integral since we have not specified the sign of the parameter λ. As soon as we indicate, using assume(), that $\lambda > 0$, Maple correctly integrates the density and finds $\int_0^\infty \lambda e^{-\lambda t}\, dt = 1$.

```
>   assume(lambda>0);
>   int(f,t=0..infinity);
```

$$1$$

Example 26 *A multiple integral.* Maple can also evaluate multiple integrals. As an example, consider a problem in probability where X and Y are independent uniform random variables distributed over the interval $(0, a)$ and we wish to evaluate the distribution $F_W(w)$ of the product $W = XY$. It can be shown (see, Sivazlian and Stanfel [177, p. 54]) that this problem can be solved by evaluating the double integral

$$F_W(w) = 1 - \int_{w/a}^{a} \int_{w/y}^{a} \frac{1}{a^2}\, dx\, dy, \quad 0 < w \le a^2.$$

We find the result using Maple's help:

```
>   restart: # DoubleIntegral.mws
>   F:=1-Int(Int(1/a^2,x=w/y..a),y=w/a..a);
    F:=unapply(value(F),w);
```

$$F := 1 - \int_{\frac{w}{a}}^{a} \int_{\frac{w}{y}}^{a} \frac{1}{a^2}\, dx\, dy$$

$$F := w \to 1 + \frac{-a^2 + w + w \ln(a) - w \ln(\frac{w}{a})}{a^2}$$

```
>   simplify(F(w));
```

$$\frac{w\left(1 + \ln(a) - \ln(\frac{w}{a})\right)}{a^2}$$

```
>   limit(F(w),w=0);  F(a^2);
```

$$0$$

$$1$$

3.3.4   Finite Sums and Infinite Series

Maple also knows how to compute finite sums and infinite series as we demonstrate in the next few examples.

```
>    restart: # SumsSeries.mws
>    Sum(a*x^k,k=0..n-1); value(%);
```

$$\sum_{k=0}^{n-1} a x^k$$

$$\frac{a x^n}{x-1} - \frac{a}{x-1}$$

The next sum is obtained correctly, but one should be careful in using it, as the result applies only when $-1 < x < 1$.

```
>    Sum(a*x^k,k=0..infinity); value(%);
```

$$\sum_{k=0}^{\infty} a x^k$$

$$-\frac{a}{x-1}$$

```
>    Sum(k^3*x^k,k=1..infinity); value(%);
```

$$\sum_{k=1}^{\infty} k^3 x^k$$

$$x \left(\frac{1}{(x-1)^2} - 6\frac{x}{(x-1)^3} + 6\frac{x^2}{(x-1)^4} \right)$$

```
>    Sum(x^k/k,k=1..infinity); value(%);
```

$$\sum_{k=1}^{\infty} \frac{x^k}{k}$$

$$-\ln(1-x)$$

```
>    factor(simplify(sum(k,k=1..n)));
```

$$\frac{1}{2} n (n+1)$$

```
>    factor(simplify(sum(k^2,k=1..n)));
```

$$\frac{1}{6} n (n+1) (2n+1)$$

```
>    factor(simplify(sum(k^3,k=1..n)));
```

$$\frac{1}{4} n^2 (n+1)^2$$

Example 27 *Geometric random variable.* The next example concerns the sum of the probabilities of the geometric random variable N such that $\Pr(N = k) = p(1 - p)^k$ for $k = 0, 1, 2, \ldots$ Provided that $0 < p < 1$, the sum $\sum_{k=0}^{\infty} p(1 - p)^k$ converges to 1. Since Maple does not check for the correct interval of convergence of these infinite sums, it is up to the user to find out about the convergence properties.

```
> Sum(rho* (1-rho)^k,k=0..infinity); value(%);
```

$$\sum_{k=0}^{\infty} p(1 - p)^k$$

$$1$$

Example 28 *Moment generating function of the binomial.* The moment generating function of the binomial random variable is defined as $\phi(t) = \sum_{k=0}^{n} e^{tk} \binom{n}{k} p^k (1 - p)^{n-k}$. Maple easily finds the result that involves the use of the binomial theorem, i.e., $(x + y)^n = \sum_{k=0}^{n} \binom{n}{k} x^k y^{n-k}$.

```
> mgf:=Sum(exp(t* k)*binomial(n,k)*p^k* (1-p)^(n-k),
  k=0..n);
```

$$mgf := \sum_{k=0}^{n} e^{(t\,k)} \text{binomial}(n,\ k)\, p^k\, (1 - p)^{(n-k)}$$

```
> assume(n,integer); simplify(value(mgf));
```

$$(1 - p + e^t\, p)^n$$

3.4 Linear Algebra

Linear algebra is probably as important as calculus in the modeling and analysis of operations research problems. In this section we demonstrate Maple's deep knowledge of linear algebra by providing examples relevant to operations research.

3.4.1 Matrix Operations

A matrix in Maple is represented as a two-dimensional array with rows and columns indexed from 1. Matrices can be input either directly, as a two-dimensional array, or using the matrix command in the linear algebra package.

Consider the following examples.

After loading the linear algebra package `linalg`, we define two matrices **A** and **B** each having a symbolic entry. Matrix algebra (such as multiplying the matrix **A** by a constant γ, adding **A** and **B**, multiplying **A** and **B**) is performed using the command `evalm()`.

```
> restart: # Matrices.mws
```

```
>   with(linalg):
```

Warning, new definition for norm

Warning, new definition for trace

```
>   A:=matrix(2,2,[ 1,3,2,a] );
```

$$A := \begin{bmatrix} 1 & 3 \\ 2 & a \end{bmatrix}$$

```
>   B:=matrix(2,2,[ b,4,3,7] );
```

$$B := \begin{bmatrix} b & 4 \\ 3 & 7 \end{bmatrix}$$

```
>   evalm(gamma*A);
```

$$\begin{bmatrix} \gamma & 3\gamma \\ 2\gamma & \gamma a \end{bmatrix}$$

```
>   transpose(A);
```

$$\begin{bmatrix} 1 & 2 \\ 3 & a \end{bmatrix}$$

```
>   evalm(A+B);
```

$$\begin{bmatrix} 1+b & 7 \\ 5 & a+7 \end{bmatrix}$$

The determinant of a matrix is computed with the det() command. We see in this example that as long as the parameter $a \neq 6$, the **A** matrix should possess an inverse, $\mathbf{A}^{-1}$, which we compute with the inverse() command. The result is checked by multiplying the inverse by the original matrix.

```
>   det(A);
```

$$a - 6$$

```
>   AI:=inverse(A);
```

$$AI := \begin{bmatrix} \dfrac{a}{a-6} & -3\dfrac{1}{a-6} \\ -2\dfrac{1}{a-6} & \dfrac{1}{a-6} \end{bmatrix}$$

```
>   simplify(evalm(A &* AI));
```

$$\begin{bmatrix} 1 & 0 \\ 0 & 1 \end{bmatrix}$$

An easy way to input an $n \times n$ identity matrix is to use the array(identity ,1..n,1..n) command.

```
>   ID := array(identity,1..2,1..2);
```

$$ID := \text{array}(identity, 1..2, 1..2, [])$$

```
>   print(ID);
```

$$\begin{bmatrix} 1 & 0 \\ 0 & 1 \end{bmatrix}$$

```
>   evalm(ID &* A);
```

$$\begin{bmatrix} 1 & 3 \\ 2 & a \end{bmatrix}$$

Vectors can be input in a manner similar to matrices. However, the shape of a vector (i.e., whether it is a row or column vector) is decided by Maple when the vector is involved in a vector/matrix multiplication.

```
>   C:=vector(2,[ 3,c] );
```

$$C := [3, \, c]$$

```
>   CA:=evalm(C &* A);
```

$$CA := [3 + 2c, \, 9 + ca]$$

```
>   AC:=evalm(A &* C);
```

$$AC := [3 + 3c, \, 6 + ca]$$

If it is important to have control over the actual dimensions of a vector, it should be input as an $n \times 1$ matrix (i.e., a column vector) or as a $1 \times n$ matrix (i.e., a row vector).

```
>   evalm(C &* C);
```

```
Error, (in evalm/evaluate) improper op or subscript
selector
```

```
>   C:=matrix(2,1,[ 3,c] );
```

$$C := \begin{bmatrix} 3 \\ c \end{bmatrix}$$

```
>   evalm(A &* C);
```

$$\begin{bmatrix} 3 + 3c \\ 6 + ca \end{bmatrix}$$

```
>   C:=matrix(1,2,[ 3,c] );
```

$$C := \begin{bmatrix} 3 & c \end{bmatrix}$$

```
>   evalm(C &* A);
```

$$\begin{bmatrix} 3 + 2c & 9 + ca \end{bmatrix}$$

The next example evaluates the product $\mathbf{C}_{1\times2}\mathbf{A}_{2\times2}\mathbf{C}_{2\times1}^T$, which results in a scalar.

```
>   evalm(C &* A &* transpose(C));
```

$$\begin{bmatrix} 9 + 6c + (9 + ca)c \end{bmatrix}$$

3.4.2   Solution of Simultaneous Linear Equations

Solving linear equations plays an important role in many branches of operations research. Consider the linear system $\mathbf{Ax} = \mathbf{b}$ where $\mathbf{A}$ is an $n \times n$ matrix, $\mathbf{b}$ is an $n \times 1$ column vector and $\mathbf{x}$ is also an $n \times 1$ column vector of unknowns. If $\mathbf{A}$ is nonsingular (i.e., if its inverse exists), then we may solve for $\mathbf{x}$ as $\mathbf{x} = \mathbf{A}^{-1}\mathbf{b}$. The computations involving $\mathbf{A}^{-1}\mathbf{b}$ can be performed in Maple with the `linsolve()` command.

Consider a simple example.

```
>   restart: # SLE.mws
>   with(linalg):
```

Warning, new definition for norm

Warning, new definition for trace

```
>   A:=matrix([[ 3,1,1] ,[ 1,2,1] ,[ 1,1,7]] );
```

$$A := \begin{bmatrix} 3 & 1 & 1 \\ 1 & 2 & 1 \\ 1 & 1 & 7 \end{bmatrix}$$

We find (before solving the problem) that the inverse of $\mathbf{A}$ exists; hence we should be able to compute the solution vector $\mathbf{x}$.

```
>   inverse(A);
```

$$\begin{bmatrix} \dfrac{13}{32} & \dfrac{-3}{16} & \dfrac{-1}{32} \\ \dfrac{-3}{16} & \dfrac{5}{8} & \dfrac{-1}{16} \\ \dfrac{-1}{32} & \dfrac{-1}{16} & \dfrac{5}{32} \end{bmatrix}$$

```
>   b:=vector([ 10,20,30] );
```

$$b := [10, 20, 30]$$

Using the `geneqns()` function, we can view the linear equations arising from the system $\mathbf{Ax} = \mathbf{b}$.

```
>   geneqns(A,[ x,y,z] ,b);
```

$$\{3x + y + z = 10,\ x + 2y + z = 20,\ x + y + 7z = 30\}$$

```
>   linsolve(A,b);
```

$$\left[\frac{-5}{8}, \frac{35}{4}, \frac{25}{8}\right]$$

Thus, we find $\mathbf{x} = (-5/8, 35/4, 25/8)$ as the solution of the system.

It is, of course, possible that some linear systems may have infinitely many solutions or no solutions at all. This is determined by computing the rank of a matrix. The rank of a matrix $\mathbf{A}$ [denoted by $r(\mathbf{A})$] is defined as follows: If there exists at least one nonsingular square submatrix in **A** of order k and if all square submatrices of order $k + i$, $i \geq 1$ (if any) are singular, then $r(\mathbf{A}) = k$.

Consider an $n \times n$ matrix **A** and an $n \times 1$ vector **b**. The **A** matrix augmented with **b** is denoted by $(\mathbf{A} \mid \mathbf{b})$. The following rules are useful in determining the nature of the solution:

- If, $r(\mathbf{A}) = r(\mathbf{A} \mid \mathbf{b}) = n$, then the system has a unique solution.

- If $r(\mathbf{A}) = r(\mathbf{A} \mid \mathbf{b}) = m < n$, then the system $\mathbf{Ax} = \mathbf{b}$ has an infinite number of solutions.

- Finally, if $r(\mathbf{A}) < r(\mathbf{A} \mid \mathbf{b})$, then the system has no solution.

These results are illustrated in the following examples.

The example with $n = 3$ had a unique solution and this is again noted by the fact that $r(\mathbf{A}) = r(\mathbf{A} \mid \mathbf{b}) = 3$.

```
>    Ab:=augment(A,b);
```

$$Ab := \begin{bmatrix} 3 & 1 & 1 & 10 \\ 1 & 2 & 1 & 20 \\ 1 & 1 & 7 & 30 \end{bmatrix}$$

```
>    rank(A); rank(Ab);
```

$$3$$

$$3$$

For the next example with $n = 2$, we see that $r(\mathbf{A}) = r(\mathbf{A} \mid \mathbf{b}) = 1 < 2$; thus the system has infinitely many solutions.

```
>    A:=matrix([[ 3,1] ,[ 6,2]]);
```

$$A := \begin{bmatrix} 3 & 1 \\ 6 & 2 \end{bmatrix}$$

```
>    b:=vector([ 4,8] );
```

$$b := [4, 8]$$

```
>    geneqns(A,[ x,y,z] ,b); linsolve(A,b);
```

$$\{3x + y = 4, 6x + 2y = 8\}$$

$$[t_1, -3\_t_1 + 4]$$

```
>    Ab:=augment(A,b);
```

$$Ab := \begin{bmatrix} 3 & 1 & 4 \\ 6 & 2 & 8 \end{bmatrix}$$

```
>    rank(A); rank(Ab);
```

$$1$$

$$1$$

The final example with $n = 2$ involves a system that has no solution. This is observed by the fact that $r(\mathbf{A}) = 1 < r(\mathbf{A} \mid \mathbf{b}) = 2$.

```
>   A:=matrix([[ 1,1] ,[ 1,1]]);
```

$$A := \begin{bmatrix} 1 & 1 \\ 1 & 1 \end{bmatrix}$$

```
>   b:=vector([ 2,4]); linsolve(A,b);
```

$$b := [2, 4]$$

```
>   geneqns(A,[ x,y] ,b); linsolve(A,b);
```

$$\{x + y = 2, \ x + y = 4\}$$

```
>   Ab:=augment(A,b);
```

$$Ab := \begin{bmatrix} 1 & 1 & 2 \\ 1 & 1 & 4 \end{bmatrix}$$

```
>   rank(A); rank(Ab);
```

$$1$$

$$2$$

Homogeneous System of Linear Equations

As we pointed out, every system of linear equations has either (i) one solution, (ii) infinitely many solutions or (iii) no solutions at all. We now consider the case of a homogeneous system of n linear equations in n unknowns, i.e., $\mathbf{Ax} = \mathbf{0}$, and discuss the conditions under which the system has solutions.

First note that if $\det(\mathbf{A}) \neq 0$, that is, if $\mathbf{A}^{-1}$ exists, then the zero vector $\mathbf{x} = \mathbf{A}^{-1}\mathbf{0} = \mathbf{0}$ is a trivial solution. However, if $\det(\mathbf{A}) = 0$, then infinitely many nontrivial solutions exist, as we see in the following example.

```
>   restart: # HomoSLE.mws
>   with(linalg):
```

Warning, new definition for norm

Warning, new definition for trace

```
>   A:=matrix([[ 1,5,3] ,[ 5,1,-1] ,[ 1,2,1]]);
```

$$A := \begin{bmatrix} 1 & 5 & 3 \\ 5 & 1 & -1 \\ 1 & 2 & 1 \end{bmatrix}$$

```
>   det(A);
```

$$0$$

```
>   basis([ col(A,1),col(A,2),col(A,3)]);
```

$$[[1, 5, 1], [5, 1, 2]]$$

```
>   b:=vector([ 0,0,0] );
```
$$b := [0, 0, 0]$$
```
>   geneqns (A,[ x,y,z] ,b);
```
$$\{x + 5y + 3z = 0,\ 5x + y - z = 0,\ x + 2y + z = 0\}$$
```
>   linsolve (A,b);
```
$$\left[-\frac{1}{2}\_t_1,\ \_t_1,\ -\frac{3}{2}\_t_1 \right]$$

Since we find that $x = -t/2$, $y = -t$ and $z = -3t/2$, the system has infinitely many nontrivial solutions.

3.4.3 Eigenvalues, Eigenvectors and Diagonalization

Let A be an arbitrary square matrix of order n, and let x be an n-dimensional nonzero vector. The vector x is called an *eigenvector* of A if Ax is a scalar multiple of x, i.e., $Ax = \lambda x$ for some scalar λ. The scalar λ is called the *eigenvalue* of A and x is said to be an eigenvector corresponding to λ. To find the eigenvalues of A we rewrite $Ax = \lambda x$ as $Ax = \lambda Ix$ (where I is the $n \times n$ identity matrix) or equivalently as

$$(A - \lambda I)x = 0. \tag{3.1}$$

For λ to be an eigenvalue, the system (3.1) must have a nonzero solution $x \neq 0$. But this implies that the columns of $(A - \lambda I)$ are linearly dependent and $(A - \lambda I)$ is singular. Hence, the scalar λ is an eigenvalue of A if and only if $\det(A - \lambda I) = 0$. Expanding this gives a polynomial in terms of λ, which must be solved to find the eigenvalues. (Some eigenvalues may be repeated.) To find an eigenvector x_i corresponding to the eigenvalue λ_i, we form the system $(A - \lambda_i I)x_i = 0$ and solve for the vector x_i.

These results are useful in determining whether the original matrix A can be diagonalized. That is, does there exist a Q matrix such that $Q^{-1}AQ = D$ is a diagonal matrix? It can be shown that the matrix A is diagonalizable if and only if A has linearly independent eigenvectors and that Q is formed using eigenvectors of A. An additional result states that if the eigenvalues of A are distinct, then A can be diagonalized; see Anton [7, p. 277]. If A can be diagonalized, then the nth power of A can be computed as $A^n = QD^nQ^{-1}$. Since D is diagonal, its nth power can be computed much more quickly than the nth power of A.

Consider the following example to illustrate these ideas.

Example 29 *Transition probabilities of a three-state Markov chain.* Suppose we have a three-state discrete-time Markov chain with the transition probability matrix A given as

$$A = \begin{pmatrix} 0 & \frac{1}{2} & \frac{1}{2} \\ \frac{1}{2} & 0 & \frac{1}{2} \\ \frac{1}{2} & \frac{1}{2} & 0 \end{pmatrix}.$$

We wish to compute, say, the 10th power of this matrix to find the probability that the chain will be in some state $j = 1, 2, 3$ after 10 transitions given that it is in some state $i = 1, 2, 3$ now. In order to do this, we find the eigenvalues and the corresponding eigenvectors of A, diagonalize it as D and form the Q matrix. Computing QD^nQ^{-1} then gives the nth power of A.

```
>    restart: # EigensMC3.mws
```

```
>    with(linalg):
```

Warning, new definition for norm

Warning, new definition for trace

```
>    A:=matrix(3,3,[ 0,1/2,1/2, 1/2,0,1/2, 1/2,1/2,0] );
```

$$A := \begin{bmatrix} 0 & \dfrac{1}{2} & \dfrac{1}{2} \\ \dfrac{1}{2} & 0 & \dfrac{1}{2} \\ \dfrac{1}{2} & \dfrac{1}{2} & 0 \end{bmatrix}$$

We could use Maple's charpoly() function to find the characteristic polynomial, i.e., $\det(A - \lambda I) = 0$, and then compute the polynomial's roots for the eigenvalues, which are obtained as $\lambda_1 = 1, \lambda_2 = \lambda_3 = -\frac{1}{2}$. (Note that the eigenvalue $-\frac{1}{2}$ is repeated; thus its *multiplicity* is 2.)

```
>    CharacteristicPolynomial:=charpoly(A,lambda);
```

$$CharacteristicPolynomial := \lambda^3 - \frac{3}{4}\lambda - \frac{1}{4}$$

```
>    solve(CharacteristicPolynomial);
```

$$1, \frac{-1}{2}, \frac{-1}{2}$$

However, it is easier to use the eigenvectors() function to find *all* eigenvalues and the corresponding eigenvectors v[1] and v[2] .

```
>    v:=[ eigenvectors(A)] ;
```

$$v := [[1, 1, \{[1, 1, 1]\}], [\frac{-1}{2}, 2, \{[-1, 0, 1], [-1, 1, 0]\}]]$$

```
>    v[ 1] ; v[ 2] ;
```

$$[1, 1, \{[1, 1, 1]\}]$$

$$[\frac{-1}{2}, 2, \{[-1, 0, 1], [-1, 1, 0]\}]$$

Note here that the list v[1] contains the information related to the eigenvalue 1 and its eigenvector: This eigenvalue's multiplicity is 1, and its eigenvector is

[1,1,1] . The list v[2] indicates that the multiplicity of eigenvalue $-1/2$ is 2 and the corresponding eigenvectors are [-1,0,1] and [-1,1,0] .

>   `Eigenvectors:=[ v[ 1][ 3,1] , v[ 2][ 3,1] , v[ 2][ 3,2]] ;`

$$Eigenvectors := [[1, 1, 1], [-1, 0, 1], [-1, 1, 0]]$$

Using the basis() function, we note that the eigenvectors are linearly independent; thus the **A** matrix is diagonalizable.

>   `basis(Eigenvectors);`

$$[[1, 1, 1], [-1, 0, 1], [-1, 1, 0]]$$

>   `Q:=transpose(matrix(3,3,Eigenvectors));`

$$Q := \begin{bmatrix} 1 & -1 & -1 \\ 1 & 0 & 1 \\ 1 & 1 & 0 \end{bmatrix}$$

>   `DiagonalizedA:=evalm(inverse(Q) &* A &* Q);`

$$DiagonalizedA := \begin{bmatrix} 1 & 0 & 0 \\ 0 & \dfrac{-1}{2} & 0 \\ 0 & 0 & \dfrac{-1}{2} \end{bmatrix}$$

Next, we evaluate $\mathbf{QD}^{10}\mathbf{Q}^{-1}$ to find the 10th power of **A**. As a check, we also compute $\mathbf{A}^{10}$ directly and see that the results agree.

>   `evalm(Q &* (DiagonalizedA)^10 &* inverse(Q));`

$$\begin{bmatrix} \dfrac{171}{512} & \dfrac{341}{1024} & \dfrac{341}{1024} \\ \dfrac{341}{1024} & \dfrac{171}{512} & \dfrac{341}{1024} \\ \dfrac{341}{1024} & \dfrac{341}{1024} & \dfrac{171}{512} \end{bmatrix}$$

>   `evalm(A^10);`

$$\begin{bmatrix} \dfrac{171}{512} & \dfrac{341}{1024} & \dfrac{341}{1024} \\ \dfrac{341}{1024} & \dfrac{171}{512} & \dfrac{341}{1024} \\ \dfrac{341}{1024} & \dfrac{341}{1024} & \dfrac{171}{512} \end{bmatrix}$$

3.4.4 Least Squares Fitting to Data

An interesting application of linear algebra is to "fitting" a curve to a set of experimentally obtained n data points (x_1, y_1), (x_2, y_2), ..., (x_n, y_n) in order to develop a mathematical relationship of the form $y = f(x)$. Based on theoretical considerations or the graphical display of data that may reveal a pattern, a general form such as $y = a + bx$ or $y = a + bx + cx^2 + dx^3$ may be used. The curve-fitting problem then involves determining the parameters (coefficients) a, b, etc., of the model using the given n data points.

Consider the simplest case of fitting a straight line $y = a + bx$ to a given set of n data points. If the data were all on the same line (i.e., collinear), then the coefficients would satisfy $y_1 = a + bx_1, \ldots, y_n = a + bx_n$, which can be written in matrix form as $\mathbf{y} = \mathbf{X}\alpha$ where

$$\mathbf{y} = \begin{pmatrix} y_1 \\ \vdots \\ y_n \end{pmatrix}_{n \times 1}, \quad \mathbf{X} = \begin{pmatrix} 1 & x_1 \\ \vdots & \vdots \\ 1 & x_n \end{pmatrix}_{n \times 2}, \quad \alpha = \begin{pmatrix} a \\ b \end{pmatrix}_{2 \times 1}.$$

Thus, for collinear data points, we would have $\mathbf{y} - \mathbf{X}\alpha = \mathbf{0}$. However, when the data are not collinear, it is impossible to find a and b that satisfy $\mathbf{y} = \mathbf{X}\alpha$. In this case one attempts to find a vector α that minimizes the Euclidean length of the difference (deviation) $\|\mathbf{y} - \mathbf{X}\alpha\|^2 = \sum_{i=1}^{n}[y_i - (a + bx_i)]^2$. It can be shown [7, Section 8.3] that the vector minimizing the deviations is given by

$$\alpha = (\mathbf{X}'\mathbf{X})^{-1}\mathbf{X}'\mathbf{y}.$$

Thus, the coefficients a and b can be found by matrix algebra that involves matrix inversion and matrix multiplications. This has been implemented by Maple, as we will demonstrate.

The same technique can be generalized to fitting a polynomial of degree m, i.e., $y = a_0 + a_1 x + \cdots + a_m x^m$ to the n data points $(x_1, y_1), (x_2, y_2), \ldots, (x_n, y_n)$. In this case the $\mathbf{X}$ becomes an $n \times (m + 1)$ matrix and α an $(m + 1) \times 1$ vector. The optimal solution is again found as $\alpha = (\mathbf{X}'\mathbf{X})^{-1}\mathbf{X}'\mathbf{y}$.

Maple's implementation of the least squares solution is provided in the fit() subpackage of the stats package. Thus both of these packages must be loaded before doing least squares computations.

```
>    restart: # LeastSquares.mws
>    with(stats); with(fit);
```

[anova, describe, fit, importdata, random, statevalf, statplots, transform]

[leastmediansquare, leastsquare]

The first example fits a straight line $y = a + bx$ to four data points with coordinates defined as lists xData and yData. Next, using the leastsquare[ ] function, we find that $y = 18.12 - 1.03x$.

```
>    xData:=[ 0,1,2,3] ;
```

$$xData := [0, 1, 2, 3]$$

> `yData:=[ 21,.3,41,4] ;`

$$yData := [21, .3, 41, 4]$$

> `leastsquare[ [ x, y] , y=a+b* x] ([ xData, yData] );`

$$y = 18.12000000 - 1.030000000\,x$$

In the next example, we fit a quadratic $y = a + bx + cx^2$ to five data points and find that $y = 1.05 - 12.03x + 35.5x^2$.

> `xData:=[ .1,.2,.3,.4,.5] ; nops(xData);`

$$xData := [.1, .2, .3, .4, .5]$$

$$5$$

> `yData:=[ .18,.31,.03,2.48,3.73] ; nops(yData);`

$$yData := [.18, .31, .03, 2.48, 3.73]$$

$$5$$

> `leastsquare[ [ x, y] , y=a+b* x+c* x^2] ([ xData, yData] );`

$$y = 1.050000000 - 12.03000000\,x + 35.50000000\,x^2$$

The next example finds the analytic solution for the coefficients a and b in linear regression.

Example 30 *Coefficients of linear regression equation.* We now directly minimize the sum $\sum_{i=1}^{n}[y_i - (a + bx_i)]^2$ by using calculus techniques.

> `restart: # LSEClosedForm.mws`

> `f:=Sum((y[ i] -a-b* x[ i] )^2,i=1..n);`

$$f := \sum_{i=1}^{n} (y_i - a - b\,x_i)^2$$

> `fa:=diff(f,a); fb:=diff(f,b);`

$$fa := \sum_{i=1}^{n} (-2\,y_i + 2\,a + 2\,b\,x_i)$$

$$fb := \sum_{i=1}^{n} (-2\,(y_i - a - b\,x_i)\,x_i)$$

> `solve ({ fa, fb} ,{ a,b} );`

$$
\left\{
\begin{aligned}
b &= \frac{(\sum\limits_{i=1}^{n} x_i\, y_i)\, (\sum\limits_{i=1}^{n} 1) - (\sum\limits_{i=1}^{n} x_i)\, (\sum\limits_{i=1}^{n} y_i)}{-(\sum\limits_{i=1}^{n} x_i)^2 + \left(\sum\limits_{i=1}^{n} x_i{}^2\right)\, (\sum\limits_{i=1}^{n} 1)}, \\[2ex]
a &= -\frac{-(\sum\limits_{i=1}^{n} y_i)\, \left(\sum\limits_{i=1}^{n} x_i{}^2\right) + (\sum\limits_{i=1}^{n} x_i)\, (\sum\limits_{i=1}^{n} x_i\, y_i)}{-(\sum\limits_{i=1}^{n} x_i)^2 + \left(\sum\limits_{i=1}^{n} x_i{}^2\right)\, (\sum\limits_{i=1}^{n} 1)}
\end{aligned}
\right.
$$

This is a rather impressive result since Maple was not given the specific value of n for the number of data points. This approach can be used even for fitting data to a higher-order polynomial such as the quadratic, cubic, etc.

3.4.5 Special Matrices

In operations research applications certain special matrices and vectors are frequently encountered. For example, the *Hessian* matrix is used to check the convexity/concavity of a multivariable function. The *Jacobian* matrix is used to find the joint probability distribution of functions of random variables. The *gradient* of a function is a vector of first partial derivatives of the function. Maple has implemented these (and other matrices) in the linalg package. The next example illustrates the use of these matrices.

```
>    restart: # SpecialMatrices.mws
>    with(linalg):
```

Warning, new definition for norm

Warning, new definition for trace

The Hessian matrix **H** of a function of three variables $f = f(x, y, z)$ is computed as

$$
\mathbf{H} = \begin{pmatrix} f_{xx} & f_{xy} & f_{xz} \\ f_{yx} & f_{yy} & f_{yz} \\ f_{zx} & f_{zy} & f_{zz} \end{pmatrix}.
$$

In our example, $f(x, y, z) = x^2 yz + 3xy^2 z$ and the Hessian is

```
>    f:=x^2* y* z+3* x* y^2* z;
```

$$
f := x^2\, yz + 3\, x\, y^2\, z
$$

```
>    hessian(f,[ x,y,z] );
```

$$
\begin{bmatrix} 2\,yz & 2\,xz + 6\,yz & 2\,xy + 3\,y^2 \\ 2\,xz + 6\,yz & 6\,xz & x^2 + 6\,xy \\ 2\,xy + 3\,y^2 & x^2 + 6\,xy & 0 \end{bmatrix}
$$

If each component of a vector **A** is a function of two variables, say, **A** = $[f(r, \theta), g(r, \theta)]$, then the Jacobian matrix is given by

$$
\begin{pmatrix}
\frac{\partial f}{\partial r} & \frac{\partial f}{\partial \theta} \\
\frac{\partial g}{\partial r} & \frac{\partial g}{\partial \theta}
\end{pmatrix}.
$$

In our example, **A** = $[r \cos(\theta), r \sin(\theta)]$ and the Jacobian is

```
>   A := vector([ r*cos(theta), r*sin(theta)] );
```

$$A := [r \cos(\theta), r \sin(\theta)]$$

```
>   jacobian(A,[ r,theta] );
```

$$
\begin{bmatrix}
\cos(\theta) & -r \sin(\theta) \\
\sin(\theta) & r \cos(\theta)
\end{bmatrix}
$$

```
>   simplify(det(%));
```

$$r$$

Using the function f, we find its gradient $\nabla f = [f_x, f_y, f_z]$ as

```
>   grad(f, [ x,y,z] );
```

$$[2 x y z + 3 y^2 z, x^2 z + 6 x y z, x^2 y + 3 x y^2]$$

3.4.6   Positive Definiteness

Maple has implemented a procedure called definite() for checking the convexity or concavity of a multivariable function—or, equivalently the positive (or negative) definiteness (or semidefiniteness) of its Hessian. This can be very useful in optimizing nonlinear functions as we see in the following example.

```
>   restart: # PositiveDefinite.mws
>   with(linalg): with(student):
```

Warning, new definition for norm

Warning, new definition for trace

The interface() command below asks Maple to indicate the variables with assumptions (such as positive, real, integer, etc.) as a list displayed at the end of an expression.

```
>   interface(showassumed=2);
>   assume(x,real, y,real, z,real):
>   f:=(x,y,z)->exp(x-y)+exp(y-x)+exp(x^2)+z^2;
```

$$f := (x, y, z) \rightarrow e^{(x-y)} + e^{(y-x)} + e^{(x^2)} + z^2$$

```
>   H:=hessian(f(x,y,z),[ x,y,z] );
```

$$H := \begin{bmatrix} e^{(x-y)} + e^{(y-x)} + 2\,e^{(x^2)} + 4\,x^2\,e^{(x^2)} & -e^{(x-y)} - e^{(y-x)} & 0 \\ -e^{(x-y)} - e^{(y-x)} & e^{(x-y)} + e^{(y-x)} & 0 \\ 0 & 0 & 2 \end{bmatrix}$$

with assumptions on x and y

```
>   definite(H,'positive_def');
```

$-e^{(x-y)} - e^{(y-x)} - 2\,e^{(x^2)} - 4\,x^2\,e^{(x^2)} < 0$ **and**

$-(e^{(x-y)} + e^{(y-x)} + 2\,e^{(x^2)} + 4\,x^2\,e^{(x^2)})\,(e^{(x-y)} + e^{(y-x)})$

$+ (-e^{(x-y)} - e^{(y-x)})^2 < 0$ **and** $-4\,e^{(x^2)}\,e^{(x-y)} - 4\,e^{(x^2)}\,e^{(y-x)}$

$-8\,x^2\,e^{(x^2)}\,e^{(x-y)} - 8\,x^2\,e^{(x^2)}\,e^{(y-x)} < 0$

with assumptions on x and y

By using the is() function, we see that these conditions are satisfied for all real x, y and z, thus making the function a convex one.

```
>   is(%);
```

true

We can now use the extrema() function to optimize (i.e., minimize) f. We find that f's optimum value is 3 and the optimal solution is at $(x, y, z) = (0, 0, 0)$.

```
>   extrema(f(x,y,z),{},{x,y,z},'sol');
```

$\{3\}$

```
>   sol;
```

$\{\{z = 0,\ x = 0,\ y = 0\}\}$
with assumptions on z, x and y

3.5   Differential Equations

Ordinary differential equations (ODE) play an important role in modeling problems where the system's behavior is time-dependent. For example, in order to find the probability that a queueing system will be in a particular state at a finite time t, one normally needs to solve a system of ordinary differential equations. Similarly, applications of optimal control theory require a good understanding of differential equations and their exact or numerical solution.

Partial differential equations (PDE) have also found applicability in operations research in the analysis of some types of queues where they are used to represent the generating function of transient probabilities. To optimize some types of continuous-time systems, one usually applies Bellman's principle of optimality, which results in the so-called Hamilton-Jacobi-Bellman partial differential equation.

In this section we describe Maple's powerful differential equation solvers and discuss some operations research applications where the solution can be obtained analytically. We also consider some problems that must be solved numerically since either an exact solution is not possible or Maple cannot find it.

3.5.1 Ordinary Differential Equations

First, we examine some problems in ordinary differential equations.

Analytic Solutions

Example 31 *Savings account.* One of the simplest ordinary differential equations arises in the modeling of growth of cash deposited in a bank account. If the initial deposit in a savings account is S_0 dollars and the savings grows at r percent per annum compounded continuously, then the first-order homogeneous linear ordinary differential equation with a constant coefficient representing the level of savings $S(t)$ at time t is given by

$$\frac{dS(t)}{dt} = rS(t), \quad S(0) = S_0 \text{ given.}$$

Maple solves this ODE as follows.

```
>    restart:  # Savings.mws
>    DE1:=diff(S(t),t)=r*S(t);
```

$$DE1 := \tfrac{\partial}{\partial t} S(t) = r\, S(t)$$

```
>    IC:=S(0)=S[ 0] ;
```

$$IC := S(0) = S_0$$

```
>    Sol1:=dsolve({ DE1, IC} ,S(t));
```

$$Sol1 := S(t) = S_0\, e^{(r\,t)}$$

Thus we find that at time t the level of savings account is given as $S(t) = S_0 e^{rt}$. The fact that the solution is correct is verified using the odetest() function, which results in a 0 signaling that Maple has solved the ODE correctly.

```
>    odetest(Sol1,DE1);
```

$$0$$

When the interest rate $r(t)$ is a function of time t, then the ODE becomes

$$\frac{dS(t)}{dt} = r(t)S(t), \quad S(0) = S_0 \text{ given.}$$

This is a slightly more challenging problem to solve since now the ODE has a variable coefficient, $r(t)$. But this creates no difficulty for Maple, which finds that $S(t) = S_0 e^{\int_0^t r(u)\,du}$.

```
>    DE2:=diff(S(t),t)=r(t)*S(t);
```

$$DE2 := \frac{\partial}{\partial t} S(t) = r(t) S(t)$$

```
>   Sol2:=dsolve({ DE2,IC} ,S(t));
```

$$Sol2 := S(t) = S_0 \, e^{\left(- \int_0^t -r(u) \, du\right)}$$

```
>   odetest(Sol2,DE2);
```

$$0$$

Numerical Solutions

In many problems, it may be impossible to solve the ODE analytically. When this is the case, Maple can still be useful by generating a numerical solution of the differential equation.

Example 32 *Van der Pol's nonlinear differential equation.* One interesting example is the well-known van der Pol equation that represents the behavior of an electronic circuit used in radars, or the behavior of the heart; see Bellman and Cooke [25, p. 128]. This nonlinear ODE is given by

$$y'' - (1 - y^2)y' + y = 0$$

where the initial conditions $y(0)$ and $y'(0)$ must be specified. We attempt to solve this ODE (with $y(0) = 1$ and $y'(0) = 0$) hoping for an analytic solution, but Maple fails to produce any results. But using the type=numeric option of dsolve() we are able to find the numerical solution for the problem.

```
>   restart: # VanDerPol.mws
>   DE:=diff(y(t),t$2)-(1-y(t)^2)*diff(y(t),t)
    +y(t)=0;
```

$$DE := (\frac{\partial^2}{\partial t^2} y(t)) - (1 - y(t)^2) (\frac{\partial}{\partial t} y(t)) + y(t) = 0$$

```
>   IC:=y(0)=1,D(y)(0)=0;
```

$$IC := y(0) = 1, \, D(y)(0) = 0$$

```
>   dsolve({ DE,IC} ,y(t));
```

The numeric solver using type=numeric returns a Maple list that consists of time t, the value of the variable $y(t)$ at time t and the value of the first derivative $y'(t)$ at time t.

```
>   F:=dsolve({ DE,IC} ,y(t),type=numeric);
```

$$F := \mathbf{proc}(rkf45\_x) \ldots \mathbf{end}$$

```
>   F(0);
```

$$[t = 0, \, y(t) = 1., \, \frac{\partial}{\partial t} y(t) = 0]$$

```
>   F(1);
```

$$[t = 1, \, y(t) = .4976154397782068,$$
$$\frac{\partial}{\partial t} y(t) = -1.044238324424431]$$

To plot the solution we convert the numerical solution into a function. Plotting the solution for $t \in [0, 15]$ in Figure 3.1 indicates the periodic nature of the solution.

```
>   Y:=t->rhs(op(2,F(t)));
```

$$Y := t \rightarrow \text{rhs(op(2, } F(t)))$$

```
>   plot(Y,0..15);
```

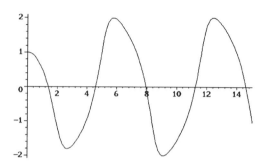

FIGURE 3.1. Solution of van der Pol's nonlinear ODE for $0 \le t \le 15$.

If it is known that the ODE does not have an analytic solution (or that Maple cannot find it), then one can use an easier method to plot the numerical solution of the ODE. This method requires loading the differential equations tools package (DEtools) and using the DEplot() function. Removing the "#" sign before DEplot() and executing the worksheet produces the same graph as obtained in Figure 3.1.

```
>   restart: # vanderPolDETools.mws
>   with(DEtools):
>   #DEplot(diff(y(t),t$2)-(1-y(t)^2)*diff(y(t),t)
    +y(t)=0,y(t),t=0..15,[[ y(0)=1,D(y)(0)=0]],
    y=-3..3,stepsize=.05);
```

Systems of Differential Equations

There are many problems in operations research where systems of ordinary differential equations arise naturally. For example, transient solutions of queueing problems are found by solving systems of ODEs. Application of the maximum principle to solve optimal control problems requires the solution of a two-point boundary value problem with several differential equations.

In some cases, Maple is successful in solving a system of ODEs analytically. When this is not possible, we can resort to a rich collection of numerical methods that Maple knows about.

Example 33 *Lancaster's equations.* As an example of a case that can be solved analytically, consider the system of ODEs

$$\frac{dx(t)}{dt} = -k\beta y(t), \quad x(0) = x_0$$

$$\frac{dy(t)}{dt} = -kax(t), \quad y(0) = y_0.$$

This system of ODEs is known as Lancaster's differential equations. F. W. Lancaster analyzed these ODEs during World War I in the context of "mathematics of warfare." For a brief description of these equations and comments on Lancaster's "square law," see Saaty [161, p. 71].

This system can be solved very easily with Maple's dsolve() function.

```
>    restart: # Lancaster.mws
>    DEx:=diff(x(t),t)=-k*beta*y(t); ICx:=x(0)=x[ 0] ;
```

$$DEx := \frac{\partial}{\partial t}x(t) = -k\beta y(t)$$

$$ICx := x(0) = x_0$$

```
>    DEy:=diff(y(t),t)=-k*alpha*x(t); ICy:=y(0)=y[ 0] ;
```

$$DEy := \frac{\partial}{\partial t}y(t) = -k a x(t)$$

$$ICy := y(0) = y_0$$

```
>    System:=DEx,DEy; ICSystem:=ICx,ICy;
```

$$System := \frac{\partial}{\partial t}x(t) = -k\beta y(t), \frac{\partial}{\partial t}y(t) = -k a x(t)$$

$$ICSystem := x(0) = x_0, y(0) = y_0$$

```
>    Sol:=dsolve({ System,ICSystem} ,{ x(t),y(t)} );
```

$$Sol := \left\{ y(t) = \cosh(\%1)\, y_0 - \frac{\sinh(\%1)\, x_0\, a}{\sqrt{\alpha\beta}}, \right.$$

$$\left. x(t) = -\frac{\sqrt{\alpha\beta}\,(\sinh(\%1)\, y_0 - \frac{\cosh(\%1)\, x_0\, a}{\sqrt{\alpha\beta}})}{a} \right\}$$

$$\%1 := k\sqrt{\alpha\beta}\, t$$

We use `odetest()` and see that the solution found does satisfy the original system of ODEs.

```
>    odetest(Sol,{ System} );
```

$$\{0\}$$

Another important class of systems of ODEs that can be solved using `dsolve()` is the two-point boundary value problem (TPBVP) arising in optimal control applications; see Kamien and Schwartz [99] and Sethi and Thompson [169].

Example 34 *An optimal control problem.* Consider the following nontrivial example adapted from Sethi and Thompson [169, Section 5.1] corresponding to the optimal control of a production/inventory system. The objective is to find the optimal production rate $u(t)$ in order to minimize the total cost of deviations from the constant target levels of inventory $(\hat{x})$ and production rate $(\hat{u})$, i.e.,

$$\int_0^T \left\{ \frac{1}{2}h[x(t) - \hat{x}]^2 + \frac{1}{2}c[u(t) - \hat{u}]^2 \right\} dt$$

subject to the state equation $\dot{x}(t) = u(t) - S(t)$, with the initial condition $x(0)$ as a given constant.[4] The parameters h and c represent the costs of deviations from the target levels and $S(t)$ is the estimated sales (demand) rate. Applying the maximum principle we find the (unconstrained) optimal production rate as $u(t) = \hat{u} + \lambda(t)$ and

$$\dot{x}(t) = \hat{x} + \frac{1}{c}\lambda(t) - S(t), \quad x(0) = x_0$$
$$\dot{\lambda}(t) = h[x(t) - \hat{x}], \quad \lambda(T) = 0.$$

where $\lambda(t)$ is the adjoint variable. This TPBVP where $x(t)$ is specified at $t = 0$ and $\lambda(t)$ is specified at $t = T$ is solved by Maple as follows.

```
>   restart: # TPBVP.mws
```

We first assign values to the parameters and specify the exact form of the demand rate $S(t)$. (The parameters uH and xH correspond to $\hat{u}$ and $\hat{x}$, respectively.)

```
>   uH:=50; xH:=25; T:=8; h:=5; c:=1; x0:=20;
```

$$uH := 50$$
$$xH := 25$$
$$T := 8$$
$$h := 5$$
$$c := 1$$
$$x0 := 20$$

```
>   S:=t->t* (t-4)* (t-8)+40; #plot(S(t),t=0..T);
```

$$S := t \to t(t-4)(t-8) + 40$$

```
>   inventory:=diff(x(t),t)=uH+lambda(t)/c-S(t);
```

$$inventory := \frac{\partial}{\partial t} x(t) = 10 + \lambda(t) - t(t-4)(t-8)$$

```
>   adjoint:=diff(lambda(t),t)=h* (x(t)-xH);
```

$$adjoint := \frac{\partial}{\partial t} \lambda(t) = 5x(t) - 125$$

```
>   sys:=inventory,adjoint;
```

---

[4]The dot $(\cdot)$ above $x(t)$ denotes the time derivative of $x(t)$.

$$sys := \frac{\partial}{\partial t} x(t) = 10 + \lambda(t) - t\,(t-4)\,(t-8),\ \frac{\partial}{\partial t}\lambda(t) = 5\,x(t) - 125$$

```
>   funcs:={ x(t),lambda(t)} ;
```

$$funcs := \{x(t),\ \lambda(t)\}$$

The `dsolve()` function succeeds in solving the TPBVP exactly.

```
>   dsolve({ sys,x(0)=x0,lambda(T)=0} ,funcs);
```

$$\{x(t) = -\frac{291}{50}\,e^{(\sqrt{5}\,t)} - \frac{291}{50}\,e^{(-\sqrt{5}\,t)} - \frac{1}{250}\,\frac{\%2\,\sqrt{5}\,e^{(-\sqrt{5}\,t)}}{\%1+1}$$

$$+\frac{1}{250}\,\frac{\%2\,\sqrt{5}\,e^{(\sqrt{5}\,t)}}{\%1+1} + \frac{3}{5}\,t^2 + \frac{791}{25} - \frac{24}{5}\,t,\ \lambda(t) = \frac{291}{50}\,\sqrt{5}\,e^{(-\sqrt{5}\,t)}$$

$$-\frac{291}{50}\,\sqrt{5}\,e^{(\sqrt{5}\,t)} + \frac{1}{50}\,\frac{\%2\,e^{(\sqrt{5}\,t)}}{\%1+1} + \frac{1}{50}\,\frac{\%2\,e^{(-\sqrt{5}\,t)}}{\%1+1} - 12\,t^2 - \frac{74}{5}$$

$$+\frac{166}{5}\,t + t^3\}$$

$$\%1 := (e^{(8\,\sqrt{5})})^2$$

$$\%2 := -291\,\sqrt{5} + 291\,\sqrt{5}\,\%1 + 260\,e^{(8\,\sqrt{5})}$$

```
>   assign(%);
```

```
>   simplify(subs(t=0,x(t)));
    simplify(subs(t=T,lambda(t)));
```

$$20$$

$$0$$

Once we find the optimal state trajectory, the optimal control trajectory can also be determined using $u(t) = \hat{u} + \lambda(t)$. The plots of the trajectories for $S(t)$, $x(t)$ and $u(t)$ can then be easily plotted with the `plot()` function. We leave this to the reader as an exercise.

```
>   #u:=unapply(uH+lambda(t)/c,t);
```

```
>   #plot({ x(t)} ,t=0..T,title='Optimal x(t)');
```

```
>   #plot({ u(t)} ,t=0..T,title='Optimal u(t)');
```

Example 35 *Numerical solution of system of ODEs.* In some cases, Maple may not be able to find the exact solution to a system of ODEs. When this happens, it may be necessary to compute the solution numerically. For example, consider the problem of solving the ODE system

$$\mathbf{P}'(t) = \mathbf{P}(t)\mathbf{Q}(t),\quad \mathbf{P}(0) = \mathbf{I}$$

where $P(t) = [P_0(t), \ldots, P_4(t)]$, $P'(t) = [P'_0(t), \ldots, P'_4(t)]$ and

$$Q(t) = \begin{pmatrix} -\lambda(t) & \lambda(t) \\ \mu & -[\lambda(t) + \mu] & \lambda(t) \\ & \mu & -[\lambda(t) + \mu] & \lambda(t) \\ & & \mu & -[\lambda(t) + \mu] & \lambda(t) \\ & & & \mu & -\mu \end{pmatrix},$$

which arises in queueing analysis of computer systems; see Kao [101, Section 2.3]. Assuming, for example, that $\lambda(t) = 100t\,(1-t)$ is the time-dependent arrival rate and that $\mu = 7$, the problem is to solve this system of five ODEs in order to compute the transient (time-dependent) probabilities of being in different states.

```
>    restart: # ComputerSystem.mws
>    lambda:=t->100*t*(1-t); mu:=7;
     #plot(lambda(t),t=0..1);
```

$$\lambda := t \rightarrow 100\,t\,(1-t)$$

$$\mu := 7$$

```
>    DE[ 0] :=diff(p0(t),t)=-p0(t)*' lambda(t)'
     +p1(t)*' mu' ;
```

$$DE_0 := \tfrac{\partial}{\partial t}\,p0(t) = -p0(t)\,\lambda(t) + p1(t)\,\mu$$

```
>    DE[ 1] :=diff(p1(t),t)=p0(t)*' lambda(t)'
     -p1(t)*' (lambda(t)+mu)' +p2(t)*'
     mu' ;
```

$$DE_1 := \tfrac{\partial}{\partial t}\,p1(t) = p0(t)\,\lambda(t) - p1(t)\,(\lambda(t) + \mu) + p2(t)\,\mu$$

```
>    DE[ 2] :=diff(p2(t),t)=p1(t)*' lambda(t)'
     -p2(t)*' (lambda(t)+mu)' +p3(t)*'
     mu' ;
```

$$DE_2 := \tfrac{\partial}{\partial t}\,p2(t) = p1(t)\,\lambda(t) - p2(t)\,(\lambda(t) + \mu) + p3(t)\,\mu$$

```
>    DE[ 3] :=diff(p3(t),t)=p2(t)*' lambda(t)'
     -p3(t)*' (lambda(t)+mu)' +p4(t)*'
     mu' ;
```

$$DE_3 := \tfrac{\partial}{\partial t}\,p3(t) = p2(t)\,\lambda(t) - p3(t)\,(\lambda(t) + \mu) + p4(t)\,\mu$$

```
>    DE[ 4] :=diff(p4(t),t)=p3(t)*' lambda(t)'
     -p4(t)*' mu' ;
```

$$DE_4 := \tfrac{\partial}{\partial t}\,p4(t) = p3(t)\,\lambda(t) - p4(t)\,\mu$$

```
>    DESystem:={ seq(DE[ i] ,i=0..4)} ;
```

$DESystem := \{\frac{\partial}{\partial t} p2(t) =$

$100\,p1(t)\,t\,(1-t) - p2(t)\,(100\,t\,(1-t)+7) + 7\,p3(t)$

$,\frac{\partial}{\partial t} p0(t) = -100\,p0(t)\,t\,(1-t) + 7\,p1(t), \frac{\partial}{\partial t} p1(t) =$

$100\,p0(t)\,t\,(1-t) - p1(t)\,(100\,t\,(1-t)+7) + 7\,p2(t)$

$,\frac{\partial}{\partial t} p4(t) = 100\,p3(t)\,t\,(1-t) - 7\,p4(t), \frac{\partial}{\partial t} p3(t) =$

$100\,p2(t)\,t\,(1-t) - p3(t)\,(100\,t\,(1-t)+7) + 7\,p4(t)$

$\}$

```
> ICSystem:={ p0(0)=1,seq(p.i(0)=0,i=1..4)};
```

$ICSystem := \{$
$p0(0) = 1,\ p1(0) = 0,\ p2(0) = 0,\ p3(0) = 0,\ p4(0) = 0$
$\}$

After trying very hard, Maple fails to find the exact solution for this problem.

```
> #dsolve(DESystem union ICSystem,
  { seq(p.i(t),i=0..4)} ); #This fails
```

But using the type=numeric option of dsolve(), we can compute the solution numerically and plot the transient probability trajectory for the loss probability $P_4(t)$ as given Figure 3.2.

```
> F:=dsolve(DESystem union
  ICSystem,[ seq(p.i(t),i=0..4)] ,type=numeric);
```

$F := \mathbf{proc}(rkf45\_x) \ldots \mathbf{end}$

```
> F(0);
```

$[t = 0,\ p0(t) = 1.,\ p1(t) = 0,\ p2(t) = 0,\ p3(t) = 0,\ p4(t) = 0]$

```
> F(.25);
```

$[t = .25,\ p0(t) = .1977703300783685,$
$p1(t) = .2605045316892459,$
$p2(t) = .2344519154105270,$
$p3(t) = .1667040464906527,$
$p4(t) = .1405691763312056]$

```
> F(.5);
```

$[t = .5,\ p0(t) = .01776293908215261,$
$p1(t) = .04255366365736456,$
$p2(t) = .09076540588829907,$
$p3(t) = .2133530755905687,$
$p4(t) = .6355649157816149]$

```
> P4:=t->rhs(op(6,F(t)));
```

$P4 := t \rightarrow \text{rhs}(\text{op}(6,\ F(t)))$

```
>   plot(P4,0..1);
```

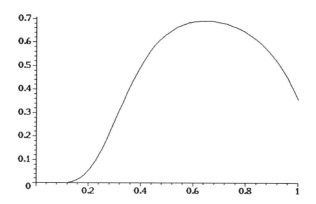

FIGURE 3.2. Numerical solution of the loss probability, $P_4(t)$, $0 \le t \le 1$, with $\lambda(t) = 100t(1 - t)$.

It is interesting to note that the loss probability trajectory $P_4(t)$ has a shape that resembles the time-dependent arrival rate function $\lambda(t) = 100t(1 - t)$, which is concave and reaching its maximum at $t = 0.5$. This resemblance is not a coincidence since the increasing arrival rates result in a higher probability of a busy system.

3.5.2 Partial Differential Equations

Partial differential equations (PDE) do not appear very frequently in operations research, but they have been used in the modeling and solution of some problems including the transient solution of ample server queue (Gross and Harris [76, p. 139]), manpower planning (Gerchak, Parlar and Sengupta [71]) and the optimal control of distributed parameter systems (Derzko, Sethi and Thompson [61]). In this section we demonstrate the solution of some PDEs using Maple's `pdsolve()` function.

Consider the PDE

$$\frac{\partial u}{\partial x} + \frac{\partial u}{\partial y} + u(x, y) = e^{x+2y}$$

with the initial condition $u(x, 0) = 0$. To solve this PDE for the unknown function $u(x, y)$ we proceed as follows.

```
>   restart: # PDE.mws
>   PDE:=diff(u(x,y),x) + diff(u(x,y),y) + u(x,y)
    = exp(x+2*y); # u(x,0)= 0 is the IC
```

$$PDE := (\tfrac{\partial}{\partial x} u(x, y)) + (\tfrac{\partial}{\partial y} u(x, y)) + u(x, y) = e^{(x+2y)}$$

> `pdsolve(PDE);`

$$u(x, y) = \frac{1}{4} \frac{e^{(x+2y)} e^x + 4\_F1(y-x)}{e^x}$$

Maple finds the general solution that involves an unknown function $\_F1(\cdot)$ that must be determined using the initial condition.

> `Sol:=combine(expand(%));`

$$Sol := u(x, y) = \frac{1}{4} e^{(x+2y)} + e^{(-x)}\_F1(y-x)$$

> `solve(subs(y=0,u(x,0)=0,Sol),_F1(-x));`

$$-\frac{1}{4} \frac{e^x}{e^{(-x)}}$$

> `_F1:=unapply(simplify(subs(x=-x,%)),x);`

$$\_F1 := x \rightarrow -\frac{1}{4} e^{(-2x)}$$

Thus, $\_F1(x) = -e^{-2x}/4$ and so $\_F1(y-x) = -e^{-2y+2x}/4$.

> `_F1(y-x);`

$$-\frac{1}{4} e^{(-2y+2x)}$$

> `combine(Sol);`

$$u(x, y) = \frac{1}{4} e^{(x+2y)} - \frac{1}{4} e^{(x-2y)}$$

It then follows that the solution is

$$u(x, y) = \frac{1}{4} e^{x+2y} - \frac{1}{4} e^{x-2y}.$$

> `assign(%); u(x,y);`

$$\frac{1}{4} e^{(x+2y)} - \frac{1}{4} e^{(x-2y)}$$

To check the result we substitute the solution in the original PDE and find that it is satisfied.

> `is(diff(u(x,y),x) + diff(u(x,y),y) + u(x,y)`
> `= exp(x+2*y));`

true

Example 36 *Ample server queue.* The following PDE arises in the transient analysis of the Markovian ample server queueing system $M/M/\infty$ with exponential interarrival times (rate λ) and exponential service times (rate μ). Defining $G(z, t) = \sum_{n=0}^{\infty} p_n(t)z^n$ as the generating function of the transient probability $p_n(t)$ of having n customers in the system at time t, it can be shown that $G(z, t)$ satisfies

$$\frac{\partial G}{\partial t} = \mu(1-z)\frac{\partial G}{\partial z} - \lambda(1-z)G(z, t)$$

with the initial condition $G(z, 0) = 1$ assuming that at time $t = 0$ the system is empty.

Maple solves this PDE as follows.

```
>    restart: # AmpleServer.mws
>    PDE:=diff(G(z,t),t)=mu* (1-z)*diff(G(z,t),z)
     -lambda* (1-z)*G(z,t);
```

$$PDE := \frac{\partial}{\partial t} G(z, t) = \mu (1 - z) (\frac{\partial}{\partial z} G(z, t)) - \lambda (1 - z) G(z, t)$$

```
>    Sol:=pdsolve(PDE,G(z,t));
```

$$Sol := G(z, t) = \_F1(-\frac{-t\,\mu + \ln(-1 + z)}{\mu}) e^{(\frac{\lambda z}{\mu})}$$

```
>    pdetest(Sol,PDE);
```

$$0$$

Using the initial condition, it can be shown that the exact solution of this PDE is

$$G(z, t) = e^{-\lambda(z-1)(e^{-t\mu}-1)/\mu}.$$

We will provide a detailed solution of this PDE in Section 9.3.2 of Chapter 9, Queueing Systems.

3.5.3   Difference (Recurrence) Equations

Maple can solve a large class of difference (i.e., recurrence) equations with the rsolve() command. This command can solve linear recurrences with constant coefficients, systems of linear recurrences with constant coefficients, many first-order linear recurrences, and some nonlinear first-order recurrences.

Example 37 *Fibonacci sequence.* The Fibonacci sequence is obtained as the solution to the difference equation $f_n = f_{n-1} + f_{n-2}$, with the initial conditions $f_0 = f_1 = 1$. Maple solves this difference equation as follows.

```
>    restart: # Fibonacci.mws
>    rsolve({ f(n)=f(n-1)+f(n-2), f(0)=1, f(1)=1}, f);
```

$$\frac{2}{5}\frac{\sqrt{5}(2\frac{1}{-1+\sqrt{5}})^n}{-1+\sqrt{5}} + \frac{2}{5}\frac{\sqrt{5}(-2\frac{1}{1+\sqrt{5}})^n}{1+\sqrt{5}}$$

If we use the ' makeproc' option, then rsolve() returns a procedure for evaluating the function defined by the difference equations.

```
>    f:=rsolve({ f(n)=f(n-1)+f(n-2), f(0)=1, f(1)=1},
     f,'makeproc');
```

$f := \textbf{proc}(n)$
$\textbf{local } i, s, t, bipow;$
$\quad bipow := \textbf{proc}(n) \ldots \textbf{end};$
$\quad \textbf{if } 1 < \text{nargs } \textbf{or not } \text{type}(n, integer) \textbf{ then } \text{'procname'}(args)$
$\quad \textbf{else}$
$\quad\quad s := bipow(n-1);$
$\quad\quad t := 0;$
$\quad\quad \textbf{for } i \textbf{ to } 2 \textbf{ do } t := t + s_{1,i} \times (\text{array}(1 .. 2,[(2)=1,(1)=1]))_i \textbf{ od};$
$\quad\quad t$
$\quad \textbf{fi}$
$\textbf{end}$

```
>  f(1); f(2); f(5); f(10);
```

$$1$$
$$2$$
$$8$$
$$89$$

Example 38 *Coin toss.* In this more complicated example, consider a biased coin for which $p = \text{Pr(Heads)}$ and $q = \text{Pr(Tails)}$. The coin is tossed repeatedly and stopped when two heads (HH) occur in succession for the first time. If we let X denote the number of such trials needed, and $a_n = \text{Pr}(X = n)$, then it follows that for $n = 1, 2$ and 3, we have $a_1 = 0$, $a_2 = p^2$, $a_3 = p^2 q$ as the initial conditions. Using probabilistic arguments as in Kao [101, Chapter 1], one can show that

$$a_n = q a_{n-1} + pq a_{n-2}, \quad n = 4, 5, \ldots$$

We now solve these difference equations with `rsolve()`.

```
>  restart: # BiasedCoin.mws
```

Maple can find the exact solution of these difference equations but the result looks unattractive. Instead, we choose to solve the problem using the ' make-proc' option and obtain a procedure for the solution a_n.

```
>  a:=rsolve({ a(n)=q* a(n-1)+p* q* a(n-2),a(1)=0,
   a(2)=p^2,a(3)=p^2* q} ,a,' makeproc' ):
```

Following are the first 10 terms of the solution.

```
>  seq([ n,simplify(a(n))] ,n=1..10);
```

$$[1, 0], [2, p^2], [3, p^2 q], [4, q p^2 (q + p)], [5, q^2 p^2 (q + 2 p)],$$
$$[6, q^2 p^2 (q^2 + 3 p q + p^2)], [7, q^3 p^2 (q + p) (q + 3 p)],$$
$$[8, q^3 p^2 (q^3 + 5 q^2 p + 6 p^2 q + p^3)],$$
$$[9, q^4 p^2 (q + 2 p) (q^2 + 4 p q + 2 p^2)],$$
$$[10, q^4 p^2 (q^4 + 7 q^3 p + 15 p^2 q^2 + 10 p^3 q + p^4)]$$

Next, we solve the same problem with the `'genfunc'` (z) option that produces the generating function $P_X(z)$ of the distribution. Since the mean $E(X)$ can be found by differentiating the generating function and substituting $z = 1$, i.e., $E(X) = P'_X(1)$, we find that the mean of X is $E(X) = (1 + p)/p^2$.

```
>   restart:
>   PXz:=rsolve({ a(n)=q*a(n-1)+p*q*a(n-2),a(1)=0,
    a(2)=p^2,a(3)=p^2*q} ,a,'genfunc'(z));
```

$$PXz := -\frac{p^2 z^2}{-1 + q z + p q z^2}$$

```
>   simplify(subs(q=1-p,normal(subs(z=1,
    diff(PXz,z)))));
```

$$\frac{1 + p}{p^2}$$

3.6 Transform Methods

Many problems in probability and stochastic processes can be successfully solved using transform techniques such as Laplace transforms (LT) and generating functions. When the direct solution of a problem—such as finding the distribution of a random variable—is not available, it may sometimes be possible to express the problem in terms of the Laplace transform (or the generating functions if the random variable is discrete). If the transform can be inverted using `invlaplace()`, then the solution to the original problem is obtained.

3.6.1 Laplace Transforms

The Laplace transform $\tilde{f}(s)$ of a real-valued function $f(t)$ is defined as

$$\tilde{f}(s) = \int_0^\infty e^{-st} f(t) \, dt$$

provided that the integral exists. When $f(t)$ is the p.d.f. of a nonnegative random variable, $\tilde{f}(s) = E(e^{-sX})$, i.e., the LT of the density is the expected value of e^{-sX}. Thus, it follows that $f^{(n)}(s) = (-1)^n E(X^n e^{-sX})$, so the nth moment of X is $E(X^n) = (-1)^n f^{(n)}(0)$.

Example 39 *Erlang density with four stages.* Consider the Erlang density with $k = 4$ stages, i.e.,

$$f(t) = \frac{\lambda e^{-\lambda t}(\lambda t)^{k-1}}{(k-1)!}, \quad t > 0.$$

```
>   restart: # ErlangLT.mws
>   with(inttrans);
```

[*addtable, fourier, fouriercos, fouriersin, hankel, hilbert, invfourier, invhilbert, invlaplace, invmellin, laplace, mellin, savetable*]

```
>   k:=4;
```

$$k := 4$$

```
>   f:=lambda*exp(-lambda*t)*(lambda*t)^('k'-1)/('k'-1)!;
```

$$f := \frac{\lambda e^{(-\lambda t)} (\lambda t)^{(k-1)}}{(k-1)!}$$

The LT of this density is found using `laplace()` as

$$\tilde{f}(s) = \left(\frac{\lambda}{\lambda + s}\right)^4.$$

```
>   fLT:=laplace(f,t,s);
```

$$fLT := \frac{\lambda^4}{(s + \lambda)^4}$$

Inverting the transform gives the original density, as expected.

```
>   invlaplace(fLT,s,t);
```

$$\frac{1}{6} \lambda^4 e^{(-\lambda t)} t^3$$

Finally, the expected value of the Erlang is obtained as $E(X) = 4/\lambda$ using the fact that $E(X) = (-1)f'(0)$.

```
>   EX:=-subs(s=0,diff(fLT,s));
```

$$EX := 4\frac{1}{\lambda}$$

Example 40 *Partial fraction expansion.* As another example, suppose that we wish to find the function $f(t)$ whose LT is

$$\tilde{f}(s) = \frac{s^3}{(s + 1)^2 (s - 1)^3}.$$

If we were solving this problem manually, we would first find the partial fraction expansion of the LT and then invert each term by referring to tables of Laplace transforms. We see below that by simply using the `invlaplace()` function on $\tilde{f}(s)$ we immediately obtain the function $f(t)$.

```
>   restart:  # LT.mws
>   with(inttrans):
>   fLT:=s^3/((s+1)^2*(s-1)^3);
```

$$fLT := \frac{s^3}{(s + 1)^2 (s - 1)^3}$$

```
>    PF:=convert(fLT,parfrac,s);
```

$$PF := \frac{1}{8}\frac{1}{(s+1)^2} - \frac{3}{16}\frac{1}{s+1} + \frac{1}{4}\frac{1}{(s-1)^3} + \frac{1}{2}\frac{1}{(s-1)^2} + \frac{3}{16}\frac{1}{s-1}$$

```
>    invlaplace(PF,s,t);
```

$$\frac{1}{8}t\,e^{(-t)} - \frac{3}{16}e^{(-t)} + \frac{1}{8}t^2\,e^t + \frac{1}{2}t\,e^t + \frac{3}{16}e^t$$

```
>    invlaplace(fLT,s,t);
```

$$\frac{1}{8}t\,e^{(-t)} - \frac{3}{16}e^{(-t)} + \frac{1}{8}t^2\,e^t + \frac{1}{2}t\,e^t + \frac{3}{16}e^t$$

Numerical Inversion of Laplace Transforms

Quite often, it may be impossible (or very difficult) to invert a Laplace transform algebraically. In such a case, numerical inversion of the Laplace transform may be a feasible alternative.

If $\tilde{f}(s)$ is the LT of $f(t)$, then the numerical inversion of $\tilde{f}(s)$ can be performed by using

$$f_{\text{Numerical}}(t) = \frac{e^{A/2}}{2t}\mathcal{R}\left(\tilde{f}\left(\frac{A}{2t}\right)\right) + \frac{e^{A/2}}{t}\sum_{k=1}^{\infty}(-1)^k\mathcal{R}\left(\tilde{f}\left(\frac{A+2k\pi i}{2t}\right)\right)$$

$$(3.2)$$

where A is a parameter related to the approximation error, i is the complex number and $\mathcal{R}(z)$ is the real part of the complex number z. This numerical method was suggested by Abate and Whitt [2] who advise using $A = v\ln(10)$ for an accuracy of 10^{-v}; see also Kao [101, p. 25]. When $|f(t)| \le 1$ for all $t \ge 0$, it can be shown that the error is bounded by

$$|f(t) - f_{\text{Numerical}}(t)| \le \frac{e^{-A}}{1 - e^{-A}},$$

so when e^{-A} is small, the error is approximately e^{-A}.

In the next example, we attempt to invert the transform

$$\tilde{f}(s) = \frac{1}{(1+\sqrt{s})(1+s)}.$$

```
>    restart: # NumericalLT.mws
>    with(inttrans):
>    fLT:=s->1/((1+sqrt(s))*(1+s));
```

$$fLT := s \to \frac{1}{(1+\sqrt{s})(1+s)}$$

```
>    f:=unapply(invlaplace(fLT(s),s,t),t);
```

$$f := t \to \int_0^t -\frac{1}{2}\frac{\sqrt{2}\left(\sqrt{2}\sqrt{\_U1}\,e^{-U1}\,\text{erfc}(\sqrt{\_U1}) - \frac{\sqrt{2}}{\sqrt{\pi}}\right)e^{(-t+\_U1)}}{\sqrt{\_U1}}\,d\_U1$$

Maple makes a valiant attempt to invert this LT but finds only an integral involving the complementary error function erfc() and the exponential function. Plotting the resulting function (which we do not display) reveals that $f(t) \to 0$ as $t \to \infty$.

```
>   v:=5.; A:=v* ln (10.);  exp (-A);
```

$$v := 5.$$

$$A := 11.51292547$$

$$.9999999950 \, 10^{-5}$$

```
>   #plot (f (t) ,t=0..10);
```

Next, we compute the triple $\left[t, \, f(t), \, \int_0^t f(u)\,du \right]$ for increasing values of t and observe that $f(t)$ *appears* to be a density of a (yet to be discovered?) random variable since $\lim_{t\to\infty} \int_0^t f(u)\,du = 1$.

```
>   for t from 5 to 40 by 5 do [ t,evalf(f(t)),
    evalf(Int(f(u),u=0..t))]  od;
```

$$[5, .03317546275, .7344982429]$$

$$[10, .009359468114, .8200628136]$$

$$[15, .004943879701, .8538194324]$$

$$[20, .003184776647, .8736012833]$$

$$[25, .002270662056, .8870247002]$$

$$[30, .001724058011, .8969068485]$$

$$[35, .001366589281, .9045757418]$$

$$[40, .001117717361, .9107517465]$$

Finally, using the numerical inversion formula in (3.2) we compare the values for $f(t)$ and $f_{\text{Numerical}}(t)$ and find that the errors resulting from the numerical approximation are quite negligible.

```
>   fNumer:=t->exp (A/2) / (2* t)* Re (fLT (A/ (2* t)))
    +exp (A/2) /t
    * sum ( (-1) ^k* Re (fLT ( (A+2* k* Pi* I) / (2* t))),
    k=1..1000);
```

$$\textit{fNumer} := t \to \frac{1}{2} \, \frac{e^{(1/2\,A)} \, \Re(\text{fLT}(\frac{1}{2}\frac{A}{t}))}{t}$$

$$+ \frac{e^{(1/2\,A)} \left(\sum_{k=1}^{1000} (-1)^k \, \Re(\text{fLT}(\frac{1}{2}\frac{A+2\,I\,k\,\pi}{t})) \right)}{t}$$

```
>   for t from 0.01 to 1 by .1 do [ t,evalf(f(t)),
    fNumer (t)]  od;
```

[.01, .1028407591, .1027793021]

[.11, .2652275434, .265022770]

[.21, .3116393936, .311356022]

[.31, .3294644327, .329119930]

[.41, .3336666867, .333270567]

[.51, .3302478027, .329806145]

[.61, .3222873512, .321804697]

[.71, .3115759949, .311055639]

[.81, .2992342356, .298678991]

[.91, .2859958883, .285407838]

3.6.2 Generating Functions

If X is a discrete-valued random variable with probability density $a_n = \Pr(X = n)$, $n = 0, 1, \ldots$, then the generating function of X is defined as $\hat{a}(z) = E(z^X) = \sum_{n=0}^{\infty} a_n z^n$. Note that $\hat{a}(1) = 1$ and the series converges for at least $|z| \leq 1$. Additionally, $E(X) = \hat{a}'(1)$, and $E(X^2) = \hat{a}''(1) + \hat{a}'(1)$ so that $\mathrm{Var}(X) = \hat{a}''(1) + \hat{a}'(1) - [\hat{a}'(1)]^2$. For an excellent review of generating functions see Kleinrock [110, Appendix I].[5]

Manipulation of (rational) generating functions in Maple is achieved by first loading the genfunc() package. Two important functions are rgf_encode() and rgf_expand(). The first function finds the generating function $\hat{a}(z)$ given the sequence a_n, and the second inverts a given generating function $\hat{a}(z)$ to obtain the original sequence a_n.

Example 41 *The use of* rgf_encode() *and* rgf_expand(). We start with the density of the geometric r.v. $a_n = \Pr(X = n) = p(1 - p)^n$, $n = 0, 1, \ldots$

```
>    restart: # Genfunc.mws

>    with(genfunc);
```

[*rgf_charseq, rgf_encode, rgf_expand, rgf_findrecur, rgf_hybrid, rgf_norm, rgf_pfrac, rgf_relate, rgf_sequence, rgf_simp, rgf_term, termscale*]

---

[5]Some authors (including Kleinrock [110, p. 327]) use the term "z-transform" for generating functions. Maple also has a procedure called ztrans(), but it should not be confused with generating functions. Maple's ztrans() is defined as $\sum_{n=0}^{\infty} a_n z^{-n}$, which is not used very frequently in operations research applications.

Maple finds the generating function $\hat{a}(z)$ easily using `rgf_encode()`. To compute the mean and variance we find $\hat{a}'(1)$ and $\hat{a}''(1)$ and see that $E(X) = (1-p)/p$ and $\text{Var}(X) = (1-p)/p^2$.

```
>   aGF:=rgf_encode(p*(1-p)^n, n, z);   #Geometric
```

$$aGF := \frac{p}{1-(1-p)z}$$

```
>   aGFz:=eval(diff(aGF,z),z=1);
```

$$aGFz := -\frac{-1+p}{p}$$

```
>   aGFzz:=eval(diff(aGF,z$2),z=1);
```

$$aGFzz := 2\frac{(-1+p)^2}{p^2}$$

```
>   Var:=normal(aGFzz+aGFz-(aGFz)^2);
```

$$Var := -\frac{-1+p}{p^2}$$

In the next example, suppose that the generating function of a r.v. is given by[6]

$$\hat{a}(z) = \frac{8}{(2-z)(3-z)^3}.$$

We now use `rgf_expand()` to invert the generating function.

```
>   aGF:=8/((2-z)*(3-z)^3);
```

$$aGF := 8\frac{1}{(2-z)(3-z)^3}$$

```
>   a:=rgf_expand(aGF,z,n);
```

$$a := 4\left(\frac{1}{2}\right)^n - \frac{8}{27}(n+1)\left(\frac{1}{2}n+1\right)\left(\frac{1}{3}\right)^n + \frac{1}{9}(-8n-8)\left(\frac{1}{3}\right)^n - \frac{8}{3}\left(\frac{1}{3}\right)^n$$

We first evaluate the result for some small values of n and see that the numbers are all between zero and one. Summing the result from zero to infinity gives one, implying that we have found the correct density.

```
>   evalf(seq(eval(a,n=k),k=0..10));
```

.1481481481, .2222222222, .2098765432, .1598079561, .1073388203, .06647233653, .03892635777, .02190183153, .01196702103, .006397479330, .003364327190

```
>   sum(a,n=0..infinity);
```

$$1$$

---

[6]It is easy to see that since $\hat{g}(1) = 1$, this *must* be the generating function of a proper discrete random variable.

Finally, we also compute the mean in two different ways and find that it equals 5/2.

```
>   eval(diff(aGF,z),z=1);
```
$$\frac{5}{2}$$

```
>   sum(a*n,n=0..infinity);
```
$$\frac{5}{2}$$

Numerical Inversion of Generating Functions

As in Laplace transforms, it may sometimes be difficult (or impossible) to invert a generating function $\hat{a}(z) = \sum_{n=0}^{\infty} a_n z^n$ of a probability density a_n. Fortunately, there is an easy-to-use numerical inversion method due to Abate and Whitt [1] that works for $n \geq 1$. If we let $a_n^{\text{Numerical}}$ denote the probability a_n, then

$$a_n^{\text{Numerical}} = \frac{1}{2nr^n}\left\{\hat{a}(r) + (-1)^n \hat{a}(-r) + 2\sum_{j=1}^{n-1}(-1)^j \mathcal{R}\left(\hat{a}\left(r\exp\left(\frac{\pi ji}{n}\right)\right)\right)\right\}$$

where $r \in (0, 1)$ and $\mathcal{R}(\zeta)$ is the real part of the complex number ζ. To have an accuracy of 10^{-v}, we let $r = 10^{-v/2n}$. It can be shown that the error bound is

$$\left|a_n - a_n^{\text{Numerical}}\right| \leq \frac{r^{2n}}{1 - r^{2n}}.$$

Example 42 *Numerical inversion of a generating function.* We invert the generating function discussed in Example 41

$$\hat{a}(z) = \frac{8}{(2 - z)(3 - z)^3}$$

and compare the results to what we already have obtained as the exact expressions.

```
>   restart: # NumericalGF.mws
>   v:=5;
```
$$v := 5$$

```
>   aGF:=z->8/((2-z)*(3-z)^3);
```
$$aGF := z \rightarrow 8\,\frac{1}{(2 - z)\,(3 - z)^3}$$

```
>   aNumer:=n-> (1/(2*n*r^n))*(aGF(r)+(-1)^n*aGF(-r)
    +2*Sum((-1)^j*Re(aGF(r*exp(Pi*j*I/n))),j=1..n-1));
```

$$aNumer :=$$

$$n \rightarrow \frac{1}{2}\,\frac{aGF(r) + (-1)^n\, aGF(-r) + 2\left(\sum_{j=1}^{n-1}(-1)^j\,\Re(aGF(r\,e^{(\frac{I\pi j}{n})}))\right)}{n\,r^n}$$

```
>   r:=10^(-v/(2*n));
```

$$r := 10^{(-5/2\frac{1}{n})}$$

We see that for this problem the numerical inversion gives results that are very close to what we had found in Example 41.

```
>   seq([n,evalf(value(aNumer(n)))],n=1..25);
```

[1, .2222238606], [2, .2098769451], [3, .1598080311], [4, .1073388311],
[5, .06647233500], [6, .03892634146], [7, .02190184246],
[8, .01196702514], [9, .006397466901], [10, .003364318742],
[11, .001747392703], [12, .0008990553030],
[13, .0004592782611], [14, .0002333670562],
[15, .0001180794478], [16, .00005955754688],
[17, .00002997095156], [18, .00001506825305],
[19, .7566997567 10^{-5}], [20, .3775759525 10^{-5}],
[21, .1885319824 10^{-5}], [22, .9688068833 10^{-6}],
[23, .4798412623 10^{-6}], [24, .2483705579 10^{-6}],
[25, .1164350634 10^{-6}]

```
>   n:=' n' :
```

We also compute the remaining probability $a_0^{\text{Numerical}} = 1 - \sum_{n=1}^{\bar{n}} a_n^{\text{Numerical}}$ where $\bar{n}$ is such that $a_n^{\text{Numerical}} \approx 0$ for $n \geq \bar{n}$. This gives $a_0^{\text{Numerical}} = 0.148146$.

```
>   sum(evalf(value(aNumer(n))),n=1..25);
```

.8518538100

```
>   evalf(1-%); # a(0)
```

.1481461900

3.7 Probability

Maple recognizes a large number of continuous and discrete random variables whose density function, cumulative distribution function and inverse cumulative distribution function can be evaluated *numerically*. These functions are loaded by entering the command `with(stats);`. By using Maple's symbolic manipulation capabilities, one can also perform certain types of probabilistic analysis such as computing expectations and finding the joint probability distribution of functions of random variables.

3.7.1 Continuous and Discrete Random Variables

Maple knows about the following 13 continuous random variables: beta, Cauchy, χ^2, exponential, F, gamma, Laplace, logistic, lognormal, normal, t, uniform and

Weibull. To evaluate (or plot) the density function of a continuous random variable, we use `statevalf[ pdf,name] ()`. To evaluate the cumulative distribution function, the first argument is replaced by `cdf`, and to evaluate the inverse cumulative distribution function, the first argument is replaced by `icdf`.

Maple also knows about the binomial, discrete uniform, hypergeometric, negative binomial and Poisson discrete random variables. To evaluate the density function of a discrete random variable, we use `statevalf[ pf,name] ()`. The first argument is replaced with `dcdf` and `idcdf` to evaluate the cumulative distribution and inverse cumulative distribution functions, respectively.

Detailed information about the evaluation of these functions can be obtained with `? stats`.

Example 43 *Normal random variable.* In this example we evaluate and plot the density, distribution function and inverse distribution function of the unit normal random variable with mean 0 and variance 1. The density

$$f(x) = \frac{1}{\sqrt{2\pi}} e^{-\frac{1}{2}x^2}, \quad -\infty < x < \infty$$

is defined as `f:=x->statevalf[ pdf,normald[ 0,1]] (x)`.
```
>    restart: # ContinuousDistributions.mws
>    with(stats);
```
[*anova, describe, fit, importdata, random, statevalf, statplots, transform*]

The distribution function $F(x) = \int_{-\infty}^{x} f(u)\, du$ and inverse distribution function $F^{-1}(x)$ are defined similarly.

Evaluating these functions, we see that $f(0) = .398$, $F(3) = .998$ and $F^{-1}(.3) = -.524$. These three functions are plotted on the same graph with the `display()` command and presented in Figure 3.3
```
>    f:=x->statevalf[ pdf,normald[ 0,1]] (x):
>    F:=x->statevalf[ cdf,normald[ 0,1]] (x):
>    iF:=x->statevalf[ icdf,normald[ 0,1]] (x):
>    f(0); F(3); iF(.3);
```
                           .3989422803

                           .9986501020

                          −.5244005127
```
>    fPlot:=plot(f(x),x=-3..3,linestyle=1,
        color=black,thickness=2):
>    FPlot:=plot(F(x),x=-3..3,linestyle=2,
        color=black,thickness=2):
>    iFPlot:=plot(iF(x),x=-1..1,linestyle=3,
        color=black,thickness=2):
>    with(plots):
>    display([ fPlot,FPlot,iFPlot] );
```

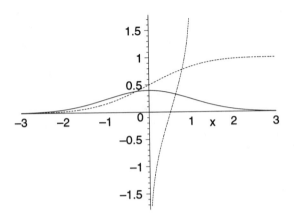

FIGURE 3.3. The density $f(x)$, cumulative distribution function $F(x)$ and inverse cumulative distribution function $F^{-1}(x)$ of the unit normal random variable drawn as a solid line, dotted line and dashed line, respectively.

Example 44 *Poisson random variable.* Next, we consider the Poisson random variable with parameter $\lambda = 3$ and density

$$p(n) = \frac{\lambda^n e^{-\lambda}}{n!}, \quad n = 0, 1, \dots.$$

This is defined in Maple as `p:=statevalf[ pf,poisson[ 3]] (n);`.

```
>    restart: # DiscreteDistributions.mws
>    with(stats);
```

[anova, describe, fit, importdata, random, statevalf, statplots, transform]

The distribution function $P(n) = \sum_{k=0}^{n} p(k)$ and the inverse distribution function $P^{-1}(n)$ are defined in a similar manner. We find, for example, that $p(5) = .1$, $P(5) = .91$ and $P^{-1}(.8) = 3$; i.e., there is a 0.10 chance that the Poisson random variable with rate $\lambda = 3$ will take the value 5 and a 0.91 chance that it will take a value less than or equal to 5. The first two functions are plotted in Figure 3.4. The inverse distribution is not included in this graph but it too can be plotted using the command `plot(iP(n),n=0..1,0..8);`.

```
>    p:=n->statevalf[ pf,poisson[ 3]] (n);
```

$$p := statevalf_{pf, poisson_3}$$

```
>    P:=n->statevalf[ dcdf,poisson[ 3]] (n);
```

$$P := statevalf_{dcdf, poisson_3}$$

```
>    iP:=n->statevalf[ idcdf,poisson[ 3]] (n);
```

$$iP := statevalf_{idcdf, poisson_3}$$

```
>    p(5); P(5); iP(.8);
```

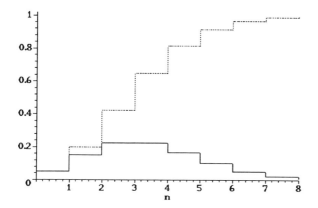

FIGURE 3.4. The density $p(n)$ and the cumulative distribution function $P(n)$ of Poisson with parameter $\lambda = 3$ drawn as a solid line and a dotted line, respectively.

.1008188134

.9160820583

3.
```
>   pPlot:=plot(p(floor(n)),n=0..8,linestyle=1,
    color=black,thickness=2):
>   PPlot:=plot(P(floor(n)),n=0..8,linestyle=2,
    color=black,thickness=2):
>   with(plots):
>   display([ pPlot,PPlot] );
>   #plot(iP(n),n=0..1,0..8):
```
In addition to the nearly 20 frequently used distributions, Maple also allows the definition of arbitrary discrete empirical distributions defined over the integers $1, \ldots, N$. This is made possible by using the empirical command.

Example 45 *A discrete empirical distribution.* Consider a discrete random variable X with density $p_X(1) = 0.2$, $p_X(2) = 0.3$, $p_X(3) = 0.5$. The probabilities are entered as an expression sequence named Probs and the density is defined simply as p:=empirical[Probs].
```
>   restart: # Empirical.mws
>   with(stats):
>   with(random);
```

[β, *binomiald, cauchy, chisquare, discreteuniform, empirical, exponential, fratio, γ, laplaced, logistic, lognormal, negativebinomial, normald, poisson, studentst, uniform, weibull*]
```
>   Probs:=.2,.3,.5;
```

$$Probs := .2, .3, .5$$

```
>  p:=empirical[ Probs] ;
```

$$p := empirical_{.2, .3, .5}$$

The density and the cumulative distribution function of X can be evaluated using the pf and dcdf functions. It is also possible to evaluate the inverse cumulative distribution function using idcdf and by supplying as input a number between 0 and 1.

```
>  statevalf[ pf,p] (1);
```

$$.2$$

```
>  statevalf[ dcdf,p] (2);
```

$$.5$$

```
>  statevalf[ idcdf,p] (.215);
```

$$1.$$

```
>  seq(statevalf[ dcdf,p] (n),n=1..3);
```

$$.2, .5, 1.0$$

Finally, it is also possible to generate random variates from this distribution using the random command.

```
>  random[ p] (10);
```

$$2.0, 3.0, 3.0, 3.0, 2.0, 1.0, 1.0, 3.0, 1.0, 3.0$$

3.7.2 Expectation

Since the computation of moments of a random variable requires evaluating integrals or sums, Maple is well suited for performing such operations.

Example 46 *Mean and the variance of the normal distribution.* For example, let us consider the normally distributed random variable X with parameters μ and σ^2 and compute its mean $E(X)$ and variance $\mathrm{Var}(X)$.

```
>  restart: # NormalMoments.mws
>  interface(showassumed=2); assume(sigma>0);
>  f:=(1/(sigma* (2* Pi)^(1/2)))* exp(-(x-mu)^2/
   (2* sigma^2));
```

$$f := \frac{1}{2} \frac{\sqrt{2}e^{(-1/2\frac{(x-\mu)^2}{\sigma^2})}}{\sigma\sqrt{\pi}}$$

with assumptions on σ

```
>  Int(f,x=-infinity..infinity);
```

$$\int_{-\infty}^{\infty} \frac{1}{2} \frac{\sqrt{2}\, e^{(-1/2\, \frac{(x-\mu)^2}{\sigma^2})}}{\sigma\, \sqrt{\pi}}\, dx$$

with assumptions on σ

Maple easily evaluates the integral of density and finds the result as 1. It also finds that $E(X) = \mu$ and $\text{Var}(X) = \sigma^2$.

>   value(%);

$$1$$

>   EX:=int(x* f,x=-infinity..infinity);

$$EX := \mu$$

>   VarX:=int((x-mu)^2* f,x=-infinity..infinity);

$$VarX := \sigma^2$$

with assumptions on σ

Example 47 *A discrete random variable taking finitely many values.* Suppose a random variable X takes the value x with probability ax, $x = 1, \ldots, N$. We wish to find the mean $E(X)$ and variance $\text{Var}(X)$ after determining the proper value of a.

>   restart: # Expectation.mws

>   f:=a* x;

$$f := a x$$

>   a:=solve(sum(a* x,x=1..N)=1,a);

$$a := 2\, \frac{1}{N\,(N+1)}$$

Thus, a is found to be $2/[N(N+1)]$.

>   f;

$$2\, \frac{x}{N\,(N+1)}$$

>   EX:=normal(sum(x* f,x=1..N));

$$EX := \frac{2}{3}\, N + \frac{1}{3}$$

>   EX2:=normal(sum(x^2* f,x=1..N));

$$EX2 := \frac{1}{2}\, N\,(N+1)$$

>   VarX:=normal(EX2-EX^2);

$$VarX := \frac{1}{18}\, N^2 + \frac{1}{18}\, N - \frac{1}{9}$$

Example 48 *Cauchy distribution.* We wish to find a constant $c(a, b)$ for which the function $f(x) = c(a, b)(1 + x^2)^{-1}$ for $a < x < b$ is a density.

```
>    restart:  # Cauchy.mws
>    f:=c/(1+x^2);
```

$$f := \frac{c}{1 + x^2}$$

```
>    p:=Int(f,x=a..b);
```

$$p := \int_a^b \frac{c}{1 + x^2}\, dx$$

To do this, we solve $\int_a^b c(a, b)(1 + x^2)^{-1}\, dx = 1$ for $c(a, b)$ and find that $c(a, b) = [\arctan(b) - \arctan(a)]^{-1}$. When $x \in (-1, 1)$, we have $c(-1, 1) = 2/\pi$ and when $x \in (-\infty, \infty)$, we obtain $c(-\infty, \infty) = 1/\pi$ which gives the Cauchy density

$$f(x) = \frac{1}{\pi(1 + x^2)}, \quad -\infty < x < \infty.$$

```
>    c:=solve(value(p)=1,c);
```

$$c := \frac{1}{\arctan(b) - \arctan(a)}$$

```
>    limit(limit(c,b=1),a=-1);
```

$$2\frac{1}{\pi}$$

```
>    limit(limit(c,b=infinity),a=-infinity);
```

$$\frac{1}{\pi}$$

The fact that the Cauchy density has no expected value can be shown as follows. We write $E(X) = \int_{-\infty}^a xf(x)\, dx + \int_a^\infty xf(x)\, dx$. The second integral evaluates to $\frac{1}{2\pi} \ln(1 + a^2)$, which diverges to infinity as $a \to \infty$. Thus, $E(X)$ does not exist.

```
>    f:=1/(Pi*(1+x^2));
```

$$f := \frac{1}{\pi(1 + x^2)}$$

```
>    Integral[2]:=Int(x* f,x=0..a);
     value(Integral[2]);
```

$$Integral_2 := \int_0^a \frac{x}{\pi(1 + x^2)}\, dx$$

$$\frac{1}{2}\frac{\ln(1 + a^2)}{\pi}$$

```
>    limit(%,a=infinity);
```

$$\infty$$

3.7.3   Jointly Distributed Random Variables

Let X and Y be jointly distributed continuous random variables with density $f_{X,Y}(x, y)$. Suppose we define two new random variables $U = u(X, Y)$ and $V = v(X, Y)$ and assume that the system of nonlinear equations $u = u(x, y)$ and $v = v(x, y)$ can be solved uniquely for $x = x(u, v)$ and $y = y(u, v)$ in terms of u and v. Then the joint density of the random vector (U, V) is given by

$$f_{U,V}(u, v) = f_{X,Y}(x(u, v), y(u, v)) \, |J(u, v)| \qquad (3.3)$$

where

$$J(u, v) = \det \begin{pmatrix} \frac{\partial x}{\partial u} & \frac{\partial x}{\partial v} \\ \frac{\partial y}{\partial u} & \frac{\partial y}{\partial v} \end{pmatrix}$$

is the determinant of the Jacobian of the transformation; see Harris [82, p. 169] and Stirzaker [181, p. 270].

Example 49 *Joint distribution of a random vector.* We assume that the joint density of the random vector (X, Y) is $f_{X,Y}(x, y) = \lambda^2 e^{-\lambda x} e^{-\lambda y}$, i.e., X and Y are individually distributed as exponential random variables with parameter λ.

```
>    restart: # JointDensity.mws
>    with(linalg):
```

Warning, new definition for norm

Warning, new definition for trace

```
>    assume(lambda>0);
>    fXY:=(x,y)->lambda*exp(-lambda*x)*lambda
     *exp(-lambda*y);
```

$$fXY := (x,\ y) \to \lambda^2\, e^{(-\lambda x)}\, e^{(-\lambda y)}$$

Using the transformation $U = X + Y$ and $V = X/(X + Y)$, we obtain the inverse transformation as $x = uv$ and $y = u - uv$ and the determinant of the Jacobian as $-u$.

```
>    uxy:=u=x+y; vxy:=v=x/(x+y);
```

$$uxy := u = x + y$$

$$vxy := v = \frac{x}{x + y}$$

```
>    solve({uxy,vxy},{x,y});
```

$$\{y = u - v\,u,\ x = v\,u\}$$

```
>    assign(%);
>    A:=evalm(vector([x,y]));
```

$$A := [v\,u,\ u - v\,u]$$

```
>    jacobian(A,[u,v]);
```

$$\begin{bmatrix} v & u \\ 1-v & -u \end{bmatrix}$$

>   J:=det(%);

$$J := -u$$

>   fXY(x,y)*abs(J);

$$\lambda^2\, e^{(-\lambda v\, u)}\, e^{(-\lambda\,(u-v\, u))}\, |u|$$

Applying (3.3) gives $f_{U,V}(u,v) = \lambda^2 e^{-\lambda u}\,|u|,\ u > 0$, which integrates to 1, as expected.

>   fUV:=simplify(%);

$$fUV := \lambda^2\, e^{(-\lambda u)}\, |u|$$

>   int(fUV,u=0..infinity);

$$1$$

3.8 Summary

Modeling and solution of many operations research problems requires a knowledge of a wide spectrum of mathematical topics such as algebra and calculus, linear algebra, differential equations, transform methods and probability theory. This chapter covered these topics and their treatment with Maple. We surveyed these mathematical topics by providing a large number of examples. With the detailed explanation of the useful mathematical techniques as presented in this chapter, we are now ready to begin our study of OR techniques and models.

3.9 Exercises

1. Find the values of x that satisfy the inequality $x^2 + x + 1 \geq 3$. Check your result by comparing the plots of $x^2 + x + 1$ and 3 on the same graph.

2. The nonlinear equation $1 - \int_0^Q f(x)\,dx = \beta$ (where $0 < \beta < 1$ and $f(x)$ is the demand density) arises in stochastic inventory theory. Solve this equation for the order quantity Q under the following assumptions:

   (a) $f(x) = 1/(b-a),\ a < b$ (uniform demand)
   (b) $f(x) = \lambda e^{-\lambda x}(\lambda x)^{n-1}/(n-1)!$, for $n = 3,\ \beta = 0.7$ and $\lambda = 0.01$ (gamma demand).

3. Find the points of intersection of the two circles $x^2 + y^2 = 1$ and $(x-1)^2 + y^2 = \frac{1}{4}$.

4. The function $C(u) = \sum_{n=0}^{\infty} e^{-anu}(k + c\lambda u)$ arises in the analysis of an inventory problem with discounted cost objective function; see Zipkin [199, p. 63] and Hadley and Whitin [81, p. 80]. Differentiate this function with respect to u and find the u^* that satisfies $C'(u^*) = 0$ assuming $c = 1$, $\lambda = 10$, $k = 10$, $a = 0.1$.

5. Consider the function $G(u) = \int_0^u \frac{1}{10}(u - x)\,dx + \int_u^{10} \frac{1}{10}(x - u)\,dx$. Find the derivative of this function in two different ways and locate the point that minimizes $G(u)$.

6. Evaluate the integral $\int_0^{\infty} x^a e^{-bx}\,dx$ assuming that a and b are strictly positive. HINT: Use the assume() statement.

7. Determine whether the linear system of equations $\mathbf{Ax} = \mathbf{b}$ with

$$\mathbf{A} = \begin{bmatrix} 1 & 1 & 3 \\ 2 & 5 & -6 \\ 3 & 6 & -9 \end{bmatrix}, \quad \mathbf{b} = \begin{bmatrix} 9 \\ 1 \\ 0 \end{bmatrix}$$

has a unique solution by comparing the ranks of $r(\mathbf{A})$ and $r(\mathbf{A} \mid \mathbf{b})$. If the system has a unique solution, find it by (i) evaluating $\mathbf{A}^{-1}\mathbf{b}$, and (ii) using linsolve().

8. Find the 10th power of the transition probability matrix

$$\mathbf{P} = \begin{bmatrix} \frac{1}{2} & \frac{1}{2} \\ \frac{2}{5} & \frac{3}{5} \end{bmatrix}$$

after diagonalizing it. Does your result agree with $\mathbf{P}^{10}$ obtained directly?

9. The second order differential equation $x''(t) = rx'(t) + \frac{1}{2}$ arises in the minimization of the functional $\int_0^T e^{-rt}\{[x'(t)]^2 + 2x(t)\}\,dt$ with conditions $x(0) = x_0$ and $x(T) = x_T$. Solve the differential equation and plot the trajectory assuming $r = 0.10$, $T = 1$, $x_0 = 10$ and $x_T = 8$.

10. Consider the system of differential equations:

$$\begin{aligned} m_1'(t) &= \lambda - \mu m_1(t), \quad m_1(0) = i \\ m_2'(t) &= \lambda + (2\lambda + \mu)m_1(t) - 2\mu m_2(t), \quad m_2(0) = i^2. \end{aligned}$$

This system arises in the analysis of an infinite server queueing system where λ and μ are positive arrival and departure rates, respectively. The functions $m_1(t)$ and $m_2(t)$ are the expected number of busy servers and mean-squared number of busy servers. Solve this system by dsolve() and use the method=laplace option. Show that $\lim_{t\to\infty} m_1(t) = \lambda/\mu$. Next, define the variance of the number of busy servers as $M_1(t) = m_2(t) - [m_1(t)]^2$ and show that $\lim_{t\to\infty} M_1(t) = \lambda/\mu$. (Ghosal et al. [72, p. 124].)

11. Solve the difference equation $m_k = 2m_{k-1} + 1$, with initial conditions $m_0 = 0, m_1 = 1$ using generating functions.

12. Some differential equations can be solved easily using Laplace transforms (LT). Consider the second order differential equation $y''(t) + 3y'(t) + 2y(t) = 0$ with $y(0) = 1$ and $y'(0) = 3$. Let $\tilde{y}(s)$ be the LT of $y(t)$. Find the LT of the individual terms of this differential equation, isolate $\tilde{y}(s)$, and invert the resulting transform to find $y(t)$. Compare your result to what Maple would find using `dsolve()`.

13. Invert the generating function

$$\hat{a}(z) = \frac{18}{(3 - z)(4 - z)^2}$$

using `rgf_expand()` to find the sequence a_n. Is the resulting sequence a probability density of a discrete random variable? If it is, find its mean using two different methods.

14. Consider two random variables X and Y with joint density given by $f(x, y) = e^{-(x+y)}$ for $0 < x < \infty$ and $0 < y < \infty$ and zero otherwise. Find the distribution function and the density of the ratio X/Y.

15. The Euclidean distance R of a random point (X, Y) from the origin is given by $R = \sqrt{X^2 + Y^2}$. If the joint density of (X, Y) is $f(x, y) = 1/(\pi u^2)$ for $x^2 + y^2 \leq u^2$ and zero otherwise, find the mean distance $E(R)$.

16. An experiment consists of n independent and identical trials where each trial can result in one of three classifications for the items tested: (i) grade A classification, (ii) grade B classification or (iii) defective with probabilities p, q and $1 - p - q$, respectively. Let X denote the number of grade A items, and Y denote the number of grade B items. Determine the joint density of X and Y, their expectations, variances and covariance.

17. Consider two random variables X and Y distributed jointly with density

$$f(x, y) = \frac{1}{2\pi\sigma^2} \exp\left(-\frac{x^2 + y^2}{2\sigma^2}\right), \quad -\infty < x < \infty, \; -\infty < y < \infty.$$

Let two new random variables (R, Θ) be related to (X, Y) as $X = R\cos(\Theta)$ and $Y = R\sin(\Theta)$. Determine the joint density $g(r, \theta)$ of (R, Θ). Show that $\int_0^\infty \int_0^{2\pi} g(r, \theta) \, d\theta \, dr = 1$.

4

Linear Programming

4.1 Introduction

Linear programming (LP) is a flexible and powerful optimization technique that is used to determine the nonnegative values of n decision variables x_j, which satisfy m linear constraints

$$\sum_{j=1}^{n} a_{ij}x_j \ \{\leq, =, \geq\}\ b_i, \quad i = 1, 2, \ldots, m$$

and maximize (or minimize) a linear objective function

$$z = \sum_{j=1}^{n} c_j x_j$$

where the parameters a_{ij}, b_i and c_j are given constants.

The origins of LP date back to the late 1940s when G. B. Dantzig [58] developed a very efficient and general algebraic method known as the "simplex method" that could solve LP problems.[1] About a decade later, Dantzig [59] published a major survey covering the theoretical developments and applications of linear programming up to the early 1960s. Since the publication of Dantzig's original paper, practically hundreds of books and thousands of research articles on LP have appeared in dozens of different languages.

---

[1] In late 1930s the Russian mathematician Kantorovich [100] had, however, studied linear programming models for production planning.

Any textbook on operations research published since the 1950s has at least one chapter describing LP. Linear programming is now a very important and established branch of operations research that has found wide applicability in business and governmental decisions.

Despite the mathematical nature of the simplex method, which requires a good understanding of linear algebra, LP has become one of the most useful operations research techniques and is frequently used by practitioners—perhaps due to the proliferation of easy-to-use PC software such as LINDO and the inclusion of the simplex algorithm in almost every spreadsheet software such as Excel and Quat-troPro. A recent survey by Lane, Mansour and Harpell [116] indicates that along with simulation and statistical methods, LP has found its way as an important technique into the practitioner's toolkit of analytical methods.

In this chapter, we will briefly describe the mathematics of the simplex method and present some examples of LP modeling and solution using Maple. Since LP problems cannot be solved symbolically (i.e., the parameters a_{ij}, b_j and c_i must be real numbers), Maple's symbolic manipulation capabilities will not be used in this chapter. However, since Maple can plot regions of linear inequalities with the command inequal(), this plotting feature will be very useful in identifying the feasible region and the optimal solution of an LP with $n = 2$ decision variables.

4.2   Graphical Solution

To motivate the discussion, we first formulate a simple LP with two decision variables and three constraints and present its solution using the graphical method.

Example 50 *A simplified production problem.* Suppose we are in the business of producing toy cars and toy trains. The accounting department has analyzed the costs and revenues and found that each car made (and immediately sold) results in a profit of \$30 and the profit from each train is \$40. We have two departments where these toys are made. The car department has a daily production capacity of 90 units and the train department 60 units. A complicating factor in the production of these toys is a special part that has to be purchased from an outside source that can provide only 600 parts per day. The engineering department has done some computations and found that each car needs 5 parts and each train needs 6 parts. The problem is to find the optimal values of the decision variables that maximize the daily profit.

In this problem it is clear that there are $n = 2$ decision variables. These are defined as x_1 : number of toy cars to produce per day and x_2 : number of toy trains to produce per day. The objective is to maximize the daily profit $z = 30x_1 + 40x_2$. The car department's production capacity indicates that the solution must satisfy $x_1 \leq 90$. Similarly for the train department the constraint is $x_2 \leq 60$. Finally, the special parts constraint is given as $5x_1 + 6x_2 \leq 600$. Since negative production is impossible, we also have the "natural" constraints $x_1 \geq 0$ and $x_2 \geq 0$. Thus, the linear programming problem for the toy manufacturer can be stated as

$$
\begin{array}{llllll}
\max & z = & 30x_1 & + & 40x_2 & \\
\text{s.t.} & & x_1 & & & \leq & 90 \\
& & & & x_2 & \leq & 60 \\
& & 5x_1 & + & 6x_2 & \leq & 600 \\
& & & & x_1, x_2 & \geq & 0
\end{array}
$$

In this simple case, it is possible to plot the feasible region F (i.e., the set of all points (x_1, x_2) satisfying all constraints) and find the point that maximizes the objective function $z = 30x_1 + 40x_2$. We do this using Maple's inequal() function that can plot linear inequalities and hence the feasible region F.

At this point we introduce the concepts of a *convex set* and an *extreme point* that play an important role in the theory of linear programming.

Definition 1 *A set $\mathcal{F}$ is said to be* convex *if, for any two points $x_1, x_2 \in \mathcal{F}$, the line segment joining the two points lies entirely within the set, i.e., $\lambda x_1 + (1 - \lambda)x_2 \in \mathcal{F}$ for $0 \leq \lambda \leq 1$.*

Intuitively, a convex set cannot have any "holes" in it and its boundaries are always "flat" or "bent away" from the set, as we will see in the current example.

Definition 2 *A point x is said to be an* extreme point *of a convex set if and only if there do not exist other points x_1, x_2 ($x_1 \neq x_2$) in the set such that $x = \lambda x_1 + (1 - \lambda)x_2$ for $0 < \lambda < 1$.*

Intuitively, an extreme point cannot be "between" any other two points of the set.

We start by introducing the constraints, which are placed in a list named CS (Constraint Set). The feasible region is plotted (but not yet displayed) using the inequal() function. We also plot—but do not yet display—three iso-profit lines corresponding to profit levels of $z = 1200$, 2400 and 3600. These lines will be useful when we attempt to identify the optimal solution. The feasible region $\mathcal{F}$ (and the iso-profit lines) are displayed with the display() command. We find that the feasible region is a convex set bounded by the lines $x_2 = 0$, $x_1 = 90$, $5x_1 + 6x_2 = 600$, $x_2 = 60$ and $x_1 = 0$ with the corner points at O, A, B, C and D.

```
>    restart: # ToyLP.mws

>    with(plots):

>    CS:=[ x[ 1] <=90,x[ 2] <=60,5* x[ 1] +6* x[ 2] <=600,
     x[ 1] >=0,x[ 2] >=0] ;
```

$$CS := [x_1 \leq 90,\ x_2 \leq 60,\ 5x_1 + 6x_2 \leq 600,\ 0 \leq x_1,\ 0 \leq x_2]$$

```
>    FeasibleRegion:=inequal(CS,  x[ 1] =0..120,
     x[ 2] =0..120,optionsfeasible=(color=white),
     optionsexcluded=(color=yellow)):
```

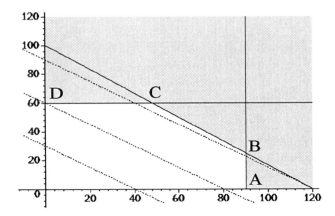

FIGURE 4.1. The feasible region with extreme points O, A, B, C and D and the iso-profit lines for $z = 1200, 2400$ and 3600.

```
>   ProfitLines:=contourplot(30*x[ 1] +40*x[ 2] ,
    x[ 1] =-10..120,x[ 2] =-10..120,
    contours=[ 1200,2400,3600] ,linestyle=2,
    numpoints=2000,color=black):
>   display([ FeasibleRegion,ProfitLines] );
```

It can be shown that in linear programming problems, such feasible regions are always convex; see Hadley [79, p. 60]. Additionally, if the convex feasible region corresponding to the linear programming problem is nonempty, it must have at least one *extreme* (corner) point. In our example, the feasible set has five extreme points (denoted by O, A, B, C and D), which are located at the intersection of the constraint boundary lines; see Figure 4.1. It is also important to add that if there is a finite optimal solution to the linear program, then this optimal solution must be an extreme point of the feasible region. For a discussion of these important properties of the solution of linear programming problems, see Luenberger [127, Chapter 2].

Returning to our problem, we observe that the extreme points of the feasible region—except the uninteresting point O at the origin where there is no production activity—are easily computed by Maple as follows.

```
>   with(simplex):
```

Warning, new definition for display

Warning, new definition for maximize

Warning, new definition for minimize

We start by solving for the intersection point of the boundaries $x_1 = 90$ and $x_2 = 0$ of the constraints CS[1] and CS[5] , respectively. This is achieved by first converting the inequality constraints $x_1 \leq 90$ and $x_2 \geq 0$ to equalities using the convert() function. The obvious solution corresponds to the corner point A on the feasible set. Substituting the solution $x_1 = 90$ and $x_2 = 0$ in z gives $z = 2700$.

```
>    solve(convert({ CS[ 1] ,CS[ 5]} ,equality));
     assign(%); z:=30* x[ 1] +40* x[ 2] ;
     x[ 1] :=′ x[ 1]′ :  x[ 2] :=′ x[ 2]′ :
```

$$\{x_2 = 0, \ x_1 = 90\}$$

$$z := 2700$$

The coordinates of the next three points B, C and D are similarly computed and the objective function is evaluated at each point.

```
>    solve(convert({ CS[ 1] ,CS[ 3]} ,equality));
     assign(%); z:=30* x[ 1] +40* x[ 2] ; x[ 1] :=′ x[ 1]′ :
     x[ 2] :=′ x[ 2]′ :
```

$$\{x_2 = 25, \ x_1 = 90\}$$

$$z := 3700$$

```
>    solve(convert({ CS[ 3] ,CS[ 2]} ,equality));
     assign(%); z:=30* x[ 1] +40* x[ 2] ; x[ 1] :=′ x[ 1]′ :
     x[ 2] :=′ x[ 2]′ :
```

$$\{x_2 = 60, \ x_1 = 48\}$$

$$z := 3840$$

```
>    solve(convert({ CS[ 4] ,CS[ 2]} ,equality));
     assign(%); z:=30* x[ 1] +40* x[ 2] ; x[ 1] :=′ x[ 1]′ :
     x[ 2] :=′ x[ 2]′ :
```

$$\{x_2 = 60, \ x_1 = 0\}$$

$$z := 2400$$

Thus, we find that the optimal extreme point is C with coordinates $(x_1, x_2) = (48, 60)$ that gives rise to the highest profit level of $z = 30 \times 48 + 40 \times 60 = 3840$ dollars per day.

Although in this example it was not too difficult to examine each extreme point individually and find the optimal solution, this method is not very efficient if computations are being performed manually. In problems with several constraints, the number of extreme points may be quite large, which makes the computation of the coordinates of the extreme points and the corresponding profit values unnecessarily cumbersome. In those cases, it would be preferable to use a visual method whereby an iso-profit line is drawn on the feasible region for an arbitrary value of z and moved in a direction to maximize the profit.

We start by arbitrarily assigning a fixed value to the daily profit z, say, $z = 1200$, and ask whether it is possible to find at least one feasible production combination (x_1, x_2) that gives rise to the value of $z = 1200$. We see in Figure 4.1 that for $z = 1200$, the lowest iso-profit line passes through the points $(40, 0)$ and $(0, 30)$, and hence every point between and including these points gives a profit of $z = 1200$.

Increasing the value of z (say, doubling it to 2400) and redrawing the line, we see that it is still possible to find some feasible production combinations. Hence, by visual inspection (perhaps by using a ruler and sliding it upward parallel to the first two iso-profit lines) we find that the optimal point must be at the intersection of the boundary lines $5x_1 + 6x_2 = 600$ and $x_2 = 60$. Since the solution of this system can be found easily, we conclude that the optimal result is at point C with coordinates $x_1 = 48$ and $x_2 = 60$ with a maximum profit of $z = 3840$. It is important to note that with the iso-profit line method it is not necessary to solve for the coordinates of every extreme point. Once we visually identify the *optimal* extreme point, its coordinates can then be found by solving a system of two linear equations in x_1 and x_2.

It should be clear that when there are only $n = 2$ decision variables, this graphical method (while cumbersome) can always be used to locate the optimal solution—when it exists. If there are $n = 3$ decision variables, the graphical method can still be used but it becomes very difficult to identify the feasible region as it becomes a solid geometrical figure (a polytope) bounded by planes in three dimensions. When there are four or more variables, the graphical method can no longer be used to find the optimal solution since it becomes impossible to plot objects in dimensions higher than three. Real-life models that may have thousands of variables and constraints are commonly solved using the simplex method.

4.3   The Simplex Method

Essentially, implementing the simplex method involves nothing more difficult than solving a sequence of system of linear *equations*. This requires the conversion of linear programming problems with a mixture of $\leq, =$ or $\geq$ constraints to the form $\mathbf{Ax} = \mathbf{b}$ where $\mathbf{A}$ is an $m \times n$ matrix of constraint coefficients (with columns denoted by $\mathbf{a}_1, \mathbf{a}_2, \ldots, \mathbf{a}_n$), $\mathbf{b}$ is an $m \times 1$ vector of right-hand-side (RHS) constants and $\mathbf{x}$ is an $n \times 1$ vector of unknown decision variables. The vector $\mathbf{x}$ contains the original set of decision variables and a new collection of variables introduced to convert the inequalities to equalities.

For example, the toy manufacturing problem with three constraints can be converted to a linear system of the form $\mathbf{Ax} = \mathbf{b}$ as follows. Since the first constraint is $x_1 \leq 90$, by introducing a *slack variable* $x_3 \geq 0$, we can write $x_1 + x_3 = 90$. Similarly, for the second and third constraints $x_2 \leq 60$ and $5x_1 + 6x_2 \leq 600$, we introduce the slack variables $x_4 \geq 0$ and $x_5 \geq 0$, respectively, and write $x_2 + x_4 = 60$ and $5x_1 + 6x_2 + x_5 = 600$. These conversions result in the following problem with $m = 3$ equality constraints and $n = 5$ variables:

$$
\begin{array}{lrcrcrcrcrcr}
\max & z & = & 30x_1 & + & 40x_2 & + & 0x_3 & + & 0x_4 & + & 0x_5 \\
\text{s.t.} & & & x_1 & & & + & x_3 & & & & & = & 90 \\
& & & & & x_2 & & & + & x_4 & & & = & 60 \\
& & & 5x_1 & + & 6x_2 & & & & & + & x_5 & = & 600 \\
& & & & & & & & & & \text{all } x_j & \geq & 0
\end{array}
$$

It is usually assumed that the slack variables make no contribution to the objective function nor do they result in any cost. Thus, in this example x_3, x_4 and x_5 appear in the objective function with zero coefficients. Of course, if there is a profit or cost associated with the slack variables it should be reflected in the objective function. See Ignizio [96, p. 115] for a more detailed discussion of this issue.

Note that the constraint equations $x_1 + x_3 = 90$, $x_2 + x_4 = 60$ and $5x_1 + 6x_2 + x_5 = 600$ form a system of $m = 3$ equations in $n = 5$ unknowns. Thus, the optimal solution of the optimization problem must be a solution to this set of linear equations. But since $m < n$, this system will have an infinite number of solutions.[2] However, in the simplex method only a finite number of these solutions (known as the *basic solution set*) will be of interest.

Given a system $\mathbf{Ax} = \mathbf{b}$, we may select any $m \times m$ nonsingular submatrix of $\mathbf{A}$ (termed the *basis matrix*), set the remaining $n - m$ *nonbasic* variables to zero and solve the resulting system for the m *basic* variables. This gives a *basic solution* denoted by $\mathbf{x}_B$.

Denoting the basis matrix by $\mathbf{B}$, the resulting system after setting $n - m$ variables to zero may be written as $\mathbf{B}\mathbf{x}_B = \mathbf{b}$ or $\mathbf{x}_B = \mathbf{B}^{-1}\mathbf{b}$. Thus, finding the basic solution $\mathbf{x}_B$ involves a matrix inversion that can be easily performed by Maple. Since the decision variables must be nonnegative, the simplex method examines only the basic *feasible* points for which $\mathbf{x}_B \geq 0$. We see that the number of basic solutions must be limited by $n!/[m!(n - m)!]$ since this is the number of combinations of n variables taken m at a time. For example, as we will show, the toy manufacturing problem with $n = 5$ and $m = 3$ has a total of eight basic solutions, which is less than the maximum possible of $5!/[3!(5 - 3)!] = 10$.

Suppose we set $x_4 = x_5 = 0$. In this case the basic variables would be x_1, x_2 and x_3 for which the basis matrix $\mathbf{B}$ and the basis $\mathbf{x}_B$ are obtained as follows.

```
>    restart: # BasicSols.mws
>    with(linalg):

Warning, new definition for norm

Warning, new definition for trace
```

---

[2]More accurately, if the system $\mathbf{Ax} = \mathbf{b}$ is consistent (i.e., the rank of the coefficient matrix, $r(\mathbf{A})$, is equal to the rank of the augmented matrix, $r(\mathbf{A} \mid \mathbf{b})$) and if $r(\mathbf{A}) < n$, then the system has an infinite number of solutions; see Ignizio [96, Chapter 3].

```
>   System:={ x[ 1] +x[ 3] =90,  x[ 2] +x[ 4] =60,
    5* x[ 1] +6* x[ 2] +x[ 5] =600} ;
```

$System := [x_1 + x_3 = 90, \ x_2 + x_4 = 60, \ 5\,x_1 + 6\,x_2 + x_5 = 600]$

```
>   A:=genmatrix(System,[ x[ 1] ,x[ 2] ,x[ 3] ,x[ 4] ,x[ 5] ] ,
    RHS) ;
```

$$A := \begin{bmatrix} 1 & 0 & 1 & 0 & 0 \\ 0 & 1 & 0 & 1 & 0 \\ 5 & 6 & 0 & 0 & 1 \end{bmatrix}$$

```
>   b:=evalm (RHS) ;
```

$$b := [90, \ 60, \ 600]$$

```
>   for j from 1 to 5 do a[ j] :=col (A,j) od:
>   B:=augment (a[ 1] ,a[ 2] ,a[ 3] ) ;
```

$$B := \begin{bmatrix} 1 & 0 & 1 \\ 0 & 1 & 0 \\ 5 & 6 & 0 \end{bmatrix}$$

```
>   xB:=evalm(inverse (B) &* b) ;
```

$$xB := [48, \ 60, \ 42]$$

Thus, $\mathbf{x}_B = (x_1, x_2, x_3) = (48, 60, 42)'$, which is a feasible point. This solution corresponds to producing 48 cars and 60 trains, leaving an idle capacity of 42 units in the car department.

But when $x_2 = x_4 = 0$, the resulting basis matrix for the columns $(\mathbf{a}_1, \mathbf{a}_3, \mathbf{a}_5)$ is

$$B = \begin{pmatrix} 1 & 1 & 0 \\ 0 & 0 & 0 \\ 5 & 0 & 1 \end{pmatrix}$$

whose inverse does not exist since $x_2 + x_4 = 0 + 0 \neq 60$.

```
>   B:=augment (a[ 1] ,a[ 3] ,a[ 5] ) ;
```

$$B := \begin{bmatrix} 1 & 1 & 0 \\ 0 & 0 & 0 \\ 5 & 0 & 1 \end{bmatrix}$$

```
>   xB:=evalm(inverse (B) &* b) ;

    Error, (in inverse) singular matrix
```

If we choose $(\mathbf{a}_1, \mathbf{a}_2, \mathbf{a}_5)$ as the columns of the basis matrix (i.e., if $x_3 = x_4 = 0$), then we find that $\mathbf{x}_B = (x_1, x_2, x_5) = (90, 60, -210)'$, which is an *infeasible* basic solution. This corresponds to producing 90 cars and 60 trains that would require a total of $90 \times 5 + 60 \times 6 = 810$ parts, which is 210 more than the available 600.

```
>   B:=augment (a[ 1] ,a[ 2] ,a[ 5] ) ;
```

$$B := \begin{bmatrix} 1 & 0 & 0 \\ 0 & 1 & 0 \\ 5 & 6 & 1 \end{bmatrix}$$

```
>   xB:=evalm(inverse(B) &* b);
```

$$xB := [90, \ 60, \ -210]$$

The following table summarizes results for all possible solutions, some of which do not exist and some of which are infeasible. The five corner points O, A, B, C and D as shown in Figure 4.1 are indicated in the first column labelled "POINTS."

POINTS↓	x_1	x_2	x_3	x_4	x_5	PROPERTY
1 (C)	48	60	42	0	0	**Feasible**
2 (B)	90	25	0	35	0	**Feasible**
3	90	60	0	0	−210	*Infeasible*
4	120	0	−30	60	0	*Infeasible*
5	?	0	?	0	?	Singular **B**
6 (A)	90	0	0	60	150	**Feasible**
7	0	100	90	−40	0	*Infeasible*
8 (D)	0	60	90	0	240	**Feasible**
9	0	?	0	?	?	Singular **B**
10 (O)	0	0	90	60	600	**Feasible**

Essentially, the simplex method locates the optimal solution by examining only a small fraction of the basic feasible solution generated using the procedure described. It is well known that most linear programming problems are solved within the range of $1.5m$ to $3m$ iterations (i.e., corner point evaluations) where m is the number of constraints; see Wagner [190, p. 116].

4.3.1   Manual Solution of the Production Problem

We now demonstrate the steps involved in the simplex method by solving the toy manufacturing problem with Maple's help.

We first define the system of three equations in three unknowns arising from the introduction of the slack variables x_3, x_4 and x_5. Next, the coefficient matrix A is generated from these equations using genmatrix(). We also extract the right-hand-side vector b and define the objective function coefficient vector c. The jth column of the A matrix is denoted by a_j, which is indicated by a[j] in Maple notation.

```
>   restart: # ToyLPManualSimplex.mws
>   with(linalg):

Warning, new definition for norm

Warning, new definition for trace
```

```
>   System:=[ x[ 1] +x[ 3] =90,  x[ 2] +x[ 4] =60,
    5* x[ 1] +6* x[ 2] +x[ 5] =600] ;
```

$$System := [x_1 + x_3 = 90, \ x_2 + x_4 = 60, \ 5x_1 + 6x_2 + x_5 = 600]$$

```
>   A:=genmatrix (System,[ x[ 1] ,x[ 2] ,x[ 3] ,x[ 4] ,x[ 5]] ,
    RHS);
```

$$A := \begin{bmatrix} 1 & 0 & 1 & 0 & 0 \\ 0 & 1 & 0 & 1 & 0 \\ 5 & 6 & 0 & 0 & 1 \end{bmatrix}$$

```
>   b:=evalm (RHS);
```

$$b := [90, \ 60, \ 600]$$

```
>   c:=vector (5,[ 30,40,0,0,0] );
```

$$c := [30, \ 40, \ 0, \ 0, \ 0]$$

Next, we form a basis matrix $\mathbf{B}$ from $\mathbf{A}$ by using the last three columns in the order $\mathbf{B} = [\mathbf{a}_3, \mathbf{a}_4, \mathbf{a}_5]$. Thus, $\mathbf{b}_1 = \mathbf{a}_3$, $\mathbf{b}_2 = \mathbf{a}_4$ and $\mathbf{b}_3 = \mathbf{a}_5$.

```
>   # a[ 3] ,  a[ 4]  and a[ 5]  are in basis
>   for j from 1 to 5 do a[ j] :=col (A,j) od:
>   B:=augment (a[ 3] ,a[ 4] ,a[ 5] );
```

$$B := \begin{bmatrix} 1 & 0 & 0 \\ 0 & 1 & 0 \\ 0 & 0 & 1 \end{bmatrix}$$

The basic solution $\mathbf{x}_B = (x_{B1}, x_{B2}, x_{B3})' = (x_3, x_4, x_5)'$ given by this basis matrix $\mathbf{B}$ is found as $\mathbf{x}_B = \mathbf{B}^{-1}\mathbf{b} = (90, 60, 600)'$. This implies that if $x_1 = x_2 = 0$, then $x_3 = 90$, $x_4 = 60$ and $x_5 = 600$; i.e., if nothing is produced, then there will be an idle capacity of 90 units and 60 units in the car department and train department, respectively. With no production, the 600 parts will also not be used. This corresponds to Point O in Figure 4.1.

The basic solution $\mathbf{x}_B$ is also associated with a $1 \times m$ vector $\mathbf{c}_B = (c_{B1}, c_{B2}, \ldots, c_{Bm})$ of the objective function coefficients of the basic variables $x_{B1}, x_{B2}, \ldots, x_{Bm}$. In this example, $\mathbf{c}_B = (0, 0, 0)$ since the coefficients of the slack variables are all zero. Hence, given a basic feasible solution $\mathbf{x}_B \geq \mathbf{0}$, the value of the objective function, z, is found as $z = \mathbf{c}_B \mathbf{x}_B$. In our case, we find $z = 0$ since $\mathbf{c}_B = (0, 0, 0)$ and $\mathbf{x}_B = (90, 60, 600)'$.

```
>   xB:=evalm (inverse (B)  &* b);
```

$$xB := [90, \ 60, \ 600]$$

```
>   cB:=vector (3,[ c[ 3] ,c[ 4] ,c[ 5]] );
```

$$cB := [0, \ 0, \ 0]$$

```
>   z:=evalm (cB &* xB);
```

$$z := 0$$

Since the basis matrix $\mathbf{B}$ consists of linearly independent vectors $\mathbf{b}_1, \mathbf{b}_2, \ldots, \mathbf{b}_m$, *any* column $\mathbf{a}_j$ of the coefficient matrix can be written as a linear combination of the columns of $\mathbf{B}$, i.e.,

$$\mathbf{a}_j = \sum_{i=1}^{m} y_{ij} \mathbf{b}_i$$

$$= \mathbf{B}\mathbf{y}_j$$

where $\mathbf{y}_j = (y_{1j}, y_{2j}, \ldots, y_{mj})'$. Since the inverse of $\mathbf{B}$ exists, we can also write $\mathbf{y}_j = \mathbf{B}^{-1}\mathbf{a}_j$. In our example we find $\mathbf{y}_1 = (1, 0, 5)'$ and $\mathbf{y}_2 = (0, 1, 6)'$. Note that the entries of the $\mathbf{y}_j$ vectors correspond to an amount by which a basic variable's value will change if one unit of a nonbasic variable x_j is introduced into the solution. For example, if we decide to produce one car (i.e., increase x_1 from 0 to 1 while keeping x_2 at 0), then $y_{13} = 5$ implies that the value of x_5 will be reduced by 5 units since

$$x_5 = 600 - 5x_1 - 6x_2$$
$$= 600 - 5x_1 - 6 \times 0$$
$$= 600 - 5x_1.$$

These values are computed in the next line.
```
>    y[ 1] :=evalm(inverse(B)  &*  a[ 1] );
     y[ 2] :=evalm(inverse(B)  &*  a[ 2] );
```

$$y_1 := [1, 0, 5]$$

$$y_2 := [0, 1, 6]$$

Using the definition of the $\mathbf{y}_j$'s provided, we now define a new scalar quantity $z_j = \sum_{i=1}^{m} y_{ij} c_{Bi} = \mathbf{c}_B \mathbf{y}_j$. This important quantity is interpreted as the decrease in the value of the objective function that will result if *one* unit of a nonbasic variable x_j is brought into the solution. At this particular stage of the solution, we find that $z_1 = (0, 0, 0) \times (1, 0, 5)' = 0$ and $z_2 = (0, 0, 0) \times (0, 1, 6)' = 0$.
```
>    z[ 1] :=evalm(cB  &*  y[ 1] );
     z[ 2] :=evalm(cB  &*  y[ 2] );
```

$$z_1 := 0$$

$$z_2 := 0$$

Now we are faced with the important question of improving the currently available solution. To do this we define $c_j - z_j$ as the net contribution resulting from introducing one unit of a nonbasic variable x_j into the solution. The x_j that gives the highest $c_j - z_j$ value is chosen as a new basic variable. (Ties can be broken arbitrarily.) In our example, we see that since $c_1 - z_1 = 30$ and $c_2 - z_2 = 40$, we decide to make x_2 a new basic variable—i.e., include the column vector $\mathbf{a}_2$ in the basis matrix $\mathbf{B}$—and assign x_2 the highest possible value in order to increase the profit z as much as possible.

```
>    c[ 1] -z[ 1] ; c[ 2] -z[ 2] ;
```
$$30$$
$$40$$
```
>    # So, a[ 2] enters
```

However, this cannot be done arbitrarily. For example, the parts constraint is $5x_1 + 6x_2 + x_5 = 600$, which can be written as $x_5 = 600 - 6x_2$ since $x_1 = 0$. Thus, the highest possible value that x_2 can take without making x_5 infeasible is $600/6 = 100$. But recalling that the train production constraint was $x_2 + x_4 = 60$, or $x_4 = 60 - x_2$, we see that the value of $x_2 = 100$ cannot be allowed. Thus, the highest value that x_2 can take would be 60 units, which is the minimum of the two ratios $60/1 = 60$ for x_4 and $600/6 = 100$ for x_5.

These results can be summarized using a table that indicates that the nonbasic variable x_2 should enter the basis and x_4 should leave the basis (i.e., the vector $\mathbf{a}_2$ should replace the vector $\mathbf{a}_4$ in the basis matrix $\mathbf{B}$).

Basis	c_j	x_1 30	x_2 40	x_3 0	x_4 0	x_5 0	b	Ratio
x_3	0	1	0	1	0	0	90	–
x_4	0	0	1	0	1	0	60	$60/1 = 60$ (out)
x_5	0	5	6	0	0	1	600	$600/6 = 100$
	z_j	0	0	0	0	0		
	$c_j - z_j$	30	40 (in)	0	0	0	0	

```
>    if y[ 2][ 1] >0 then print(xB[ 1] /y[ 2][ 1] ) else
     print("No Ratio") fi;
```
"No Ratio"
```
>    if y[ 2][ 2] >0 then print(xB[ 2] /y[ 2][ 2] ) else
     print("No Ratio") fi;
```
$$60$$
```
>    if y[ 2][ 3] >0 then print(xB[ 3] /y[ 2][ 3] ) else
     print("No Ratio") fi;
```
$$100$$
```
>    # So, a[ 4] leaves
```

This example suggests that the column (say, $\mathbf{b}_r$) of the basis matrix $\mathbf{B}$ that should be replaced by $\mathbf{a}_j$ is the one that satisfies

$$\frac{x_{Br}}{y_{rj}} = \min_i \left(\frac{x_{Bi}}{y_{ij}}, \quad y_{ij} > 0 \right). \tag{4.1}$$

The rest of the Maple worksheet implements the simplex procedure until the optimal solution is reached. Now with $(\mathbf{a}_3, \mathbf{a}_2, \mathbf{a}_5)$ in the basis, we find that $\mathbf{x}_B = (x_3, x_2, x_5) = (90, 60, 240)'$ with an objective function value as $z = 30 \times 0 + 40 \times 60 = 2400$ since $x_1 = 0$ and $x_2 = 60$. This corresponds to Point D in Figure 4.1.

```
> B:=augment(a[ 3] ,a[ 2] ,a[ 5] );
```

$$B := \begin{bmatrix} 1 & 0 & 0 \\ 0 & 1 & 0 \\ 0 & 6 & 1 \end{bmatrix}$$

```
> xB:=evalm(inverse(B) &* b);
```

$$xB := [90, 60, 240]$$

```
> cB:=vector(3,[ c[ 3] ,c[ 2] ,c[ 5]] );
```

$$cB := [0, 40, 0]$$

```
> z:=evalm(cB &* xB);
```

$$z := 2400$$

```
> y[ 1] :=evalm(inverse(B) &* a[ 1] );
  y[ 3] :=evalm(inverse(B) &* a[ 3] );
```

$$y_1 := [1, 0, 5]$$

$$y_3 := [1, 0, 0]$$

```
> z[ 1] :=evalm(cB &* y[ 1] );
  z[ 3] :=evalm(cB &* y[ 3] );
```

$$z_1 := 0$$

$$z_3 := 0$$

```
> c[ 1] -z[ 1] ;    c[ 3] -z[ 3] ;
```

$$30$$

$$0$$

In this iteration we find that a_1 should enter and a_5 should leave the basis.

```
> # So, a[ 1]  enters
> if y[ 1][ 1] >0 then print(xB[ 1] /y[ 1][ 1] ) else
  print("No Ratio") fi;
```

$$90$$

```
> if y[ 1][ 2] >0 then print(xB[ 2] /y[ 1][ 2] ) else
  print("No Ratio") fi;
```

"No Ratio"

```
> if y[ 1][ 3] >0 then print(xB[ 3] /y[ 1][ 3] ) else
  print("No Ratio") fi;
```

$$48$$

```
> # So, a[ 5]  leaves
```

Now, with (a_3, a_2, a_1) in the basis, the basis vector is found as $\mathbf{x}_B = (x_3, x_2, x_1)$ $= (42, 60, 48)$, which gives rise to an objective function value of $z = 30 \times 48 + 40 \times 60 = 3840$ since $x_1 = 48$ and $x_2 = 60$. This corresponds to Point C in Figure 4.1.

```
> B:=augment(a[ 3] ,a[ 2] ,a[ 1] );
```

$$B := \begin{bmatrix} 1 & 0 & 1 \\ 0 & 1 & 0 \\ 0 & 6 & 5 \end{bmatrix}$$

```
>   xB:=evalm(inverse(B) &* b);
```

$$xB := [42, 60, 48]$$

```
>   cB:=vector(3,[ c[ 3] ,c[ 2] ,c[ 1]]);
```

$$cB := [0, 40, 30]$$

```
>   z:=evalm(cB &* xB);
```

$$z := 3840$$

```
>   y[ 4] :=evalm(inverse(B) &* a[ 4] );
    y[ 5] :=evalm(inverse(B) &* a[ 5] );
```

$$y_4 := \left[\frac{6}{5}, 1, \frac{-6}{5}\right]$$

$$y_5 := \left[\frac{-1}{5}, 0, \frac{1}{5}\right]$$

```
>   z[ 4] :=evalm(cB &* y[ 4] );
    z[ 5] :=evalm(cB &* y[ 5] );
```

$$z_4 := 4$$

$$z_5 := 6$$

We also find that $c_4 - z_4 = -4$ and $c_5 - z_5 = -6$. This implies that we have found the optimal solution since introducing any of the nonbasic variables (x_4 or x_5) back into the basis would lower the objective function.

```
>   c[ 4] -z[ 4] ;  c[ 5] -z[ 5] ;
```

$$-4$$

$$-6$$

```
>   # Optimal solution reached!
```

In summary, the optimal solution is to produce $x_1 = 48$ cars and $x_2 = 60$ trains that would result in a maximum profit of $z = \$3840$. The final simplex table follows.

Basis	c_j	x_1 30	x_2 40	x_3 0	x_4 0	x_5 0	b
x_3	0	0	0	1	6/5	−1/5	42
x_2	40	0	1	0	1	0	60
x_1	30	1	0	0	−6/5	1/5	48
	z_j	30	40	0	4	6	
	$c_j - z_j$	0	0	0	−4	−6	3840

4.3.2   Solution Using Maple's `simplex` Package

Fortunately, it is not necessary to solve LP problems "manually." Maple has an implementation of the simplex algorithm that can be used either to maximize or minimize a linear objective function subject to linear constraints. (Maple's implementation of the simplex algorithm is based on the initial chapters of Chvátal's *Linear Programming* [46].)

We now return to the toy manufacturing problem and solve it using Maple's `simplex` package. In order to see the steps used by Maple, we insert the command `infolevel[ simplex] :=5` before loading `simplex`. Note that it is not necessary to introduce the slack variables as Maple automatically adds them to each constraint.

```
>    restart: # ToyLPSimplex.mws
>    infolevel[ simplex] :=5;
```

$$infolevel_{simplex} := 5$$

```
>    with(simplex):
```

Warning, new definition for maximize

Warning, new definition for minimize

```
>    z:=30* x[ 1] +40* x[ 2] ;
```

$$z := 30x_1 + 40x_2$$

```
>    Const[ 1] :=x[ 1] <=90; Const[ 2] :=x[ 2] <=60;
     Const[ 3] :=5* x[ 1] +6* x[ 2] <=600;
```

$$Const_1 := x_1 \le 90$$

$$Const_2 := x_2 \le 60$$

$$Const_3 := 5x_1 + 6x_2 \le 600$$

The optimal solution is found as $x_1 = 48$, $x_2 = 60$ and $z = \$3840$, as before.

```
>    Sol:=maximize(z,{ seq(Const[ i] ,i=1..3)} ,
     NONNEGATIVE);
```

simplex[feasible]   verify that constraints are linear

simplex[feasible]   the variables are of type NONNEGATIVE

simplex[feasible]   initial equations:

[_SL1 = 90-x[1] , _SL2 = 60-x[2] , _SL3 = 600-5* x[1] -6* x[2]]

simplex[feasible]   initial equations already feasible

maximize: eqns =

$$[\_SL1 = 90 - x_1, \_SL2 = 60 - x_2, \_SL3 = 600 - 5x_1 - 6x_2]$$

maximize: trans =

$$\{\}$$

```
simplex/max:   eqns =    [ _SL1 = 90-x[ 1] , _SL2 = 60-x[ 2] ,
_SL3 =600-5* x[ 1] -6* x[ 2] ]
```

```
simplex/max:  # [ _AR, _SL1, _SL2, _SL3, x[ 1] , x[ 2] ]   #
```

$$eqns =, [\_SL1 = 90 - x_1, \_SL2 = 60 - x_2, \_SL3 = 600 - 5x_1 - 6x_2]$$

$$obj =, 30x_1 + 40x_2$$

$$pivot\ variable =, x_1$$

$$pivot\ equation =, \_SL1 = 90 - x_1$$

```
simplex[ pivot]    working on
```

$$[\_SL1 = 90 - x_1, \_SL2 = 60 - x_2, \_SL3 = 600 - 5x_1 - 6x_2]$$

$$x_1$$

$$\_SL1 = 90 - x_1$$

```
simplex/max:  # [ _AR, _SL1, _SL2, _SL3, x[ 1] , x[ 2] ]   #
```

$$eqns =, [x_1 = -\_SL1 + 90, \_SL2 = 60 - x_2, \_SL3 = 150 + 5\_SL1 - 6x_2]$$

$$obj =, -30\_SL1 + 2700 + 40x_2$$

$$pivot\ variable =, x_2$$

$$pivot\ equation =, \_SL3 = 150 + 5\_SL1 - 6x_2$$

```
simplex[ pivot]    working on
```

$$[x_1 = -\_SL1 + 90, \_SL2 = 60 - x_2, \_SL3 = 150 + 5\_SL1 - 6x_2]$$

$$x_2$$

$$\_SL3 = 150 + 5\_SL1 - 6x_2$$

```
simplex/max:  # [ _AR, _SL1, _SL2, _SL3, x[ 1] , x[ 2] ]   #
```

$$eqns =, [x_1 = -\_SL1 + 90, \_SL2 = 35 + \frac{1}{6}\_SL3 - \frac{5}{6}\_SL1,$$

$$x_2 = -\frac{1}{6}\_SL3 + 25 + \frac{5}{6}\_SL1]$$

$$obj =, \frac{10}{3}\_SL1 + 3700 - \frac{20}{3}\_SL3$$

$$pivot\ variable =, \_SL1$$

$$pivot\ equation =, \_SL2 = 35 + \frac{1}{6}\_SL3 - \frac{5}{6}\_SL1$$

```
simplex[ pivot]    working on
```

$$[x_1 = -\_SL1 + 90, \ \_SL2 = 35 + \frac{1}{6}\_SL3 - \frac{5}{6}\_SL1,$$

$$x_2 = -\frac{1}{6}\_SL3 + 25 + \frac{5}{6}\_SL1]$$

$$\_SL1$$

$$\_SL2 = 35 + \frac{1}{6}\_SL3 - \frac{5}{6}\_SL1$$

`simplex/max:    ###### final description ######`

$$eqns = , [x_1 = \frac{6}{5}\_SL2 + 48 - \frac{1}{5}\_SL3, \ \_SL1 = -\frac{6}{5}\_SL2 + 42 + \frac{1}{5}\_SL3,$$

$$x_2 = 60 - \_SL2]$$

$$obj = , \ -4\_SL2 + 3840 - 6\_SL3$$

$$Sol := \{x_1 = 48, x_2 = 60\}$$

`>   assign(Sol); z;`

$$3840$$

It is interesting to note that Maple follows a slightly different path in finding the optimal solution for this problem. As an exercise, we leave it to the reader to follow Maple's steps and determine the points examined by Maple. It is worth noting that Maple also informs the user that in the optimal solution the final basis consists of x_1, x_2 and x_3 (which is denoted as _SL1 by Maple), and hence the nonbasic variables are x_4 (denoted by _SL2) and x_5 (denoted by _SL2). Using the information under the section of the output titled "final description," we can easily reconstruct the final simplex table. The last set of equations found in this section are given as

$$x_1 = \frac{6}{5}\_SL2 + 48 - \frac{1}{5}\_SL3$$

$$\_SL1 = -\frac{6}{5}\_SL2 + 42 + \frac{1}{5}\_SL3$$

$$x_2 = 60 - \_SL2.$$

Collecting the variables on the left-hand side and writing _SL1, _SL2 and _SL3 as x_3, x_4, and x_5, we get

$$x_1 - \frac{6}{5}x_4 + \frac{1}{5}x_5 = 48$$

$$x_2 + x_4 = 60$$

$$x_3 + \frac{6}{5}x_4 - \frac{1}{5}x_5 = 42.$$

This is exactly the same set of equations that is implied by the information presented in the final table at the end of Section 4.3.1. Moreover, information about the objective function that is given as $obj =, -4\_SL2 + 3840 - 6\_SL3$ corresponds to the $c_j - z_j$ row in the final simplex table: If $\_SL2(= x_4)$ is made a basic variable, the objective function decreases by \$4 and if $\_SL3(= x_5)$ is made a basic variable, the objective function decreases by \$6. Since $\_SL2 = 0$ and $\_SL3 = 0$ in the optimal solution, the maximum value of the objective is $z = \$3840$.

4.3.3   A Problem with Mixed Constraints

LP problems with a mixture of less-than-or-equal-to, greater-than-or-equal-to and/ or equality constraints require special care before they can be solved with the simplex method. While each less-than-or-equal-to constraint can easily be converted to an equality constraint by the inclusion of a slack variable, the other constraint types necessitate additional steps.

Consider, for example, the following problem with two variables and three constraints:

$$
\begin{aligned}
\min \quad z = \quad & 2x_1 + 4x_2 \\
\text{s.t.} \quad & x_1 + 5x_2 \le 80 \\
& 4x_1 + 2x_2 \ge 20 \\
& x_1 + x_2 = 10 \\
& x_1, x_2 \ge 0
\end{aligned}
$$

Using the graphical method, it can be shown that this problem has the optimal solution where $x_1 = 10, x_2 = 0$ with the minimum value of the objective function $z = 20$.

If we want to solve this problem using the "manual" version of the simplex method as demonstrated in Section 4.3.1, we need to convert the inequalities to equalities. The first constraint is converted by introducing a slack variable, say, $x_3 \ge 0$, which results in $x_1 + 5x_2 + x_3 = 80$. By introducing a "surplus" variable, say, $x_4 \ge 0$, and *subtracting* it from the left-hand side of the second constraint, we obtain $4x_1 + 2x_2 - x_4 = 20$. Since the last constraint is already an equality, we obtain the linear system and the objective function as

$$
\begin{aligned}
\min \quad z = \quad & 2x_1 + 4x_2 + 0x_3 + 0x_4 \\
\text{s.t.} \quad & x_1 + 5x_2 + x_3 \qquad\qquad = 80 \\
& 4x_1 + 2x_2 \qquad - x_4 = 20 \\
& x_1 + x_2 \qquad\qquad = 10 \\
& \text{all } x_j \ge 0
\end{aligned}
$$

Although we now have a system of equations, the **A** matrix of coefficients does not contain an immediately obvious starting feasible basis. To obtain a convenient starting basis that is also feasible, we introduce two new ("artificial") variables and add them to the left-hand side of the second and third constraints. This results in the following coefficient matrix A with columns $(\mathbf{a}_1, \mathbf{a}_2, \mathbf{a}_3, \mathbf{a}_4, \mathbf{a}_5, \mathbf{a}_6)$.

$$
\begin{array}{rcrcrcrcrcrcr}
x_1 & + & 5x_2 & + & x_3 & & & & & & & = & 80 \\
4x_1 & + & 2x_2 & & & - & x_4 & + & x_5 & & & = & 20 \\
x_1 & + & x_2 & & & & & & & + & x_6 & = & 10
\end{array}
$$

It is now easy to see that the initial basis matrix can be chosen as $\mathbf{B} = (\mathbf{a}_3, \mathbf{a}_5, \mathbf{a}_6)$, which results in the initial solution of $\mathbf{x}_B = \mathbf{B}^{-1}\mathbf{b} = (x_3, x_5, x_6) = (80, 20, 10)$.

But we note that the artificial variables must eventually be eliminated from the basis, i.e., they must be forced to take a value of 0 in the optimal solution. In order to assure this outcome, we assign a very high cost of keeping these variables in the solution. That is, we rewrite the objective function as

$$
\min z = 2x_1 + 4x_2 + 0x_3 + 0x_4 + Mx_5 + Mx_6
$$

where M is a "big" number.[3]

Since minimizing a function is equivalent to maximizing its negative, we may also convert the objective function to

$$
\max -z = -2x_1 - 4x_2 - 0x_3 - 0x_4 - Mx_5 - Mx_6
$$

and solve this problem as a maximization. We leave the solution of this problem to the reader as an exercise.

Fortunately, if Maple is used to solve this problem with the `simplex` package, there is obviously no need to introduce the slack, surplus or the artificial variables as Maple takes care of this step automatically. Following is the solution using `simplex` where we have chosen to minimize the original objective function $\min z = 2x_1 + 4x_2$.

```
>    restart: # MixedConstraints.mws
>    with(simplex):
```

Warning, new definition for maximize

Warning, new definition for minimize

```
>    z:=2* x[ 1] +4* x[ 2] ;
```
$$
z := 2x_1 + 4x_2
$$
```
>    Const[ 1] :=x[ 1] +5* x[ 2] <=80;
     Const[ 2] :=4* x[ 1] +2* x[ 2] >=20;
     Const[ 3] :=x[ 1] +x[ 2] =10;
```
$$
Const_1 := x_1 + 5x_2 \le 80
$$
$$
Const_2 := 20 \le 4x_1 + 2x_2
$$
$$
Const_3 := x_1 + x_2 = 10
$$

---

[3] The "big" M could be approximately ten times larger than the largest parameter in the problem.

```
>   Sol:=minimize(z,{ seq(Const[ i] ,i=1..3)},
    NONNEGATIVE);
```

$$Sol := \{x_2 = 0, x_1 = 10\}$$

```
>   assign(Sol); z;
```

$$20$$

4.4   Special Cases and Difficulties

Some linear programming problems may have unusual characteristics resulting in certain complications in formulation and/or solution. We now examine these special cases and attempt their solution using Maple.

4.4.1   Infeasibility

It is sometimes possible to formulate LP problems that do not have any feasible points! As an example, consider the following formulation and an attempt to solve the problem using Maple.

```
>   restart: # Infeasible.mws
```

```
>   with(simplex):
```

Warning, new definition for maximize

Warning, new definition for minimize

```
>   z:=2* x[ 1] +3* x[ 2] ;
```

$$z := 2x_1 + 3x_2$$

```
>   Const[ 1] :=x[ 1] +x[ 2] <=10; Const[ 2] :=x[ 1] +x[ 2] >=20;
```

$$Const_1 := x_1 + x_2 \le 10$$

$$Const_2 := 20 \le x_1 + x_2$$

```
>   Sol:=maximize(z,{ seq(Const[ i] ,i=1..2)},
    NONNEGATIVE);
```

$$Sol := \{\}$$

```
>   assign(Sol); z;
```

$$2x_1 + 3x_2$$

Clearly, since it is impossible to satisfy both the constraints $x_1 + x_2 \le 10$ and $x_1 + x_2 \ge 20$, Maple cannot find a solution to this problem and gives the "solution" as Sol:={ } ;—an empty set!

4.4.2 Unbounded Solutions

Another type of unusual case is when the objective function can be maximized at infinity. An LP model that gives rise to such a case cannot be the representation of a real system. In this situation the analyst who modeled the system must return to the drawing board and correct the faulty formulation.

Here is a simple example that produces an unbounded solution.

```
>   restart: # Unbounded.mws
>   #infolevel[ simplex] :=5;
>   with(simplex):
```

Warning, new definition for maximize

Warning, new definition for minimize

```
>   z:=-2* x[ 1] +x[ 2] ;
```

$$z := -2x_1 + x_2$$

```
>   Const[ 1] :=5* x[ 1] -x[ 2] <=20; Const[ 2] :=x[ 1] <=5;
```

$$Const_1 := 5x_1 - x_2 \leq 20$$

$$Const_2 := x_1 \leq 5$$

```
>   Sol:=maximize(z,{ seq(Const[ i] ,i=1..2)} ,
    NONNEGATIVE);
```

$$Sol :=$$

```
>   assign(Sol); z;
```

$$-2x_1 + x_2$$

In this case Maple does not even attempt to make a statement about the existence of the solution and gives `Sol:=`. If the # symbol before `infolevel[ simplex]` statement is removed and the problem is solved again, one would see that Maple correctly identifies the unboundedness of the solution and flags it.

4.4.3 Degeneracy

Normally, in an LP problem with m constraints, the optimal basis consists of m basic feasible variables all assuming *strictly positive* values. In some "degenerate" cases, the optimal basis may contain fewer than m positive basic variables as the following example demonstrates.

```
>   restart: # Degeneracy.mws
>   with(simplex):
```

Warning, new definition for maximize

Warning, new definition for minimize

```
> infolevel[ simplex] :=2:
> z:=2* x[ 1] +6* x[ 2] ;
```

$$z := 2x_1 + 6x_2$$

```
> Const[ 1] :=x[ 1] +2* x[ 2] <=20;
  Const[ 2] :=2* x[ 1] +3* x[ 2] <=30;
  Const[ 3] :=x[ 2] <=10;
```

$$Const_1 := x_1 + 2x_2 \leq 20$$

$$Const_2 := 2x_1 + 3x_2 \leq 30$$

$$Const_3 := x_2 \leq 10$$

```
> Sol:=maximize(z,{ seq(Const[ i] ,i=1..3)} ,
  NONNEGATIVE);
```

```
simplex/max:    ###### final description ######
```

$$eqns = , [\_SL1 = \frac{2}{3}\_SL2 + \frac{1}{3}x_1, x_2 = -\frac{2}{3}x_1 - \frac{1}{3}\_SL2 + 10,$$

$$\_SL3 = \frac{2}{3}x_1 + \frac{1}{3}\_SL2]$$

$$obj = , -2\_SL2 + 60 - 2x_1$$

$$Sol := \{x_1 = 0, x_2 = 10\}$$

```
> assign(Sol); z;
```

$$60$$

Thus, in this case with $m = 3$ constraints, we would expect three basic variables taking strictly *positive* values. But as we see in the final simplex table, the optimal basis consists of the variables $(\_SL1, x_2, \_SL3) = (0, 10, 0)$. Thus, two basic variables are equal to zero, implying degeneracy.

Unlike the cases of infeasibility and unboundedness, problems having a degenerate solution do not present any serious difficulties. In the simplex method of solution, degeneracy arises when the ratio test (4.1) produces a result whereby the minimum ratio is not unique and the variable that is to leave the basis is not immediately obvious. In such a case an arbitrary choice is made and the ties are broken according to a specific rule such as choosing the variable with largest y_{ij} or selecting the variable with the smallest i index. For a good discussion of this issue, see the classical text by Hadley [79, p. 113].

4.5   Other Examples

In this section we present a few more LP formulations and their solution with Maple's simplex package. Unlike the examples in the previous sections where we considered problems with only a few variables and constraints, the examples in this section involve problems that can potentially have a very large number of variables and constraints.

4.5.1 The Transportation Problem

The transportation problem arises frequently in the optimal distribution of goods from several supply points (e.g., factories, plants) to demand points (e.g., warehouses, customers, retail outlets). The objective is usually to find the best distribution plan to minimize the total cost of shipments from the supply points to demand points.

Example 51 *Ski Shipments.* We illustrate this model by considering a problem faced by SunSno, a multi-national company that operates three factories in (1) Jasper, Canada, (2) Seoul, Korea, and (3) Toronto, Canada. SunSno ships skis to four company owned warehouses in (1) Frankfurt, Germany, (2) New York, USA, (3) Paris, France, and (4) Yokohama, Japan. The weekly production capacities s_i, $i = 1, 2, 3$ of the $m = 3$ factories and the weekly demands d_j, $j = 1, \ldots, 4$ of the $n = 4$ warehouses are given in the following table where we also indicate the unit transportation costs c_{ij}, $i = 1, \ldots, 3$ and $j = 1, \ldots, 4$ in bold face.

From↓ \To→	Frankf't (1)	NY (2)	Paris (3)	Yok'ma (4)	SUPPLY
	19	**7**	**13**	**8**	
Jasper (1)					100
	15	**21**	**18**	**6**	
Seoul (2)					300
	11	**3**	**12**	**20**	
Toronto (3)					200
DEMAND	150	100	200	150	600

To solve this problem as a linear program, we define x_{ij} as the number of units shipped from factory i to warehouse j for factories $i = 1, \ldots, m$ and warehouses $j = 1, \ldots, n$. Thus, the LP model of the transportation problem takes the following form:

$$\min \sum_{i=1}^{m} \sum_{j=1}^{n} c_{ij} x_{ij}$$

subject to

$$\sum_{j=1}^{n} x_{ij} = s_i, \quad i = 1, \ldots, m$$

$$\sum_{j=1}^{m} x_{ij} = d_j, \quad j = 1, \ldots, n$$

with $x_{ij} \geq 0, i = 1, \ldots, m$ and $j = 1, \ldots, n$.

Specializing this to the SunSno company's decision problem, we obtain the following formulation.

$$\min z = 19x_{11} + 7x_{12} + 13x_{13} + 8x_{14} + +15x_{21} + 21x_{22} + 18x_{23} + 6x_{24}$$
$$+ 11x_{31} + 3x_{32} + 12x_{33} + 20x_{34}$$

subject to

$$
\begin{aligned}
x_{11} + x_{12} + x_{13} + x_{14} &= 100 \quad \text{(Jasper capacity)} \\
x_{21} + x_{22} + x_{23} + x_{24} &= 300 \quad \text{(Seoul capacity)} \\
x_{31} + x_{32} + x_{33} + x_{34} &= 200 \quad \text{(Toronto capacity)} \\
x_{11} + x_{21} + x_{31} &= 150 \quad \text{(Frankfurt demand)} \\
x_{12} + x_{22} + x_{32} &= 100 \quad \text{(N.Y. demand)} \\
x_{13} + x_{23} + x_{33} &= 200 \quad \text{(Paris demand)} \\
x_{14} + x_{24} + x_{34} &= 150 \quad \text{(Yokohoma demand)} \\
\text{All } x_{ij} &\geq 0.
\end{aligned}
$$

Note that, in general, formulating the transportation problem as a linear program results in $m \times n$ decision variables and $m + n$ constraints. However, although there is a total of $m + n$ constraints in transportation problems, the total number of basic variables is never more than $(m + n - 1)$ due to the redundancy in the constraints—if a solution is found that satisfies the first $m + n - 1$ constraints, it must automatically satisfy the last one since the total supply equals the total demand.

In this problem we have $3 \times 4 = 12$ variables and $3 + 4 = 7$ constraints. Also, since the total supply of $\sum_{i=1}^{3} s_i = 600$ units equals the total demand $\sum_{i=1}^{4} d_j = 600$, the factories will be shipping all that they can produce and the demands of the warehouses will be satisfied exactly. Transportation problems where $\sum_{i=1}^{m} s_i = \sum_{j=1}^{n} d_j$ are said to be *balanced*.

Formulation and solution of this problem using Maple is now a straightforward exercise in entering the objective function and the constraints. Once that is done, we again use simplex to find the optimal solution.

```
>   restart: # Transportation.mws
>   m:=3: n:=4:
>   c:=matrix(m,n,[[ 19,7,13,8] ,[ 15,21,18,6] ,
    [ 11,3,12,20]] );
```

$$
c := \begin{bmatrix} 19 & 7 & 13 & 8 \\ 15 & 21 & 18 & 6 \\ 11 & 3 & 12 & 20 \end{bmatrix}
$$

```
>   z:=sum(sum(c[ i,j] * x[ i,j] ,j=1..n) ,i=1..m) :
>   s:=array([ 100,300,200] ); sum(s[ i] ,i=1..m);
```

$$s := [100, 300, 200]$$

$$600$$

```
>   d:=array([ 150,100,200,150] ); sum(d[ j] ,j=1..n);
```

$$d := [150, 100, 200, 150]$$

$$600$$

```
>   for i from 1 to m do
    Supply[ i] :=sum(x[ i,j] ,j=1..n)=s[ i]  od:
>   i:=' i' :
>   for j from 1 to n do
    Demand[ j] :=sum(x[ i,j] ,i=1..m)>=d[ j]  od:
>   j:=' j' :
>   constraints:={ seq(Supply[ i] ,i=1..m)}  union
    { seq(Demand[ j] ,j=1..n)} :
>   with(simplex):

Warning, new definition for maximize

Warning, new definition for minimize

>   Sol:=minimize(z,constraints,NONNEGATIVE);
```

$Sol := \{x_{1,1} = 0, x_{1,2} = 0, x_{3,4} = 0, x_{1,4} = 0, x_{2,2} = 0, x_{1,3} = 100, x_{3,3} = 100,$
$x_{3,2} = 100, x_{2,3} = 0, x_{3,1} = 0, x_{2,1} = 150, x_{2,4} = 150\}$

```
>   assign(Sol):
>   z;
```

$$5950$$

Thus, the optimal solution is obtained with a minimum cost of $z = \$5,950$. We summarize the optimal solution in the following table.

From↓ \To→	Frankf't (1)		NY (2)		Paris (3)		Yok'ma (4)		SUPPLY
		19		7		13		8	
Jasper (1)	0		0		100		0		100
		15		21		18		6	
Seoul (2)	150		0		0		150		300
		11		3		12		20	
Toronto (3)	0		100		100		0		200
DEMAND	150		100		200		150		600

In some cases, total supply may not equal total demand. If total supply exceeds total demand, i.e., when $\sum_{i=1}^{m} s_i > \sum_{j=1}^{n} d_j$, then the only modification necessary is to rewrite the supply constraints as $\sum_{j=1}^{n} x_{ij} \leq s_i$, $i = 1, \ldots, m$. In this case excess supply would appear as a slack, which should be interpreted as the unused supply that is not shipped to any destination.

When total supply is less than total demand, i.e., when $\sum_{i=1}^{m} s_i < \sum_{j=1}^{n} d_j$, then the LP formulation would be infeasible. In this case we would introduce a *dummy* origin with an artificial supply equal $\sum_{j=1}^{n} d_j - \sum_{i=1}^{m} s_i$, the difference between the total demand and total supply. The optimal "shipment" amounts from the dummy origin to a particular destination would then correspond to the amount by which the destination's demand is unsatisfied.

When the number of origins and destinations is not excessively large, moderate-size transportation problems can be solved using the `simplex` package. But for a realistic problem with, say, $m = 200$ origins and $n = 500$ destinations, we would have $m + n = 700$ constraints (less one due to redundancy) and $m \times n = 100000$ variables! Thus, to solve these problems one would need more efficient procedures. Such special-purpose procedures for solving transportation problems indeed exist, but they have not been implemented by Maple. For a discussion of these procedures see, for example, Hillier and Lieberman [92, Chapter 7].

4.5.2   Two-Person Zero-Sum Games

Competitive situations are characterized by the presence of two or more players (decision makers) whose decisions influence the objective functions of all players. A zero-sum (or a constant sum) game is one in which the gains of one player (or players) are equivalent to the losses of another player (or players).

As an example, consider the simple *coin-matching* game with two players P1 and P2 who select a head (H) or a tail (T). If the outcomes match, i.e., both players select H or T, player P1 wins \$1 from player P2. Otherwise, P1 loses \$1 to P2. In this game, the 2×2 payoff matrix, expressed in terms of P1's payoff, is as follows.

		Player P2	
		H	T
Player P1	H	1	−1
	T	−1	1

A solution to a zero-sum game is obtained when each player determines his "best" strategy without knowing, *a priori*, what the other player will do. In general, the best strategy for a player is found by assigning a probability distribution over the player's set of strategies. Mathematically, we let

$$x_i \quad : \quad \text{probability that P1 will use strategy } i = 1, \ldots, m$$
$$y_j \quad : \quad \text{probability that P2 will use strategy } j = 1, \ldots, n$$

where m and n are the number of strategies available to P1 and P2, respectively, and $\sum_{i=1}^{m} x_i = 1$ and $\sum_{j=1}^{n} y_j = 1$. The vectors $\mathbf{x} = (x_1, \ldots, x_m)$ and $\mathbf{y} = (y_1, \ldots, y_n)$ are referred to as *mixed strategies*. But if some $x_i = 1$ (and all other probabilities are zero), the player P1 is said to use a *pure strategy*.

A generally agreed-upon objective is the expected payoff resulting from applying the mixed strategies. Denoting p_{ij} as the payoff to P1 if P1 uses strategy $i = 1, \ldots, m$ and P2 uses strategy $j = 1, \ldots, n$, we find P1's expected payoff E_1

as

$$E_1 = \sum_{i=1}^{m} x_i y_j p_{ij}.$$

We now assume that P1 will choose $x_1, \ldots, x_m$ in such a way that regardless of how P2 chooses $y_1, \ldots, y_n$, the gain to P1 is at least equal to some value v where v is to be made as large as possible. Thus, P1 should choose $x_1, \ldots, x_m$ such that

$$\sum_{i=1}^{m} x_i p_{ij} \geq v, \quad j = 1, \ldots, n$$

and that v is maximized. Note that the variable v is, in general, unrestricted in sign. Additionally, since the $x_1, \ldots, x_m$ constitute a probability distribution, we must have $\sum_{i=1}^{m} x_i = 1$ with $x_i \geq 0, i = 1, \ldots, m$.

For example, in the case of the coin-matching game, P1's problem would be to

$$\begin{aligned}
\max \quad & v \\
\text{s.t.} \quad & x_1 - x_2 \geq v \\
& -x_1 + x_2 \geq v \\
& x_1 + x_2 = 1 \\
& x_1, x_2 \geq 0
\end{aligned} \tag{4.2}$$

where v is unrestricted in sign. Solving this problem (as we show below using Maple) gives the optimal solution as $x_1 = x_2 = \frac{1}{2}$ and $v = 0$.

Now considering P2's problem, we see that by using the same arguments employed for P1's strategy, P2 would choose $y_1, \ldots, y_n$ so that the gain to P1 is at most equal to some value u where u is to be made as small as possible where variable u is, in general, unrestricted in sign. (Note that minimizing the gain to P1 maximizes P2's gain in zero-sum games.) Thus, P2 should choose $y_1, \ldots, y_n$ such that

$$\sum_{j=1}^{n} y_j p_{ij} \leq u, \quad i = 1, \ldots, m$$

and that u is minimized. As before, since the $y_1, \ldots, y_n$ constitute a probability distribution, we must have $\sum_{j=1}^{n} y_j = 1$ with $y_j \geq 0, j = 1, \ldots, n$.

For the case of the coin-matching game, P2's problem would be to

$$\begin{aligned}
\min \quad & u \\
\text{s.t.} \quad & y_1 - y_2 \leq u \\
& -y_1 + y_2 \leq u \\
& y_1 + y_2 = 1 \\
& y_1, y_2 \geq 0
\end{aligned} \tag{4.3}$$

where u is unrestricted in sign. Solving this problem (as follows) gives the optimal solution as $y_1 = y_2 = \frac{1}{2}$ and $u = 0$. Hence, for this game the maximum P1 can expect to gain is equal to the minimum P2 can expect to lose, that is, $u = v$.

The following Maple worksheet solves this simple coin-matching game for each player. Note that the variables v and u in the first problem and second problem, respectively, *are* allowed to take any positive or negative value since they are not restricted. Normally, such unrestricted variables must be represented as the difference of two nonnegative variables in a problem formulation, e.g., $u = u' - u''$ where $u' \geq 0$ and $u'' \geq 0$. If the option NONNEGATIVE is not included in simplex, all variables are assumed to be unrestricted in sign by default—unless otherwise indicated as in { seq(x[i] >=0, i=1..m)}.

```
>   restart: # CoinTossGame.mws
>   with(simplex):
```

Warning, new definition for maximize

Warning, new definition for minimize

```
>   P:=matrix(2,2,[ 1,-1,-1, 1] );
```

$$P := \begin{bmatrix} 1 & -1 \\ -1 & 1 \end{bmatrix}$$

```
>   m:=2; n:=2;
```

$$m := 2$$
$$n := 2$$

```
>   # Player 1's problem
>   for j from 1 to n do
    Player1[ j] :=add(x[ i] * P[ i,j] ,i=1..m)>=v od;
```

$$Player1_1 := v \leq x_1 - x_2$$
$$Player1_2 := v \leq -x_1 + x_2$$

```
>   Unity:=add(x[ i] ,i=1..m)=1;
```

$$Unity := x_1 + x_2 = 1$$

```
>   maximize(v,{ seq(Player1[ j] ,j=1..n)} union
    { Unity}  union { seq(x[ i] >=0,i=1..m)} );
```

$$\{x_2 = \frac{1}{2}, x_1 = \frac{1}{2}, v = 0\}$$

```
>   # Player 2's problem
>   for i from 1 to m do
    Player2[ i] :=add(y[ j] * P[ i,j] ,j=1..n)<=u od;
```

$$Player2_1 := y_1 - y_2 \leq u$$
$$Player2_2 := -y_1 + y_2 \leq u$$

```
>   Unity:=add(y[ j] ,j=1..n)=1;
```

$$Unity := y_1 + y_2 = 1$$

```
>   minimize(u,{ seq(Player2[ i] ,i=1..m)} union
    { Unity}  union { seq(y[ j] >=0,j=1..n)} );
```

$$\{u = 0, y_2 = \frac{1}{2}, y_1 = \frac{1}{2}\}$$

We now consider another example of a competitive situation with a larger number of strategies available to each player.

Example 52 *Advertising policies.* Consider a marketing problem where two firms are about to launch competitive products [60, p. 34]. Firm P1 has enough funds in its advertising budget to buy three one-hour blocks of TV time and firm P2 has enough money for two such blocks. We assume that the day is divided into three periods: morning (M), afternoon (A) and evening (E). Each firm must purchase its TV time in advance but naturally keeps the information confidential. It has been determined that 20% of the TV audience watch in the morning, 30% in the afternoon and the remaining 50% in the evening. To keep the analysis simple, we assume that no person watches in more than one period.

Past experience indicates that if a firm buys more time in any period than its competitor, it captures the entire audience in that period. If both firms buy the same number of blocks in any period (including zero blocks), then each gets half the audience. For example, suppose firm P1 chooses two morning hours and one evening hour (MME) and firm P2 chooses one morning hour and one evening hour (ME). In this case, P1 captures all morning audience of 20% as it bought two Ms. Since each firm has no afternoon hours they each capture half the afternoon audience of 30%. Finally, since firm each has one E block, they again capture half the evening audience of 50%. This means that P1's percentage of the market is $1.0 \times 20\% + 0.5 \times 30\% + 0.5 \times 50\% = 60\%$. The market share (i.e., the payoff) matrix for P1 where it has a total of 10 strategies and P2 has a total of 6 strategies is given in the Table 4.1.

		MM	AM	AA	EM	EA	EE
		y_1	y_2	y_3	y_4	y_5	y_6
MMM	x_1	60	45	45	35	20	35
AMM	x_2	65	60	45	50	35	50
AAM	x_3	55	65	60	40	50	50
AAA	x_4	55	55	65	30	40	40
EMM	x_5	75	70	70	60	45	35
EAM	x_6	80	75	70	65	60	50
EAA	x_7	80	80	75	55	65	40
EEM	x_8	65	60	70	75	70	60
EEA	x_9	80	65	60	80	75	65
EEE	x_{10}	65	50	60	65	60	75

TABLE 4.1. The market share matrix for player P1.

Note that this is a *constant-sum* (not a zero-sum) game since the sum of both players' market shares equals 100%. However, as both players still have diametrically opposed interests the game can be reduced to a zero-sum game and solved using the procedure described.

A careful examination of the payoff table reveals that EEM (x_8) dominates the three strategies AMM (x_2), MMM (x_1) and AAA (x_4), and EAM (x_6) dominates EMM (x_5) and AAM (x_3). Thus, we could discard P1's first five strategies. If this is done, then the P2's strategy AM (y_2) dominates MM (y_1). The game could then be reduced to one with only five strategies for each player. We leave this as an exercise and solve the complete problem with ten strategies for P1 and six strategies for P2.

```
>    restart: # MarketingGame.mws

>    with(simplex):
```

Warning, new definition for maximize

Warning, new definition for minimize

```
>    m:=10: n:=6:
>    P:=matrix(m,n,[
     [ 60,45,45,35,20,35] ,
     [ 65,60,45,50,35,50] ,
     [ 55,65,60,40,50,50] ,
     [ 55,55,65,30,40,40] ,
     [ 75,70,70,60,45,35] ,
     [ 80,75,70,65,60,50] ,
     [ 80,80,75,55,65,40] ,
     [ 65,60,70,75,70,60] ,
     [ 80,65,60,80,75,65] ,
     [ 65,50,60,65,60,75]] ):

>    # Player 1's problem
>    for j from 1 to n do
     Player1[ j] :=add(x[ i] * P[ i,j] ,i=1..m)>=v od:

>    Unity:=add(x[ i] ,i=1..m)=1:
>    maximize(v,{ seq(Player1[ j] ,j=1..n)} union
     { Unity}  union { seq(x[ i] >=0,i=1..m)} );
```

$$\{x_1 = 0,\ x_2 = 0,\ x_3 = 0,\ x_4 = 0,\ x_5 = 0,\ x_7 = 0,\ x_{10} = 0,\ x_6 = 0,$$
$$x_9 = \frac{2}{3},\ x_8 = \frac{1}{3},\ v = \frac{190}{3}\}$$

```
>    # Player 2's problem
>    for i from 1 to m do
     Player2[ i] :=add(y[ j] * P[ i,j] ,j=1..n)<=u od:

>    Unity:=add(y[ j] ,j=1..n)=1:
>    minimize(u,{ seq(Player2[ i] ,i=1..m)} union
     { Unity}  union { seq(y[ j] >=0,j=1..n)} );
```

$$\{y_5 = 0, \ y_6 = \frac{2}{5}, \ y_2 = \frac{4}{15}, \ y_3 = \frac{1}{3}, \ y_1 = 0, \ y_4 = 0, \ u = \frac{190}{3}\}$$

Clearly, the dominated states $i = 1, \ldots, 5$ for P1 and $j = 1$ for P2 do not appear in the solution indicating that they would never be used. As in the coin-tossing game, the maximum P1 can expect to gain is equal to the minimum P2 can expect to lose, i.e., $u = v = 190/3$.

4.5.3 A Linear Program with Randomly Generated Data

In this section we present a Maple worksheet that randomly generates the vector $\mathbf{c}$ of objective function coefficients, the matrix $\mathbf{A}$ of constraints coefficients and the vector $\mathbf{b}$ of RHS constants of a linear program.

We set $m = 4$ and $n = 6$ and generate the integer entries of the $\mathbf{c}$ vector distributed uniformly between 0 and 10 using the command c:=randvector (n,entries=rand(0..10));. Using similar statements, we also generate the $\mathbf{A}$ matrix and the $\mathbf{b}$ vector. Once the objective function and constraints are defined using the multiply() command, the randomly generated problem is solved in the usual way. In the optimal solution we find that $x_1 = x_2 = x_3 = x_5 = 0$, but $x_4 = \frac{6213}{9679}$ and $x_6 = \frac{16362}{9679}$.

```
>    restart: # RandomLP.mws
>    m:=4;
```

$$m := 4$$

```
>    n:=6;
```

$$n := 6$$

```
>    with(linalg): with(simplex):
```

Warning, new definition for norm

Warning, new definition for trace

Warning, new definition for basis

Warning, new definition for maximize

Warning, new definition for minimize

Warning, new definition for pivot

```
>    readlib(randomize); randomize();
```

proc(n)
description '*Reset the seed for the random number generator*'
    . . .
end

```
>   c:=randvector(n,entries=rand(0..10));
```
$$c := [3, 2, 3, 7, 7, 8]$$
```
>   A:=randmatrix(m,n,entries=rand(0..150));
```
$$A := \begin{bmatrix} 102 & 116 & 57 & 125 & 51 & 4 \\ 118 & 57 & 127 & 6 & 64 & 24 \\ 14 & 97 & 47 & 134 & 15 & 77 \\ 85 & 23 & 83 & 49 & 97 & 79 \end{bmatrix}$$
```
>   b:=randvector(m,entries=rand(0..1000));
```
$$b := [87, 864, 565, 165]$$
```
>   x:=vector(n);
```
$$x := \text{array}(1..6, [])$$
```
>   z:=multiply(c,x):
>   LHS:=multiply(A,x):
>   constraints:={ seq(LHS[ k] <=b[ k] ,k=1..m)} :
>   Sol:=maximize(z,constraints,NONNEGATIVE);
```
$$Sol := \{x_1 = 0,\ x_2 = 0,\ x_3 = 0,\ x_5 = 0,\ x_6 = \frac{16362}{9679},\ x_4 = \frac{6213}{9679}\}$$
```
>   assign(Sol); evalf(z);
```
$$18.01704722$$

The reader can change the parameters m and n and solve different random problems with the same worksheet. However, since every element of $\mathbf{A}$ is positive, for large problems (e.g., with $m = 200$ and $n = 400$) one should expect Maple to take several minutes to find the optimal solution.

4.6   Sensitivity Analysis and Duality

Very often, after solving an LP problem completely, it is advantageous to investigate if changing any of the parameters affects the optimal solution. This could be important since the values of parameters such as demand, revenue, or cost used as inputs to the problem formulation are only estimates that are likely to deviate from their true values when the solution is actually implemented. Such sensitivity (postoptimality) analysis is an important part of any application of linear programming. We now discuss several issues related to sensitivity analysis.

4.6.1   Changes in the RHS Values and Dual Prices

Let us return to the toy manufacturing problem introduced in Section 4.2 and investigate if it is worthwhile to acquire additional resources. We know from the solution to the original formulation that it is optimal to produce $x_1 = 48$ toy cars and $x_2 = 60$ toy trains that would result in a daily profit of $z = \$3840$. With this solution the capacity in the train department is utilized completely since the constraint $x_2 \leq 60$ is satisfied as an equality. Thus, we now wish to determine if it would be desirable to acquire additional resources and increase the production capacity in the train department.

Let us suppose that the train production capacity is increased by 10 units to 70. Re-solving the problem using Maple gives

```
>   restart: # ToyLPRHS70.mws

>   with(simplex):
Warning, new definition for maximize

Warning, new definition for minimize

>   z:=30*x[ 1] +40*x[ 2] ;
```
$$z := 30\,x_1 + 40\,x_2$$
```
>   Const[ 1] :=x[ 1] <=90; Const[ 2] :=x[ 2] <=70;
    Const[ 3] :=5* x[ 1] +6* x[ 2] <=600;
```
$$Const_1 := x_1 \leq 90$$
$$Const_2 := x_2 \leq 70$$
$$Const_3 := 5\,x_1 + 6\,x_2 \leq 600$$
```
>   Sol:=maximize(z,{ seq(Const[ i] ,i=1..3)} ,
    NONNEGATIVE);
```
$$Sol := \{x_1 = 36,\ x_2 = 70\}$$
```
>   assign(Sol); z;
```
$$3880$$

With an extra capacity of 10 units in the train department, it is now optimal to produce $x_1 = 36$ cars and $x_2 = 70$ trains with a maximum daily profit of $z = \$3880$. Hence, the maximum profit increases by an amount of $40. This implies that the *marginal* increase in profit for a *unit* increase in the train production capacity is $40/10 = \$4$. This $4 is known as the *dual price* of the train production department capacity. Naturally, if the marginal cost of increasing the capacity in this department is less than the marginal benefit of $4, then management should increase the capacity since the net benefit is positive.

For small increases in the available capacity in the train department, we observe that the dual price is $4. Does this result apply regardless of the amount of capacity added to the train department? Increasing the capacity in increments of ten units and resolving the problem gives the following.

Train Department

Production Capacity	x_1	x_2	z (\$)
60	48	60	3840
70	36	70	3880
80	24	80	3920
90	12	90	3960
100	0	100	4000
110	0	100	4000
120	0	100	4000

We see that if, for example, the new capacity is $b_2 = 120$, the optimal solution is found as $x_1 = 0$, $x_2 = 100$ and $z = \$4000$. Thus, for an increase of 60 units in the capacity, the maximum profit increases by $\$4000 - \$3840 = \$160$, implying that the marginal benefit per unit is now only $\$160/60 = \$2.67$ instead of $\$4$. Why is the dual price no longer equal to $\$4$? Using the graph of the feasible region as in Figure 4.1, the reader can easily show that for such a substantial increase in the train department capacity, this department's constraint becomes redundant and thus increases in this capacity no longer add to the profit beyond a certain threshold level.

If this is the case, there must be some range of values for the train department capacity for which the dual price would still be $\$4$. This is indeed the case as we now demonstrate using Maple's help.

First, recall from the analysis of the problem in Section 4.3.1 that in the optimal solution, the basic variables are $x_1 = 48$, $x_2 = 60$ and $x_3 = 42$ with the basis matrix $\mathbf{B} = (\mathbf{a}_1, \mathbf{a}_2, \mathbf{a}_3)$. Now, let us introduce a parameter Δ_2 corresponding to the change in the train department production capacity so that the RHS vector $\mathbf{b}$ assumes the form $\mathbf{b} = (90, 60 + \Delta_2, 600)'$.

With this change in the train production capacity, Maple tells us that the new values of the optimal basic variables are now $\mathbf{x}_B = \mathbf{B}^{-1}\mathbf{b} = (48 - \frac{6}{5}\Delta_2, 60 + \Delta_2, 42 + \frac{6}{5}\Delta_2)'$ provided that $\mathbf{x}_B \geq 0$. It is an easy matter to solve the resulting set of inequalities for Δ_2, which gives $-35 \leq \Delta_2 \leq 40$ as we shall see. This result implies that as long as the train department capacity is not increased by more than 40 units or is not decreased by more than 35 units, the optimal basis (x_1, x_2, x_3) remains optimal. Hence, for this range of Δ_2 values, the dual price of $\$4$ is still valid. Equivalently, this implies that as long as the new capacity in the train department is between $25(= 60-35)$ and $100(= 60+40)$ units, the optimal basis does not change and the dual price is still $\$4$. These results are presented in the following Maple worksheet.

```
>    restart: # ToyLPRHSDelta2.mws

>    with(linalg):

Warning, new definition for norm

Warning, new definition for trace
```

```
>  System:=[ x[ 1] +x[ 3] =90,  x[ 2] +x[ 4] =60+Delta[ 2] ,
   5* x[ 1] +6* x[ 2] +x[ 5] =600] ;
```

$System := [x_1 + x_3 = 90, \ x_2 + x_4 = 60 + \Delta_2, \ 5x_1 + 6x_2 + x_5 = 600]$

```
>  A:=genmatrix(System,[ x[ 1] ,x[ 2] ,x[ 3] ,x[ 4] ,x[ 5]] ,
   RHS) :
>  b:=evalm(RHS);
```

$$b := [90, \ 60 + \Delta_2, \ 600]$$

```
>  for j from 1 to 5 do a[ j] :=col(A,j) od:
>  B:=augment(a[ 1] ,a[ 2] ,a[ 3] ) :
>  xB:=evalm(inverse(B)  &* b);
```

$$xB := \left[48 - \frac{6}{5}\Delta_2, \ 60 + \Delta_2, \ 42 + \frac{6}{5}\Delta_2 \right]$$

```
>  solve({ seq(xB[ i] >=0,i=1..3)} );
```

$$\{-35 \leq \Delta_2, \ \Delta_2 \leq 40\}$$

Using the approach described, it can also be shown that the dual price for the toy car department production capacity is \$0 and for the special parts constraint is \$6.[4] Moreover, it is easy to show that for the car production capacity, the permissible range of values for the change Δ_1 is $-42 \leq \Delta_1$, and for the special parts the range is $-240 \leq \Delta_3 \leq 210$. These findings are summarized in the following table.

Constraints	Dual Price	Allowable Decrease in RHS	Current RHS	Allowable Increase in RHS
Car	0	42	90	∞
Train	4	35	60	40
Parts	6	240	600	210

4.6.2 Changes in the Objective Function Coefficients

The second problem in sensitivity analysis is concerned with the effects of changes in the solution that result from changes in the objective function coefficients of the decision variables. For example, suppose that due to cost savings in the toy car production department, the unit profit from each car increases to $c_1 = \$35$ (up from \$30). Does this change the original optimal solution? It is easy to show that the optimal solution now moves to a different corner point (to B in Figure 4.1) where $x_1 = 90$, $x_2 = 25$ and $z = \$4150$. Thus, for this increase of \$5 in the unit profit c_1, the optimal solution changes. But if the increase in c_1 were small, say, only \$2, would the solution still have changed? Or more generally, for what

---

[4]Note that if there is any unused capacity in the car department (in this case, $x_3 = 42$ units), adding more capacity would not improve the profit since the new units would also not be used.

range of c_1 values (the *range of optimality*) does the original optimal solution *not* change?

Suppose that the profit from each car is $30 + \theta_1$ dollars where θ_1 is the amount by which the unit profit changes. Thus, the objective coefficient vector now becomes $\mathbf{c} = (30 + \theta_1, 40, 0, 0, 0)$ and since (x_1, x_2, x_3) constitute the optimal basis, we have $\mathbf{c}_B = (30 + \theta_1, 40, 0)$. This change in $\mathbf{c}_B$ affects the $z_j = \mathbf{c}_B \mathbf{B}^{-1} \mathbf{a}_j$ values and hence the $c_j - z_j$ row in the final simplex table. Indeed, as the following Maple worksheet demonstrates, we find that $c_4 - z_4 = -4 + \frac{6}{5}\theta_1$ and $c_5 - z_5 = -6 - \frac{1}{5}\theta_1$. In order for the original solution to remain optimal, we must of course have $c_4 - z_4 \le 0$ and $c_5 - z_5 \le 0$ for these nonbasic variables x_4 and x_5. Maple easily solves the inequalities and finds that for $-30 \le \theta_1 \le \frac{10}{3}$, the original basis remains optimal.

```
>    restart: # ToyLPCostthetal.mws
>    with(linalg):
```

Warning, new definition for norm

Warning, new definition for trace

```
>    System:=[ x[ 1] +x[ 3] =90,  x[ 2] +x[ 4] =60,
     5* x[ 1] +6* x[ 2] +x[ 5] =600] :
>    A:=genmatrix(System,[ x[ 1] ,x[ 2] ,x[ 3] ,x[ 4] ,x[ 5]] ,
     RHS) :
>    b:=evalm(RHS) :
>    c:=vector(5,[ 30+theta[ 1] ,40,0,0,0] );
```

$$c := [30 + \theta_1, 40, 0, 0, 0]$$

```
>    for j from 1 to 5 do a[ j] :=col(A,j) od:
>    B:=augment(a[ 1] ,a[ 2] ,a[ 3] ) :
>    xB:=evalm(inverse(B) &* b);
```

$$xB := [48, 60, 42]$$

```
>    cB:=vector(3,[ 30+theta[ 1] ,40,0] );
```

$$cB := [30 + \theta_1, 40, 0]$$

```
>    for j from 1 to 5 do
     y[ j] :=evalm(inverse(B) &* a[ j] );
     z[ j] :=evalm(cB &* y[ j] ) od:
>    seq(c[ j] -z[ j] <=0,j=1..5);
```

$$0 \le 0, \ 0 \le 0, \ 0 \le 0, \ -4 + \frac{6}{5}\theta_1 \le 0, \ -6 - \frac{1}{5}\theta_1 \le 0$$

```
>    solve({ seq(c[ j] -z[ j] <=0,j=1..5)} );
```

$$\{-30 \le \theta_1, \ \theta_1 \le \frac{10}{3}\}$$

Next, writing $40 + \theta_2$ as the new value of the unit profit from each toy train (where θ_2 is the change in this unit profit), it can be shown that as long as $-4 \le \theta_2$, the original optimal solution ($x_1 = 48$, $x_2 = 60$ and $z = \$3840$) remains optimal. These results are summarized in the following table.

Variable	Allowable Decrease	Current Coefficient	Allowable Increase
x_1 (Car)	30	30	3.33
x_2 (Train)	4	40	∞

4.6.3 Addition of a New Decision Variable

We now examine a third important issue in sensitivity analysis. Suppose that the company is considering adding a new product (say, toy bulldozers denoted by x_p) to their product line where the profit for each bulldozer is $c_p = \$50$. Each bulldozer needs some of the resources available in the car and train department. In particular, each bulldozer needs labor and machine time resources from the truck department equivalent to $1/2$ car, labor and machine time resources from the tractor department equivalent to 1 train, and 10 special parts. Thus, the column vector $\mathbf{a}_p$ that would correspond to the new variable x_p is $\mathbf{a}_p = (\frac{1}{2}, 1, 10)'$.

Given this information, should the company should produce the bulldozers? Although this problem could be formulated and solved with the addition of the new variable x_p, it is often easier to consider the effect of the new variable on the optimal solution.

To do this, we first find the updated vector $\mathbf{y}_p$ corresponding to x_p by the transformation $\mathbf{y}_p = \mathbf{B}^{-1}\mathbf{a}_p$, which gives $\mathbf{y}_p = (\frac{4}{5}, 1, -\frac{3}{10})'$. Recall that the entries of the $\mathbf{y}_p$ vector correspond to the amounts by which a basic variable's value will change if one unit of the nonbasic variable x_p, is introduced into the solution. This is the vector that would have appeared in the final solution table if we had initially introduced the x_p into the formulation. Hence, $z_p = \mathbf{c}_B \mathbf{y}_p = (30, 40, 0) \times (\frac{4}{5}, 1, -\frac{3}{10})' = 64$, which is the amount by which the objective function value would decrease if one unit of x_p is introduced into the solution. This gives $c_p - z_p = 50 - 64 = -14$, implying that it is *not* beneficial to introduce the new product into the solution. We see that profit from each bulldozer would have to increase by $\$14$ before it would be worthwhile to produce a bulldozer. The Maple commands that perform the above computations follow.

```
>   restart: # ToyLPNewVar.mws
>   with(linalg):

Warning, new definition for norm

Warning, new definition for trace
>   System:=[ x[ 1] +x[ 3] =90,  x[ 2] +x[ 4] =60,
    5* x[ 1] +6* x[ 2] +x[ 5] =600] :
>   A:=genmatrix(System,[ x[ 1] ,x[ 2] ,x[ 3] ,x[ 4] ,x[ 5]] ,
    RHS):
```

```
>   b:=evalm(RHS):
>   for j from 1 to 5 do a[ j] :=col(A,j) od:
>   B:=augment(a[ 1] ,a[ 2] ,a[ 3] ):
>   c[ p] :=50;
```

$$c_p := 50$$

```
>   a[ p] :=vector(3,[ 1/2,1,10] );
```

$$a_p := \left[\frac{1}{2},\ 1,\ 10\right]$$

```
>   y[ p] :=evalm(inverse(B) &* a[ p] );
```

$$y_p := \left[\frac{4}{5},\ 1,\ \frac{-3}{10}\right]$$

```
>   cB:=vector(3,[ 30,40,0] );
```

$$cB := [30,\ 40,\ 0]$$

```
>   z[ p] :=evalm(cB &* y[ p] );
```

$$z_p := 64$$

```
>   c[ p] -z[ p] ;
```

$$-14$$

These arguments also imply that while it is not optimal to produce bulldozers, if for some reason the management has to produce one unit (i.e., if the constraint $x_p \geq 1$ is introduced, which would result in $x_p = 1$), then the optimal profit would decrease by \$14. This amount is also known as the *reduced cost* associated with the nonbasic variable x_p.

4.6.4   Duality

In Section 4.6.1 we learned that the dual prices of the car and train production department resources and the special parts resource were \$0, \$4 and \$6, respectively. As we discussed, these dual prices can be found by increasing the right-hand side of a particular constraint by one unit (thereby increasing the associated resource by one unit) and re-solving the problem. The change in the objective function with the higher value of the resource was the dual price for that resource constraint.

We now follow a different route and attempt to compute the dual prices as the optimal solution to a linear programming problem ("dual program") that is intimately related to the original problem ("primal program.")

Example 53 *Furniture production.* In this example, we consider a simplified version of a furniture manufacturing problem where only tables (x_1) and chairs (x_2) are made by using wood and labor as resources. The problem data are summarized in the following table.

| | Unit Requirements | | Amount |
Resource	Table	Chair	Available
Wood (board feet)	5	4	240
Labor (hours)	2	5	130
Profit/unit	10	5	

Following is the *primal* program for this problem and its solution using Maple.

$$
\begin{array}{rrcrcl}
\max & z_P & = & 10x_1 & + & 5x_2 \\
\text{s.t.} & & & 5x_1 & + & 4x_2 \le 240 \\
& & & 2x_1 & + & 5x_2 \le 130 \\
& & & & & x_1, x_2 \ge 0
\end{array}
$$

```
>   restart: # FurnitureDual.mws
>   infolevel[ simplex] :=2;
```
$$infolevel_{simplex} := 2$$
```
>   with(simplex):
```
Warning, new definition for maximize

Warning, new definition for minimize
```
>   zPrimal:=10* x[ 1] +5* x[ 2] ;
```
$$zPrimal := 10x_1 + 5x_2$$
```
>   ConstraintsP:=[ 5* x[ 1] +4* x[ 2] <=240,
    2* x[ 1] +5* x[ 2] <=130] ;
```
$$ConstraintsP := [5x_1 + 4x_2 \le 240, \ 2x_1 + 5x_2 \le 130]$$
```
>   PrimalSol:=maximize(zPrimal,ConstraintsP,
    NONNEGATIVE) ;
```

simplex/max: ###### final description ######

$$eqns =, \ [x_1 = -\frac{1}{5}\_SL1 + 48 - \frac{4}{5}x_2, \ \_SL2 = 34 + \frac{2}{5}\_SL1 - \frac{17}{5}x_2]$$

$$obj =, \ -2\_SL1 + 480 - 3x_2$$

$$PrimalSol := \{x_2 = 0, \ x_1 = 48\}$$

Note that by inserting the command `infolevel[ simplex] :=2;` we have requested Maple to provide detailed information on the solution. In order to conserve space, we deleted the output until the `final description`, which gives the final simplex table. The final stage in the solution indicates that the two basic variables are x_1 and $\_SL2$, which is the slack variable for the second constraint. (Using our standard notation, $\_SL2$ would be denoted by x_4.) Writing out the equations after collecting the variables of the constraints on the left, we obtain the following final simplex table for the primal problem.

Basis	c_j	x_1 10	x_2 5	$SL1(=x_3)$ 0	$SL2(=x_4)$ 0	b
x_1	10	0	4/5	1/5	0	48
$SL2(=x_4)$	0	1	17/5	−2/5	1	34
	$c_j − z_j$	0	−3	−2	0	480

Thus, the optimal solution to the primal program is to produce $x_1 = 48$ tables and $x_2 = 0$ chairs resulting in a maximum profit of $z_P = \$480$. It is easy to show that if the wood availability increases by 1 board foot, then the optimal profit increases by \$2, which is the dual price for the wood resource. On the other hand, increasing the labor availability does not increase the profit, implying that the dual price for labor is \$0.

In the development of the dual program we assume that the dual prices of the resources are not yet known, but they are the decision variables in a different problem. Thus, for each of the two constraints in the primal program we introduce unknown variables defined as u_1 : dual price for the wood capacity resource and u_2 : dual price for the labor capacity resource. We now generate a new linear program whose solution gives us the values of the dual prices.

Suppose that management wishes to know the total value of the two resources available. This value is $240u_1 + 130u_2$ since there are 240 board feet of wood and 130 hours of labor available. Suppose now that management decides to *reduce* table production (x_1) by one unit, which would release the wood and labor resources for alternative use. Specifically, 5 board feet of wood and 2 hours of labor would be released, which could then be sold in the marketplace. How much would company management be willing to sell these resources for in order to be at least as well off as before?

Since the value of released resources is $5u_1 + 2u_2$ and the since the profit foregone (by reducing the table production by one unit) is \$10, the constraint

$$5u_1 + 2u_2 \geq 10 \qquad (4.4)$$

must be satisfied in order for the furniture company to be at least as well off as before. Similarly, if the chair production is reduced by one unit, then the constraint

$$4u_1 + 5u_2 \geq 5 \qquad (4.5)$$

must be satisfied in order for the furniture company to be at least as well off as before.

Finally, let us suppose that a prospective customer has been found who is willing to buy the resources from the company. This customer would pay u_1 and u_2 dollars for the wood and labor resources, respectively, while minimizing

$$z_D = 240u_1 + 130u_2$$

and satisfying the two constraints (4.4) and (4.5).[5]

---

[5] If the constraints are not satisfied, company management would not sell the resources.

These arguments lead to the formulation of the dual program given by

$$\begin{aligned} \min \quad z_D = \quad & 240u_1 + 130u_2 \\ \text{s.t.} \quad\quad & 5u_1 + 2u_2 \geq 10 \\ & 4u_1 + 5x_2 \geq 5 \\ & u_1, u_2 \geq 0 \end{aligned}$$

Maple has implemented a procedure (dual) that simplifies the generation (and hence the solution) of the dual program. This is illustrated below. Note that we enclosed the original constraints in a *list* (in square brackets [ ] rather than the curly brackets { } of a *set*) since the order in which the constraints appear is important.

```
>   assign(PrimalSol); zPrimal;
```
$$480$$
```
>   x[ 1] :=' x[ 1]' : x[ 2] :=' x[ 2]' :
>   z[ P] :=10* x[ 1] +5* x[ 2] ;
```
$$z_P := 10x_1 + 5x_2$$
```
>   DualProgram:=dual(zPrimal,ConstraintsP,u);
```

DualProgram $:= 240\,u1 + 130\,u2, [10 \leq 5\,u1 + 2\,u2, 5 \leq 4\,u1 + 5\,u2]$
```
>   DualSol:=minimize(DualProgram[ 1] ,DualProgram[ 2] ,
    NONNEGATIVE);
```

simplex/max: ###### final description ######

$$eqns = , [u1 = 2 + \frac{1}{5}\_SL1 - \frac{2}{5}u2, \_SL2 = 3 + \frac{4}{5}\_SL1 + \frac{17}{5}u2]$$

$$obj = , -480 - 48\_SL1 - 34\,u2$$

$$DualSol := \{u2 = 0, \ u1 = 2\}$$
```
>   assign(DualSol); DualProgram[ 1] ;
```
$$480$$

Maple finds that the dual prices are $u_1 = 2$ and $u_2 = 0$ with the minimum value of the dual objective function $z_D = \$480$.[6] It is interesting to note that without solving the primal problem we could have found its optimal solution by using the results for the optimal solution for the dual. The final table for the dual can be constructed as follows.

Basis	c_j	u_1 240	u_2 130	$\_SL1$ 0	$\_SL2$ 0	b
$\_SL2$	0	0	$-17/5$	$-4/5$	1	3
u_1	240	1	$8/20$	$-1/5$	0	2
	$c_j - z_j$	0	-34	-48	0	480

---

[6]It is not clear why Maple gives *obj* $=, -480 - 48\_SL1 - 34u2$, which would incorrectly imply that the minimum value of the dual objective is $z_D = -480$ instead of $+480$.

This example demonstrates several important properties of the primal and dual problems that are true in general:

- The optimal objective function value for both problems is the same, i.e., $z_P = z_D = 480$.

- The optimal values of the primal decision variables (x_j) are the negatives of $c_j - z_j$ entries for the _$SL.i$ variables in the final dual table. For the primal, $x_1 = 48$ and $x_2 = 0$. These values could also be deduced from the information in the final table in the dual: $c_j - z_j = -48$ for _$SL1$ so that $x_1 = -(-48) = 48$, and $c_j - z_j = 0$ for _$SL2$ so that $x_2 = -0 = 0$.

- The optimal values of the dual variables (u_i) are the negatives of $c_j - z_j$ entries for the _$SL.i$ variables in the final primal table. These values could also be deduced from the information in the final table in the primal: $c_j - z_j = -2$ for _$SL1$ so that $u_1 = -(-2) = 2$, and $c_j - z_j = 0$ for _$SL2$ so that $u_2 = -0 = 0$.

We can now generalize these ideas by presenting the concept of duality in a formal manner. We first assume that the primal problem has been written in the form of maximization with $\leq$ constraints. This is always possible since a $\geq$ constraint can be written as a $\leq$ constraint by multiplying both sides of the former by -1, for example, $x_1 + x_2 \geq 3$ can be written as $-x_1 - x_2 \leq -3$. Similarly, an equality constraint can be converted to two inequality constraints; $x_1 + 2x_2 = 1$ can be written equivalently as $x_1 + 2x_2 \leq 1$ and $x_1 + 2x_2 \geq 1$ (or, $-x_1 - 2x_2 \leq -1$).

The primal program is stated as

$$\begin{aligned} \max \quad & z_P = \mathbf{cx} \\ \text{s.t.} \quad & \mathbf{Ax} \leq \mathbf{b} \\ & \mathbf{x} \geq \mathbf{0}. \end{aligned}$$

With this form of the primal, the dual will always be

$$\begin{aligned} \min \quad & z_D = \mathbf{b'u} \\ \text{s.t.} \quad & \mathbf{A'u} \geq \mathbf{c'} \\ & \mathbf{u} \geq \mathbf{0}. \end{aligned}$$

We now present some of the main theoretical results associated with the pair of primal/dual programs.

- Suppose $\mathbf{x}$ and $\mathbf{u}$ are feasible solutions to the primal and dual programs, respectively. Given $\mathbf{x}$ and $\mathbf{u}$, the value of the primal objective function is always less than or equal to the value of the dual objective function. This result easily follows by noting that the inequality $\mathbf{Ax} \leq \mathbf{b}$ can be written as $\mathbf{x'A'} \leq \mathbf{b'}$. Multiplying both sides on the right by any feasible $\mathbf{u} \geq \mathbf{0}$ gives $\mathbf{x'A'u} \leq \mathbf{b'u} = z_D$. Now, writing $\mathbf{A'u} \geq \mathbf{c'}$ as $\mathbf{c'} \leq \mathbf{A'u}$ and multiplying both sides on the left by $\mathbf{x} \geq \mathbf{0}$, we get $z_P = \mathbf{x'c'} \leq \mathbf{x'A'u}$. But since $\mathbf{x'A'u} \leq \mathbf{b'u} = z_D$, we obtain $z_P \leq z_D$.

As an example, consider the furniture production example. The vector $\mathbf{x} = (10, 10)'$ is a feasible solution for the primal problem and at that point the primal objective function is $z_P = \mathbf{cx} = (10, 5) \times (10, 10)' = 150$. Now consider the vector $\mathbf{u} = (5, 5)'$, which is also feasible for the dual for which the dual objective function is $z_D = \mathbf{b'u} = (240, 130) \times (5, 5)' = 1850 > z_P = 150$.

- Another important property in duality theory that is a corollary to the previous result is that if $\hat{\mathbf{x}}$ is feasible for the primal and $\hat{\mathbf{u}}$ is feasible for the dual such that $\mathbf{c\hat{x}} = \mathbf{b'\hat{u}}$, then $\hat{\mathbf{x}}$ is an optimal solution for the primal and $\hat{\mathbf{u}}$ is an optimal solution for the dual.

As an example, consider again the furniture production problem. For $\hat{\mathbf{x}} = (48, 0)'$ and $\hat{\mathbf{u}} = (2, 0)$, we have $\mathbf{c\hat{x}} = (10, 5) \times (48, 0)' = 480$ and $\mathbf{b'\hat{u}} = (240, 130) \times (2, 0) = 480$. But, the $\hat{\mathbf{x}}$ and $\hat{\mathbf{u}}$ vectors are optimal for the primal and the dual, respectively.

- A third important property concerns the existence of the optimal solutions for the primal and dual: If either the primal or the dual has an optimal solution, then the other has an optimal solution, too, and the corresponding values of the objective functions are equal. Moreover, if $\mathbf{B}$ is the basis matrix for the primal for an optimal solution $\mathbf{x}$ and if $\mathbf{c}_B$ is the objective function coefficient vector for the basic variables, then $\mathbf{u'} = \mathbf{c}_B \mathbf{B}^{-1}$. Note that $z_j = \mathbf{c}_B \mathbf{y}_j = \mathbf{c}_B \mathbf{B}^{-1} \mathbf{a}_j$. Since $c_j - z_j \leq 0$ (when the primal is optimal), then $\mathbf{c} - \mathbf{c}_B \mathbf{B}^{-1} \mathbf{A} \leq \mathbf{0}$, which becomes $\mathbf{c} - \mathbf{u'A} \leq \mathbf{0}$ (after substituting $\mathbf{u'} = \mathbf{c}_B \mathbf{B}^{-1}$), or $\mathbf{A'u} \geq \mathbf{c'}$, which is, of course, the dual problem.

Returning again to the furniture example, we see that the optimal primal basis is $\mathbf{x}_B = (x_1, \_SL2) = (x_1, x_4)$. The solution for the dual variables $\mathbf{u'} = \mathbf{c}_B \mathbf{B}^{-1} = (u_1, u_2) = (2, 0)$ is given by the following Maple worksheet.

```
>    restart: # DualFormula.mws
>    cB:=vector(2,[ 10,0] );
```
$$cB := [10,\ 0]$$
```
>    B:=matrix(2,2,[ 5,0,2,1] );
```
$$B := \begin{bmatrix} 5 & 0 \\ 2 & 1 \end{bmatrix}$$
```
>    with(linalg):
```
Warning, new definition for norm

Warning, new definition for trace
```
>    u:=evalm(cB &* inverse(B));
```
$$u := [2,\ 0]$$

Example 54 *Dual of the problem with mixed constraints.* As another example, consider the problem with mixed constraints discussed in Section 4.3.3:

$$
\begin{array}{rlrcrcl}
\min & z & = & 2x_1 & + & 4x_2 & \\
\text{s.t.} & & & x_1 & + & 5x_2 & \leq & 80 \\
& & & 4x_1 & + & 2x_2 & \geq & 20 \\
& & & x_1 & + & x_2 & = & 10 \\
& & & & & x_1, x_2 & \geq & 0
\end{array}
$$

This problem must first be converted to the standard form max $z_P = \mathbf{c}\mathbf{x}$ subject to $\mathbf{A}\mathbf{x} \leq \mathbf{b}$ and $\mathbf{x} \geq \mathbf{0}$. After converting the objective to maximization, we let Maple take over and do the necessary conversions of the constraints using the `convert(,stdlc)` command. This gives rise to a problem in standard form with four $\leq$ constraints where the last two are obtained from the equality constraint $x_1 + x_2 = 10$.

Maple finds that $u_1 = u_2 = u_3 = 0$ and $u_4 = 2$ where the last dual price corresponds to $-x_1 - x_2 \leq -10$. The minimum value of the dual program's objective function is found as $z_D = -20$. The fact that $u_1 = u_2 = 0$ implies that increasing the RHS of the first two constraints in the primal program would not affect the optimal value of the objective z_P. But $u_3 = 0$ and $u_4 = 2$ imply that increasing the RHS of $-x_1 - x_2 \leq -10$ by 1 to $-10 + 1 = -9$ would increase the objective function $z_P = -20$ by 2 units to $-20 + 2 = -18$. Translating these to the original problem statement, we find that if $x_1 + x_2 = 10$ were written as $x_1 + x_2 = 9$, then the optimal value of $2x_1 + 4x_2$ that is to be minimized would be lower and reach a level of 18. The Maple worksheet that produces these results follows.

```
>    restart: # MixedConstraintsDual.mws
>    with(simplex):
```

Warning, new definition for maximize

Warning, new definition for minimize

```
>    zPrimal:=-2*x[ 1] -4* x[ 2] ;
```

$$zPrimal := -2x_1 - 4x_2$$

```
>    ConstraintsP:={ x[ 1] +5* x[ 2] <=80,4* x[ 1] +2* x[ 2] >=20,
     x[ 1] +x[ 2] =10] ;
```

$$ConstraintsP := [x_1 + 5x_2 \leq 80,\ 20 \leq 4x_1 + 2x_2,\ x_1 + x_2 = 10]$$

```
>    ConstraintsPStandard:=
     convert(ConstraintsP,stdle);
```

$$
\begin{aligned}
&ConstraintsPStandard := \\
&[x_1 + 5x_2 \leq 80,\ -4x_1 - 2x_2 \leq -20,\ x_1 + x_2 \leq 10,\ -x_1 - x_2 \leq -10]
\end{aligned}
$$

```
>    PrimalSol:=maximize(zPrimal,ConstraintsPStandard,
     NONNEGATIVE);
```

$$PrimalSol := \{x_2 = 0,\ x_1 = 10\}$$

```
>   assign(PrimalSol); zPrimal;
```

$$-20$$

```
>   x[ 1] :=' x[ 1]' :  x[ 2] :=' x[ 2]' :
>   zPrimal;
```

$$-2x_1 - 4x_2$$

```
>   DualProgram:=dual(zPrimal,ConstraintsPStandard,u);
```

$$DualProgram := 80\,u1 - 20\,u2 + 10\,u3 - 10\,u4,$$
$$[-2 \le u1 - 4\,u2 + u3 - u4, \ -4 \le 5\,u1 - 2\,u2 + u3 - u4]$$

```
>   DualSol:=minimize(DualProgram[ 1] ,DualProgram[ 2] ,
    NONNEGATIVE);
```

$$DualSol := \{u1 = 0, \ u2 = 0, \ u3 = 0, \ u4 = 2\}$$

```
>   assign(DualSol); DualProgram[ 1] ;
```

$$-20$$

Example 55 *Duality in game theory.* In Section 4.5.2 we presented two examples of the LP formulations of zero-sum games and solved them using Maple. In both examples, the minimum value of the second player's objective and the maximum value of the first player's objective were found to be equal: In the first example we found min u = max v = 0, and in the second example min u = max v = 190/3. The fact that the minimum value of the objective of one problem equals the maximum value of the objective of another problem would suggest that there may be a duality-type relationship between the two problems. This is indeed the case and we now demonstrate it for the case of the coin-tossing example.

Without loss of generality, we will suppose that the second player's LP problem is the primal program since its constraints are in $\le$ form. Recalling that in general, the variable u is unrestricted in sign, we let $u = u' - u''$ where $u', u'' \ge 0$ and rewrite the problem as

$$
\begin{array}{ll}
\max & -(u' - u'') \\
s.t. & y_1 - y_2 \le u' - u'' \\
& -y_1 + y_2 \le u' - u'' \\
& y_1 + y_2 = 1 \\
& y_1, y_2 \ge 0
\end{array}
$$

where the original objective function min u = min $(u' - u'')$ is converted to a maximization objective as max $-(u' - u'')$.

At this stage, we let Maple take over and find the dual of Player 2's problem.

```
>   restart: # CoinTossDual.mws
>   with(simplex):

Warning, new definition for maximize
```

```
     Warning, new definition for minimize

   >  P:=matrix(2,2,[ 1,-1,-1,1] );
```

$$P := \begin{bmatrix} 1 & -1 \\ -1 & 1 \end{bmatrix}$$

```
   >  m:=2; n:=2;
```

$$m := 2$$
$$n := 2$$

```
   >  # Player 2's problem
   >  for i from 1 to m do
      Player2[ i] :=add(y[ j] * P[ i,j] ,j=1..n)<=up-upp od;
```

$$Player2_1 := y_1 - y_2 \le up - upp$$

$$Player2_2 := -y_1 + y_2 \le up - upp$$

```
   >  Unity:=add(y[ j] ,j=1..n)=1;
```

$$Unity := y_1 + y_2 = 1$$

```
   >  Constraints:=[ seq(Player2[ i] ,i=1..m),Unity]  ;
```

$$Constraints := [y_1 - y_2 \le up - upp, -y_1 + y_2 \le up - upp, y_1 + y_2 = 1]$$

```
   >  ConstraintsSTDLE:=convert(Constraints,stdle);
```

$$ConstraintsSTDLE := [y_1 - y_2 - up + upp \le 0, -y_1 + y_2 - up + upp \le 0, y_1 + y_2 \le 1, -y_1 - y_2 \le -1]$$

```
   >  dual(-up+upp,ConstraintsSTDLE,x);
```

$$x3 - x4, [0 \le x1 - x2 + x3 - x4, 0 \le -x1 + x2 + x3 - x4, -1 \le -x1 - x2, 1 \le x1 + x2]$$

Maple finds the dual as follows:

$$\begin{aligned} &\min & &(x_3 - x_4) \\ &\text{s.t.} & x_1 - x_2 + (x_3 - x_4) &\ge 0 \\ & & -x_1 + x_2 + (x_3 - x_4) &\ge 0 \\ & & x_1 + x_2 &\le 1 \\ & & x_1 + x_2 &\ge 1. \end{aligned}$$

Letting $v = x_3 - x_4$ and noting that the two constraints $x_1 + x_2 \le 1$ and $x_1 + x_2 \ge 1$ imply $x_1 + x_2 = 1$, we obtain the problem for Player 1 as given in (4.2).

Thus, since the two players' problems are the duals of each other, it is sufficient to solve only one player's problem and deduce the solution for the other player using the $c_j - z_j$ values from the final optimal table.

4.7   Integer Linear Programming

In many realistic applications of linear programming, the optimal values of decision variables must be integer. For example, when a decision problem requires finding the optimal number of employees to be assigned to jobs or number of aircraft to purchase, the decision variables must be obviously limited to integer values. Since the simplex method does not necessarily produce integer optimal solutions, alternative means must be found to compute the optimal integer solutions.

As a simple example of linear programming requiring integer values for the decision variables, consider the following.

Example 56 *Aircraft purchase.* Air NoFrills is a small airline company that has decided to acquire new aircraft for its operations. After some preliminary analysis, management has decided that they will limit their choice to Boeings and Airbuses. The relevant information on profitability, price and maintenance requirements are as follows.

	Boeing	**Airbus**
Profit/year ($000)	200	100
Price/unit ($000,000)	5	4
Maintenance time/year (hours)	200	500

The company has a budget of $24,000,000 available for the purchases of these aircraft, and the annual maintenance time available per year is 1,300 hours. The problem is to find the optimal number of Boeings (x_1) and Airbuses (x_2) to purchase in order to maximize the annual profit.

We first solve this problem as an ordinary linear program using Maple's $\texttt{simplex}$ package and find that the solution is *not* integer since $x_1 = 4.8$ and $x_2 = 0$ with $z = 960$.

```
>   restart: # AircraftLP.mws
>   with(simplex):

Warning, new definition for maximize

Warning, new definition for minimize

>   z:=200*x[ 1] +100*x[ 2] ;
```
$$z := 200\,x_1 + 100\,x_2$$
```
>   Constraints:= { 5* x[ 1] +4* x[ 2] <=24.,
    200* x[ 1] +500* x[ 2] <=1300} ;
```
$$Constraints := \{5\,x_1 + 4\,x_2 \le 24., \ 200\,x_1 + 500\,x_2 \le 1300\}$$
```
>   maximize(z,Constraints,NONNEGATIVE);
```
$$\{x_1 = 4.800000000, \ x_2 = 0\}$$
```
>   assign(%); z;
```

$$960.0000000$$

At this stage one may be tempted to round the noninteger solution up and down to the nearest integer with the hope of obtaining an optimal integer result. This usually does not work, as we see in the following lines. Rounding $x_1 = 4.8$ up to 5, we find an infeasible solution, and rounding down to 4 gives an integer result that is not optimal!

```
>    x[ 1] :=5;  x[ 2] :=0;  z;  Constraints;
```

$$x_1 := 5$$

$$x_2 := 0$$

$$1000$$

$$\{25 \leq 24., \ 1000 \leq 1300\}$$

```
>    x[ 1] :=4;  x[ 2] :=0;  z;  Constraints;
```

$$x_1 := 4$$

$$x_2 := 0$$

$$800$$

$$\{20 \leq 24., \ 800 \leq 1300\}$$

Although Maple has not implemented any algorithms to solve integer linear programs, it includes a share package written by Anu Pathria in its share folder.[7] This package is named ilp() and it can be loaded by using the commands with(share); and with(ilp);.

The ilp() function's syntax is identical to that used in simplex as can be seen in the following Maple worksheet. Note that ilp() maximizes the objective function in an integer programming problem. For minimization problems, the user must first multiply the original objective function by -1 and then maximize the resulting function subject to the usual constraints. Once the optimal solution is found, taking the negative value of the objective function would give the minimum of the original objective.

```
>    restart: # AircraftIP.mws
>    with(share): with(ilp);
```

```
See ?share and ?share,contents for information about the
share library

Share Library:    ilp

Author: Pathria, Anu.
```

---

[7]Anu Pathria's ilp() is available with Release 5.1 and it implements the branch-and-bound algorithm for solving integer programming problems. A good reference on this algorithm is Zionts [198, Chapter 15].

Description: an integer linear programming algorithm for maximizing a linear function given a set of linear constraints

["ilp"]

>   z:=200* x[1] +100* x[2] ;

$$z := 200 x_1 + 100 x_2$$

>   Constraints:= { 5* x[1] +4* x[2] <=24.,
>   200* x[1] +500* x[2] <=1300} ;

$$Constraints := \{5 x_1 + 4 x_2 \leq 24., \ 200 x_1 + 500 x_2 \leq 1300\}$$

>   ilp(z,Constraints,NONNEGATIVE);

$$\{x_1 = 4, \ x_2 = 1\}$$

>   assign(%); z; Constraints;

$$900$$
$$\{24 \leq 24., \ 1300 \leq 1300\}$$

We thus see that it is optimal to purchase $x_1 = 4$ Boeings and $x_2 = 1$ Airbuses resulting in a maximum profit of $z = \$900,000$.

4.8   Summary

Linear programming (LP) is one of the oldest and most thoroughly studied OR technique. We started this chapter by introducing a simple example of LP modeling with two decision variables and solved the resulting problem using the graphical method. For larger problems with four or more variables the graphical method fails since plotting the feasible set becomes geometrically impossible. In that case, the (algebraic) simplex method can be used to locate the optimal solution. We provided a step-by-step description of the simplex method that involves inverting matrices. This was made relatively painless by using Maple's inverse() command. We also explained the implementation of Maple's own simplex package and discussed certain types of problems that may have unusual characteristics such as infeasibility and unbounded solutions. Sensitivity analysis is always an important component of any application of LP. This topic was also discussed in some detail. The chapter ended with a very brief description of a share package known as ilp() that can be used to solve integer linear programming problems.

4.9   Exercises

1. Consider the following LP problem with two decision variables.

$$
\begin{array}{rrl}
\max & 20x_1 + 30x_2 & \\
\text{s.t.} & \tfrac{1}{2}x_1 + x_2 & \leq \quad 8 \\
& x_1 + x_2 & \leq \quad 10 \\
& x_2 & \geq \quad 4 \\
& x_1 & \geq \quad 0
\end{array}
$$

   (a) Solve this problem graphically using Maple's inequality plotting fa-
       cilities.

   (b) Solve the same problem *manually* (but with Maple's help) using the
       simplex method.

   (c) Now solve the problem using Maple's simplex.

   (d) Formulate and solve the dual problem and interpret its solution.

2. The Kingdom of Albricol's national oil company that produces oil special-
   ity additives purchases three grades of petroleum distillates A, B and C.
   The company then combines the three according to King Peter the First's
   specifications of the maximum and minimum percentages of grades B and
   C, respectively, in each blend.

Blend	Max % Allowed for B	Min % Allowed for C	Selling Price of Mixture ($/L)
Royal Deluxe	60%	10%	3.75
Princely Standard	20%	30%	3.10
Pauper's Special	1%	65%	1.99

   The supply of the three additives and their costs are:

Distillate	Maximum Quantity Available per day (L)	Cost of Distillate ($/L)
A	16000	.37
B	19000	.29
C	12000	.24

   Formulate a linear program to determine the daily production policy that
   will maximize profits for King Peter and his Kingdom. Solve using simplex.

3. The Claus Cookie Manufacturing Society packages three types of cookies
   into two different one pound assortment packages. Each week, 1000 pounds
   of chocolate chip cookies, 2000 pounds of peanut butter cookies and 1500
   pounds of coconut crispies are available. Assortment A contains 50 percent
   chocolate chip cookies, 25 percent peanut butter cookies and 25 percent

coconut crispies. Assortment B contains 50 percent peanut butter cookies, 30 percent coconut crispies and 20 percent chocolate chip cookies. The net profit contribution for each pound of Assortment A cookies is fifty cents, while for Assortment B cookies it is forty cents.

(a) Formulate an appropriate linear program for Claus Cookie Manufacturing Society to find its optimal mix and solve using `simplex`.

(b) Formulate and solve the dual problem and interpret its solution.

4. Captain's Boats Inc. makes three different kinds of boats. All of these boats can be made profitably, but the company's monthly production is constrained by the limited amount of wood, screws, and labor available each month. The director will choose the combination of boats that maximizes his revenue in view of the following information:

Requirements per boat

Input	ROW BOAT	CANOE	KAYAK	Available/month
Labour (hr.)	12	7	9	1,260 hrs.
Wood (board ft.)	22	18	16	19,008 bd. ft.
Screws (lb.)	2	4	3	396 lbs.
Selling Price	$400	$200	$500	

Formulate this problem and solve using Maple's `simplex`.

(a) How many boats of each type will be produced and what will be the resulting revenue?

(b) How much wood will be used to make all of these boats? If extra wood were to become available, how much would Captain's Boats be willing to pay for some?

(c) You will find that very few canoes are made because their selling price is so small. Suppose someone offered to buy canoes for $x each. For what value of x, would Captain's Boats produce any canoes?

(d) Suppose Kayaks were sold for $y per unit rather than $500. For what value of y, fewer Kayaks would be produced?

(e) What is the highest hourly wage Captain's Boats would be willing to pay for extra workers?

5. Smart Bucks, Inc., an investment agency, has been asked to advise one of its clients from the province of Kebek on how to invest all of his $100,000 among the 3 assets described below:

Assets	Units of Risk per Dollar Invested	Rate of Return
1. Northern Mines Shares	4	0.13
2. Royal Bank of Commerce Shares	3	0.12
3. Kebek Savings Bonds	1	0.14

The client would like as high an annual return as is possible to receive while incurring an average of no more than 2.9 risk units per dollar invested. Kebek Savings Bonds unfortunately are available only up to a value of $15,000 per Kebek resident.

Formulate and solve this problem using Maple's `simplex` and answer the following questions.

(a) How much of each asset will be bought and what will be the average rate of return on these?

(b) How much risk is being incurred on average? What is its shadow (dual) price?

(c) Obviously Kebek Savings Bonds are paying an excessive rate of return in view of their relatively low level of risk. How low a rate of return could the province of Kebek have offered for its bonds without this investor having invested differently than he did?

(d) If the Royal Bank of Commerce raised the rate of return on its shares to 12.9%, would it be able to increase its sale of shares to the particular investor we are discussing? What would be the new average rate of return on this investor's portfolio?

(e) Formulate and solve the dual problem and interpret its solution.

6. Solve the following integer programming problem using Maple's `ilp()` share package if this package is available in your system.

$$\begin{aligned} \min \quad & 12x_1 + 9x_2 + 8x_3 + 6x_4 \\ \text{s.t.} \quad & 4x_1 + x_2 - x_4 \geq 16 \\ & x_1 + x_2 + x_3 \geq 20 \\ & 3x_1 + 3x_3 + x_4 \geq 36 \\ & x_1, x_2, x_3, x_4 \geq 0 \text{ and integers} \end{aligned}$$

7. An airplane with a weight capacity of 40,000 pounds and a volume capacity of 7,000 cubic feet will be sent to a disaster area. The weight and volume of the container for each of the five items is given below. Each item has an associated utility index that depends on the number of people who will need these items.

Item	Weight per Container (lb.)	Volume per Container (ft³)	Utility Index
Blood	80	25	7
Drugs	150	30	80
Supplies	130	40	1
Food	40	7	2
Water	70	5	5

Formulate the problem to find the optimal number of containers to ship in order to maximize the total utility index. Assume that at least 50 containers of blood, 15 containers of medical drugs, and 100 containers of water must be shipped. Use the `ilp()` share package to solve this problem if this package is available in your system.

5
Nonlinear Programming

5.1 Introduction

As we discussed in Chapter 4, in linear programming (LP) applications, our pur-
pose is to optimize (i.e., maximize or minimize) a *linear* objective function subject
to *linear* constraints. Although a large number of practical decision problems can
be accurately modeled as linear programs, it should be emphasized that in some
cases using LP to formulate an inherently nonlinear decision problem (i.e., forcing
the world to fit the model) may not provide an accurate representation of reality.
In many OR problems due to economies/diseconomies of scale (e.g., quantity
discounts), interaction among decision variables (e.g., variance of a portfolio) or
transformations that result in nonlinearities (e.g., a linear stochastic programming
formulation transformed into an equivalent nonlinear deterministic one), an LP
formulation becomes impossible to use. In these cases, we would need to utilize
the techniques of nonlinear programming (NLP) that can deal with both nonlinear
objective functions and nonlinear constraints.

A general NLP problem with both inequality and equality constraints is expressed as follows:

$$
\begin{aligned}
\min \quad & f(\mathbf{x}) \\
\text{s.t.} \quad & g_i(\mathbf{x}) \le 0, \ i = 1, \ldots, m \\
& h_i(\mathbf{x}) = 0, \ i = 1, \ldots, p \\
& \mathbf{x} \in S.
\end{aligned}
$$

Here, $f(\mathbf{x})$ is the objective function with $\mathbf{x} = (x_1, \ldots, x_n)'$ as the column
vector of n decision variables; $g_i(\mathbf{x}) \le 0$, $i = 1, \ldots, m$, are the m inequality

constraints, $h_i(\mathbf{x}) = 0$, $i = 1, \ldots, p$ are the p equality constraints and S is a subset of the n-dimensional Euclidean space R^n. (Of course, since max $f(\mathbf{x}) = -\min[-f(\mathbf{x})]$, an NLP problem can be stated as a maximization problem. Also, inequality constraints of the type $g_i(\mathbf{x}) \geq 0$, $i = 1, \ldots, m$ can be transformed to this form if both sides of these constraints are multiplied by -1.)

Example 57 *Portfolio selection.* As a simple example of the formulation of a nonlinear programming problem, consider a portfolio selection problem where we wish to invest all of $\$10{,}000$ in two stocks. Assume W_i is the random annual return on $\$1$ invested in stock $i = 1, 2$ with the expected returns $E(W_1) = 0.10$, $E(W_2) = 0.05$, variances $\text{Var}(W_1) = 0.15$, $\text{Var}(W_2) = 0.06$, and the covariance $\text{cov}(W_1, W_2) = 0.07$. If the objective is to minimize the variance of the portfolio with an expected annual return of at least 9%, we can define decision variables $x_i = $ amount invested in stock $i = 1, 2$ and find the expected annual rate of return as

$$\frac{E(W_1) \cdot x_1 + E(W_2) \cdot x_2}{10000} = \frac{0.10x_1 + 0.05x_2}{10000}$$

which must be at least equal to 0.09, that is, $0.10x_1 + 0.05x_2 \geq 900$. The variance of the portfolio is computed as

$$\begin{aligned}
\text{Var}(W_1x_1 + W_2x_2) &= \text{Var}(W_1) \cdot x_1^2 + \text{Var}(W_2) \cdot x_2^2 + 2\text{cov}(W_1, W_2) \cdot x_1x_2 \\
&= 0.15x_1^2 + 0.06x_2^2 + 0.14x_1x_2
\end{aligned}$$

Since the amounts invested must be nonnegative and we wish to invest all of $\$10{,}000$, this nonlinear programming problem can be formulated as follows.

$$\begin{aligned}
\min \quad & f(x_1, x_2) = 0.15x_1^2 + 0.06x_2^2 + 0.14x_1x_2 \\
\text{s.t.} \quad & g(x_1, x_2) = 900 - 0.10x_1 - 0.05x_2 \leq 0 \\
& h(x_1, x_2) = 10000 - x_1 - x_2 = 0 \\
& x_1 \geq 0, x_2 \geq 0.
\end{aligned}$$

The optimal solution for this problem is obtained as $x_1 = \$8{,}000$, $x_2 = \$2{,}000$ and $f = 1.208 \times 10^4$ with the standard deviation of the portfolio as $\sqrt{1.208 \times 10^4} \simeq 3475$; see Example 68.

The next section introduces the concept of convexity (of sets and functions), which plays a very important role in the study of nonlinear programming.

5.2   Convexity of Sets and Functions

Let R^n be the n-dimensional Euclidean space. A nonempty set $S \subset R^n$ is said to be a **convex set** if the line segment joining any two points of S also belongs to S. Stated mathematically, if $\mathbf{x}_1, \mathbf{x}_2 \in S$, then the convex combination of $\mathbf{x}_1$ and $\mathbf{x}_2$ also belongs to S, i.e., $\lambda\mathbf{x}_1 + (1 - \lambda)\mathbf{x}_2 \in S$ where $\lambda \in [0, 1]$.

Let $f : S \rightarrow R^1$ with S as a convex set. The function $f(\mathbf{x})$ is said to be a **convex function** on S if $f(\lambda \mathbf{x}_1 + (1-\lambda)\mathbf{x}_2) \le \lambda f(\mathbf{x}_1) + (1-\lambda)f(\mathbf{x}_2)$ for each $\mathbf{x}_1, \mathbf{x}_2 \in S$ where $\lambda \in [0,1]$. Geometrically, this has the interpretation that when the function f is convex, its value at points on the line $\lambda \mathbf{x}_1 + (1-\lambda)\mathbf{x}_2$ is less than or equal to the height of the cord joining the points $(\mathbf{x}_1, f(\mathbf{x}_1))$ and $(\mathbf{x}_2, f(\mathbf{x}_2))$. Note that if $\lambda \in (0,1)$ and if the inequality is satisfied strictly, then f is a **strictly convex function**. Also, if f is (strictly) convex, then $-f$ is (**strictly**) **concave**.

Example 58 *Convexity of $x^2 - 4x - 8$.* Using the foregoing definition of convexity, we now use Maple's `assume()` facility and `is()` function to check whether the function $f(x) = x^2 - 4x - 8$ is convex. We let $x_1 = a$ and $x_2 = b$.

```
>    restart: # convex.mws
>    assume(lambda>=0,lambda<=1,lambda,real,
     a,real,b,real);
>    f:=x->x^2-4*x-8;
```

$$f := x \rightarrow x^2 - 4x - 8$$

```
>    LHS:=f(lambda*a+(1-lambda)*b);
```

$$LHS := (\lambda a + (1-\lambda)b)^2 - 4\lambda a - 4(1-\lambda)b - 8$$

```
>    RHS:=lambda*f(a)+(1-lambda)*f(b);
```

$$RHS := \lambda(a^2 - 4a - 8) + (1-\lambda)(b^2 - 4b - 8)$$

```
>    is(LHS<=RHS);
```

true

Maple indicates that the condition $LHS \le RHS$ is always true indicating that $f(\lambda a + (1-\lambda)b) \le \lambda f(a) + (1-\lambda)f(b)$, for all $\lambda \in (0,1)$ and all real a and b. This proves the convexity of $f(x) = x^2 - 4x - 8$.

The following intuitive result is concerned with the sum of convex functions.

Theorem 1 *Let the functions $f_i(\mathbf{x})$, $i = 1, \ldots, k$ be convex over some convex set $S \subset R^n$. Then the function $f(\mathbf{x}) = \sum_{i=1}^{k} f_i(\mathbf{x})$ is also convex over S.* ∎

A useful theorem that relates the convexity of functions to the convexity of sets is the following.

Theorem 2 *[80, p. 86] If $f(\mathbf{x})$ is a convex function over the nonnegative orthant of R^n, then if the set of points S satisfying $f(\mathbf{x}) \le b$ and $\mathbf{x} \ge 0$ is not empty, S is a convex set.* ∎

For example, since $f(\mathbf{x}) = x_1^2 + x_2^2$ is a convex function, the set $S = \{\mathbf{x} : f(\mathbf{x}) \le 1, \mathbf{x} \ge 0\}$—which corresponds to the portion of the unit circle in the first quadrant—is also convex.

Let the nonempty set S be a subset of R^n and let $f : S \rightarrow R^1$. If the function f is differentiable at some point $\bar{x} \in S$, then its **gradient** vector at $\bar{x}$ is

$$\nabla f(\bar{x}) = \left(\frac{\partial f(\bar{x})}{\partial x_1}, \dots, \frac{\partial f(\bar{x})}{\partial x_n} \right)'$$

where $\partial f(\bar{x})/\partial x_i$ is the partial derivative of f with respect to x_i at the point $\bar{x}$.

Maple excels in computing the partial derivatives—and hence the gradient of a function—symbolically. In order to compute the gradient of an expression (or function) using the `grad()` function, the linear algebra package must be loaded with the command `with(linalg)`.

Example 59 *Terrain depth.* Consider the following function f of two variables adapted from Schmidt and Davis [167, p. 161] that is purported to describe the depth of the terrain beneath a harbor's waters. With x and y as measured in miles due east and due north, respectively, from the center of the harbor, the function $f(x, y)$ is given as

$$f(x, y) = -(5/10)x^2 + (1/10)y^3 - (2/10)y^2 + (4/100)xy + (1/100)x + (3/100)y - (5/10).$$

We find the gradient vector of the function as follows.

```
>    restart: # HARBOUR1.MWS
>    with(linalg):
```

Warning, new definition for norm

Warning, new definition for trace

```
>    f:=-(5/10)*x^2+(1/10)*y^3-(2/10)*y^2
     +(4/100)*x*y+(1/100)*x+(3/100)*y-(5/10);
```

$$f := -\frac{1}{2}x^2 + \frac{1}{10}y^3 - \frac{1}{5}y^2 + \frac{1}{25}xy + \frac{1}{100}x + \frac{3}{100}y - \frac{1}{2}$$

```
>    grad(f,[ x,y] );
```

$$\left[-x + \frac{1}{25}y + \frac{1}{100}, \; \frac{3}{10}y^2 - \frac{2}{5}y + \frac{1}{25}x + \frac{3}{100} \right]$$

For a twice differentiable function f, we define the **Hessian** matrix $\mathbf{H}(\mathbf{x})$ of f at $\mathbf{x}$ as a square matrix with the ith row and jth column entry as the second partial derivative $\partial^2 f(\mathbf{x})/\partial x_i \partial x_j$:

$$\mathbf{H}(\mathbf{x}) = \begin{bmatrix} \dfrac{\partial^2 f(\mathbf{x})}{\partial x_1^2} & \cdots & \dfrac{\partial^2 f(\mathbf{x})}{\partial x_1 \partial x_n} \\ \vdots & \ddots & \vdots \\ \dfrac{\partial^2 f(\mathbf{x})}{\partial x_n \partial x_1} & \cdots & \dfrac{\partial^2 f(\mathbf{x})}{\partial x_n^2} \end{bmatrix}.$$

Provided that $\partial f(\mathbf{x})/\partial x_i$, $i = 1, \ldots, n$ and $\partial^2 f(\mathbf{x})/\partial x_i \partial x_j$, $i, j = 1, \ldots, n$ are continuous, the Hessian is a symmetric matrix; see [8, p. 121].

Maple has a hessian() function that simplifies the computation of the Hessian. We demonstrate its use on the function f:

```
>   H:=hessian(f,[ x,y] );
```

$$H := \begin{bmatrix} -1 & \dfrac{1}{25} \\ \dfrac{1}{25} & \dfrac{3}{5}y - \dfrac{2}{5} \end{bmatrix}$$

5.2.1 Positive and Negative Definite Matrices

Now, let $\mathbf{Q}$ be an $n \times n$ symmetric matrix. The matrix $\mathbf{Q}$ is called **positive definite** if the **quadratic form** $\mathbf{x}'\mathbf{Q}\mathbf{x} > 0$ for all *nonzero* $\mathbf{x} \in R^n$ and **positive semidefinite** if $\mathbf{x}'\mathbf{Q}\mathbf{x} \geq 0$ for *all* $\mathbf{x} \in R^n$. When the inequalities are reversed, we obtain the definitions for **negative definite**ness and **negative semidefinite**ness, respectively. The matrix $\mathbf{Q}$ is called **indefinite** if $\mathbf{x}'\mathbf{Q}\mathbf{x}$ is positive for some values of $\mathbf{x}$ and negative for others. These definitions are summarized in the following table.

	Sign of the Quadratic Form $\mathbf{x}'\mathbf{Q}\mathbf{x}$				
	> 0 for $\mathbf{x} \neq \mathbf{0}$	≥ 0 for all $\mathbf{x}$	< 0 for $\mathbf{x} \neq \mathbf{0}$	≤ 0 for all $\mathbf{x}$	Mixed
Property of $\mathbf{Q}$	Positive Definite	Positive Semidefinite	Negative Definite	Negative Semidefinite	Indefinite

Example 60 *Quadratic forms.* The quadratic form $f(x_1, x_2) = x_1^2 + x_2^2$ is positive definite since f is positive except when $x = 0$, and the quadratic form $f(x_1, x_2) = (x_1 + x_2)^2$ is positive semidefinite since f is positive except for $x_1 = -x_2$. But the quadratic form $f(x_1, x_2) = x_1^2 - x_2^2$ is indefinite since f is positive for $|x_1| > |x_2|$ and negative for $|x_1| < |x_2|$.

It is usually difficult to check the positive and negative definiteness properties of a symmetric matrix using the stated definitions. We will now present an easy-to-use operational result that makes use of the signs of certain types of submatrices of $\mathbf{Q}$ for examining these properties.

Theorem 3 *(Sylvester's Theorem) [148, p. 16] For the $n \times n$ symmetric $\mathbf{Q}$ matrix, we define Δ_k to be the determinant of the upper left-hand corner $k \times k$ submatrix of $\mathbf{Q}$ for $k = 1, \ldots, n$, and call it the kth principal minor of $\mathbf{Q}$. Then, $\mathbf{Q}$ is positive definite if and only if $\Delta_k > 0$ for $k = 1, \ldots, n$; and negative definite if and only if $(-1)^k \Delta_k > 0$ for $k = 1, \ldots, n$. Otherwise, $\mathbf{Q}$ is indefinite. These results are summarized in the following table.*

	Sign of the Δ_k's, $k = 1, \ldots, n$		
	$\Delta_k > 0$	$(-1)^k \Delta_k > 0$	Mixed
Property of	Positive	Negative	Indefinite
Q	Definite	Definite	

These results must be interpreted and used with care since merely replacing the $>$ sign by $\geq$ (or $<$ by $\leq$) does *not* mean that the **Q** matrix is positive (or negative) *semi*definite. For example, if

```
>   restart: # notsemi.mws
>   Q:=matrix(3,3,[ 2,2,2,
>   2,2,2,
>   2,2,1] );
```

$$Q := \begin{bmatrix} 2 & 2 & 2 \\ 2 & 2 & 2 \\ 2 & 2 & 1 \end{bmatrix}$$

then it is easy to show that $\Delta_1 > 0$, $\Delta_2 = 0$ and $\Delta_3 = 0$. But for a nonzero vector such as $\mathbf{x} = (\frac{1}{2}, \frac{1}{2}, -1)'$, i.e.,

```
>   x:=array([ 1/2,1/2,-1] );
```

$$x := \left[\frac{1}{2}, \frac{1}{2}, -1 \right],$$

we obtain $\mathbf{x}'\mathbf{Q}\mathbf{x}$ as

```
>   evalm(x &* Q &* x);
```

$$-1$$

implying that **Q** is *not* positive semidefinite.

There is, however, a special case where it can be shown that if $\Delta_k > 0$, $k = 1, \ldots, n-1$, and $\Delta_n = 0$, then **Q** is **positive semidefinite** [148, p. 19]. Similarly, if $(-1)^k \Delta_k > 0$, $k = 1, \ldots, n-1$ but $\Delta_n = 0$, then **Q** is **negative semidefinite**.

In general, in order to determine the *positive semidefiniteness* of a symmetric matrix, the following procedure is used: Let $\mathbf{Q}_r$ be the matrix obtained by removing the first r rows and r columns of the original matrix $\mathbf{Q}$, $r = 0, \ldots, n-1$ (with $\mathbf{Q}_0 \equiv \mathbf{Q}$). Define Δ_{rk} to be the kth principal minor of $\mathbf{Q}_r$. Then, **Q** is **positive semidefinite** if and only if $\Delta_{rk} \geq 0$ for $r = 0, \ldots, n-1$, $k = 1, \ldots, n-r$; and **negative semidefinite** if and only if $(-1)^k \Delta_{rk} \geq 0$ for $r = 0, \ldots, n-1$, $k = 1, \ldots, n-r$. For example, if $n = 3$ and

$$\mathbf{Q} = \begin{bmatrix} q_{11} & q_{12} & q_{13} \\ q_{21} & q_{22} & q_{23} \\ q_{31} & q_{32} & q_{33} \end{bmatrix},$$

then for positive semidefiniteness we would need to show that

$$\Delta_{01} = q_{11} \geq 0, \quad \Delta_{02} = \det \begin{bmatrix} q_{11} & q_{12} \\ q_{21} & q_{22} \end{bmatrix} \geq 0, \quad \Delta_{03} = \det \mathbf{Q} \geq 0,$$

$$\Delta_{11} = q_{22} \geq 0, \quad \Delta_{12} = \det \begin{bmatrix} q_{22} & q_{23} \\ q_{32} & q_{33} \end{bmatrix} \geq 0, \quad \text{and} \quad \Delta_{21} = q_{33} \geq 0.$$

Note that this method—which is actually implemented by Maple—of checking the positive semidefiniteness of symmetric matrices requires the computation of $n + (n - 1) + \cdots + 2 + 1 = n(n + 1)/2$ determinants. Although it would be cumbersome to implement this test manually, for Maple it is straightforward, as we demonstrate in the next example.

Example 61 *Maple's test for semi-definiteness with* `definite()`. *We let* $n =$ 3 and ask Maple to find the conditions under which a general (i.e., symbolic) symmetric 3 × 3 symmetric matrix is positive semidefinite.

```
>    restart: # semid33.mws
>    with(linalg):

Warning, new definition for norm

Warning, new definition for trace

>    H:=array(1..3,1..3,'symmetric');
```

$$H := \text{array}(symmetric, \ 1..3, \ 1..3, \ [])$$

```
>    definite(H,'positive_semidef');
```

$-H_{1,1} \leq 0$ **and** $- H_{1,1} H_{2,2} + H_{1,2}{}^2 \leq 0$ **and** $- H_{1,1} H_{2,2} H_{3,3} + H_{1,1} H_{2,3}{}^2$ $+ H_{1,2}{}^2 H_{3,3} - 2 H_{1,2} H_{1,3} H_{2,3} + H_{1,3}{}^2 H_{2,2} \leq 0$ **and** $- H_{2,2} \leq 0$ **and** $- H_{2,2} H_{3,3} + H_{2,3}{}^2 \leq 0$ **and** $- H_{3,3} \leq 0$

There is another method that makes use of the **eigenvalues** and provides information on all aspects of positive and negative definiteness and semidefiniteness of a symmetric matrix **Q**. We summarize this result in the following theorem.

Theorem 4 *[148, p. 30] The symmetric matrix* **Q** *is positive (negative) definite if and only if all the eigenvalues of* **Q** *are positive (negative) and is positive (negative) semidefinite if and only if all the eigenvalues of* **Q** *are nonnegative (nonpositive). However, matrix* **Q** *is* **indefinite** *if and only if* **Q** *has at least one positive eigenvalue and at least one negative eigenvalue.*

	Sign of the Eigenvalues of Q				
	All > 0	All ≥ 0	All < 0	All ≤ 0	Mixed
Property of **Q**	Positive Definite	Positive Semidefinite	Negative Definite	Negative Semidefinite	Indefinite

Provided that the eigenvalues' signs can be easily determined, this test is simpler to implement than the principal minor test. We should also note that although, in general, the eigenvalues may assume complex values, due to the symmetric nature of the **Q** matrix that we are assuming, it can be shown that its eigenvalues are always real [91, p. 32].

5.2.2   Convexity of a Function and Definiteness of its Hessian

There is a very important relationship between the **convexity** of a function f and the **positive semidefiniteness** of its Hessian matrix. For a proof of the following theorem on this relationship, see Bazaraa and Shetty [18, Section 3.3].

Theorem 5 *Let $f : S \to R^1$ be a twice differentiable function on a nonempty open convex set $S \subset R^n$. Then f is **convex** if and only if the Hessian is **positive semidefinite** at each point of S. Also, if the Hessian is **positive definite** at each point of S, then the function f is **strictly convex**. But, if f is strictly convex, then the Hessian is positive semidefinite at each point in S.* ■

The fact that strict convexity of f does *not* imply positive definiteness can be seen if we take the strictly convex function $f(x) = \frac{1}{12}x^4$. The Hessian is $H(x) = f''(x) = x^2$, which is positive *semi*definite at $x = 0$.

As demonstrated, Maple can be very helpful in identifying the positive or negative definiteness or semidefiniteness properties of a symmetric matrix. Since the Hessian **H** is symmetric, Maple's `definite()` function automates the task of determining the definiteness properties of **H** and hence the convexity or concavity of the function f.

5.2.3   Examples of Definiteness/Convexity

Example 62 *Terrain depth (continued)*. Consider again the terrain depth function f discussed in Example 59. We first check to see if the Hessian is positive (semi)definite. This being the irregular terrain beneath the water, we would not expect such a property to hold true.

```
>   restart: # HARBOUR2.MWS
>   with(linalg):
```

Warning, new definition for norm

Warning, new definition for trace

```
>   f:=-(5/10)*x^2+(1/10)*y^3-(2/10)*y^2
    +(4/100)*x*y+(1/100)*x+(3/100)*y-(5/10);
```

$$f := -\frac{1}{2}x^2 + \frac{1}{10}y^3 - \frac{1}{5}y^2 + \frac{1}{25}xy + \frac{1}{100}x + \frac{3}{100}y - \frac{1}{2}$$

```
>   H:=hessian(f,[x,y]):
```

Now let us examine the negative (semi)definiteness of f. In this case, we would expect the terrain to have this property, but due to the cubic term in f, the function may be concave only over some regions:

```
>   definite(H,'positive_def');
    definite(H,'positive_semidef');
```

false

false

```
>   definite(H,' negative_def' );
    definite(H,' negative_semidef' );
```

$$125\, y - 83 < 0$$

$$125\, y - 83 \leq 0 \text{ and } 3\, y - 2 \leq 0$$

Thus, we see that only in some regions, the terrain is negative definite (or negative semidefinite), and hence f is strictly concave (or concave). We can ask Maple to solve these inequalities to obtain the region over which the terrain is, say, negative semidefinite:

```
>   solve({ definite(H,' negative_semidef' )} ,y);
```

$$\{y \leq \frac{83}{125}\}$$

Since the `definite()` function would not be directly useful to determine where the Hessian is indefinite, (i.e., where f is neither convex nor concave) we make use of the eigenvalue test and obtain the eigenvalues of **H** as

```
>   s:=eigenvalues(H);
```

$$s := -\frac{7}{10} + \frac{3}{10}\, y + \frac{1}{50}\, \sqrt{229 + 450\, y + 225\, y^2},$$

$$-\frac{7}{10} + \frac{3}{10}\, y - \frac{1}{50}\, \sqrt{229 + 450\, y + 225\, y^2}$$

The two eigenvalues are given in terms of the variable y; thus, we ask Maple to solve for the region of values of this variable for which the eigenvalues assume opposite signs:

```
>   solve({ s[ 1] >0, s[ 2] <0} ,y);
```

$$\{\frac{83}{125} < y\}$$

This result implies that for $\{(x, y) : y > 83/125\}$, the terrain of the surface beneath the water resembles a part of a saddle without any convex or concave regions. See Figure 5.1 for the plot of the surface of the terrain of function f produced by the following Maple three-dimensional plotting command.

```
>   plot3d(f,x=-15..15,y=-15..15,axes=boxed,
    shading=none,orientation=[ -33,76] );
```

Example 63 *Nonconvexity of the backorder function in continuous-review (Q, r) inventory model.* We now use the above results and discuss an incorrect statement made by Hadley and Whitin in their classic text on inventory theory [81, p. 221, problem 4-6]. Under suitable assumptions, Hadley and Whitin (H-W) find the average number of backorders per year for the (Q, r) model as

$$B(Q, r) = \frac{\lambda}{Q} \int_r^\infty (x - r)k(x)\, dx$$

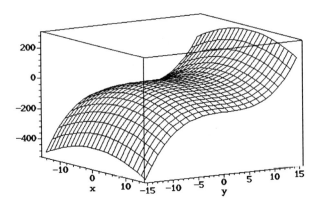

FIGURE 5.1. Surface plot of the terrain depth function $f(x, y)$.

where Q is the order quantity, r is the reorder point, λ is the average demand per year, and $k(x)$ is the density of demand during the leadtime. It is easy to show that the $B(Q, r)$ function is separately convex in Q and convex in r, but despite H-W's claim it is not jointly convex in both Q and r. Let us use Maple to demonstrate this fact.

```
>    restart: # backord.mws
>    B:=lambda*int((x-r)*k(x),x=r..infinity)/Q;
```

$$B := \frac{\lambda \int_r^\infty (x - r)\, k(x)\, dx}{Q}$$

```
>    with(linalg):
```

Warning, new definition for norm

Warning, new definition for trace

```
>    H:=hessian(B,[ Q, r] );
```

$$H := \begin{bmatrix} 2\, \dfrac{\lambda \int_r^\infty (x - r)\, k(x)\, dx}{Q^3} & -\dfrac{\lambda \int_r^\infty -k(x)\, dx}{Q^2} \\[2em] -\dfrac{\lambda \int_r^\infty -k(x)\, dx}{Q^2} & \dfrac{\lambda\, k(r)}{Q} \end{bmatrix}$$

In order for B to be positive semidefinite, the diagonal elements and the determinant must be nonnegative. The diagonal elements are nonnegative, but the determinant is computed as

```
>    det(H);
```

$$\frac{\lambda^2 \left(2 \int_r^\infty -(-x+r)\,k(x)\,dx\,k(r) - (\int_r^\infty -k(x)\,dx)^2\right)}{Q^4}$$

Since $\lambda^2/Q^4 > 0$, we see that if at $r = 0$ the density function has the property that $k(0) = 0$, the determinant assumes a negative value as it reduces to $-(\lambda^2/Q^4)\left[-\int_r^\infty k(x)\,dx\right]^2 < 0$. Thus, contrary to H-W's claim, it is *not*, in general, true that the backorder cost function is always convex.

However, as has been shown by Brooks and Lu [38], if ξ is a number such that $k(x)$ is nonincreasing for $x \geq \xi$, then $B(Q,r)$ is convex in the region $\{(Q,r) : 0 < Q < \infty \text{ and } \xi \leq r < \infty\}$.

Example 64 *Convexity of $f(a, b) = \sqrt{a^2 + b^2}$.* The Euclidean distance function $f(a, b) = \sqrt{a^2 + b^2}$ (also known as the ℓ_2 distance function) plays a very important role in location theory. In this example we will show using Maple that this function is convex over the real plane R^2 except at the origin where it has no derivative.

```
>    restart: # sqrt.mws
>    f:=sqrt(a^2+b^2);
```

$$f := \sqrt{a^2 + b^2}$$

```
>    diff(f,a);
```

$$\frac{a}{\sqrt{a^2 + b^2}}$$

```
>    limit(%,{ a=0,b=0} );
```

$$\textit{undefined}$$

```
>    with(linalg):

Warning, new definition for norm

Warning, new definition for trace

>    H:=hessian(f,[ a,b] );
```

$$H := \begin{bmatrix} -\dfrac{a^2}{(a^2 + b^2)^{(3/2)}} + \dfrac{1}{\sqrt{a^2 + b^2}} & -\dfrac{b\,a}{(a^2 + b^2)^{(3/2)}} \\[2em] -\dfrac{b\,a}{(a^2 + b^2)^{(3/2)}} & -\dfrac{b^2}{(a^2 + b^2)^{(3/2)}} + \dfrac{1}{\sqrt{a^2 + b^2}} \end{bmatrix}$$

```
>    H11:=normal(H[ 1,1] ,expanded);
     H22:=normal(H[ 2,2] ,expanded);
```

$$H11 := \frac{b^2}{(a^2 + b^2)^{(3/2)}}$$

$$H22 := \frac{a^2}{(a^2 + b^2)^{(3/2)}}$$

```
>   assume(a<0,a>0,b<0,b>0);
>   is(H11>0); is(H22>0);
```

$$true$$

$$true$$

```
>   det(H);
```

$$0$$

Since the first principal minor of the Hessian is $\Delta_1 = H_{11} > 0$ for $\Omega = \{(a, b) : a \in R^1 \setminus \{0\}, b \in R^1 \setminus \{0\}\}$ and the determinant of the Hessian is zero, it follows that the Hessian is positive semidefinite over Ω. Thus, the function $f(a, b) = \sqrt{a^2 + b^2}$ is convex in Ω.

5.2.4 Cholesky Factorization

In closing this section we will mention a simple but effective algorithm for testing if a symmetric matrix Q is positive definite, positive semidefinite or indefinite. This method is based on Cholesky factorization and it is briefly described in Gill, Murray and Wright [73] but more thoroughly covered in Vickson [189].

In a nutshell, the Cholesky factorization of a positive definite or a positive semidefinite matrix Q is $Q = LL'$ where L is a real lower-triangular matrix. When this factorization exists, Q is positive definite or positive semidefinite; but if L fails to exist or if it contains some complex numbers then Q is indefinite. A simple recursive algorithm has been developed based on these arguments and Maple has implemented a portion of this algorithm that works only on positive definite matrices: If Q is positive definite, Maple indicates that this is the case by producing the L matrix; when Q is positive semidefinite or indefinite, Maple produces an error message indicating that the matrix is not positive definite.

Example 65 *Maple's implementation of the Cholesky factorization.* A simple example with the Hessians of positive definite, positive semidefinite and indefinite quadratic forms $f = x^2 + y^2$, $g = (x + y)^2$ and $h = x^2 - y^2$, respectively, is given below.

```
>   restart: # cholesky.mws
>   with(linalg):
```

Warning, new definition for norm

Warning, new definition for trace

```
>   f:=x^2+y^2; g:=(x+y)^2; h:=x^2-y^2;
```

$$f := x^2 + y^2$$

$$g := (x + y)^2$$

$$h := x^2 - y^2$$

```
>   Hf:=hessian(f,[ x,y] ); Hg:=hessian(g,[ x,y] );
    Hh:=hessian(h,[ x,y] );
```

$$Hf := \begin{bmatrix} 2 & 0 \\ 0 & 2 \end{bmatrix}$$

$$Hg := \begin{bmatrix} 2 & 2 \\ 2 & 2 \end{bmatrix}$$

$$Hh := \begin{bmatrix} 2 & 0 \\ 0 & -2 \end{bmatrix}$$

```
>   cholesky(Hf);
```

$$\begin{bmatrix} \sqrt{2} & 0 \\ 0 & \sqrt{2} \end{bmatrix}$$

```
>   cholesky(Hg);
```

Error, (in cholesky) matrix not positive definite

```
>   cholesky(Hh);
```

Error, (in cholesky) matrix not positive definite

In the next example, Maple is able to compute the lower triangular matrix **L** of a 3 × 3 symmetric matrix **Q**, which, by implication, is positive definite.

```
>   Q:=matrix([[ 2,-1,-3] , [ -1,2,4] , [ -3,4,9]] );
```

$$Q := \begin{bmatrix} 2 & -1 & -3 \\ -1 & 2 & 4 \\ -3 & 4 & 9 \end{bmatrix}$$

```
>   L:=cholesky(Q);
```

$$L := \begin{bmatrix} \sqrt{2} & 0 & 0 \\ -\frac{1}{2}\sqrt{2} & \frac{1}{2}\sqrt{6} & 0 \\ -\frac{3}{2}\sqrt{2} & \frac{5}{6}\sqrt{6} & \frac{1}{3}\sqrt{3} \end{bmatrix}$$

As a check of the results, when we ask Maple whether **LL**′ is equal to the original matrix **Q**, the answer is obtained in the affirmative.

```
>   equal(L &* transpose(L), Q);
```

true

5.3   Unconstrained Optimization

In unconstrained optimization, the objective is to optimize the function $f(\mathbf{x})$ without any constraints on the decision vector $\mathbf{x}$. In operations research applications, such problems arise infrequently since in many business and industrial models the decision variables are required to assume nonnegative values (e.g., the number of units of production, order quantity, number of people to hire, amount of money to invest). Additionally, these decision variables are usually constrained to take values within some feasible region due to resource or budget limitations.

However, there are a few cases (such as the single-facility location problem to be discussed) where the decision variables do not have to be constrained. Moreover, the optimality conditions for the constrained problems are natural extensions of the conditions for unconstrained problems, which provides additional justification for the study of unconstrained optimization.

We first give some definitions that will be useful in the analysis of general optimization problems.

Definition 3 *Let $f(\mathbf{x})$ be a function defined on $S \subset R^n$. A point $\bar{\mathbf{x}} \in S$ is*

*(a) a **global minimizer** of $f(\mathbf{x})$ on S if $f(\bar{\mathbf{x}}) \leq f(\mathbf{x})$ for all $\mathbf{x} \in S$,*

*(b) a **local minimizer** of $f(\mathbf{x})$ if there exists an $\varepsilon > 0$ such that $f(\bar{\mathbf{x}}) \leq f(\mathbf{x})$ for all $\mathbf{x} \in S$ for which $\mathbf{x} \in N_\varepsilon(\bar{\mathbf{x}}) = \{\mathbf{y} : \|\mathbf{y} - \bar{\mathbf{x}}\| \leq \varepsilon\}$.*

Note that replacing the inequality sign ($\leq$) by the strict inequality sign ($<$) gives the definitions for **strict global minimizer** and **strict local minimizer**.

The next two theorems, which give the necessary and sufficient conditions for the optimality, are proved in Bazaraa and Shetty [18, Section 4.1].

Theorem 6 *Suppose $f(\mathbf{x})$ is differentiable at $\bar{\mathbf{x}}$. If $\bar{\mathbf{x}}$ is a local minimum, then the first order condition on the gradient[1] is*

$$\nabla f(\bar{\mathbf{x}}) = \left(\frac{\partial f(\bar{\mathbf{x}})}{\partial x_1}, \ldots, \frac{\partial f(\bar{\mathbf{x}})}{\partial x_n}\right)' = \mathbf{0}.$$

Also, if $f(\mathbf{x})$ is twice differentiable and $\bar{\mathbf{x}}$ is a local minimum, a second-order condition is that the Hessian $\mathbf{H}(\bar{\mathbf{x}})$ is positive semidefinite. ∎

This theorem provides the **necessary conditions** for the local optimality of $\bar{\mathbf{x}}$; these conditions are true for every local optimum, but a point satisfying these conditions does not have to be a local minimum, the classical example being $f(x) = x^3$. The following theorem provides a **sufficient condition** for the local minimum.

---

[1]The points of $f(\mathbf{x})$ obtained as a solution of $\nabla f(\bar{\mathbf{x}}) = \mathbf{0}$ are called the **critical** (or **stationary**) **points**.

Theorem 7 *Suppose $f(\mathbf{x})$ is twice differentiable at $\bar{\mathbf{x}}$. If $\nabla f(\bar{\mathbf{x}}) = 0$ and Hessian $\mathbf{H}(\bar{\mathbf{x}})$ is positive definite, then $\bar{\mathbf{x}}$ is a local minimum.* ■

Let us now illustrate the necessary and sufficient conditions using a specific function.

Example 66 *Analysis of $f(x) = (x - 4)^4(x + 3)^3$.* The following steps in Maple compute the derivative of this seventh-degree polynomial, compute its stationary points, which are obtained from the necessary condition $f'(x) = 0$, and evaluate the Hessian $f''(x)$ at each stationary point.

```
>   restart: # necsuf.mws
>   f:=x-> (x-4)^4* (x+3)^3;
```

$$f := x \rightarrow (x - 4)^4 (x + 3)^3$$

```
>   solve(diff(f(x),x),x);
```

$$0, -3, -3, 4, 4, 4$$

```
>   H:=diff(f(x),x$2);
```

$$H := 12(x - 4)^2(x + 3)^3 + 24(x + 3)^2(x - 4)^3 + 6(x - 4)^4(x + 3)$$

```
>   subs(x=0,H); subs(x=-3,H); subs(x=4,H);
```

$$-4032$$

$$0$$

$$0$$

The function is negative definite at $x = 0$, but it is negative semidefinite at $x = -3$ and $x = 4$. Using Maple's is() function, we can check the characteristics of these two points. We hypothesize that both of these stationary points are local minima, which should imply that the value of the function at these points is lower than the neighboring points, i.e., $f(-3 - \varepsilon) > f(-3)$ and $f(-3) < f(-3 + \varepsilon)$, and $f(4 - \varepsilon) > f(4)$ and $f(4) < f(4 + \varepsilon)$.

```
>   epsilon:=.01;
```

$$\varepsilon := .01$$

```
>   is( f(-3-epsilon)>f(-3) );
    is( f(-3)<f(-3+epsilon) );
```

false

true

```
>   is( f(4-epsilon)>f(4) );
    is( f(4)<f(4+epsilon) );
```

true

true

This demonstrates that $x = -3$ is neither a local minimum nor a local maximum and that $x = 4$ is a local minimum. See Figure 5.2 depicting the graph

of the function $f(x) = (x - 4)^4 (x + 3)^3$ obtained using the Maple command
`plot(f(x),x=-4..5).`

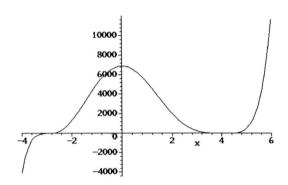

FIGURE 5.2. Plot of the function $f(x) = (x - 4)^4 (x + 3)^3$.

Remark 1 *The results obtained for the stationary points $x = -3$ and $x = 4$
using the* `is ()` *function of Maple could have been shown more easily using the
following result: If the function of a single variable $f(x)$ and its first n derivatives
are continuous, then $f(x)$ has a (local) maximum or minimum at $\bar{x}$ if and only
if n is even, where n is the order of the first nonvanishing derivative at $\bar{x}$. The
function has a maximum or minimum at $\bar{x}$ according to whether $f^{(n)}(\bar{x}) < 0$
or > 0, respectively [77, p. 29]. Maple can conveniently compute the higher-
order derivatives and evaluate them at the stationary points -3 and 4. Thus, the
following commands and their output indicate that the former point is neither a
local maximum nor a local minimum since $f^{(3)}(-3) > 0$ and that the latter point
is a local minimum since $f^{(4)}(4) > 0$.*

```
>   subs(x=-3,diff(f(x),x$2));
    subs(x=-3,diff(f(x),x$3));
    subs(x=-3,diff(f(x),x$4));
```

$$0$$

$$14406$$

$$-32928$$

```
>   subs(x=4,diff(f(x),x$2)); subs(x=4,diff(f(x),x$3));
    subs(x=4,diff(f(x),x$4));
```

$$0$$

$$0$$

$$8232$$

```
>   plot(f(x),x=-4..6);
```

Example 67 *Solution of the single facility location problem.* Location theory deals with finding the optimal geographic location of one or more sources (e.g., plants or factories) to meet the requirements of destinations (e.g., warehouses or markets). Suppose that there are n destinations located at the coordinates (x_i, y_i), $i = 1, \ldots, n$, and we wish to locate optimally a source at a point with coordinates (a, b). The Euclidean distance from the source to destination i is given as $d_i = \sqrt{(a - x_i)^2 + (b - y_i)^2}$. We assume that c_i is the cost of shipping one unit per kilometer ($/unit/distance) from the source to destination i and m_i is the requirement (in units) at destination i. Thus, the total cost of shipments as a function of c_i, m_i and the distance is $S(a, b) = \sum_{i=1}^{n} c_i m_i \sqrt{(a - x_i)^2 + (b - y_i)^2}$, which we wish to minimize. Unfortunately, due to the presence of the square root function, it is impossible to find a closed-form solution for this problem; see Love, Morris and Wesolowsky [126, Chapter 2]. We demonstrate this using Maple.

```
>    restart: # locgen.mws
>    S:=sum(c[ i] *m[ i] *' sqrt' ((a-x[ i] )^2+(b-y[ i] )^2),
     i=1..n);
```

$$S := \sum_{i=1}^{n} c_i \, m_i \, \sqrt{(a - x_i)^2 + (b - y_i)^2}$$

```
>    Sa:=diff(S,a);
```

$$Sa := \sum_{i=1}^{n} \left(\frac{1}{2} \frac{c_i \, m_i \, (2\,a - 2\,x_i)}{\sqrt{a^2 - 2\,a\,x_i + x_i{}^2 + b^2 - 2\,b\,y_i + y_i{}^2}} \right)$$

```
>    Sb:=diff(S,b);
```

$$Sb := \sum_{i=1}^{n} \left(\frac{1}{2} \frac{c_i \, m_i \, (2\,b - 2\,y_i)}{\sqrt{a^2 - 2\,a\,x_i + x_i{}^2 + b^2 - 2\,b\,y_i + y_i{}^2}} \right)$$

```
>    solve({ Sa,Sb} ,{ a,b} );
```

Understandably, Maple finds no solution!

The decision variables a and b appear in every term both in the numerator and inside the square root in the denominator of the necessary conditions $\partial S/\partial a \equiv Sa = 0$ and $\partial S/\partial b \equiv Sb = 0$, which makes it impossible to solve these equations analytically. But we can use Maple to easily obtain a *numerical* solution that minimizes the total cost for this unconstrained optimization problem.

We know that the function $S(a, b)$ must be convex, since we have shown that the distance function $\sqrt{a^2 + b^2}$ is convex. As the sum of convex functions is also convex, any numerical solution we find for the necessary conditions $\partial S/\partial a = \partial S/\partial b = 0$ will be the global minimum for this problem.

Consider a problem with $n = 10$ destinations and shipment costs c_i, demands m_i, and the coordinates (x_i, y_i) of the destinations given as follows.

```
>    restart: # location.mws
>    c:=[ 21,1,3,2,2,3,4,3,2,15] ;
```

$$c := [21, 1, 3, 2, 2, 3, 4, 3, 2, 15]$$

```
>   m:=[ 40,70,56,38,59,83,38,57,86,19] ;
```

$$m := [40, 70, 56, 38, 59, 83, 38, 57, 86, 19]$$

```
>   x:=[ 1,6,9,13,8,17,11,6,8,9] ;
```

$$x := [1, 6, 9, 13, 8, 17, 11, 6, 8, 9]$$

```
>   y:=[ 8,9,4,13,8,6,1,13,9,10] ;
```

$$y := [8, 9, 4, 13, 8, 6, 1, 13, 9, 10]$$

The objective function S is written conveniently with Maple's sum() function:

```
>   S:=sum(c[ i] *m[ i] *sqrt((a-x[ i] )^2+(b-y[ i] )^2),
    i=1..10);
```

$$
\begin{aligned}
S := {}& 840\sqrt{a^2 - 2a + 65 + b^2 - 16b} + 70\sqrt{a^2 - 12a + 117 + b^2 - 18b} \\
&+ 168\sqrt{a^2 - 18a + 97 + b^2 - 8b} \\
&+ 76\sqrt{a^2 - 26a + 338 + b^2 - 26b} \\
&+ 118\sqrt{a^2 - 16a + 128 + b^2 - 16b} \\
&+ 249\sqrt{a^2 - 34a + 325 + b^2 - 12b} \\
&+ 152\sqrt{a^2 - 22a + 122 + b^2 - 2b} \\
&+ 171\sqrt{a^2 - 12a + 205 + b^2 - 26b} \\
&+ 172\sqrt{a^2 - 16a + 145 + b^2 - 18b} \\
&+ 285\sqrt{a^2 - 18a + 181 + b^2 - 20b}
\end{aligned}
$$

```
>   Sa:=diff(S,a): Sb:=diff(S,b):
```

Next, we use Maple's floating point solver function fsolve() to compute the optimal solution.

```
>   fsolve({ Sa,Sb} ,{ a,b} );
```

$$\{a = 6.966982752, \ b = 8.539483288\}$$

```
>   sol:=assign(%):
>   subs(sol,S);
```

$$12216.13251$$

The optimal location of the source is thus found to be at the coordinates $(a, b) = (6.96, 8.53)$ with a total minimum cost of $S_{min} = \$12,216.13$.

5.4 Inequality and Equality Constrained Optimization

We now extend the above results to the following general nonlinear programming problem with inequality and equality constraints:

Problem P : min $f(\mathbf{x})$
          s.t. $g_i(\mathbf{x}) \leq 0, i = 1, \ldots, m$
                 $h_i(\mathbf{x}) = 0, i = 1, \ldots, p$
                 $\mathbf{x} \in S.$

With the inclusion of these constraints, the optimization problem becomes substantially more difficult to solve. Nevertheless, by using the theory developed by H.W. Kuhn and A.W. Tucker [114], we can develop necessary (and sufficient) conditions to solve these problems.

Theorem 8 *Kuhn-Tucker (K-T) Necessary Conditions For the Problem P let $f(\mathbf{x})$, $g_i(\mathbf{x})$, $i = 1, \ldots, m$ and $h_i(\mathbf{x})$, $i = 1, \ldots, p$ be continuously differentiable. Suppose that $\bar{\mathbf{x}}$ is a feasible point that solves P locally and that the gradient vectors $\nabla g_i(\bar{\mathbf{x}})$ for $i \in B = \{i : g_i(\bar{\mathbf{x}}) = 0\}$ and $\nabla h_i(\bar{\mathbf{x}})$ are linearly independent. Then there exist scalars λ_i, $i = 1, \ldots, m$ and μ_i, $i = 1, \ldots, p$ such that*

$$\nabla f(\bar{\mathbf{x}}) + \sum_{i=1}^{m} \lambda_i \nabla g_i(\bar{\mathbf{x}}) + \sum_{i=1}^{p} \mu_i \nabla h_i(\bar{\mathbf{x}}) \;\; = \;\; \mathbf{0} \tag{5.1}$$

$$\lambda_i g_i(\bar{\mathbf{x}}) \;\; = \;\; 0, \; i = 1, \ldots, m \tag{5.2}$$

$$\lambda_i \;\; \geq \;\; 0, \; i = 1, \ldots, m, \tag{5.3}$$

with the original structural constraints

$$g_i(\bar{\mathbf{x}}) \;\; \leq \;\; 0, \; i = 1, \ldots, m, \tag{5.4}$$

$$h_i(\bar{\mathbf{x}}) \;\; = \;\; 0, \; i = 1, \ldots, p. \tag{5.5}$$

∎

The scalars λ_i, $i = 1, \ldots, m$ and μ_i, $i = 1, \ldots, p$ are known as **Lagrange multipliers**, and they have an important economic interpretation that will be discussed in Section 5.4.4 below. The conditions $\lambda_i g_i(\bar{\mathbf{x}}) = 0$, for $i = 1, \ldots, m$ are known as the **complementary slackness** conditions, stating that if the constraint $g_i(\bar{\mathbf{x}})$ is not binding, i.e., if $g_i(\bar{\mathbf{x}}) < 0$, then $\lambda_i = 0$. Similarly, if $\lambda_i > 0$, then the constraint must be binding, i.e., $g_i(\bar{\mathbf{x}}) = 0$. Finally, note that there is no sign restriction on the multipliers μ_i that correspond to the equality constraints.

Theorem 8 provides necessary conditions for the optimal solution. For a discussion of the sufficient conditions that involve the forms of the functions, i.e., pseudoconvexity of f and quasiconvexity of g_i and h_i, see Bazaraa and Shetty [18, Section 4.3].

In general, the solution of the conditions is not a trivial matter since it involves solving a *system of nonlinear equalities and inequalities*. But Maple's `solve()` function can deal with such a system and compute the values of all the variables exactly. We demonstrate this with the portfolio selection problem.

Example 68 *Solution of the portfolio problem using Maple.* Recall that in this problem the objective was to minimize $f(x_1, x_2) = 0.15x_1^2 + 0.06x_2^2 + 0.14x_1x_2$

subject to the constraints $g(x_1, x_2) = 900 - 0.10x_1 - 0.05x_2 \le 0$, $h(x_1, x_2) = 10000 - x_1 - x_2 = 0$ and $x_1 \ge 0, x_2 \ge 0$.

After inputting the data for the objective function and the constraints, we check and see that f is strictly convex since its Hessian is positive definite. This assures that the solution that will be found by Maple is the global optimum.

```
>   restart: # portfoli.mws
>   with(linalg):
```

Warning, new definition for norm

Warning, new definition for trace

```
>   f:=(15/100)*x[ 1] ^2+(6/100)*x[ 2] ^2
    +(14/100)*x[ 1] *x[ 2] ;
```

$$f := \frac{3}{20} x_1{}^2 + \frac{3}{50} x_2{}^2 + \frac{7}{50} x_1 x_2$$

```
>   g:=900-(1/10)*x[ 1] -(5/100)*x[ 2] ;    # g <= 0
```

$$g := 900 - \frac{1}{10} x_1 - \frac{1}{20} x_2$$

```
>   h:=10000-x[ 1] -x[ 2] ;   # h = 0
```

$$h := 10000 - x_1 - x_2$$

```
>   vars:=[ x[ 1] ,x[ 2]] :
>   H:=hessian(f,vars); definite(H,' positive_def' );
```

$$H := \begin{bmatrix} \dfrac{3}{10} & \dfrac{7}{50} \\[2mm] \dfrac{7}{50} & \dfrac{3}{25} \end{bmatrix}$$

true

```
>   grad_f:=grad(f,vars):
>   grad_g:=grad(g,vars):
>   grad_h:=grad(h,vars):
>   eq[ 1] :=grad_f[ 1] +lambda* grad_g[ 1] +mu* grad_h[ 1] ;
```

$$eq_1 := \frac{3}{10} x_1 + \frac{7}{50} x_2 - \frac{1}{10} \lambda - \mu$$

```
>   eq[ 2] :=grad_f[ 2] +lambda* grad_g[ 2] +mu* grad_h[ 2] ;
```

$$eq_2 := \frac{3}{25} x_2 + \frac{7}{50} x_1 - \frac{1}{20} \lambda - \mu$$

```
>   comp_slack:=lambda* g=0;
```

$$comp\_slack := \lambda \left(900 - \frac{1}{10} x_1 - \frac{1}{20} x_2 \right) = 0$$

```
>   structural[ 1] :=g<=0;
```

$$structural_1 := 900 - \frac{1}{10}x_1 - \frac{1}{20}x_2 \leq 0$$

```
> structural[ 2] :=h=0;
```

$$structural_2 := 10000 - x_1 - x_2 = 0$$

```
> solve({ eq[ 1] ,eq[ 2] ,comp_slack,lambda>=0,
  structural[ 1] ,structural[ 2] ,
  x[ 1] >=0,x[ 2] >=0} );
```

$$\{\lambda = 26400, \ \mu = 40, \ x_1 = 8000, \ x_2 = 2000\}$$

```
> sol:=assign(%):
```

```
> subs(sol,f);
```

$$12080000$$

The results indicate that it is optimal to invest $x_1 = \$8000$ in the first stock and $x_2 = \$2000$ in the second stock, giving rise to a portfolio variance of 12080000.

5.4.1   Geometric Interpretation of Kuhn-Tucker Conditions

There is an interesting geometric interpretation of the K-T conditions $\nabla f(\bar{x}) + \sum_{i=1}^{m} \lambda_i \nabla g_i(\bar{x}) + \sum_{i=1}^{p} \mu_i \nabla h_i(\bar{x}) = 0$. Defining $B = \{i : g_i(\bar{x}) = 0\}$ as the set of binding inequality constraints and rewriting this condition, we have

$$\sum_{i \in B} \lambda_i \nabla g_i(\bar{x}) + \sum_{i=1}^{p} \mu_i \nabla h_i(\bar{x}) = -\nabla f(\bar{x}).$$

In other words, the (negative) of the gradient vector $\nabla f(\bar{x})$ of the objective function f is a *linear combination* of the gradient vectors $\nabla g_i(\bar{x})$, $i \in B$ of the binding inequality constraints and the gradient vectors $\nabla h_i(\bar{x})$ of the equality constraints. (If any of the inequality constraints were *not* binding, then the corresponding multiplier would be zero due to the complementary slackness conditions.)

For the portfolio problem, at the optimal solution $\bar{x} = (8000, 2000)$ we have

$$\nabla g(\bar{x}) = \left[\frac{-1}{10}, \frac{-1}{20}\right]', \quad \nabla h(\bar{x}) = [-1, -1]' \quad \text{and} \quad \nabla f(\bar{x}) = [2680, 1360]'.$$

Thus, the values we have found for $\lambda = 26400$ and $\mu = 40$ will satisfy

$$\lambda \left[\begin{array}{c} \frac{-1}{10} \\ \frac{-1}{20} \end{array} \right] + \mu \left[\begin{array}{c} -1 \\ -1 \end{array} \right] = - \left[\begin{array}{c} 2680 \\ 1360 \end{array} \right],$$

implying that $\bar{x} = (8000, 2000)$ is a Kuhn-Tucker point. Geometrically, then, the vector $-\nabla f(\bar{x})$ is in the n-dimensional cone spanned by the vectors $\nabla g(\bar{x})$ and $\nabla h(\bar{x})$.

5.4.2 Constraint Qualification

Are there nonlinear programming problems whose optimal solutions do not satisfy the Kuhn-Tucker conditions? Consider the following problem from Kuhn and Tucker [114].

$$
\begin{aligned}
\min \quad & f(x,y) = -x \\
\text{s.t.} \quad & g_1(x,y) = -(1-x)^3 + y \le 0 \\
& g_2(x,y) = -y \le 0.
\end{aligned}
$$

As shown on the graph of the feasible region and the objective function in Figure 5.3, the optimal solution for this problem is clearly at $x = 1$, $y = 0$. But let us attempt to solve this problem using Maple.

```
>   restart: # CONSQUAL.MWS
>   f:=-x; #Minimize
```

$$f := -x$$

```
>   g1:=-(1-x)^3+y; #g1 <= 0
```

$$g1 := -(1-x)^3 + y$$

```
>   g2:=-y;          #g2 <= 0
```

$$g2 := -y$$

```
>   with(linalg):
```

Warning, new definition for norm

Warning, new definition for trace

```
>   vars:=[ x,y] :
>   grad_f:=grad(f,vars): #Gradients
>   grad_g1:=grad(g1,vars):
>   grad_g2:=grad(g2,vars):
>   eq[ 1] :=grad_f[ 1] +lambda[ 1] * grad_g1[ 1]
    +lambda[ 2] * grad_g2[ 1] :
    # Linear independence of gradient vectors ...
>   eq[ 2] :=grad_f[ 2] +lambda[ 1] * grad_g2[ 1]
    +lambda[ 2] * grad_g2[ 2] :
>   comp_slack[ 1] :=lambda[ 1] * g1:
    # Complementary slackness ...
>   comp_slack[ 2] :=lambda[ 2] * g2:
>   structural[ 1] :=g1<=0:
    # Structural constraints ...
>   structural[ 2] :=g2<=0:
>   solve({ eq[ 1] ,eq[ 2] ,comp_slack[ 1] ,comp_slack[ 2] ,
    lambda[ 1] >=0,lambda[ 2] >=0
    } );
```

$$\{\lambda_2 = 0, \ y = 1 - 3x + 3x^2 - x^3, \ \lambda_1 = \frac{1}{3}\frac{1}{(x-1)^2}, \ x = x\}$$

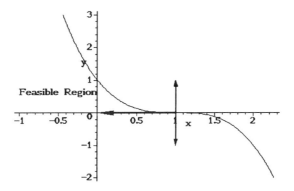

FIGURE 5.3. Optimal solution is at $\bar{x} = (\bar{x}, \bar{y}) = (1, 0)$ where constraint qualification is not satisfied since at $\bar{x}$ the vectors $\nabla g_1(\bar{x}) = [0, 1]'$ (pointing up) and $\nabla g_2(\bar{x}) = [0, -1]'$ (pointing down) are linearly dependent.

As we see, Maple tries its best to find a solution and produces one. But due to the unusual nature of this problem, the solution found by Maple (using Kuhn-Tucker conditions) does not seem right, especially when we note that at $x = 1$, the multiplier $\lambda_1 = \infty$. What went wrong?

What happened in this problem was the *lack* of linear independence of the gradient vectors $\nabla g_i(\bar{x})$, $i = 1, 2$ at the point $\bar{x} = (1, 0)$. Since

$$\nabla g_1(\bar{x}) = [0, 1]', \quad \nabla g_2(\bar{x}) = [0, -1]' \quad \text{and} \quad \nabla f(\bar{x}) = [-1, 0]'.$$

the system

$$\lambda_1 \begin{bmatrix} 0 \\ 1 \end{bmatrix} + \lambda_2 \begin{bmatrix} 0 \\ -1 \end{bmatrix} = -\begin{bmatrix} -1 \\ 0 \end{bmatrix}$$

has no solution, as can be seen from the following Maple output:

```
>   sol:={ x=1,y=0} ; # This is known to be the optimal
solution
```

$$sol := \{y = 0, \ x = 1\}$$

```
>   grad_g1:=subs(sol,evalm(grad_g1));
```

$$grad\_g1 := [0, 1]$$

```
>   grad_g2:=subs(sol,evalm(grad_g2));
```

$$grad\_g2 := [0, -1]$$

```
>   grad_f:=subs(sol,evalm(grad_f));
```

$$grad\_f := [-1, 0]$$

```
>   vector_space:=stackmatrix(grad_g1,grad_g2);
```

$$vector\_space := \begin{bmatrix} 0 & 1 \\ 0 & -1 \end{bmatrix}$$

```
>   linsolve(transpose(vector_space),-grad_f);
```

Maple finds no solution!

The fact that the vectors $[0, 1]'$ and $[0, -1]'$ are linearly *dependent* is shown in the next set of commands where Maple indicates that the matrix formed by these vectors has rank 1 and the solution of the linear system for λ_1, and λ_2 gives $\lambda_1 = \lambda_2 = \_t_1$:

```
>   rank(vector_space);
```

$$1$$

```
>   linsolve(transpose(vector_space),[ 0,0] );
```

$$[\_t_1, \_t_1]$$

Thus, at the optimal solution $\bar{x} = (1, 0)$, the linear independence **constraint qualification** is *not* satisfied, which is the reason for the failure of the Kuhn-Tucker conditions at this optimal point.

The graph of the feasible set in Figure 5.3 can be obtained with the following Maple commands.

```
>   with(plots): with(plottools):
>   Delg1:=arrow([ 1,0] , [ 1,1] ,.01,.05,.2,
    color=black):
>   Delg2:=arrow([ 1,0] ,[ 1,-1] ,.01,.05,.2,
    color=black):
>   Delf:= arrow([ 1,0] , [ 0,0] ,.03,.15,.2,
    color=black):
>   fr:=textplot([ -1,1/2, 'Feasible Region'] ,
    align={ ABOVE,RIGHT} ):
>   g1plot:=implicitplot(g1,x=-2..3,y=-2..3,
    color=black,numpoints=5000,th
    ickness=2):
>   g2plot:=implicitplot(g2,x=-1..2,y=-1..2,
    color=black,numpoints=5000):
>   display({ fr,g1plot,g2plot,Delg1,
    Delg2,Delf} );
```

5.4.3   An Equivalent Formulation Using the Lagrangian

It would be convenient to introduce the **Lagrangian** function L associated with the original constrained problem as

$$L(\mathbf{x}, \boldsymbol{\lambda}, \boldsymbol{\mu}) = f(\mathbf{x}) + \sum_{i=1}^{m} \lambda_i g_i(\mathbf{x}) + \sum_{i=1}^{p} \mu_i h_i(\mathbf{x})$$

where $\lambda = (\lambda_1, \ldots, \lambda_m)$ and $\mu = (\mu_1, \ldots, \mu_p)$. The K-T conditions can now be expressed in terms of the Lagrangian as

$$\frac{\partial L(\mathbf{x}, \lambda, \mu)}{\partial x_j} = \frac{\partial f(\mathbf{x})}{\partial x_j} + \sum_{i=1}^{m} \lambda_i \frac{\partial g_i(\mathbf{x})}{\partial x_j} + \sum_{i=1}^{p} \mu_i \frac{\partial h_i(\mathbf{x})}{\partial x_j} = 0, j = 1, \ldots, n$$

$$\lambda_i \frac{\partial L(\mathbf{x}, \lambda, \mu)}{\partial \lambda_i} = \lambda_i g_i(\mathbf{x}) = 0, i = 1, \ldots, m$$

$$\lambda_i \geq 0, i = 1, \ldots, m$$

$$\frac{\partial L(\mathbf{x}, \lambda, \mu)}{\partial \lambda_i} = g_i(\mathbf{x}) \leq 0, i = 1, \ldots, m$$

$$\frac{\partial L(\mathbf{x}, \lambda, \mu)}{\partial \mu_i} = h_i(\mathbf{x}) = 0, i = 1, \ldots, p.$$

Using this equivalent formulation, we now present another constrained minimization example that can be solved more conveniently with Maple.

Example 69 *A problem with a quadratic objective function and mixed constraints.* Suppose we wish to minimize $f(x_1, x_2) = (x_1 - 3)^2 + (x_2 - 2)^2$, subject to two mixed (i.e., nonlinear and linear) inequality constraints $g_1(x_1, x_2) = x_1^2 + x_2^2 - 5 \leq 0$, and $g_2(x_1, x_2) = x_1 + 2x_2 - 4 \leq 0$. We enter the problem data as

```
>  restart:  # KTMIN.MWS
>  f:=(x[ 1] -3)^2+(x[ 2] -2)^2;  # Minimize
```

$$f := (x_1 - 3)^2 + (x_2 - 2)^2$$

```
>  g[ 1] :=x[ 1] ^2+x[ 2] ^2-5;  # g[ 1]  <= 0
```

$$g_1 := x_1{}^2 + x_2{}^2 - 5$$

```
>  g[ 2] :=x[ 1] +2* x[ 2] -4;   # g[ 2]  <= 0
```

$$g_2 := x_1 + 2x_2 - 4$$

```
>  vars:=[ x[ 1] ,x[ 2]] ;
```

$$vars := [x_1, x_2]$$

and generate the K-T conditions with the following commands:

```
>  L:=f+sum(lambda[ k] * g[ k] ,k=1..2);
```

$$L := (x_1 - 3)^2 + (x_2 - 2)^2 + \lambda_1 (x_1{}^2 + x_2{}^2 - 5) + \lambda_2 (x_1 + 2x_2 - 4)$$

```
>  L_x:=seq(diff(L,x[ k] ),k=1..2);
```

$$L\_x := 2x_1 - 6 + 2\lambda_1 x_1 + \lambda_2, 2x_2 - 4 + 2\lambda_1 x_2 + 2\lambda_2$$

```
>  eq:=seq(lambda[ k] * diff(L,lambda[ k] ),k=1..2);
```

$$eq := \lambda_1 (x_1{}^2 + x_2{}^2 - 5), \lambda_2 (x_1 + 2x_2 - 4)$$

```
>  ineq_lambda:=seq(lambda[ k] >=0,k=1..2);
```

$$ineq\_lambda := 0 \leq \lambda_1, 0 \leq \lambda_2$$

```
> ineq_g:=seq(diff(L,lambda[ k] )<=0,k=1..2);
```

$$ineq\_g := x_1^2 + x_2^2 - 5 \le 0, \ x_1 + 2x_2 - 4 \le 0$$

```
> L_x, ineq_g;
```

$$2x_1 - 6 + 2\lambda_1 x_1 + \lambda_2, \ 2x_2 - 4 + 2\lambda_1 x_2 + 2\lambda_2, \ x_1^2 + x_2^2 - 5 \le 0,$$
$$x_1 + 2x_2 - 4 \le 0$$

Next, the `solve()` command is used to find the solution to the K-T conditions:

```
> sol:=solve({ L_x,eq,ineq_g,
  ineq_lambda} ,{ x[ 1] ,x[ 2] ,lambda[ 1] ,lambda[ 2] } );
```

$$sol := \{x_2 = 1, \ x_1 = 2, \ \lambda_1 = \frac{1}{3}, \ \lambda_2 = \frac{2}{3}\}$$

At the optimal solution, the objective function takes the minimum value $f = 2$:

```
> evalf(subs(sol,f));
```

$$2.$$

The following commands are used to show that the constraint qualification is satisfied at the optimal solution:

```
> with(linalg):
```

Warning, new definition for norm

Warning, new definition for trace

```
> grad_f:=grad(f,vars);
```

$$grad\_f := [2x_1 - 6, \ 2x_2 - 4]$$

```
> grad_g[ 1] :=grad(g[ 1] ,vars);
```

$$grad\_g_1 := [2x_1, \ 2x_2]$$

```
> grad_g[ 2] :=grad(g[ 2] ,vars);
```

$$grad\_g_2 := [1, \ 2]$$

```
> Opt:={ x[ 1] =2,x[ 2] =1} ;
```

$$Opt := \{x_2 = 1, \ x_1 = 2\}$$

```
> grad_f:=subs(Opt,evalm(grad_f));
```

$$grad\_f := [-2, \ -2]$$

```
> grad_g[ 1] :=subs(Opt,evalm(grad_g[ 1] ));
```

$$grad\_g_1 := [4, \ 2]$$

```
> grad_g[ 2] :=subs(Opt,evalm(grad_g[ 2] ));
```

$$grad\_g_2 := [1, \ 2]$$

```
> A:=transpose(matrix(2,2,[ grad_g[ 1] ,grad_g[ 2] ] ));
```

$$A := \begin{bmatrix} 4 & 1 \\ 2 & 2 \end{bmatrix}$$

```
>  linsolve(A,-grad_f);
```

$$\begin{bmatrix} \dfrac{1}{3}, & \dfrac{2}{3} \end{bmatrix}$$

The final result indicates that the gradient of the objective function can be expressed as a linear combination of the gradients of the constraints evaluated at $\bar{x} = (2, 1)$ when $\lambda = (\frac{1}{3}, \frac{2}{3})$. The graph of the feasible set, the objective function and the optimal solution presented in Figure 5.4 are plotted using the next set of Maple commands.

```
>  with(plots):
>  f:=subs({ x[ 1] =x,x[ 2] =y} ,f);
```

$$f := (x - 3)^2 + (y - 2)^2$$

```
>  g1:=subs({ x[ 1] =x,x[ 2] =y} ,g[ 1] );
```

$$g1 := x^2 + y^2 - 5$$

```
>  g2:=subs({ x[ 1] =x,x[ 2] =y} ,g[ 2] );
```

$$g2 := x + 2y - 4$$

```
>  solve({ g1,g2} );
```

$$\{y = 1,\ x = 2\},\ \{y = \frac{11}{5},\ x = \frac{-2}{5}\}$$

```
>  feas_g1:=implicitplot(g1,x=2..3,y=-1..1):
>  feas_g2:=implicitplot(g2,x=-1..2,y=1..3):
>  obj:=contourplot(f,x=-2..6,y=-2..3,contours=
   [ 0.5,1,1.5,2,2.5] ,coloring=[ white,blue] ):
>  display({ feas_g1,feas_g2,obj} ,
   scaling=constrained);
```

Example 70 *An application in microeconomic theory: The Cobb-Douglas production function.* Consider a simple production function that is defined as $q = f(x_1, x_2)$ where $x_1 > 0$ and $x_2 > 0$ are the variable inputs (e.g., labor and capital) and q is the quantity of the output. A special class of these functions known as the Cobb-Douglas (CD) production function [48] is defined as $q = Ax_1^{a_1} x_2^{a_2}$ where $a_1 > 0$ and $a_2 > 0$. The CD function has an interesting homogeneity property: If the inputs are each increased by a factor of θ, we have $q_\theta = A(\theta x_1)^{a_1}(\theta x_2)^{a_2} = A\theta^{(a_1+a_2)}x_1^{a_1}x_2^{a_2}$. Hence, if $a_1 + a_2 = 1$, the function is linear homogeneous and we have constant returns to scale; if $a_1 + a_2 > 1 (< 1)$ we have increasing (decreasing) returns to scale [87], [94].

In this example, we discuss an optimization problem that involves the CD production function. Suppose that the desired output is specified as q and that each unit of input x_1 costs $\$c_1$ and each unit of input x_2 costs $\$c_2$. Thus, we wish

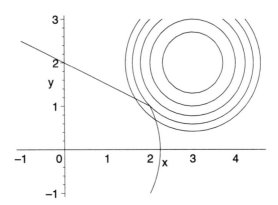

FIGURE 5.4. The feasible set, objective function and the optimal solution of the problem with a quadratic objective function and mixed constraints.

to minimize the total cost $C = c_1x_1 + c_2x_2$ of producing q units subjcct to the constraint that $Ax_1^{a_1}x_2^{a_2} = q$.

We first form the Lagrangian as $L = c_1x_1 + c_2x_2 + \mu(q - Ax_1^{a_1}x_2^{a_2})$. Using Maple's implicit differentiation function, we observe that the curve $q = Ax_1^{a_1}x_2^{a_2}$ in the (x_1, x_2) plane is decreasing and strictly convex. Since the cost function $C = c_1x_1 + c_2x_2$ is linear, this implies that the solution found from the application of the Kuhn-Tucker theorem will give the unique minimizing point.

```
>  restart: # cobbdouglas.mws
>  L:=c[ 1] * x[ 1] +c[ 2] * x[ 2]
   +mu* (q-A* x[ 1] ^a[ 1] * x[ 2] ^a[ 2] ) ;
```

$$L := c_1 x_1 + c_2 x_2 + \mu (q - A x_1^{a_1} x_2^{a_2})$$

The first-order implicit derivative is obtained as a negative quantity implying that x_2 is decreasing in x_1,

```
>  implicitdiff(q-A* x[ 1] ^a[ 1] * x[ 2] ^a[ 2] ,
   x[ 1] ,x[ 2] ) ;
```

$$-\frac{a_2 x_1}{a_1 x_2}$$

whereas the second order implicit derivative is always positive implying that x_2 is a strictly convex function of x_1.

```
>  implicitdiff(q-A* x[ 1] ^a[ 1] * x[ 2] ^a[ 2] ,
   x[ 1] ,x[ 2] $2) ;
```

$$\frac{a_2 x_1 (a_2 + a_1)}{a_1^2 x_2^2}$$

```
>  e1:=diff(L,x[ 1] )=0;
```

$$e1 := c_1 - \frac{\mu A x_1^{a_1} a_1 x_2^{a_2}}{x_1} = 0$$

```
>   e2:=diff(L,x[ 2] )=0;
```

$$e2 := c_2 - \frac{\mu \, A \, x_1{}^{a_1} \, x_2{}^{a_2} \, a_2}{x_2} = 0$$

```
>   e3:=diff(L,mu)=0;
```

$$e3 := q - A \, x_1{}^{a_1} \, x_2{}^{a_2} = 0$$

The necessary conditions give rise to a system of three nonlinear equations in three unknowns, which is solved easily by Maple's solve() command.

```
>   Sol:=solve({ e1,e2,e3} ,{ x[ 1] ,x[ 2] ,mu} );
```

$$Sol := \left\{ x_1 = e^{\left(\frac{\ln(\frac{c_2 \, a_1}{a_2 \, c_1}) a_2 + \ln(\frac{q}{A})}{a_1 + a_2} \right)}, \quad \mu = \frac{c_2 \, e^{\left(\frac{-\ln(\frac{q}{A}) + a_1 \ln(\frac{c_2 \, a_1}{a_2 \, c_1})}{a_1 + a_2} \right)}}{q \, a_2}, \right.$$

$$\left. x_2 = e^{\left(-\frac{-\ln(\frac{q}{A}) + a_1 \ln(\frac{c_2 \, a_1}{a_2 \, c_1})}{a_1 + a_2} \right)} \right\}$$

Expanding to clear the logarithms and simplifying gives the analytic solution for the optimal values of the inputs x_1 and x_2 and the Lagrange multiplier μ.

```
>   expand(simplify(Sol,power,symbolic));
```

$$\left\{ x_2 = \frac{q^{(\frac{1}{a_1 + a_2})} \, a_2^{(\frac{a_1}{a_1 + a_2})} \, c_1^{(\frac{a_1}{a_1 + a_2})}}{A^{(\frac{1}{a_1 + a_2})} \, c_2^{(\frac{a_1}{a_1 + a_2})} \, a_1^{(\frac{a_1}{a_1 + a_2})}}, \quad \mu = \frac{c_2 \, q^{(\frac{1}{a_1 + a_2})} \, a_2^{(\frac{a_1}{a_1 + a_2})} \, c_1^{(\frac{a_1}{a_1 + a_2})}}{A^{(\frac{1}{a_1 + a_2})} \, c_2^{(\frac{a_1}{a_1 + a_2})} \, a_1^{(\frac{a_1}{a_1 + a_2})} \, q \, a_2}, \right.$$

$$\left. x_1 = \frac{c_2^{(\frac{a_2}{a_1 + a_2})} \, a_1^{(\frac{a_2}{a_1 + a_2})} \, q^{(\frac{1}{a_1 + a_2})}}{a_2^{(\frac{a_2}{a_1 + a_2})} \, c_1^{(\frac{a_2}{a_1 + a_2})} \, A^{(\frac{1}{a_1 + a_2})}} \right\}$$

5.4.4   Economic Interpretation of the Lagrange Multipliers

So far we have treated the Lagrange multipliers as a purely algebraic device to convert a constrained problem into an unconstrained one. The solution of the problem assigned a nonnegative value to these multipliers, but we did not provide a discussion of the economic interpretation of Lagrange multipliers.

In the portfolio selection problem of Example 68 we had found that $\lambda = 26400$ and $\mu = 40$, and in Example 69 with a quadratic objective function with mixed constraints we had $\lambda_1 = 1/3$ and $\lambda_2 = 2/3$. One may justifiably ask whether these values have an economic meaning related to the constraints they are attached to. The answer, as in linear programming, is in the affirmative and the Lagrange multipliers provide information on the *approximate improvement* in the objective function if a particular constraint's RHS constant is increased by a *small amount*.

If, for example, in the portfolio problem we were to increase the RHS of the first constraint by one unit, i.e., $900 - 0.10x_1 - 0.05x_2 \leq 1$ or, equivalently, $g_1(x_1, x_2) = 899 - 0.10x_1 - 0.05x_2 \leq 0$, and solve the problem again, we would obtain the optimal solution as $x_1 = 7980$, $x_2 = 2020$, and $f = 12053628$, a reduction (improvement) in the objective function (i.e., portfolio variance) of $12,080,000 - 12,053,628 = 26,372$ units. This difference is very close to the value of the Lagrange multiplier $\lambda = 26400$ found as the solution of the original problem.

We formalize this important result in the following theorem.

Theorem 9 *[127, p. 317] Consider the problem*
$$\text{Problem } P: \quad \min \quad f(\mathbf{x})$$
$$\text{s.t.} \quad g_i(\mathbf{x}) \leq a_i, \, i = 1, \ldots, m$$
$$h_i(\mathbf{x}) = b_i, \, i = 1, \ldots, p.$$
Suppose that for $\mathbf{a} = (a_1, \ldots, a_m) = \mathbf{0}$, $\mathbf{b} = (b_1, \ldots, b_p) = \mathbf{0}$ *there is a solution* $\bar{\mathbf{x}}$ *that satisfies the constraint qualification with the Lagrange multipliers* $\lambda_i \geq 0$, $i = 1, \ldots, m$, *and* μ_i, $i = 1, \ldots, p$. *Then for every* $(\mathbf{a}, \mathbf{b}) \in R^{m+p}$ *in the neighborhood of* $(\mathbf{0}, \mathbf{0})$ *there is a solution* $\mathbf{x}(\mathbf{a}, \mathbf{b})$ *depending continuously on* $(\mathbf{a}, \mathbf{b})$ *such that* $\mathbf{x}(\mathbf{0}, \mathbf{0}) = \bar{\mathbf{x}}$ *and such that* $\mathbf{x}(\mathbf{a}, \mathbf{b})$ *is a relative minimum of the problem P. Moreover,*

$$\left. \frac{\partial f}{\partial a_i}(\mathbf{x}(\mathbf{a}, \mathbf{b})) \right|_{0,0} = -\lambda_i, \, i = 1, \ldots, m$$

$$\left. \frac{\partial f}{\partial b_i}(\mathbf{x}(\mathbf{a}, \mathbf{b})) \right|_{0,0} = -\mu_i, \, i = 1, \ldots, p.$$

∎

Thus, in Example 69 (ktmin.mws), if the first constraint is written as $x_1^2 + x_2^2 - 5 \leq 0.1$ or equivalently as $g_1(x_1, x_2) = x_1^2 + x_2^2 - 5.1 \leq 0$, the value $\Delta a_1 \times \lambda_1 = (0.1) \times 1/3 = 1/30$ should be the *approximate improvement* (i.e., reduction) in the optimal value of the objective function $f(x_1, x_2)$. This increase in the RHS of the first constraint by 0.1 units enlarges the feasible set and should improve the objective function by *approximately* 0.033 units as we see in the following Maple output. (The *actual* improvement is equal to $2 - 1.968468797 = 0.031531203$ units.)

```
>   restart: # KTMin2.mws
>   f:=(x[ 1] -3)^2+(x[ 2] -2)^2: # Minimize
>   g[ 1] :=x[ 1] ^2+x[ 2] ^2-5.1; # g[ 1]  <= 0
```

$$g_1 := x_1^2 + x_2^2 - 5.1$$

```
>   g[ 2] :=x[ 1] +2* x[ 2] -4:   # g[ 2]  <= 0
>   vars:=[ x[ 1] ,x[ 2] ] :
>   L:=f+sum(lambda[ k] * g[ k] ,k=1..2):
>   L_x:=seq(diff(L,x[ k] ),k=1..2):
```

```
>   eq:=seq(lambda[ k] *diff(L,lambda[ k] ),k=1..2):
>   ineq_lambda:=seq(lambda[ k] >=0,k=1..2):
>   ineq_g:=seq(diff(L,lambda[ k] )<=0,k=1..2):
>   L_x, ineq_g:
>   sol:=solve({ L_x,eq,ineq_g,
    ineq_lambda} ,{ x[ 1] ,x[ 2] ,lambda[ 1] ,lambda[ 2]} );
```

$$sol := \{\lambda_1 = .2977713690, \ x_2 = .9835585997, \ \lambda_2 = .7235658095,$$
$$x_1 = 2.032882801\}$$

```
>   evalf(subs(sol,f));
```

$$1.968468797$$

5.5   Lagrangian Duality

In Section 5.4.3 we presented the Lagrangian as

$$L(\mathbf{x}, \lambda, \mu) = f(\mathbf{x}) + \sum_{i=1}^{m} \lambda_i g_i(\mathbf{x}) + \sum_{i=1}^{p} \mu_i h_i(\mathbf{x})$$

without providing any geometric interpretation for this function. In this section we show that the Lagrangian has the important geometric property that it has a *saddle point* at the optimal solution $(\bar{\mathbf{x}}, \bar{\lambda}, \bar{\mu})$. To motivate the discussion, we first present a very simple example and plot the surface of L to observe its three-dimensional shape.

Consider the problem of minimizing $f(x) = (x - 2)^2 + 1$ subject to $g(x) = x - 1 \leq 0$ for which the obvious optimal solution is $\bar{x} = 1$. The Lagrangian is formed as $L(x, \lambda) = (x - 2)^2 + 1 + \lambda(x - 1)$, which is a function of two variables (x, λ). Using Maple's plot3d() function, we plot the surface of the Lagrangian for the range x=-2..3 and lambda=0..4 and obtain the graph of the Lagrangian depicted in Figure 5.5.[2]

```
>   restart: # saddle.mws
>   L:=(x-2)^2+1+lambda* (x-1);
```

$$L := (x - 2)^2 + 1 + \lambda(x - 1)$$

```
>   plot3d(L,x=-2..3,lambda=0..4,axes=boxed,
    shading=none,orientation=[ -152,70] ,
    labels=[ "","",""] );
```

This graph suggests that the Lagrangian appears to have a saddle point at around $(x, \lambda) = (1, 2)$, i.e., that $L(1, \lambda) \leq L(1, 2) \leq L(x, 2)$. Indeed, when

---

[2]Maple cannot use different fonts (such as the Roman x and Greek λ) to display the labels in the same graph. For this reason, the axis labels in Figure 5.5 are suppressed. Note, however, that x takes values between -2 and 3 and λ between 0 and 4.

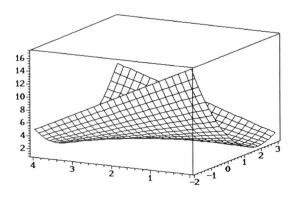

FIGURE 5.5. The Lagrangian function $L = (x - 2)^2 + 1 + \lambda(x - 1)$ has a saddle point at $(x, \lambda) = (1, 2)$.

we apply the eigenvalue test to the Hessian of L, we obtain one positive and one negative eigenvalue, implying that the Lagrangian does have a saddle point:

```
>    with(linalg):
```

Warning, new definition for norm

Warning, new definition for trace

```
>    H:=hessian(L,[ x,lambda] );
```

$$H := \begin{bmatrix} 2 & 1 \\ 1 & 0 \end{bmatrix}$$

```
>    eigenvalues(H);
```

$$1 + \sqrt{2}, \ 1 - \sqrt{2}$$

We should now be able to compute the saddle point of the Lagrangian, i.e., the x value that minimizes L and the λ value that maximizes L. We do this using the following Maple commands:

```
>    restart: # SADDUAL.MWS
>    f:=(x-2)^2+1; # Minimize
```

$$f := (x - 2)^2 + 1$$

```
>    g:=x-1; # g <= 0
```

$$g := x - 1$$

```
>    L:=f+lambda*g;
```

$$L := (x - 2)^2 + 1 + \lambda (x - 1)$$

```
>    diff(L,x);
```

$$2x - 4 + \lambda$$

```
>    xsol:=solve(%,x);
```

$$xsol := 2 - \frac{1}{2}\lambda$$

```
>    subs(x=xsol,L);
```

$$\frac{1}{4}\lambda^2 + 1 + \lambda\left(1 - \frac{1}{2}\lambda\right)$$

```
>    diff(%,lambda);
```

$$1 - \frac{1}{2}\lambda$$

```
>    lambdasol:=solve(%,lambda);
```

$$lambdasol := 2$$

```
>    xsol:=subs(lambda=lambdasol,xsol);
```

$$xsol := 1$$

These computations show that $L(1, \lambda) \le L(1, 2) \le L(x, 2)$.

To compare the results obtained by our graph based on heuristic arguments, we re-solve the same problem using Kuhn-Tucker conditions (5.1)–(5.5):

```
>    restart: # SADDLEKT.MWS
>    f:=(x-2)^2+1; # Minimize
```

$$f := (x - 2)^2 + 1$$

```
>    g:=x-1; # g <= 0
```

$$g := x - 1$$

```
>    L:=f+lambda*g;
```

$$L := (x - 2)^2 + 1 + \lambda(x - 1)$$

```
>    L_x:=diff(L,x);
```

$$L\_x := 2x - 4 + \lambda$$

```
>    eq:=lambda*diff(L,lambda);
```

$$eq := \lambda(x - 1)$$

```
>    ineq_lambda:=lambda>=0;
```

$$ineq\_lambda := 0 \le \lambda$$

```
>    ineq_g:=diff(L,lambda)<=0;
```

$$ineq\_g := x - 1 \le 0$$

```
>    sol:=solve({ L_x,eq,ineq_g, ineq_lambda},
     { x,lambda} );
```

$$sol := \{\lambda = 2, x = 1\}$$

Interestingly, for this problem the solution obtained using the saddle-point analysis of the Lagrangian and the Kuhn-Tucker conditions agree. This is *not* a coincidence and when the objective function and the constraint satisfy certain regularity conditions (e.g., convexity of the functions involved), there is a valid intimate relationship between the saddle-point criteria and the K-T conditions. This is stated in the following theorem:

Theorem 10 *[18, Section 6.2] The feasible point $\bar{x}$ for Problem P and the Lagrange multipliers $\bar{\lambda} \geq 0$ and $\bar{\mu}$ satisfy the Kuhn-Tucker conditions (5.1)–(5.5), if and only if $(\bar{x}, \bar{\lambda}, \bar{\mu})$ satisfy the saddle-point conditions $L(\bar{x}, \lambda, \mu) \leq L(\bar{x}, \bar{\lambda}, \bar{\mu}) \leq L(x, \bar{\lambda}, \bar{\mu})$.* ∎

In the simple problem of minimizing $f(x) = (x - 2)^2 + 1$ subject to $g(x) = x - 1 \leq 0$, we used Maple to differentiate $L(x, \lambda)$ with respect to x for any given value of λ. This gave $x(\lambda) = 2 - \lambda/2$ as the optimal value of x in terms of λ. When we substituted $x(\lambda)$ into the Lagrangian we obtained a new function $\ell(\lambda) = \lambda^2/4 + 1 + \lambda(1 - \lambda/2)$. Maximizing ℓ with respect to $\lambda \geq 0$ gave $\lambda = 2$ which, when substituted back into $x(\lambda)$, produced the optimal value of $x = 1$. Thus, at the optimal values found by this approach, we obtained $f(x = 1) = 2 = \ell(\lambda = 2)$, i.e., $\min f = \max \ell$.

Although the approach we used to find the optimal solution via a derived function such as $\ell(\lambda)$ may seem arbitrary, it has wide-ranging implications in theoretical and computational aspects of nonlinear programming: The function $\ell(\cdot)$ is known as the **Lagrangian dual**, and its maximum and the minimum of the original (primal) problem's objective function $f(x)$ are intimately related.

We define the Lagrangian dual D problem as

$$\text{max} \quad \ell(\lambda, \mu) = \min \left[f(x) + \sum_{i=1}^{m} \lambda_i g_i(x) + \sum_{i=1}^{p} \mu_i h_i(x) : x \in S \right]$$
$$\text{s.t.} \quad \lambda \geq 0.$$

The relationship between the values of the functions $f(x)$ and $\ell(\lambda, \mu)$ of Problems P (the **primal** problem) and D (the **dual** problem) is given in the following theorem.

Theorem 11 *Duality Theorem* *[18, Section 6.2] Under certain regularity conditions (i.e., convexity of the functions f and g_i, linearity of h_i in terms of x, and a constraint qualification), the following result holds:*

$$\min \left[f(x) : x \in S, \ g_i(x) \leq 0, \ i = 1, \ldots, m, \ h_i(x), i = 1, \ldots, p \right]$$
$$= \ \max \left[\ell(\lambda, \mu) : \lambda \geq 0 \right]$$

Moreover, the function $\ell(\lambda, \mu)$ is concave over any convex subset of its domain.

∎

This theorem can be very useful in solving constrained nonlinear programming problems where the decision vector x can be expressed in terms of (λ, μ) after minimizing the Lagrangian $L(x, \lambda, \mu)$. If this is possible, then $x = x(\lambda, \mu)$ can be substituted back into L to obtain the Lagrangian dual function $\ell(\lambda, \mu)$, which is shown to be concave; hence its global maximum can be found using the standard methods. Finally, once the optimal $(\bar{\lambda}, \bar{\mu})$ are computed, they can be substituted back into $\mathbf{x}$ to generate the optimal value of the decision vector.

As an application of the Lagrangian duality, we now discuss the solution of a discrete-time optimal control problem with quadratic objective and linear state dynamics.

Example 71 *A discrete linear quadratic control problem.* The problem is to minimize the objective

$$J = \sum_{i=0}^{N-1} (q_i x_i^2 + r_i u_i^2) + q_N x_N^2$$

subject to the state dynamics

$$x_{i+1} = a_i x_i + b_i u_i, \quad i = 0, \ldots, N-1$$

where x_0, $\{q_i\}_{i=0}^{N}$, $\{r_i\}_{i=0}^{N-1}$, $\{a_i\}_{i=0}^{N-1}$, $\{b_i\}_{i=0}^{N-1}$ are given constants. To solve this control problem as a nonlinear programming problem, we treat all of the state variables x_i and control variables u_i as decision variables, i.e., we define the decision vector as $x = (x_1, \ldots, x_N, u_0, \ldots, u_{N-1})$.

The Maple commands to solve this problem with $N = 2$ are as follows.

```
>    restart: # LQ.MWS
```

The next set of commands defines the objective and the state dynamics.

```
>    for i from 0 to 1 do c[ i] :=q[ i] * x[ i] ^2
+r[ i] * u[ i] ^2 od;
```

$$c_0 := q_0 x_0^2 + r_0 u_0^2$$

$$c_1 := q_1 x_1^2 + r_1 u_1^2$$

```
>    f:=sum(c[ n] ,n=0..1)+q[ 2] * x[ 2] ^2;
```

$$f := q_0 x_0^2 + r_0 u_0^2 + q_1 x_1^2 + r_1 u_1^2 + q_2 x_2^2$$

```
>    for j from 1 to 2 do h[ j] :=x[ j] -x[ j-1] -u[ j-1] od;
```

$$h_1 := x_1 - x_0 - u_0$$

$$h_2 := x_2 - x_1 - u_1$$

The Lagrangian L is introduced as

```
>    L:=f+sum(mu[ n] * h[ n] ,n=1..2);
```

$$L := q_0 x_0^2 + r_0 u_0^2 + q_1 x_1^2 + r_1 u_1^2 + q_2 x_2^2 + \mu_1 (x_1 - x_0 - u_0)$$
$$+ \mu_2 (x_2 - x_1 - u_1)$$

and differentiated with respect to all decision variables in the decision vector $\mathbf{x}$.

```
>    Lu[ 0] :=diff(L,u[ 0] );
```

$$Lu_0 := 2 r_0 u_0 - \mu_1$$

```
>   Lx[ 1] :=diff(L,x[ 1] );
```

$$Lx_1 := 2\,q_1\,x_1 + \mu_1 - \mu_2$$

```
>   Lu[ 1] :=diff(L,u[ 1] );
```

$$Lu_1 := 2\,r_1\,u_1 - \mu_2$$

```
>   Lx[ 2] :=diff(L,x[ 2] );
```

$$Lx_2 := 2\,q_2\,x_2 + \mu_2$$

The decision variables are computed in terms of the two Lagrange multipliers $\mu_i, i = 1, 2$.

```
>   solxu:=solve({ Lx[ 1] ,Lx[ 2] ,Lu[ 0] ,Lu[ 1]} ,
    { u[ 0] ,x[ 1] ,u[ 1] ,x[ 2]} ) ;
```

$$solxu := \{u_1 = \frac{1}{2}\frac{\mu_2}{r_1},\ x_1 = -\frac{1}{2}\frac{\mu_1 - \mu_2}{q_1},\ u_0 = \frac{1}{2}\frac{\mu_1}{r_0},\ x_2 = -\frac{1}{2}\frac{\mu_2}{q_2}\}$$

```
>   assign(solxu);
```

The Lagrangian dual $\ell(\mu_1, \mu_2)$ is computed and maximized after substituting $\mathbf{x}(\mu_1, \mu_2)$ in L.

```
>   el:=subs(solxu,L);
```

$$el := q_0\,x_0^2 + \frac{1}{4}\frac{\mu_1^2}{r_0} + \frac{1}{4}\frac{(\mu_1 - \mu_2)^2}{q_1} + \frac{1}{4}\frac{\mu_2^2}{r_1} + \frac{1}{4}\frac{\mu_2^2}{q_2}$$
$$+ \mu_1\left(-\frac{1}{2}\frac{\mu_1 - \mu_2}{q_1} - x_0 - \frac{1}{2}\frac{\mu_1}{r_0}\right) + \mu_2\left(-\frac{1}{2}\frac{\mu_2}{q_2} + \frac{1}{2}\frac{\mu_1 - \mu_2}{q_1} - \frac{1}{2}\frac{\mu_2}{r_1}\right)$$

```
>   elmu[ 1] :=diff(el,mu[ 1] );
```

$$elmu_1 := -x_0 + \mu_1\left(-\frac{1}{2}\frac{1}{q_1} - \frac{1}{2}\frac{1}{r_0}\right) + \frac{1}{2}\frac{\mu_2}{q_1}$$

```
>   elmu[ 2] :=diff(el,mu[ 2] );
```

$$elmu_2 := \frac{1}{2}\frac{\mu_1}{q_1} + \mu_2\left(-\frac{1}{2}\frac{1}{q_2} - \frac{1}{2}\frac{1}{q_1} - \frac{1}{2}\frac{1}{r_1}\right)$$

The Lagrange multipliers are found in terms of the model parameters as follows.

```
>   solmu:=solve({ elmu[ 1] ,elmu[ 2]} ,{ mu[ 1] ,mu[ 2]} ) ;
```

$$solmu := \{\mu_1 = -2\frac{x_0\,r_0\,(q_1\,r_1 + q_2\,r_1 + q_2\,q_1)}{q_1\,r_1 + q_2\,r_1 + r_1\,r_0 + q_2\,r_0 + q_2\,q_1},$$
$$\mu_2 = -2\frac{x_0\,r_0\,q_2\,r_1}{q_1\,r_1 + q_2\,r_1 + r_1\,r_0 + q_2\,r_0 + q_2\,q_1}\}$$

```
>   assign(solmu);
```

Finally, the optimal solution for the complete problem, i.e., the optimal values of u_0, x_1, u_1 and x_2, are computed in terms of model parameters.

```
>   uOpt[ 0] :=simplify(u[ 0] );
```

$$uOpt_0 := -\frac{x_0\,(q_1\,r_1 + q_2\,r_1 + q_2\,q_1)}{q_1\,r_1 + q_2\,r_1 + r_1\,r_0 + q_2\,r_0 + q_2\,q_1}$$

>   xOpt[1] :=simplify(x[1]);

$$xOpt_1 := \frac{x_0\,r_0\,(r_1 + q_2)}{q_1\,r_1 + q_2\,r_1 + r_1\,r_0 + q_2\,r_0 + q_2\,q_1}$$

>   uOpt[1] :=simplify(u[1]);

$$uOpt_1 := -\frac{x_0\,r_0\,q_2}{q_1\,r_1 + q_2\,r_1 + r_1\,r_0 + q_2\,r_0 + q_2\,q_1}$$

>   xOpt[2] :=simplify(x[2]);

$$xOpt_2 := \frac{x_0\,r_0\,r_1}{q_1\,r_1 + q_2\,r_1 + r_1\,r_0 + q_2\,r_0 + q_2\,q_1}$$

Although the solution obtained is of an "open-loop" variety, a careful look at the structure of the optimal decision vector reveals that the optimal controls u_i are linear functions of the optimal states x_i at each time; i.e., in this problem we also have a closed-loop feedback control.

>   uOpt[0] /x[0] ;

$$-\frac{q_1\,r_1 + q_2\,r_1 + q_2\,q_1}{q_1\,r_1 + q_2\,r_1 + r_1\,r_0 + q_2\,r_0 + q_2\,q_1}$$

>   uOpt[1] /xOpt[1] ;

$$-\frac{q_2}{r_1 + q_2}$$

For another application of this duality theorem, consider a discrete-time vector optimal control problem with *piecewise-quadratic* objective function containing *dead-zones*, i.e.,

$$\min J = Q_N(\mathbf{x}_N, \hat{\mathbf{x}}_N) + \sum_{i=0}^{N-1}[Q_i(\mathbf{x}_i, \hat{\mathbf{x}}_i) + (\mathbf{u}_i - \hat{\mathbf{u}}_i)'R_i(\mathbf{u}_i - \hat{\mathbf{u}}_i)]$$

subject to

$$\mathbf{x}_{i+1} = \mathbf{A}_i\mathbf{x}_i + \mathbf{B}_i\mathbf{u}_i + \boldsymbol{\xi}_i, i = 0, \ldots, N-1$$

where

$$Q_{ij}(x_{ij}, \hat{x}_{ij}) = \begin{cases} Q_{ij}^-[x_{ij} - (\hat{x}_{ij} - \alpha_{ij})]^2, & \text{if } x_{ij} \leq \hat{x}_{ij} - \alpha_{ij} \\ 0, & \text{if } x_{ij} \in [\hat{x}_{ij} - \alpha_{ij}, \hat{x}_{ij} + \beta_{ij}] \\ Q_{ij}^+[x_{ij} - (\hat{x}_{ij} + \beta_{ij})]^2, & \text{if } x_{ij} \geq \hat{x}_{ij} + \beta_{ij}. \end{cases}$$

In this multidimensional optimal control problem $\mathbf{x}_i$ are $n \times 1$ state vectors, $\mathbf{u}_i$ are $m \times 1$ control vectors, $\boldsymbol{\xi}_i$ are $n \times 1$ "disturbance" vectors, $\mathbf{A}_i$ are $n \times n$ matrices, $\mathbf{B}_i$ are $n \times m$ matrices, $\mathbf{R}_i$ are positive-definite $m \times m$ matrices, Q_{ij}^-, Q_{ij}^+ are strictly positive scalars, and α_{ij} and β_{ij} are nonnegative scalars. This nonstandard control problem with the piecewise-quadratic cost that adds increased generality to the well-known linear quadratic control model has been solved as a nonlinear programming problem using the Lagrangian duality. For details of the solution method and numerical examples with $N = 25$ periods, we refer the reader to the paper by Parlar and Vickson [146].

5.6   Summary

Many realistic decision problems are nonlinear in the sense that the objective function and/or some of the constraints may not be linear. In those cases one would need to use nonlinear programming (NLP) techniques to model and solve the problem. This chapter started with an introduction of the fundamental concepts of NLP such as the convexity of sets and functions. This was followed by a discussion of unconstrained nonlinear optimization. Next, the important problem of optimizing a nonlinear objective function subject to nonlinear constraint(s) was presented. Kuhn-Tucker conditions, constraint qualification and the economic interpretation of the Lagrange multipliers were described with several examples. Once more, Maple's ability to manipulate symbolic quantities was found useful in presenting and solving relevant problems.

5.7   Exercises

1. Let the functions $f_i(\mathbf{x})$, $i = 1, \ldots, k$ be convex over some convex set $S \subset R^n$. Prove that the function $f(\mathbf{x}) = \sum_{i=1}^{k} f_i(\mathbf{x})$ is also convex over S.

2. Consider the function $f(x, y) = xy$. Show that f is not convex over R^2 but it is convex over a line passing through the origin.

3. The average number of backorders function in the continuous-review (Q, r) inventory model is given by

$$B(Q, r) = \frac{\lambda}{Q} \int_r^{\infty} (x - r)k(x)\, dx.$$

   Show that if ξ is a number such that $k(x)$ is nonincreasing for $x \geq \xi$, then $B(Q, r)$ is convex in the region $\{(Q, r) : 0 < Q < \infty \text{ and } \xi \leq r < \infty\}$.

4. Optimize the function $f(x) = 1 - e^{-x^2}$.

5. Consider the function $f(x, y) = x^3 + y^3 + 2x^2 + 4y^2$. Find the stationary points of this function and identify them (as local minimum, local maximum or saddle point).

6. Maximize the function $f(x, y, z) = xyz$ subject to the constraint $(x/a)^2 + (y/b)^2 + (z/c)^2 = 1$ where a, b and c are positive constants. Can you find a geometric interpretation for this problem?

7. Consider the objective function $f(x, y) = x^2 + 4y^3$. Find all the points that are local maxima and local minima for f subject to the equality constraint $x^2 + 2y^2 = 1$.

8. Is the constraint qualification satisfied for the problem of minimizing the distance from the origin to the implicitly defined curve $y^2 = (x - 1)^3$?

9. Consider the following problem.

$$\begin{array}{ll} \min & (x - 1)^2 + (y - 1)^2 \\ \text{s.t.} & \\ & -x + y = \tfrac{1}{2} \\ & x + y \leq 1 \\ & x, y \geq 0 \end{array}$$

(a) Solve this problem graphically.

(b) Check your answer using the Kuhn-Tucker theory.

(c) Find the dual function $\ell(\lambda, \mu)$ and show that it is concave.

10. Minimize

$$J = \sum_{i=0}^{N} (qx_i^2 + ru_i^2) + qx_3^2$$

subject to the state dynamics

$$x_{i+1} = ax_i + bu_i, \quad i = 0, \ldots, N - 1$$

where $N = 2, q = 2, r = 4, a = 1$ and $b = 0.5$.

6
Dynamic Programming

6.1   Introduction

Dynamic programming (DP) is a simple yet powerful approach for solving certain types of sequential optimization problems. Most real-life decision problems are sequential (dynamic) in nature since a decision made now usually affects future outcomes and payoffs. An important aspect of optimal sequential decisions is the desire to balance present costs with the future costs. A decision made now that minimizes the current cost only without taking into account the future costs may not necessarily be the optimal decision for the complete multiperiod problem. For example, in an inventory control problem it may be optimal to order more than the current period's demand and incur high inventory carrying costs now in order to lower the costs of potential shortages that may arise in the future. Thus, in sequential problems it may be optimal to have some "short-term pain" for the prospect of "long-term gain."

Dynamic programming is based on relatively few concepts. The **state** variables of a dynamic process completely specify the process and provide information on all that needs to be known in order to make a decision. For example, in an inventory control problem the state variable x_t may be the inventory level of the product at the start of period t. Additionally, if the supplier of the product is not always available, then the supplier's availability status may also be another state variable describing the inventory system.

In the inventory control example, ordering decisions are made at certain times. Such times are called **stages** and they may be discrete or continuous. If the inventory is managed using a periodic-review system where decisions are made, say, once every week, then the stage variable would be discrete and it may be sequentially numbered by the integers $t = 0, 1, 2, \ldots$. If, however, a continuous-review (transactions reporting) system is in place, then the order decisions may be made at any time and the stage variable would be continuous.

Having defined the state and stage variables, we now introduce the concept of a **decision**, which is an opportunity to change the state variable. For example, if the inventory level at time t is x_t units, then ordering u_t units is a decision that increases the current inventory to $x_t + u_t$ units. The decision to order u_t units in period t results in a **cost** in the current period and depending on the demand w_t, period t may end with negative or positive inventory resulting in shortages or surpluses that may give rise to additional costs for the current period t. If the process continues for, say, N periods into the future, the optimization problem is to find the best order policy that will minimize the total cost of the inventory system. In general, the total cost incurred over N periods can be written as $\sum_{t=0}^{N-1} L_t(x_t, u_t, w_t) + L_N(x_N)$ where $L_t(\cdot)$ is the cost incurred in period t.

The processes that are studied by dynamic programming pass from stage to stage and thus the states undergo a **transformation** represented by the equation $x_{t+1} = f(x_t, u_t, w_t)$. In the inventory example, after a demand realization of w_t units in period t, the next period's state is computed from $x_{t+1} = x_t + u_t - w_t$ where the transformation function takes the form $f(x_t, u_t, w_t) = x_t + u_t - w_t$. Depending on the nature of the demand w_t, the transformation may be deterministic or stochastic. In the former case, the next period's state variable x_{t+1} is known with certainty, but in the latter case x_{t+1} is a random variable whose distribution can be obtained as a function of x_t, u_t, and w_t.

Finally, we define an optimal **policy** as a decision rule that computes an optimal decision u_t for each conceivable value of the state variable x_t at stage $t = 0, 1, \ldots, N - 1$, i.e., $\pi = [\mu_0(x_0), \mu_1(x_1), \ldots, \mu_{N-1}(x_{N-1})]$. Using the optimal policy π, the optimal decision is computed from $u_t = \mu_t(x_t)$. For example, in the inventory problem, an optimal policy may be shown to be of the form

$$\mu_t(x_t) = \begin{cases} S_t - x_t & \text{if } x_t < S_t, \\ 0 & \text{otherwise} \end{cases}$$

where S_t is the "base-stock" level computed using the problem parameters [166]. At any stage, the optimal decision (a number) is computed using the above policy. If at the start of period 4 we have $x_4 = 3$ and, say, $S_4 = 10$, then the optimal order quantity u_4 in period 4 is obtained as $u_4 = \mu_4(3) = 10 - 3 = 7$ units.

Using the "principle of optimality"[1] that was first enunciated by Richard Bellman [21], [23, p. 83], dynamic programming is used to "divide and conquer" a sequential optimization problem by decomposing it into a series of smaller problems. Once the smaller problems are solved, they are combined to obtain the solution of the complete problem.

For example, consider a three-period inventory problem with $t = 0$ as the beginning of December, $t = 1$ as the beginning January and $t = 2$ as the beginning February. Using the "backward-recursion" approach of DP, we would first find the optimal solution for February—the last stage—and compute the optimal decision $u_2 = \mu_2(x_2)$ as a function of any possible value of the inventory x_2 at the start of February. With the information on the optimal decision available, we would then compute the minimum cost $V_2(x_2)$ for February. Stepping back one month, the optimal policy $\mu_1(x_1)$ for January would be found by adding the costs for that month to the minimum cost $V_2(x_2)$ that was found for February. This solution would then be used to compute the minimum cost $V_1(x_1)$ ("cost to go") for periods 1 and 2 (January and February). Finally, the optimal solution for December would be computed taking into account the costs for that month plus the cost to go $V_1(x_1)$ for the remaining months of January and February. Once the optimal policy is found for each month as a function of the entering inventory for that month, as time progresses and state variables x_t are observed, actual decisions would be computed using the optimal policy functions $\mu_t(x_t)$, $t = 0, 1, 2$.

As we indicated, the three-month inventory control example uses the backward-recursion approach of DP. For an interesting discussion of a puzzle and a mathematical proof using the backward approach, see Nemhauser [138, pp. 19–22]. When the process studied is deterministic, it may also be possible to use the "forward-recursion" approach, which can sometimes be more intuitive. For details of the forward approach, we refer the reader to Bertsekas [31, p. 52], Cooper and Cooper [49, p. 114–122], Danø [57, pp. 150–155] and Larson and Casti [117, p. 233].

Sequential optimization problems with discrete stages can be conveniently formulated using the state-equation approach as in Bertsekas [31]. The state transformation of the basic problem is represented by a discrete-time dynamic system

$$x_{t+1} = f_t(x_t, u_t, w_t), \quad t = 0, 1, \ldots, N - 1 \tag{6.1}$$

where x_t is the state (with x_0 a given constant), u_t is the control and w_t is the uncontrollable exogenous disturbance parameter that may be deterministic or random. There may be constraints on the control variables such as $u_t \geq 0$ so that u_t may have to be an element of a constraint space C_t. When the problem is deterministic, there may also be constraints on the state variables so that x_t may have to be an element of the constraint space S_t. If $L_t(x_t, u_t, w_t)$ is the cost incurred

---

[1]"An optimal policy has the property that whatever the initial state and initial decision are, the remaining decisions must constitute an optimal policy with regard to the state resulting from the first decision."

during period t, then the objective would be to minimize

$$\sum_{t=0}^{N-1} L_t(x_t, u_t, w_t) + L_N(x_N), \tag{6.2}$$

which is the total cost function for the N-stage sequential problem. Note that if the disturbance w_t is a random variable, then the proper statement of the objective would involve expectations with respect to all the random variables, i.e.,

$$E_{w_0, w_1, \dots, w_{N-1}} \sum_{t=0}^{N-1} L_t(x_t, u_t, w_t) + L_N(x_N).$$

To solve the sequential optimization problems given by (6.1)–(6.2), we will first define the **cost to go** (or **value**) function $V_t(x_t)$ as the minimum (expected) cost that can be obtained by using an optimal sequence of decisions for the remainder of the process starting from an arbitrary state x_t in an arbitrary stage $t = 0, 1, \dots, N - 1, N$. This function can be written as

$$V_t(x_t) = \min_{u_t, \dots, u_{N-1}} \sum_{i=t}^{N-1} L_i(x_i, u_i, w_i) + L_N(x_N), \quad t = 0, 1, \dots, N - 1$$

$$V_N(x) = L_N(x_N)$$

where the decision u_i must be chosen from the feasible set C_i, $i = t, \dots, N - 1$. Using the separability properties of this additive cost function and the principle of optimality, it can be shown that (see, for example, Bertsekas [31, p. 19], Nemhauser [138, pp. 28–31], Larson and Casti [117, pp. 45–47]) the value function $V_t(x_t)$ is the solution of a functional equation given as

$$V_t(x_t) = \min_{u_t \in C_t} \{L_t(x_t, u_t, w_t) + V_{t+1}[f_t(x_t, u_t, w_t)]\}, t = 0, 1, \dots, N - 1$$

$$V_N(x_N) = L_N(x_N).$$

We note here that when w_t is random, we minimize the expected value of the term inside the brackets, i.e., $E_{w_t}\{L_t(x_t, u_t, w_t) + V_{t+1}[f_t(x_t, u_t, w_t)]\}$, where, of course, the expectation is taken with respect to the probability distribution of w_t. In this case $V_t(x_t)$ is defined as the minimum *expected* cost to go for stages $t, t + 1, \dots, N$.

If the objective function is in the multiplicative form, i.e., if we are trying to optimize $\prod_{t=0}^{N} L_t(x_t, u_t, w_t)$, then the functional equations for the value function are obtained as

$$V_t(x_t) = \min_{u_t \in C_t} \{L_t(x_t, u_t, w_t) \cdot V_{t+1}[f_t(x_t, u_t, w_t)]\}, t = 0, 1, \dots, N - 1$$

$$V_N(x_N) = \min_{u_N \in C_N} L_N(x_N, u_N, w_N).$$

For details of this problem, see Larson and Casti [117, pp. 48–49] and Nemhauser [138, p. 37]. The multiplicative objective was originally used by Bellman and Dreyfus [26] to model some types of reliability problems.

We will now discuss some examples of sequential decision problems and present their analytic solution using Maple. We will see that Maple's ability to manipulate symbolic expressions makes it an ideal tool for extracting the optimal policies for these problems.

6.2   Stagecoach Problem

Our first example is a simple but illustrative deterministic dynamic programming problem that is known in the operations research literature as the "stagecoach problem." It deals with a hypothetical 19th-century stagecoach company that transports passengers from California to New York. Although the starting point (California) and the destination (New York) are fixed, the company can choose the intermediate **states** to visit in each **stage** of the trip.

We assume that the trip is completed in four stages (legs) where stage 1 starts in California, stage 2 starts in one of three states in the Mountain Time Zone (say, Arizona, Utah or Montana), stage 3 starts in one of three states in the Central Time Zone (say, Oklahoma, Missouri or Iowa) and stage 4 starts in one of two states in the Eastern Time Zone (North Carolina or Ohio). When stage 4 ends, the stagecoach reaches New York, which is the final destination.

Since in those days travel by stagecoach was rather dangerous because of attacks by roaming criminals, life insurance was offered to the traveling passengers. Naturally, the cost of the insurance **policy** was higher on those portions of the trip where there was more danger. The stagecoach company thus faced the problem of choosing a route that would be cheapest and thus safest for its passengers.

TABLE 6.1. Data for the stagecoach problem.

The nodes in the diagram in Table 6.1 are denoted by (i, j) where i corresponds to a stage ($i = 1, 2, \ldots, 5$) and j corresponds to a state in state i. Given the state j in stage i, the decision to travel to state k in the next stage $i + 1$ results in a cost (i.e., insurance premium) of $c_{ij}(k)$. These costs are indicated next to the arrows corresponding to each decision. For example, $c_{11}(2) = 4$ and $c_{21}(3) = 6$. The problem is to find the route that results in the minimum cost.

For this problem, the value function V_{ij} is the minimum cost from state (i, j) to the final state $(5, 1)$ using the optimal policy. Thus, the dynamic programming recursive equations are written as

$$V_{51} = 0$$

$$V_{ij} = \min_k \{c_{ij}(k) + V_{i+1,k}\}$$

where the expression inside the parenthesis is minimized by a suitable choice of the decision variable k.

In the Maple worksheet that follows we enter the cost data as lists, e.g., c[1, 1] :=[3, 4, 2] . This means that if we are now in state $(1, 1)$, the cost of going to states $(2,1)$, $(2,2)$ and $(2,3)$ are 3, 4 and 2, respectively.

```
>   restart: # StageCoach.mws
>   c[ 1,1] :=[ 3,4,2] ;
```

$$c_{1,1} := [3, 4, 2]$$

```
>   c[ 2,1] :=[ 4,1,6] ;  c[ 2,2] :=[ 4,2,3] ;
    c[ 2,3] :=[ 8,4,7] ;
```

$$c_{2,1} := [4, 1, 6]$$

$$c_{2,2} := [4, 2, 3]$$

$$c_{2,3} := [8, 4, 7]$$

```
>   c[ 3,1] :=[ 3,3] ;  c[ 3,2] :=[ 3,6] ;  c[ 3,3] :=[ 4,1] ;
```

$$c_{3,1} := [3, 3]$$

$$c_{3,2} := [3, 6]$$

$$c_{3,3} := [4, 1]$$

```
>   c[ 4,1] :=[ 4] ;  c[ 4,2] :=[ 5] ;
```

$$c_{4,1} := [4]$$

$$c_{4,2} := [5]$$

For stage 5, the boundary condition for the value function is trivially obtained as 0.

```
>   i:=5;
```

$$i := 5$$

```
>   V[ i,1] :=0;
```

$$V_{5,1} := 0$$

Stage 4 computations are also trivial since, once in stage 4, the stagecoach must travel to the last state.

```
>   i:=4;
```

$$i := 4$$

```
>   for j from 1 to 2 do
>   V[ i,j] :=min(seq(c[ i,j][ k] +V[ i+1,k] ,k=1..1));
>   TC[ i,j] :=[ seq(c[ i,j][ k] +V[ i+1,k] ,k=1..1)] ;
>   for k from 1 to 1 do
    u[ i,j,k] :=is(TC[ i,j][ k] =V[ i,j] ) od od;
```

$$V_{4,1} := 4$$

$$TC_{4,1} := [4]$$

$$V_{4,2} := 5$$

$$TC_{4,2} := [5]$$

```
>   for j from 1 to 2 do for k from 1 to 1 do
    print([ i,j,k,u[ i,j,k] ,TC[ i,j][ k]] ) od od;
```

$$[4, 1, 1, true, 4]$$

$$[4, 2, 1, true, 5]$$

For stage 3, we find the value function for each state and the corresponding optimal decision. For example, $[3, 1, 1, true, 7]$ means that in state $[3, 1]$ it is optimal to choose $k = 1$, which results in a minimum overall cost of 7. Thus, the word *true* next to a decision (the third component in the list) implies the optimality of that decision. On the other hand, the word *false* implies that the corresponding decision is not optimal.

```
>   i:=3;
```

$$i := 3$$

```
>   for j from 1 to 3 do
>   V[ i,j] :=min(seq(c[ i,j][ k] +V[ i+1,k] ,k=1..2));
>   TC[ i,j] :=[ seq(c[ i,j][ k] +V[ i+1,k] ,k=1..2)] ;
>   for k from 1 to 2 do
    u[ i,j,k] :=is(TC[ i,j][ k] =V[ i,j] ) od od;
```

$$V_{3,1} := 7$$

$$TC_{3,1} := [7, 8]$$

$$V_{3,2} := 7$$

$$TC_{3,2} := [7, 11]$$

$$V_{3,3} := 6$$

$$TC_{3,3} := [8, 6]$$

```
>   for j from 1 to 3 do for k from 1 to 2 do
    print([ i,j,k,u[ i,j,k] ,TC[ i,j][ k]]) od od;
```

$$[3, 1, 1, \textit{true}, 7]$$

$$[3, 1, 2, \textit{false}, 8]$$

$$[3, 2, 1, \textit{true}, 7]$$

$$[3, 2, 2, \textit{false}, 11]$$

$$[3, 3, 1, \textit{false}, 8]$$

$$[3, 3, 2, \textit{true}, 6]$$

Stage 2 and stage 1 calculations are performed in a similar manner:

```
>   i:=2;
```

$$i := 2$$

```
>   for j from 1 to 3 do
>   V[ i,j] :=min(seq(c[ i,j][ k] +V[ i+1,k] ,k=1..3));
>   TC[ i,j] :=[ seq(c[ i,j][ k] +V[ i+1,k] ,k=1..3)] ;
>   for k from 1 to 3 do
    u[ i,j,k] :=is(TC[ i,j][ k] =V[ i,j] ) od od;
```

$$V_{2,1} := 8$$

$$TC_{2,1} := [11, 8, 12]$$

$$V_{2,2} := 9$$

$$TC_{2,2} := [11, 9, 9]$$

$$V_{2,3} := 11$$

$$TC_{2,3} := [15, 11, 13]$$

```
>   for j from 1 to 3 do for k from 1 to 2 do
    print([ i,j,k,u[ i,j,k] ,TC[ i,j][ k]]) od od;
```

$$[2, 1, 1, \textit{false}, 11]$$

$$[2, 1, 2, \textit{true}, 8]$$

$$[2, 2, 1, \textit{false}, 11]$$

$$[2, 2, 2, \textit{true}, 9]$$

$$[2, 3, 1, \textit{false}, 15]$$

$$[2, 3, 2, \textit{true}, 11]$$

```
>   i:=1;
```

$$i := 1$$

```
>   for j from 1 to 1 do
```

```
>   V[ i,j] :=min(seq(c[ i,j][ k] +V[ i+1,k] ,k=1..3));
>   TC[ i,j] :=[ seq(c[ i,j][ k] +V[ i+1,k] ,k=1..3)] ;
>   for k from 1 to 3 do
    u[ i,j,k] :=is(TC[ i,j][ k] =V[ i,j] ) od od;
```

$$V_{1,1} := 11$$

$$TC_{1,1} := [11, 13, 13]$$

```
>   for j from 1 to 1 do for k from 1 to 3 do
    print([ i,j,k,u[ i,j,k] ,TC[ i,j][ k] ] ) od od;
```

$$[1, 1, 1, \textit{true}, 11]$$

$$[1, 1, 2, \textit{false}, 13]$$

$$[1, 1, 3, \textit{false}, 13]$$

We can now extract the optimal solution from the results. From the initial state (1,1), it is optimal to travel to (2,1) at a one-stage cost of 3 since we find [1,1,1,*true*,11].[2] Next, from (2,1) it is optimal to go to (3,2) at a cost of 1 since [2,1,2,*true*,8].[3] From (3,2), the stagecoach should travel to (4,1) at a cost of 3, and finally from (4,1) it will go to (5,1) at a cost of 4. Adding the one-stage costs we find $3 + 1 + 3 + 4 = 11$ as we found from [1,1,1,*true*,11].

6.3 Models with a Linear System and Quadratic Cost

Now, consider the sequential problem with the cost function $\sum_{t=0}^{2}(x_t^2 + u_t^2) + x_3^2$ and the system equations $x_{t+1} = x_t + u_t, t = 0, 1, 2$ with x_0 given as a constant. In this deterministic problem with quadratic costs and linear system dynamics, it is desirable to bring the state as close as possible to the origin in the cheapest possible way. Deviations from the origin for both the state x_t and the decision u_t are penalized using quadratic terms. We will assume that $x_0 \neq 0$ (otherwise the solution would be trivial) and that there are no constraints on the state and the control.

Although this model with linear system and quadratic cost appears simple, its generalizations have found applications in economics and operations research; see Bensoussan, Hurst and Näslund [29, Chapter 3], Holt, Modigliani, Muth and Simon [93], Parlar [140], Parlar and Gerchak [144] and Parlar and Rempała [145].

Forming the DP functional equation, we have

$$V_t(x_t) = \min_{u_t} [x_t^2 + u_t^2 + V_{t+1}(x_t + u_t)], \quad t = 0, 1, 2$$

$$V_3(x_3) = x_3^2.$$

---

[2]This also indicates that the overall minimum cost is 11.

[3]If the trip had started in state (2,1), the overall minimum cost to New York would be 8.

We will now use Maple to solve this functional equation and generate the optimal policy for the problem.

```
>    restart: # LQ2.mws
```

We begin by informing Maple that the boundary condition is $V_3(x) = x^2$.

```
>    V[ 3] :=x->x^2;
```

$$V_3 := x \rightarrow x^2$$

Now, for period 2 the cost expression that should be minimized is $c_2 = x_2^2 + u_2^2 + V_3(x_2 + u_2)$, i.e., the sum of the current cost and the future minimum cost with $x_3 = x_2 + u_2$ as the state for stage 3.

```
>    c[ 2] :=x^2+u^2+V[ 3] (x+u);
```

$$c_2 := x^2 + u^2 + (x + u)^2$$

Differentiating this expression, equating the result to zero and solving gives the optimal policy in period 2 as $\mu_2 = -\frac{1}{2}x$.

```
>    deriv[ 2] :=diff(c[ 2] ,u);
```

$$deriv_2 := 4u + 2x$$

```
>    mu[ 2] :=solve(deriv[ 2] ,u);
```

$$\mu_2 := -\frac{1}{2}x$$

Using these results, the value function $V_2(x)$ for period 2 is found as a quadratic function of the state variable.

```
>    V[ 2] :=unapply(subs(u=mu[ 2] ,c[ 2] ),x);
```

$$V_2 := x \rightarrow \frac{3}{2}x^2$$

Repeating the same procedure for the other stages, the optimal decision and the value function is computed explicitly as a function of the state variable in each stage:

```
>    c[ 1] :=x^2+u^2+V[ 2] (x+u);
```

$$c_1 := x^2 + u^2 + \frac{3}{2}(x + u)^2$$

```
>    deriv[ 1] :=diff(c[ 1] ,u);
```

$$deriv_1 := 5u + 3x$$

```
>    mu[ 1] :=solve(deriv[ 1] ,u);
```

$$\mu_1 := -\frac{3}{5}x$$

```
>    V[ 1] :=unapply(subs(u=mu[ 1] ,c[ 1] ),x);
```

$$V_1 := x \rightarrow \frac{8}{5}x^2$$

```
>    c[ 0] :=x^2+u^2+V[ 1] (x+u);
```

$$c_0 := x^2 + u^2 + \frac{8}{5}(x+u)^2$$

```
>   deriv[ 0] :=diff(c[ 0] ,u);
```

$$deriv_0 := \frac{26}{5}u + \frac{16}{5}x$$

```
>   mu[ 0] :=solve(deriv[ 0] ,u);
```

$$\mu_0 := -\frac{8}{13}x$$

```
>   V[ 0] :=unapply(subs(u=mu[ 0] ,c[ 0] ),x);
```

$$V_0 := x \rightarrow \frac{21}{13}x^2$$

The results (to three significant digits) are summarized in the table below.

t	$\mu_t(x)$	$V_t(x)$
0	$-0.615x$	$1.615x^2$
1	$-0.6x$	$1.6x^2$
2	$-0.5x$	$1.5x^2$
3		x^2

Thus, if the initial value of the state is, say, $x_0 = 1$, then using the optimal decisions prescribed by the policy

$$\pi = (\mu_0(x), \mu_1(x), \mu_2(x)) = (-0.615x, -0.6x, -0.5x),$$

one would obtain a total minimum cost of $V_0(1) = 1.615$ for the periods $0, 1, 2$ and 3. For this case, the numerical values of the optimal decisions are obtained as follows:

t	x_t	u_t
0	1	-0.615
1	0.385	-0.231
2	0.154	$-0.077.$
3	0.077	

Now suppose that the initial decision is not made optimally and instead of $u_0 = -0.615$ we use, say, $u_0 = -0.5$, which brings the system to $x_1 = 1 - 0.5 = 0.5$. What is the best decision now at the start of period 1 given this value of the state? As the principle of optimality states, "An optimal policy has the property that whatever the initial state and initial decision are, the remaining decisions must constitute an optimal policy with regard to the state resulting from the first decision." Thus, in period 1 we make the optimal decision for that period using $u_1 = \mu_1(0.5) = -0.6 \times 0.5 = -0.3$. This is the power and elegance of the dynamic programming approach: Regardless of what may have transpired in the past, at the start of any period we know exactly what to do since DP makes the optimal policy available for all periods and for any value of the state.

In this example we can see another aspect of the power and flexibility of dynamic programming. Now consider the following case: After applying the correct decision $u_0 = -0.615$, what happens if for some reason there is an unexpected random disturbance on the system and we find $x_1 = 0.4$ (instead of 0.385)? The answer is fortunately very simple. Since we have the optimal policy available to us as $\mu_1(x_1) = -0.6x_1$, we just compute the new (optimal) decision as $u_1 = -0.6 \times 0.4 = -0.24$. Even though the correct initial decision $u_0 = -0.615$ resulted in an unexpected value of the state $x_1 = 0.4$, we can still make the next decision in stage 1 optimally using the optimal policy $\mu_1(x_1)$ for that stage.

Solution as a Nonlinear Programming Problem

We note that the sequential optimization model described above can also be solved as a standard nonlinear programming (NLP) problem with six decision variables $(\mathbf{x}, \mathbf{u}) = (x_1, x_2, x_3, u_0, u_1, u_2)$, the objective function $f(\mathbf{x}, \mathbf{u}) = \sum_{t=0}^{2}(x_t^2 + u_t^2) + x_3^2$ and three equality constraints $h_1 = x_1 - x_0 - u_0 = 0$, $h_2 = x_2 - x_1 - u_1 = 0$ and $h_3 = x_3 - x_2 - u_2 = 0$ with x_0 as a given constant. The methods discussed in Chapter 5, Nonlinear Programming, can be applied and using the method of Lagrange multipliers, the optimal solution can be found.

Writing the Lagrangian as

$$\mathcal{L}(\mathbf{x}, \mathbf{u}, \boldsymbol{\theta}) = f(\mathbf{x}, \mathbf{u}) + \sum_{t=1}^{3} \theta_t h_t(\mathbf{x}, \mathbf{u})$$

where $\boldsymbol{\theta} = (\theta_1, \theta_2, \theta_3)$ are the Lagrange multipliers, we obtain the solution for this problem using Maple as follows.

```
>    restart: # LQ2NLP.mws
>    x[ 0] :=1:
>    f:=sum(x[ t] ^2+u[ t] ^2,t=0..2)+x[ 3] ^2:
>    h[ 1] :=x[ 1] -x[ 0] -u[ 0] :
>    h[ 2] :=x[ 2] -x[ 1] -u[ 1] :
>    h[ 3] :=x[ 3] -x[ 2] -u[ 2] :
>    X:=seq(x[ t] ,t=1..3):
>    U:=seq(u[ t] ,t=0..2):
>    L:=f+sum(theta[ k] *h[ k] ,k=1..3):
>    L_x:=seq(diff(L,x[ k] ),k=1..3):
>    L_u:=seq(diff(L,u[ k] ),k=0..2):
>    L_theta:=seq(diff(L,theta[ k] ),k=1..3):
>    Digits:=3:
>    evalf(solve({ L_x,L_u,L_theta} ,{ X,U,theta[ 1] ,
       theta[ 2] ,theta[ 3] } ));
```

$\{u_2 = -.0769,\ \theta_1 = -1.23,\ x_1 = .385,\ u_1 = -.231,\ \theta_2 = -.462,\ x_2 = .154,$
$x_3 = .0769,\ \theta_3 = -.154,\ u_0 = -.615\}$

Naturally, the solution here is the same as the one we found before. However, note that the NLP solution is in some sense "frozen" because unlike the DP solution it cannot directly deal with a situation where after applying $u_0 = -0.615$ there is an unexpected random disturbance on the system and we find $x_1 = 0.4$ (instead of 0.385). To answer this question, the NLP problem must be re-solved for stages 1, 2 and 3 with $x_1 = 0.4$ as the initial value of the state in stage 1. This shows that the DP approach is, in general, much more powerful than a competing optimization technique since the DP solution is obtained in terms of dynamic policies rather than the static—frozen—decisions produced by the competing techniques.[4]

The simple model described and solved using DP can easily be generalized using Maple. For example, one may assume that there are ideal (i.e., target) levels $\{\hat{x}_t, t = 1, 2, \ldots, N\}$ and $\{\hat{u}_t, t = 0, 1, \ldots, N-1\}$ for the states and decisions, respectively, and that deviations from these trajectories are to be penalized quadratically at a cost of q_t for the states and r_t for the decisions. If the disturbances also follow a particular trajectory $\{w_t, t = 0, \ldots, N-1\}$, then the sequential optimization problem can be written

$$\min \sum_{t=0}^{N-1} \{q_t(x_t - \hat{x}_t)^2 + r_t(u_t - \hat{u}_t)^2\} + q_N(x_N - \hat{x}_N)^2$$

subject to the linear constraints

$$x_{t+1} = a_t x_t + b_t u_t + w_t, \quad t = 0, 1, \ldots, N-1$$

with given a_t and b_t, $t = 0, 1, \ldots, N-1$. Following is a Maple program that automates the solution of this more general problem via DP where $N = 3$ with the arrays xHat, q, uHat, r, a, b and w having the obvious meanings.

```
>    restart: # LQ2Auto.mws
>    N:=3: #Assume N=3
>    xHat:=array(0..N,[ 4,3,2,1] ):
     q:=array(0..N,[ 1,2,1,2] ):
>    uHat:=array(0..N-1,[ 1,2,3] ):
     r:=array(0..N-1,[ 2,2,3] ):
>    a:=array(0..N-1,[ 1,7,2] ):
     b:=array(0..N-1,[ 4,2,1] ):
>    w:=array(0..N-1,[ 2,-3,1] ):
>    Digits:=3:
>    V[ N] :=unapply(q[ N] * (x-xHat[ N] )^2,x);
```

---

[4]In engineering terminology, nonlinear programming would produce "open-loop" decisions whereas dynamic programming gives rise to "closed-loop" policies; see Bertsekas [31, p. 4] and Kirk [108, pp. 14–16].

$$V_3 := x \to 2\,(x-1)^2$$

```
>    for t from N-1 by -1 to 0 do
>    c[ t] :=q[ t] * (x-xHat[ t] )^2+r[ t] * (u-uHat[ t] )^2
     +V[ t+1] (a[ t] *x+b[ t] *u+w[ t]
     );
>    deriv[ t] :=diff(c[ t] ,u);
>    mu[ t] :=evalf(solve(deriv[ t] ,u));
>    V[ t] :=
     unapply(evalf(simplify(subs(u=mu[ t] ,c[ t] )))),x);
>    od;
```

$$c_2 := (x-2)^2 + 3\,(u-3)^2 + 2\,(2\,x+u)^2$$

$$deriv_2 := 10\,u - 18 + 8\,x$$

$$\mu_2 := 1.80 - .800\,x$$

$$V_2 := x \to 5.80\,x^2 + 10.4\,x + 14.8$$

$$c_1 := 2\,(x-3)^2 + 2\,(u-2)^2 + 5.80\,(7\,x+2\,u-3)^2 + 72.8\,x + 20.8\,u - 16.4$$

$$deriv_1 := 50.4\,u - 56.8 + 162.\,x$$

$$\mu_1 := 1.13 - 3.21\,x$$

$$V_1 := x \to 24.6\,x^2 + .192\,x + 29.8$$

$$c_0 := (x-4)^2 + 2\,(u-1)^2 + 24.6\,(x+4\,u+2)^2 + .192\,x + .768\,u + 30.2$$

$$deriv_0 := 791.\,u + 391. + 197.\,x$$

$$\mu_0 := -.494 - .249\,x$$

$$V_0 := x \to 1.12\,x^2 - 6.51\,x + 50.3$$

The results for the optimal policy and the value function for each period $t = 0, \ldots, 3$ are summarized in the following table.

t	$\mu_t(x)$	$V_t(x)$
0	$-0.249x - 0.494$	$1.12x^2 - 6.51x + 50.3$
1	$-3.21x + 1.13$	$24.6x^2 + 0.192x + 29.8$
2	$-0.8x + 1.8$	$5.80x^2 + 10.4x + 14.8$
3		$2(x-1)^2$

It is worth noting that in this more general problem the optimal policy is linear and the value function is quadratic in the state variable. This result would be valid even when the states and the controls are multidimensional and the disturbances are random provided that the cost is quadratic and the system equations are linear with no constraints on the states and controls. For a discussion of this issue, see Bertsekas [31, Section 4.1].

6.3.1   The Infinite-Stage Problem

In many realistic sequential optimization problems arising in business and economics, there may either be a very large number or infinitely many stages. For example, a long-term planning problem involving the next 10 years where a decision has to be made each month would have 120 stages. Although this number is finite, it is large enough so that approximating the problem with an infinite number of stages may be useful in understanding the structure of the optimal policy.

As an example, consider the simple sequential problem discussed above. If we now assume that $N \to \infty$, the problem can be written as

$$\min \sum_{t=0}^{\infty} (x_t^2 + u_t^2)$$

subject to

$$x_{t+1} = x_t + u_t, \quad t = 0, 1, \ldots .$$

For such a problem, the functional equations $V_t(x_t)$ for, say, $t = 100$ and $t = 99$ would be given by

$$V_{100}(x_{100}) = \min_{u_{100}} [x_{100}^2 + u_{100}^2 + V_{101}(x_{101})]$$

and

$$V_{99}(x_{99}) = \min_{u_{99}} [x_{99}^2 + u_{99}^2 + V_{100}(x_{100})].$$

Since there is some regularity in the system equations $x_{t+1} = f_t(x_t, u_t) = x_t + u_t$ and the cost function $L_t = x_t^2 + u_t^2$ (i.e., that t does not appear explicitly in f_t or L_t), under certain conditions the functional equation for the value function can be written as

$$V(x) = \min_u \{L(x, u) + V[f(x, u)]\},$$

which can be solved (usually iteratively) for the unknown function $V(x)$.[5]

Successive Approximations

One of the methods that can be used to solve this problem is known as "successive approximations" or "approximation in function space." The method starts with a

---

[5]In order for $V(x)$ to remain finite, three conditions are required: (i) $L(x, u)$ must be bounded for finite x and u, (ii) a state $\bar{x}$ and control $\bar{u}$ must exist such that $L(\bar{x}, \bar{u}) = 0$ and (iii) the specified state $\bar{x}$ must be reachable from any admissible state by applying a finite number of admissible controls; see, Larson and Casti [117, p. 222].

guess $V^{(0)}(x)$ for the solution $V(x)$. Using this guess, a new value of the function $V^{(1)}(x)$ is computed from

$$V^{(1)}(x) = \min_u \{L(x, u) + V^{(0)}[f(x, u)]\} \qquad (6.3)$$

where the corresponding policy $\mu^{(1)}(x)$ is found as the value of u that minimizes the right-hand-side of (6.3) for each x. The successive approximations method continues in this manner while computing the value function's nth approximation $V^{(n)}$ from $V^{(n-1)}$ using $V^{(n)}(x) = \min_u \{L(x, u) + V^{(n-1)}[f(x, u)]\}$.
Here, $\mu^{(n-1)}$ is determined as the value of u that minimizes the right-hand side $L(x, u) + V^{(n-1)}[f(x, u)]$ for each x. This is a relatively easy method to apply using Maple.

```
>    restart: # LQInf.mws
>    Digits:=4;
```

$$Digits := 4$$

In the finite horizon version of this problem we had observed that the value function was quadratic in the state variable x. Armed with this information, we make the initial guess for the value function as $V^{(0)} = 2x^2$. The following lines automate the successive approximation of the value function and the policy μ.

```
>    L:=(x,u)->x^2+u^2;
```

$$L := (x, u) \to x^2 + u^2$$

```
>    V[ 0] :=x->2.*x^2;
```

$$V_0 := x \to 2. x^2$$

```
>    for n from 0 to 5 do
>    c[ n] :=(x,u)->L(x,u)+V[ n] (x+u) :
>    diff(c[ n] (x,u),u) :
>    uApp:=solve(%,u);
>    subs(u=uApp,c[ n] (x,u)) :
>    V[ n+1] :=unapply(simplify(%),x);
>    od;
```

$$c_0 := (x, u) \to L(x, u) + V_n(x + u)$$

$$6. u + 4. x$$

$$uApp := -.6667 x$$

$$1.667 x^2$$

$$V_1 := x \to 1.667 x^2$$

$$c_1 := (x, u) \to L(x, u) + V_n(x + u)$$

$$5.334 u + 3.334 x$$

$$uApp := -.6250\,x$$

$$1.625\,x^2$$

$$V_2 := x \to 1.625\,x^2$$

$$c_2 := (x,\ u) \to L(x,\ u) + V_n(x + u)$$

$$5.250\,u + 3.250\,x$$

$$uApp := -.6190\,x$$

$$1.619\,x^2$$

$$V_3 := x \to 1.619\,x^2$$

$$c_3 := (x,\ u) \to L(x,\ u) + V_n(x + u)$$

$$5.238\,u + 3.238\,x$$

$$uApp := -.6182\,x$$

$$1.618\,x^2$$

$$V_4 := x \to 1.618\,x^2$$

$$c_4 := (x,\ u) \to L(x,\ u) + V_n(x + u)$$

$$5.236\,u + 3.236\,x$$

$$uApp := -.6180\,x$$

$$1.618\,x^2$$

$$V_5 := x \to 1.618\,x^2$$

$$c_5 := (x,\ u) \to L(x,\ u) + V_n(x + u)$$

$$5.236\,u + 3.236\,x$$

$$uApp := -.6180\,x$$

$$1.618\,x^2$$

$$V_6 := x \to 1.618\,x^2$$

We thus see that, after about five iterations, the solution quickly converges to $V(x) = 1.618x^2$ and $\mu(x) = -0.618x$.

A Closed-Form Solution for a More General Cost Function

Now consider a more general problem where we wish to minimize $\sum_{t=0}^{\infty}(Qx_t^2 + Ru_t^2)$ subject to the usual state dynamics $x_{t+1} = x_t + u_t$. Here, deviations from the origin for the state and control variables are penalized differently.

In this case the functional equation for the value function assumes the form $V(x) = \min_u [Qx^2 + Ru^2 + V(x + u)]$. Using the information obtained for the structure of the optimal policy and the form of the value function, we assume that $V(x) = Ax^2$ and attempt to find a closed-form formula for the coefficient A. The symbolic manipulation of the expressions to solve this problem is relegated to Maple, which finds the solution easily.

```
>   restart: # LQClosedForm.mws
>   V:=x->A*x^2;
```

$$V := x \to A x^2$$

```
>   c:=Q*x^2+R*u^2+V(x+u);
```

$$c := Q x^2 + R u^2 + A (x + u)^2$$

```
>   cu:=diff(c,u);
```

$$cu := 2 R u + 2 A (x + u)$$

```
>   usol:=solve(cu,u);
```

$$usol := -\frac{A x}{R + A}$$

```
>   RHS:=normal(subs(u=usol,c));
```

$$RHS := \frac{x^2 (A Q + R A + Q R)}{R + A}$$

```
>   Asol:=solve(V(x)=RHS,A);
```

$$Asol := \frac{1}{2} Q + \frac{1}{2} \sqrt{Q^2 + 4 Q R}, \ \frac{1}{2} Q - \frac{1}{2} \sqrt{Q^2 + 4 Q R}$$

```
>   subs(A=Asol[ 1] ,usol);
```

$$-\frac{(\frac{1}{2} Q + \frac{1}{2} \sqrt{Q^2 + 4 Q R}) x}{R + \frac{1}{2} Q + \frac{1}{2} \sqrt{Q^2 + 4 Q R}}$$

Note that since $Q - \sqrt{Q^2 + 4QR} < 0$ and since the minimum cost $V(x)$ must be nonnegative, we choose the first solution Asol[1] for A in the above computations.

```
>   V:=subs(A=Asol[ 1] ,V(x));
```

$$V := (\frac{1}{2} Q + \frac{1}{2} \sqrt{Q^2 + 4 Q R}) x^2$$

To summarize, the optimal policy and the value function for this general case are obtained as $\mu(x) = -Ax/(R + A)$ and $V(x) = Ax^2$ where

$$A = \frac{1}{2} \left(Q + \sqrt{Q^2 + 4QR} \right).$$

Naturally, for the special case when $Q = R = 1$, we obtain $A = 1.618$ so that $V(x) = 1.618x^2$ and $\mu(x) = -0.618x$ as we had found previously.

The Model with a Discount Factor

In most practical problems with long planning horizons, it would be important to discount the future cash flows to account for the concept of time value of money. Thus, if $\beta = 1/(1 + r)$ is the discount factor with r as the interest rate and $\beta \in (0, 1)$, then the objective would be to minimize the net present value of the future costs, which would now be written as $\min \sum_{t=0}^{\infty} \beta^t (x_t^2 + u_t^2)$. With this objective, the functional equation for the value function assumes the form $V(x) = \min_u \{L(x, u) + \beta V[f(x, u)]\}$, which can be solved for $V(x)$ in the usual way. Naturally, in this case the value function $V(x)$ would be defined as the minimum *discounted* cost that can be obtained by using an optimal sequence of decisions for the remainder of the process starting from an arbitrary state x in an arbitrary stage.

Although we do not present it here, it would be a simple matter to apply the successive approximation method to solve a more general problem with discounted cost and stationary target values, i.e.,

$$\min \sum_{t=0}^{\infty} \beta^t \{Q(x_t - \hat{x})^2 + R(u_t - \hat{u})^2\}$$

subject to

$$x_{t+1} = x_t + u_t, \quad t = 0, 1, \ldots.$$

For example, assuming $Q = 2$, $\hat{x} = 1$, $R = 3$, $\hat{u} = 1.5$, and $\beta = 0.9$ and starting with the initial guess of $V^{(0)} = 4x^2 + 2x + 60$, after about seven iterations the method converges to the value function $V(x) = 3.55x^2 + 1.15x + 61.6$ and to policy $\mu(x) = 0.642 - 0.516x$.

6.4 Continuous-Time Dynamic Programming

A different type of infinite-stage sequential optimization problem arises when the system evolves continuously over time and a decision must be made at each instant. In this case, even if the decision horizon is finite, the continuous nature of the process results in an infinite number of decisions.

Let us now suppose that the system evolves according to the differential equation $\dot{x}(t) = f[x(t), u(t), t]$, $0 \le t \le T$ with the initial value of the state $x(0) = x_0$ given as a constant. This differential equation is the continuous-time analogue of the discrete-time state equations that were written as $x_{t+1} = f(x_t, u_t, t)$, $t = 0, 1, \ldots, N - 1$ in Section 6.3. We wish to minimize the cost functional $\int_0^T L[x(t), u(t), t] \, dt + S[x(T), T]$, which is the continuous-time analogue of the sum $\sum_{t=0}^{N-1} L_t(x_t, u_t, w_t) + L_N(x_N)$.

Similar to the discrete-time formulation, let us define the value function $V(t, x)$ as the minimum cost that can be obtained starting in state $x(t)$ at time $t \in [0, T]$.

The value function can then be written as

$$V(t, x) = \min_u \int_t^T L(x, u, \tau)\, d\tau + S[x(T), T] \qquad (6.4)$$

subject to

$$\dot{x} = f(x, u, \tau), \quad x(t) = x$$

with the condition that at the final time $t = T$, we have $V[x(T), T] = S[x(T), T]$.
    If we write the integral in (6.4) as

$$V(t, x) = \min_u \left(\int_t^{t+\Delta t} L\, d\tau + \int_{t+\Delta t}^T L\, d\tau + S \right)$$

where Δt is an infinitesimally small interval, then by the principle of optimality
we obtain

$$V(t, x) = \min_{\substack{u \\ t \le \tau \le t+\Delta t}} \left[\int_t^{t+\Delta t} L\, d\tau + \left(\min_{\substack{u \\ t+\Delta t \le \tau \le T}} \int_{t+\Delta t}^T L\, d\tau + S \right) \right] \qquad (6.5)$$

subject to

$$\dot{x} = f(x, u, \tau), \quad x(t + \Delta t) = x + \Delta x.$$

This follows from the principle of optimality because the control $u(\tau)$ for $t+\Delta t \le \tau \le T$ is optimal for the problem starting at time $t + \Delta t$ in state $x(t + \Delta t) = x + \Delta x$.
    Now using the definition of $V(t, x)$, we can rewrite (6.5) as

$$V(t, x) = \min_{\substack{u \\ t \le \tau \le t+\Delta t}} \left[\int_t^{t+\Delta t} L\, d\tau + V(t + \Delta t, x + \Delta x) \right].$$

Assuming $V(x, t)$ to be twice differentiable and expanding $V(t + \Delta t, x + \Delta x)$
using Taylor's theorem gives

$$\begin{aligned}
V(t, x) &= \min_u [L(x, u, t)\Delta t + V(t, x) + V_t(t, x)\Delta t \\
&\quad + V_x(t, x)\Delta x + \text{higher order terms}].
\end{aligned}$$

Subtracting $V(t, x)$ from both sides, dividing by Δt and letting $\Delta t \to 0$ we obtain

$$\begin{aligned}
0 &= \min_u [L(x, u, t) + V_t(t, x) + V_x(t, x)\dot{x}] \\
&= \min_u [L(x, u, t) + V_t(t, x) + V_x(t, x)f(x, u, t)].
\end{aligned}$$

Since the term $V_t(t, x)$ does not involve the decision u, we move it to the left and
obtain the partial differential equation (PDE) for the value function $V(t, x)$ as

$$-V_t(t, x) = \min_u [L(x, u, t) + V_x(t, x)f(x, u, t)]$$

with the boundary condition $V[x(T), T] = S[x(T), T]$. The resulting PDE is known as the *Hamilton-Jacobi-Bellman* (HJB) equation since it was originally developed by Bellman [22] to solve problems in the calculus of variations. For discussions of the development of this PDE and examples of its solution, see Bertsekas [31, p. 91–92], Bryson and Ho [39, p. 135], Kamien and Schwartz [99, pp. 238–240], Kirk [108, pp. 86–90] and Sage and White [162, pp. 76–77].

A Remark on Discounting

Note that if we have an infinite horizon problem and if continuous discounting is applied with a discount factor e^{-rt}, then the optimization problem would be to minimize $\int_0^\infty e^{-rt} L(x, u, t)\, dt$ subject to the state equation $\dot{x} = f(x, u)$ with the initial state as $x(0) = x_0$. In this case, the HJB equation would be written as

$$-V_t(t, x) = \min_u [e^{-rt} L(x, u, t) + V_x(t, x) f(x, u, t)]$$

without any boundary condition on the value function. Here the value function is defined as the minimum cost *discounted to time zero* that can be obtained starting at time t in state x and using the optimal policy. However, it can be shown that for "autonomous" systems where time does not appear in the cost function (except for the discount term) and in the system equations, a similar *ordinary* differential equation (ODE) can be found for the HJB problem where the value function $V(x)$ would be defined as the minimum cost discounted to the *current time*.

In autonomous systems, the problem is to minimize $\int_0^\infty e^{-rt} L(x, u)\, dt$ subject to $\dot{x} = f(x, u)$. For this problem we write the HJB equation where discounting is to the current time (rather than to time zero) as

$$V(t, x) = \min_u [L(x, u) + e^{-r\Delta t} V(t + \Delta t, x + \Delta x)].$$

Since $e^{-r\Delta t} \approx 1 - r\Delta t$ and the value function at $(t + \Delta t, x + \Delta x)$ is expanded as

$$V(t + \Delta t, x + \Delta x) \approx V(t, x) + V_t \Delta t + V_x \Delta x,$$

after simplifications and usual limiting operations the HJB equation is obtained as an ODE in terms of $V(x)$ as follows:[6]

$$rV(x) = \min_u [L(x, u) + V'(x) f(x, u)].$$

For a more detailed discussion of this issue, see Beckmann [19, Part IV] and Kamien and Schwartz [99, pp. 241–242].

We now consider two examples where the HJB partial differential equation can be solved explicitly to find the value function and the optimal policy.

---

[6]Note that in the limiting operations we have $\lim_{\Delta t \to 0} \Delta x / \Delta t = \dot{x} = f(x, u)$.

6.4.1   A Problem with Linear System and Quadratic Cost

Consider the following continuous-time version of the sequential optimization problem with quadratic cost and linear state equation adapted from Kamien and Schwartz [99, pp. 240–241]. We wish to minimize

$$\int_0^\infty e^{-rt}(Qx^2 + Ru^2)\,dt$$

subject to $\dot{x} = u$ with the initial condition $x(0) = x_0 > 0$. We now solve the HJB equation $-V_t = \min_u[e^{-rt}(Qx^2 + Ru^2) + V_x u]$ for this problem using Maple.

```
>    restart:  # LQPDE.mws
>    L:=(x,u)->exp(-r*t)*(Q*x^2+R*u^2);
```

$$L := (x, u) \rightarrow e^{(-rt)}(Qx^2 + Ru^2)$$

```
>    f:=u;
```

$$f := u$$

```
>    RHS:=L(x,u)+diff(V(t,x),x)*f;
```

$$RHS := e^{(-rt)}(Qx^2 + Ru^2) + (\tfrac{\partial}{\partial x}V(t, x))\,u$$

The optimal policy is obtained in terms of V_x as follows.

```
>    uSol:=combine(solve(diff(RHS,u),u),exp);
```

$$uSol := -\frac{1}{2}\frac{(\tfrac{\partial}{\partial x}V(t, x))\,e^{(rt)}}{R}$$

Using this policy we develop the HJB partial differential equation and attempt to solve it using Maple's pde() function.

```
>    HJB:=-diff(V(t,x),t)=L(x,uSol)
     +diff(V(t,x),x)*uSol;
```

$$HJB := -(\tfrac{\partial}{\partial t}V(t, x)) =$$
$$e^{(-rt)}\left(Qx^2 + \frac{1}{4}\frac{(\tfrac{\partial}{\partial x}V(t, x))^2\,(e^{(rt)})^2}{R}\right) - \frac{1}{2}\frac{(\tfrac{\partial}{\partial x}V(t, x))^2\,e^{(rt)}}{R}$$

```
>    pdesolve(HJB, V(t,x));
```

```
Error, (in pdesolve/exact/charac) dsolved returned
multiple answers,
[{t(_F1) = - _F1+_C4}, {P[ 2] (_F1) =
DESol({ (diff(diff(_Y(_F1), _F1), _F1)*R
+r*diff(t(_F1), _F1)*diff(_Y(_F1),
 _F1)*R+Q*_Y(_F1))/R} ,{_Y(_F1)} )
* _C3, x(_F1) =-1/2*exp(r*t(_F1))/Q
*diff(DESol({ (diff(diff(_Y(_F1), _F1), _F1)*R
+r*diff(t(_F1), _F1)*diff(_Y(_F1), _F1)*R
+Q*_Y(_F1))/R} ,{_Y(_F1)} ), _F1)*_C3},
{ P[ 1] (_F1) =Int(1/4*r*(4*exp(-2*r*t(_F1))
*Q*x(_F1)^2*R+P[ 2] (_F1)^2)*exp(r*t(_F1))/
```

```
R, _F1) + _C2} , { U( _F1) =
Int((-P[ 1] ( _F1)*R+1/2* P[ 2] ( _F1)^2
* exp(r* t ( _F1)))/R, _F1) + _C1} ]
```

Unfortunately, Maple fails to find a solution to this PDE and returns an error message. Thus, using our intuition obtained from the discrete-time version of this type of problem with quadratic cost and linear system dynamics, we try a solution in the form $V(t, x) = e^{-rt} A x^2$ where the constant A is to be determined.

```
>   V:=(t,x)->exp(-r* t)*A* x^2; Vt:=diff(V(t,x),t);
    Vx:=diff(V(t,x),x);
```

$$V := (t, x) \rightarrow e^{(-rt)} A x^2$$

$$Vt := -r\, e^{(-rt)} A x^2$$

$$Vx := 2\, e^{(-rt)} A x$$

```
>   HJB;
```

$$r\, e^{(-rt)} A x^2 =$$
$$e^{(-rt)} (Q x^2 + \frac{(e^{(-rt)})^2 A^2 x^2 (e^{(rt)})^2}{R}) - 2\frac{(e^{(-rt)})^2 A^2 x^2 e^{(rt)}}{R}$$

```
>   simplify(HJB);
```

$$r\, e^{(-rt)} A x^2 = -\frac{e^{(-rt)} x^2 (-Q R + A^2)}{R}$$

Substituting and simplifying, we finally reduce the problem to the solution of a quadratic equation in terms of the unknown A:

```
>   ASol:=solve(HJB,A);
```

$$ASol := -\frac{1}{2} r R + \frac{1}{2}\sqrt{r^2 R^2 + 4 Q R},\ -\frac{1}{2} r R - \frac{1}{2}\sqrt{r^2 R^2 + 4 Q R}$$

Since the value function must be positive, we choose the first solution of ASol so that we have $A = -\frac{1}{2}r R + \frac{1}{2}\sqrt{r^2 R^2 + 4 Q R}$. The optimal policy is then obtained as $u = -(A/R)x$.

```
>   uSol:=combine(uSol);
```

$$uSol := -\frac{A x}{R}$$

6.4.2 A Problem with Quadratic and Linear Costs

Many of the examples discussed are somewhat standard because they involve a quadratic cost function with linear state equations. We now describe the solution of a more difficult problem where the cost function consists of a *mixture* of quadratic and linear terms. We will see that Maple is again very helpful in manipulating expressions that give rise to the value function and the optimal policy for the model.

Consider the following problem that appears as an exercise in Kamien and Schwartz [99, p. 242]. We wish to minimize the cost functional $\int_0^T (c_1 u^2 + c_2 x)\, dt$ subject to $\dot{x} = u$ with $x(0) = 0$ and $x(T) = B$. Kamien and Schwartz suggest using $V(t, x) = a + bxt + hx^2/t + kt^3$ as the trial form for the value function. Keeping this hint in mind, we proceed with the development of the HJB partial differential equation.

```
>    restart: # LPDE.mws

>    L:=(x,u)->c[ 1] *u^2+c[ 2] *x;
```

$$L := (x, u) \to c_1 u^2 + c_2 x$$

```
>    f:=u;
```

$$f := u$$

The right-hand side of the HJB equation that will be minimized is $L(x, u) + V_x f(x, u)$. Differentiating this expression, equating to zero and solving for u gives the optimal policy function in terms of the unknown V_x.

```
>    RHS:=L(x,u)+diff(V(t,x),x)*f;
```

$$RHS := c_1 u^2 + c_2 x + (\tfrac{\partial}{\partial x} V(t, x)) u$$

```
>    uSol:=solve(diff(RHS,u),u);
```

$$uSol := -\frac{1}{2} \frac{\frac{\partial}{\partial x} V(t, x)}{c_1}$$

```
>    HJB:=-diff(V(t,x),t)=L(x,uSol)
     +diff(V(t,x),x)*uSol;
```

$$HJB := -(\tfrac{\partial}{\partial t} V(t, x)) = -\frac{1}{4} \frac{(\frac{\partial}{\partial x} V(t, x))^2}{c_1} + c_2 x$$

An attempt to solve the PDE explicitly using Maple fails.

```
>    pdesolve(HJB, V(t,x));
```

```
Error, (in pdesolve/exact/charac) dsolved returned
multiple answers,
[{ P[ 1] (_F1) = _C4} , { P[ 2] (_F1) = -c[ 2] * _F1+_C3} ,
{ U(_F1) =Int((-P[ 1] (_F1)*c[ 1]
+1/2* P[ 2] (_F1)^2)/c[ 1] , _F1)+ _C2} , { x(_F1) =
Int(1/2/c[ 1] * P[ 2] (_F1), _F1)+_C1} ,
{ t(_F1) = - _F1+_C5}]
```

At this stage we make use of the trial form of the value function and compute its derivatives with respect to t and x.

```
>    V:=(t,x)->a+b*x*t+h*x^2/t+k*t^3;
     Vt:=diff(V(t,x),t); Vx:=diff(V(t,x),x);V(T,B);
```

$$V := (t, x) \to a + bxt + \frac{hx^2}{t} + kt^3$$

$$Vt := b\,x - \frac{h\,x^2}{t^2} + 3\,k\,t^2$$

$$Vx := b\,t + 2\,\frac{h\,x}{t}$$

$$a + b\,B\,T + \frac{h\,B^2}{T} + k\,T^3$$

```
>   HJB:=lhs(HJB)-rhs(HJB);
```

$$HJB := -b\,x + \frac{h\,x^2}{t^2} - 3\,k\,t^2 + \frac{1}{4}\frac{(b\,t + 2\,\frac{h\,x}{t})^2}{c_1} - c_2\,x$$

The HJB equation is now reduced to a nonlinear algebraic equation in terms of x, t^2 and x^2/t^2 with unknown constants A, B, H and K, which we now compute analytically.

The coefficients of x and t^2 are easily extracted using the Maple `coeff()` function.

```
>   cx:=coeff(HJB,x);
```

$$cx := -b + \frac{b\,h}{c_1} - c_2$$

```
>   ct2:=coeff(HJB,t^2);
```

$$ct2 := -3\,k + \frac{1}{4}\frac{b^2}{c_1}$$

To extract the coefficient of x^2/t^2, we use a somewhat obscure Maple function `algsubs()` that performs more general algebraic substitutions than the substitutions `subs()` function can make. This substitution will have a temporary effect of defining x^2/t^2 as z^2, thus making the expression a polynomial so that the `coeff()` function can be applied.

```
>   x2t2:=algsubs(x^2/t^2=z^2,expand(HJB));
```

$$x2t2 := -\frac{(b\,c_1 - b\,h + c_2\,c_1)x}{c_1} - \frac{1}{4}\frac{(12\,k\,c_1 - b^2)t^2}{c_1} + \frac{h\,z^2\,(c_1 + h)}{c_1}$$

```
>   cx2t2:=coeff(x2t2,z^2);
```

$$cx2t2 := \frac{h\,(c_1 + h)}{c_1}$$

To make sure that we have not missed any terms while extracting the coefficients, we perform a simple check and observe that everything is in order.

```
>   simplify(cx*(x)+ct2*(t^2)+cx2t2*(x^2/t^2)-HJB);
```

$$0$$

Now that we have the algebraic expressions for the coefficients of x, t^2 and x^2/t^2, (i.e., `cx`, `ct2` and `cx2t2`) and the final time condition that $V[T, x(T)] =$

$V(T, B) = 0$, we can solve the resulting system of four nonlinear equations in the four unknowns a, b, h and k.

```
>    solve({ cx,ct2,cx2t2,V(T,B)} ,{ a,b,h,k} );
```

$$\{h = 0, \ k = \frac{1}{12}\frac{c_2^2}{c_1}, \ a = -\frac{1}{12}\frac{c_2 T (-12 B c_1 + c_2 T^2)}{c_1}, \ b = -c_2\}, \{h = -c_1,$$

$$k = \frac{1}{48}\frac{c_2^2}{c_1}, \ b = -\frac{1}{2}c_2, \ a = -\frac{1}{48}\frac{-24 c_2 B T^2 c_1 - 48 c_1^2 B^2 + c_2^2 T^4}{T c_1}\}$$

Maple finds two solutions but we use the one where the solution is nonzero.

```
>    assign(%[ 2] );
>    a; b; h; k;
```

$$-\frac{1}{48}\frac{-24 c_2 B T^2 c_1 - 48 c_1^2 B^2 + c_2^2 T^4}{T c_1}$$

$$-\frac{1}{2}c_2$$

$$-c_1$$

$$\frac{1}{48}\frac{c_2^2}{c_1}$$

The value function $V(t, x)$ and the optimal production rate $\dot{x} = u$ are now easily computed as functions of the stage t and state x.

```
>    V(t,x); # Value function
```

$$-\frac{1}{48}\frac{-24 c_2 B T^2 c_1 - 48 c_1^2 B^2 + c_2^2 T^4}{T c_1} - \frac{1}{2}c_2 x t - \frac{c_1 x^2}{t} + \frac{1}{48}\frac{c_2^2 t^3}{c_1}$$

```
>    Vx;
```

$$-\frac{1}{2}c_2 t - 2\frac{c_1 x}{t}$$

```
>    expand(simplify(uSol)); # Optimal production rate
```

$$\frac{1}{4}\frac{t c_2}{c_1} + \frac{x}{t}$$

To obtain an analytic expression for the optimal production rate $u(t)$ and for the optimal inventory level $x(t)$, we solve $\dot{x}(t) = u(t)$ using the form of the optimal policy just determined.

```
>    dsolve({ diff(x(t),t)=(t* c[ 2] /(4* c[ 1] ))+x(t)/t,
         x(0)=0, x(T)=B} ,x(t));
```

$$x(t) = \frac{1}{4}\frac{t^2 c_2}{c_1} - \frac{1}{4}\frac{t (c_2 T^2 - 4 B c_1)}{T c_1}$$

```
>    assign(%);
```

```
>   xOpt:=unapply(normal(x(t)),t);
    uOpt:=unapply(normal(diff(x(t),t)),t);
```

$$xOpt := t \rightarrow -\frac{1}{4} \frac{t\,(-c_2\,t\,T + c_2\,T^2 - 4\,B\,c_1)}{T\,c_1}$$

$$uOpt := t \rightarrow -\frac{1}{4} \frac{-2\,c_2\,t\,T + c_2\,T^2 - 4\,B\,c_1}{T\,c_1}$$

6.5   A Constrained Work Force Planning Model

Many of the models discussed so far were relatively simple since the costs were quadratic and there were no constraints on the state or the decision variables. We now present a more complicated sequential decision problem of work force planning with constraints on both state and decision variables. For a discussion of similar models, see Bensoussan, Hurst and Näslund [29, pp. 58–72], Hillier and Lieberman [92, pp. 345–350], Holt, Modigliani, Muth and Simon [93] and Sasieni, Yaspan and Friedman [164, p. 280].

The work force requirements in a particular factory fluctuate due to seasonality of the demand. The estimated work force requirements during each of the four seasons for the indefinite future are as follows:

Season	Summer	Fall	Winter	Spring
Requirements (r_t)	$r_1 = 210$	$r_2 = 250$	$r_3 = 190$	$r_4 = 260$

Due to the undesirability—and costliness—of hiring and firing of the workers, management wishes to minimize fluctuations in work force levels. It is estimated that the total cost of changing the work force level from one season to the next is $s = \$100$ times the square of the difference in the levels between the two seasons. For example, if the Summer employment level were 215 people and if an additional 10 workers were hired for the Fall, the cost of this change in the work force would be $\$100 \times (225 - 215)^2 = \$10000$. Any employment level that is above the levels given in the table is considered a "waste" that costs the company approximately $w = \$1000$/person/season. For example, a Summer level of 215 would result in a waste of $\$1000 \times (215 - 210) = \$5000$ for that season.

Although this is a problem with an infinite number of stages, each year starts an identical cycle. Since the cost data are assumed to be stationary, we can consider only one cycle of four seasons that end with Spring. We will assume that the Spring production is the most important and that the management has decided to keep the Spring employment level at $u_4 = 260$. Also, note that since the estimate of 260 for the Spring is the highest, it would not be optimal to increase the employment level at any season above 260.

We denote the initial Summer season by $t = 1$ and the other periods as $t = 2, \ldots, 4$. If we define the decision variables $u_t, t = 1, \ldots, 4$ as the employment levels used in period t (with $u_4 = 260$), it is easy to see that this problem can be formulated as a nonlinear programming problem. For this problem the objective function would be

$$\min \sum_{t=1}^{4}[100(u_t - u_{t-1})^2 + 1000(u_t - r_t)]$$

subject to the constraints $210 \le u_1 \le 260, 250 \le u_2 \le 260, 190 \le u_3 \le 260$ (and $u_4 = 260$). If this problem were solved using the techniques of nonlinear programming, we would obtain the *open-loop* decisions as $u_1 = 252.5$, $u_2 = 250$, $u_3 = 252.5$ and $u_4 = 260$ with a minimum total cost of \$117,500. Fractional levels such 252.5 are assumed to be acceptable since they correspond to the employment of part-time workers.

We will now solve this problem with Maple's help using dynamic programming. Since in the current stage the state (i.e., the available information about the system) is the previous stage's decision (i.e., the employment level chosen), we write $x_{t+1} = u_t, t = 1, 2, 3$ as the state equations with $x_1 = 260$. This gives rise to the constraints on the states x_t, i.e., $210 \le x_2 \le 260, 250 \le x_3 \le 260$, and $190 \le x_4 \le 260$. Defining $V_t(x_t)$ as the value function for stage t—the minimum cost obtainable for periods $t, t+1, \ldots, 4$—when the optimal policy is implemented, the DP functional equation becomes

$$V_t(x_t) = \min_{r_t \le u_t \le r_4} [100(u_t - x_t)^2 + 1000(u_t - r_t) + V_{t+1}(x_{t+1})], t = 1, \ldots, 4$$

with the boundary condition

$$\begin{aligned} V_4(x_4) &= 100(u_4 - x_4)^2 + 1000(u_4 - r_4) \\ &= 100(260 - x_4)^2 \end{aligned}$$

since in stage 4 we must have $u_4 = r_4 = 260$.

We start by assigning values to the seasonal work force requirements and compute the maximum requirement as $R = \max_t \{r_t\} = 260$.

```
>    restart: # Manpower.mws
>    with(linalg):

Warning, new definition for norm

Warning, new definition for trace

>    s:=100; w:=1000;
```

$$s := 100$$

$$w := 1000$$

```
>    r:=array(1..4,[ 210,250,190,260] );
     R:=max(seq(r[ i] ,i=1..4));
```

$$r := [210, 250, 190, 260]$$

$$R := 260$$

For stage 4 the value function $V_4(x_4)$ is defined as the boundary condition.

```
>    V[ 4] :=unapply(s* (r[ 4] -x)^2,x);
```

$$V_4 := x \to 100\,(260 - x)^2$$

For stage 3 the recursive relationship is

$$V_3(x_3) = \min_{190 \le u_3 \le 260} c_3(x_3, u_3)$$

where $c_3(x_3, u_3) = 100(u_3 - x_3)^2 + 1000(u_3 - 190) + V_4(x_4)$. We show that c_3 is convex in u_3 given x_3.

Differentiating c_3 with respect to $u_3 = u$, equating the result to 0 and solving gives an expression for u_3 (or the policy μ_3) in terms of the state variable $x_3 = x$.

Recall that u_3 is feasible in the interval from $r_3 = 190$ to $r_4 = R = 260$. Evaluating the policy μ_3 at the end points of x_3, we find $\mu_3(x_3 = r_2) = 252.5$ and $\mu_3(x_3 = R) = 257.5$. This shows that the policy μ_3 is in the feasible interval for any x_3; i.e., we have $\mu_3 = \frac{1}{2}x_3 + \frac{255}{2}$ for $250 \le x_3 \le 260$.

With this information about the optimal u_3 we easily compute the value function $V_3(x)$ for stage 3.

```
>    c[ 3] :=s* (u-x)^2+w* (u-r[ 3] )+V[ 4] (u);
```

$$c_3 := 100\,(u - x)^2 + 1000\,u - 190000 + 100\,(260 - u)^2$$

```
>    cp[ 3] :=diff(c[ 3] ,u); cpp[ 3] :=diff(c[ 3] ,u$2);
```

$$cp_3 := 400\,u - 200\,x - 51000$$

$$cpp_3 := 400$$

```
>    mu[ 3] :=solve(cp[ 3] ,u);
```

$$\mu_3 := \frac{1}{2}x + \frac{255}{2}$$

```
>    # u[ 3] is feasible from r[ 3] =190 to R=260
>    evalf(subs(x=r[ 2] ,mu[ 3] )); evalf(subs(x=R,mu[ 3] ));
```

$$252.5000000$$

$$257.5000000$$

```
>    V[ 3] :=unapply(subs(u=mu[ 3] ,c[ 3] ),x);
```

$$V_3 := x \to 100\,(-\frac{1}{2}x + \frac{255}{2})^2 + 500\,x - 62500 + 100\,(\frac{265}{2} - \frac{1}{2}x)^2$$

For stage 2, the recursive equation becomes

$$V_2(x_2) = \min_{250 \le u_2 \le 260} c_2(x_2, u_2)$$

where $c_2(x_2, u_2) = 100(u_2 - x_2)^2 + 1000(u_2 - 250) + V_3(x_3)$. Here the analysis becomes more complicated.

Differentiating c_2 with respect to u_2 we get $\partial c_2/\partial u_2 = 300u_2 - 200x_2 + 24500$. Next, equating this result to zero and solving for u_2 we find the candidate policy as $\mu_2 = \frac{2}{3}x_2 + \frac{245}{3}$. Now, recall that u_2 is feasible in the interval from $r_2 = 250$ to $r_4 = R = 260$, but evaluating the policy μ_2 at the extreme points of x_2 we find $\mu_2(x_2 = r_1 = 210) = 221.67$ (infeasible) and $\mu_2(x_2 = R = 260) = 255$ (feasible).

To find the range of x_2 for which u_2 is feasible, we solve $\mu_2 = 250$ for x_2 and obtain $x_2 = 252.5$. This shows that the policy μ_2 applies only in the interval for x_2 from 252.5 to 260. For a given $x_2 \leq 252.5$, we find that c_2 is convex and increasing in u_2. Hence, we set $\mu_2 = 250$ for $x_2 \leq 252.5$. This gives a piecewise function for the policy μ_2; see Figure 6.1.

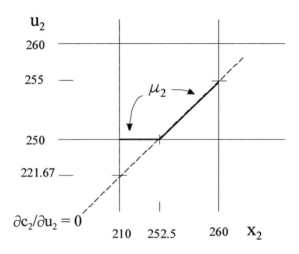

FIGURE 6.1. The form of the policy $\mu_2(x_2)$.

The value function $V_2(x_2)$ and c_1 are easily computed in terms of the `piecewise()` function. But we suppress the output of these functions in order to conserve space.

```
>    c[ 2] :=s* (u-x)^2+w* (u-r[ 2] )+V[ 3] (u);
```

$$c_2 := 100\,(u-x)^2 + 1500\,u - 312500 + 100\,(-\frac{1}{2}u + \frac{255}{2})^2$$
$$+ 100\,(\frac{265}{2} - \frac{1}{2}u)^2$$

```
>    cp[ 2] :=diff(c[ 2] ,u); cpp[ 2] :=diff(c[ 2] ,u$2);
```

$$cp_2 := 300\,u - 200\,x - 24500$$

$$cpp_2 := 300$$

```
>    mu[ 2] :=solve(cp[ 2] ,u);
```

$$\mu_2 := \frac{2}{3}x + \frac{245}{3}$$

```
>    # u[ 2]  is feasible from r[ 2] =250 to R=260
>    evalf(subs(x=r[ 1] ,mu[ 2] )); evalf(subs(x=R,mu[ 2] ));
```

$$221.6666667$$

$$255.$$

```
>    xLow:=evalf(solve(mu[ 2] =250,x));
```

$$xLow := 252.5000000$$

```
>    xHigh:=evalf(min(R,solve(mu[ 2] =260,x)));
```

$$xHigh := 260.$$

```
>    uLow:=subs(x=xLow,mu[ 2] );
```

$$uLow := 250.0000000$$

```
>    uHigh:=subs(x=xHigh,mu[ 2] );
```

$$uHigh := 255.0000000$$

```
>    mu[ 2] :=piecewise(x<=xLow,r[ 2] ,x<R,mu[ 2] );
```

$$\mu_2 := \begin{cases} 250 & x \le 252.5000000 \\ \dfrac{2}{3}x + \dfrac{245}{3} & x < 260 \end{cases}$$

```
>    cp[ 2] ;
```

$$300\,u - 200\,x - 24500$$

```
>    assume(210<x,x<252.5,250<u,u<260); is(cp[ 2] >0);
```

true

```
>    x:=' x' ; u:=' u' ;
```

$$x := x$$

$$u := u$$

```
>    V[ 2] :=unapply(subs(u=mu[ 2] ,c[ 2] ),x):
>    c[ 1] :=s* (u-r[ 4] )^2+w* (u-r[ 1] )+V[ 2] (u):
```

Although we do not present the graph of the c_1 function here, we include a
Maple command plot(c[1] ,u=190..260) that can be used to note that c_1
is convex in u_1. Thus differentiating this function using the Maple diff() command and solving gives the optimal numerical value of the first period's decision
as $u_1 = 252.5$. The minimum total cost for all four periods is now easily found as
$V_1 = 117500$ by substituting the optimal u_1 into c_1.

```
>    #plot(c[ 1] ,u=190..260,discont=true);
>    cp[ 1] :=diff(c[ 1] ,u):
```

```
>   mu[ 1] :=evalf(solve(cp[ 1] ,u)); u[ 1] :=mu[ 1] ;
```

$$\mu_1 := 252.5000000$$

$$u_1 := 252.5000000$$

```
>   V[ 1] :=evalf(subs(u=mu[ 1] ,c[ 1] ));
```

$$V_1 := 117500.0000$$

The optimal employment levels for the other periods are easily computed as $u_2 = 250$, $u_3 = 252.5$ and $u_4 = 260$.

```
>   x:=mu[ 1] ; u[ 2] :=mu[ 2] ;
```

$$x := 252.5000000$$

$$u_2 := 250$$

```
>   x:=mu[ 2] ; u[ 3] :=evalf(mu[ 3] );
```

$$x := 250$$

$$u_3 := 252.5000000$$

```
>   x:=evalf(mu[ 3] ); u[ 4] :=r[ 4] ;
```

$$x := 252.5000000$$

$$u_4 := 260$$

The following table summarizes the solution.

t	1	2	3	4
r_t	210	250	190	260
u_t	252.5	250	252.5	260

6.6 A Gambling Model with Myopic Optimal Policy

In his seminal paper entitled "A New Interpretation of Information Rate," Kelly [103] considered an unscrupulous gambler who was receiving prior information concerning the outcomes of sporting events over a noisy communication channel (e.g., a phone line) before the results became common knowledge. After receiving the word "win" or "lose"—which he could hear incorrectly due to communication difficulties—the gambler would place his bet (a nonnegative amount up to his present fortune) on the original odds. With p as the probability of correct transmission and $q = 1 - p$ the probability of an incorrect transmission, Kelly showed using calculus techniques that if $p > q$, the gambler's optimal wager u (i.e., the fraction of his capital bet each time) was simply $u = p - q$. If $p \leq q$, then $u = 0$. He also showed that the maximum growth of capital occurred at a rate equal to the capacity of the channel given as $G_{max} = \log_2 2 + p \log_2 p + q \log_2 q$. (Interestingly, this was the same result obtained by Shannon [171] from considerations of coding of information.)

Kelly's work attracted the attention of a large number of researchers including economists, e.g., Arrow [10], psychologists, e.g., Edwards [63] and applied mathematicians such as Bellman and Kalaba [27], [28], [24] who extended and reinterpreted Kelly's results. In particular, Bellman and Kalaba re-solved the same problem after formulating it as a dynamic program with a logarithmic utility function for the terminal wealth of the gambler. They assumed that the gambler is allowed N plays (bets) and that at each play of the gamble, he could bet any nonnegative amount up to his present fortune.

In the formulation of the gambling problem using DP we define $V_n(x_n)$ as the maximum expected return if the gambler has a present fortune of x_n and has n gambles left. (This is a slight change in notation compared to what we had used before. In this model $n = N - t$ is the number of gambles *left* where $n = 0, 1, \ldots, N$.) At each play, the gambler could bet any nonnegative amount up to his present fortune and win the gamble with a probability of p. The objective is to maximize the terminal utility of his wealth which, is assumed to be logarithmic; i.e., when there are no gambles left, the gambler's utility of his final wealth is $V_0(x_0) = \log(x_0)$. When n gambles are left, we define u_n as the fraction of the current wealth to gamble with $0 \le u_n \le 1$. Then the state equation assumes the form $x_{n-1} = f_n(x_n, u_n, w_n) = x_n + u_n x_n w_n$, $n = 1, \ldots, N$ where the random variable w_n takes the values 1 or -1 with probability p and $q = 1 - p$, respectively. Thus, the DP functional equation is written as

$$
\begin{aligned}
V_n(x_n) &= \max_{0 \le u_n \le 1} E_{w_n} V_{n-1}(x_n + u_n x_n w_n) \\
&= \max_{0 \le u_n \le 1} [p V_{n-1}(x_n + u_n x_n) + q V_{n-1}(x_n - u_n x_n)]
\end{aligned}
$$

with the boundary condition $V_0(x_0) = \log(x_0)$.

We now describe the DP solution of Kelly's problem using Maple. Our notation and exposition closely follow those of Ross [159, pp. 2–4].

First, we show that when $p \le \frac{1}{2}$, it is optimal not to bet ($u = 0$) and that $V_n(x_n) = \log(x_n)$. Starting with $n = 1$, we see that the c_1 function is monotone decreasing in u, so the optimal decision is to bet 0. (Actually when $p = \frac{1}{2}$, setting $u = 2p - 1$ makes c_1 constant. But since $u = 2p - 1 = 0$, the statement about the optimal decision is still valid.) With $\mu_1(x_1) = 0$, we obtain $V_1(x_1) = \log(x_1)$. For other values of $n > 1$, the same solution can be shown to be valid proving that when $p \le \frac{1}{2}$, $\mu_n(x_n) = 0$ and $V_n(x_n) = \log(x_n)$.

```
>    restart: # Gamble.mws
>    assume(0<=p,p<=1/2): additionally(0<=u,u<=1):
>    q:=1-p;
```

$$q := 1 - p$$

```
>    c[1]:=p*log(x+u*x)+q*log(x-u*x);
```

$$c_1 := p \ln(x + u x) + (1 - p) \ln(x - u x)$$

```
>    c[1,1]:=normal(diff(c[1],u));
```

$$c_{1,1} := -\frac{2p - 1 - u}{(1 + u)(-1 + u)}$$

```
>    is(c[ 1,1]<=0);
```

true

```
>    V[ 1] :=simplify(subs(u=0,c[ 1] ));
```

$$V_1 := \ln(x)$$

We now turn to the case where $p > \frac{1}{2}$.

```
>    restart: # Gamble.mws (Part 2)
>    assume(x>0,p>0);
>    q:=1-p;
```

$$q := 1 - p$$

The boundary condition is the utility of the final fortune, which is defined as the logarithmic function in x.

```
>    V[ 0] :=x->log(x);
```

$$V_0 := \log$$

When there is one play left, we find that it is optimal to bet $\mu_1 = 2p - 1 = p - (1 - p) = p - q$ fraction of the current wealth:

```
>    c[ 1] :=p* log(x+u* x)+q* log(x-u* x);
```

$$c_1 := p\ln(x + ux) + (1 - p)\ln(x - ux)$$

```
>    deriv[ 1] :=diff(c[ 1] ,u);
```

$$deriv_1 := \frac{px}{x + ux} - \frac{(1 - p)x}{x - ux}$$

```
>    mu[ 1] :=solve(deriv[ 1] ,u);
```

$$\mu_1 := 2p - 1$$

The value function $V_1(x)$ is obtained in terms of logarithms that can be written as $V_1(x) = C + \log(x)$ where $C = \log(2) + p\log(p) + \log(1 - p) - p\log(1 - p) = \log(2) + p\log(p) + q\log(q)$.

```
>    V[ 1] :=unapply(expand(simplify(
     subs(u=mu[ 1] ,c[ 1] ))),x);
```

$$V_1 := x \rightarrow p\ln(p) + \ln(2) + \ln(x\sim) + \ln(1 - p) - p\ln(1 - p)$$

When there are two plays left, the optimal fraction to bet is again $\mu_2 = p - q$ with the value function assuming the form $V_2(x) = 2C + \log(x)$.

```
>    c[ 2] :=p* V[ 1] (x+u* x)+q* V[ 1] (x-u* x);
```

$$c_2 := p\,(p\ln(p) + \ln(2) + \ln(x + ux) + \ln(1 - p) - p\ln(1 - p))$$
$$+ (1 - p)\,(p\ln(p) + \ln(2) + \ln(x - ux) + \ln(1 - p) - p\ln(1 - p))$$

```
>    deriv[ 2] :=diff(c[ 2] ,u);
```

$$deriv_2 := \frac{p\,x}{x+u\,x} - \frac{(1-p)\,x}{x-u\,x}$$

```
>    mu[ 2] :=solve(deriv[ 2] ,u);
```

$$\mu_2 := 2\,p - 1$$

```
>    V[ 2] :=unapply(expand(simplify(
     subs(u=mu[ 2] ,c[ 2] ))),x);
```

$$V_2 := x \rightarrow 2\,p\ln(p) - 2\,p\ln(1-p) + 2\ln(2) + \ln(x\tilde{\ }) + 2\ln(1-p)$$

This procedure can be automated to produce similar results.

```
>    for n from 3 to 4 do:
>    c[ n] :=p* V[ n-1] (x+u* x)+q* V[ n-1] (x-u* x):
>    deriv[ n] :=diff(c[ n] ,u):
>    mu[ n] :=solve(deriv[ n] ,u):
>    V[ n] :=unapply(expand(simplify(
     subs(u=mu[ n] ,c[ n] ))),x):
>    od;
```

$c_3 := p\,(2\,p\ln(p) - 2\,p\ln(1-p) + 2\ln(2) + \ln(x+u\,x) + 2\ln(1-p)) + (1-p)$
$(2\,p\ln(p) - 2\,p\ln(1-p) + 2\ln(2) + \ln(x-u\,x) + 2\ln(1-p))$

$$deriv_3 := \frac{p\,x}{x+u\,x} - \frac{(1-p)\,x}{x-u\,x}$$

$$\mu_3 := 2\,p - 1$$

$$V_3 := x \rightarrow 3\,p\ln(p) - 3\,p\ln(1-p) + 3\ln(2) + \ln(x\tilde{\ }) + 3\ln(1-p)$$

$c_4 := p\,(3\,p\ln(p) - 3\,p\ln(1-p) + 3\ln(2) + \ln(x+u\,x) + 3\ln(1-p)) + (1-p)$
$(3\,p\ln(p) - 3\,p\ln(1-p) + 3\ln(2) + \ln(x-u\,x) + 3\ln(1-p))$

$$deriv_4 := \frac{p\,x}{x+u\,x} - \frac{(1-p)\,x}{x-u\,x}$$

$$\mu_4 := 2\,p - 1$$

$$V_4 := x \rightarrow 4\,p\ln(p) - 4\,p\ln(1-p) + 4\ln(2) + \ln(x\tilde{\ }) + 4\ln(1-p)$$

Using induction, it can then be shown that when $p > \frac{1}{2}$ the optimal policy is to bet $\mu_n = p - q$ fraction of the current wealth at any stage of the game. This gives rise to a value function that is in the form $V_n(x) = nC(p) + \log(x)$ where $C(p) = q\ln(q) + p\ln(p) + \ln(2)$.

One of the most interesting features of this problem is the nature of its solution: The optimal strategy is myopic (invariant) in the sense that regardless of the number of bets left to place (n) and the current wealth (x), the optimal (nonnegative) fraction to bet in each period is the same, i.e., $u = p - q$ when $p > \frac{1}{2}$ and $u = 0$

when $p \leq \frac{1}{2}$. Optimal myopic policies of this type are usually difficult to obtain, but they exist for some models. For example, in some periodic-review inventory problems with random demand it has been shown that provided that some terminal cost is appended to the objective function, the optimal order quantity is the same for all periods; i.e., the policy is myopic (Veinott [188], Heyman and Sobel [89, Chapter 3]). In some dynamic portfolio problems (Mossin [133], Bertsekas [31, pp. 152–157]) myopic policy is also optimal provided that the amount invested is not constrained and the investor's utility function satisfies certain conditions.

In a recent paper, Çetinkaya and Parlar [41] argue that the assumption of the simple logarithmic function in the Kelly-Bellman-Kalaba model is somewhat unrealistic and that the problem should ideally be solved with a more general logarithmic utility function. They thus assume that, in general, the gambler's terminal utility is given as $V_0(x) = \log(b + x)$ where b is a positive constant. This generalization results in the disappearance of the myopic nature of the solution and the optimal strategy assumes a form that *depends* on the stage (n) and the state (x) of the gambling process.

6.7   Optimal Stopping Problems

Let us suppose that a professor has N days to leave a country where he has spent the past year on sabbatical. Before leaving the country he wants to sell his car and hence places an advertisement in the local newspaper.

After the publication of the advertisement each day he receives random offers $w_0, w_1, \ldots, w_{N-1}$, which are independent and identically distributed. If the professor accepts an offer, then the process ends; however if he rejects the offer, he must wait until the next day to evaluate the next offer. We assume that offers that are rejected in the past are not retained—the potential buyer goes elsewhere. (For the case where past offers are retained, see Bertsekas [31, p. 173].) The problem is to find a policy that will maximize the professor's revenue when he leaves the country.

Problems of this type are known as *stopping-rule* problems and their origin can be traced back to Wald's research on sequential analysis in the 1940s [192]. For later surveys of stopping-rule problems see Breiman [37] and Leonardz [121]. The second part of Howard's delightful review of dynamic programming [95] discusses an elementary stopping-rule problem (called an "action-timing" problem by Howard). For a rigorous exposition of the theory of stopping rules, Chow, Robbins and Siegmund [43] can be consulted. Applications in economics of job search are discussed in Lippman and McCall [123], [124].

We let $x_t = w_{t-1}$ be the state variable at the beginning of period t (or, equivalently, at the end of period $t - 1$). Defining $V_t(x_t)$ as the value function, the DP functional equation is obtained as

$$V_t(x_t) = \max \begin{cases} x_t & \text{if the decision is to accept the offer of } w_{t-1} \\ E[V_{t+1}(w_t)] & \text{if the decision is to reject the offer of } w_{t-1}. \end{cases}$$

Note that if the last offer w_{t-1} is accepted, then at the start of period t, the professor will have a revenue of x_t and the process will "stop." If the offer is rejected, then $V_t(x_t)$ is equal to the expected value of continuing optimally in periods $t+1, t+2, \ldots, N$, i.e., $E[V_{t+1}(w_t)] = \int_0^\infty V_{t+1}(w)\, dF(w)$ where $F(w)$ is the distribution function of the offer w.

Thus, the optimal policy is to accept the offer $x_t = w_{t-1}$ in period t if x_t exceeds a critical threshold level of $\theta_t = E[V_{t+1}(w_t)]$ and reject the offer if it falls below θ_t, i.e.,

$$\mu_t(x_t) = \begin{cases} \text{Accept} & \text{if } x_t > \theta_t \\ \text{Reject} & \text{if } x_t < \theta_t, \end{cases} \tag{6.6}$$

with either acceptance or rejection being optimal if $x_t = \theta_t$. Hence the value function is $V_t(x_t) = \max(x_t, \theta_t)$. Note that since $x_0 \equiv 0$, we have $V_0(0) = \max(0, \theta_0) = \theta_0$.

As the threshold value θ_t at time t is obtained in terms of $E[V_{t+1}(w_t)]$, we can develop a difference equation in terms of this value and compute θ_t, $t = 0, 1, \ldots, N$ recursively using Maple.

To obtain the difference equation, we write

$$\begin{aligned} \theta_t &= E[V_{t+1}(w_t)] = E[\max(x_{t+1}, \theta_{t+1})] = E[\max(w_t, \theta_{t+1})] \\ &= \theta_{t+1} \int_0^{\theta_{t+1}} dF(w) + \int_{\theta_{t+1}}^\infty w\, dF(w) \\ &= \theta_{t+1} F(\theta_{t+1}) + \int_{\theta_{t+1}}^\infty w\, dF(w) \end{aligned}$$

with the boundary condition being $\theta_N = 0$ (since the professor would be willing to accept *any* offer on his last day). Thus, the threshold levels are found as the solution of the nonlinear difference equation

$$\theta_t = \theta_{t+1} F(\theta_{t+1}) + A(\theta_{t+1}), \qquad t = 0, 1, \ldots, N-1$$

with $A(\theta_{t+1}) = \int_{\theta_{t+1}}^\infty w\, dF(w)$ and $\theta_N = 0$. We now discuss some special cases where the distributions of the random variables assume specific forms.

As a simple example, assume first that the professor has $N = 5$ days to leave the country and his departure is scheduled for next Friday morning. Thus the process of receiving offers starts on day $t = 0$ (Sunday) and the first decision is made on Monday morning. If we assume that the offers are distributed uniformly between 0 and 1, then their density is $f(w) = 1$ for $0 \leq w \leq 1$. With these data we find $F(\theta_{t+1}) = \theta_{t+1}$ and $A(\theta_{t+1}) = \frac{1}{2} - \frac{1}{2}\theta_{t+1}^2$ which gives rise to the following difference equation:

$$\theta_t = \frac{1}{2}(1 + \theta_{t+1}^2), \qquad t = 0, 1, \ldots, 4, \qquad \theta_N = 0.$$

Unfortunately, Maple is unable to solve this nonlinear difference equation in closed form with the `rsolve()` function:

```
>    restart: # UniformDP.mws
>    rsolve({ theta(t) = (1/2)*(1+(theta(t+1)^2)),
     theta(5)=0} ,
     theta);
```

$$\text{rsolve}(\{\theta(t) = \frac{1}{2} + \frac{1}{2}\theta(t+1)^2, \ \theta(5) = 0\}, \ \theta)$$

However, the θ_t values for $t = 0, 1, \ldots, N-1$ can easily be computed numerically as follows.

```
>    restart: # UniformDP.mws  (Part 2)
>    Digits:=4:
>    f:=w->1; # Uniform w
```

$$f := 1$$

```
>    N:=5; theta[ N] :=0;
```

$$N := 5$$

$$\theta_5 := 0$$

```
>    for t from N-1 by -1 to 0 do:
>    F[ t+1] :=int(f,w=0..theta[ t+1] ):
>    A[ t+1] :=int(w* f,w=theta[ t+1] ..1):
>    theta[ t] :=evalf(theta[ t+1] * F[ t+1] +A[ t+1] ):
>    od:
>    seq(theta[ t] ,t=0..N-1);
```

$$.7751, \ .7417, \ .6953, \ .6250, \ .5000$$

Thus, we see that at the start of the process, the professor expects to gain 0.7751 monetary units if he uses the optimal policy $\mu_t(x_t)$ that is prescribed in (6.6).

6.7.1 The Infinite-Stage Problem

What happens if the horizon N is large? Does the sequence θ_t approach a limit for large N? We now examine these issues.

First note that for large N, we must discount the future rewards, or the threshold levels would approach the maximum possible value attainable by the random variable representing the offer. For example, for the uniform example we would have $\theta_t \to 1$ as N gets large. This would be a strange result since in that case the optimal policy would *not* allow accepting any offers.

Now using discounting and assuming that $1/(1+r)$ is the discount factor per period, we can write $V_t(x_t) = \max\{x_t, (1+r)^{-1}E[V_{t+1}(w_t)]\}, t = 0, 1, \ldots, N-1$, with $V_N(x_N) = x_N$, which gives $\theta_t = (1+r)^{-1}E[V_{t+1}(w_t)]$. It can then be shown using induction that [31, p. 161]

$$V_t(x) \geq V_{t+1}(x) \quad \text{for all } x \geq 0 \text{ and } t.$$

Since $\theta_t = (1+r)^{-1}E[V_{t+1}(w)]$, we obtain $\theta_t \geq \theta_{t+1}$. This means that as the end of horizon gets one period closer, the professor would be willing to accept a lower offer since he has one less chance of getting an improved offer. For the discounted case, the nonlinear difference equation assumes the form

$$\theta_t = (1+r)^{-1}F(\theta_{t+1}) + (1+r)^{-1}A(\theta_{t+1}), \quad t = 0, 1, \ldots, N-1.$$

Now, $0 \leq F(\theta) \leq 1$ for all $\theta \geq 0$ and $0 \leq \int_{\theta_{t+1}}^{\infty} w\,dF(w) \leq E(\text{offer}) < \infty$ for all t. Thus, using the property that $\theta_t \geq \theta_{t+1}$, we obtain an algebraic equation in terms of a constant $\bar{\theta}$ whose solution would give us the limiting value of θ:

$$(1+r)\bar{\theta} = \bar{\theta}F(\bar{\theta}) + \int_{\bar{\theta}}^{\infty} w\,dF(w).$$

For the uniformly distributed offers with a discount rate of $r = 0.10$, the limiting result is obtained simply as the solution of a quadratic equation:

```
>    restart: # UniformDP.mws (Part 2)
>    r:=0.10;
```

$$r := .10$$

```
>    F:=theta;
```

$$F := \theta$$

```
>    A:=int(w,w=theta..1);
```

$$A := \frac{1}{2} - \frac{1}{2}\theta^2$$

```
>    Asy:=(1+r)*theta=theta*F+A;
```

$$Asy := 1.10\,\theta = \frac{1}{2}\theta^2 + \frac{1}{2}$$

```
>    solve(Asy,theta);
```

$$.6417424305, \ 1.558257570$$

Thus, we obtain $\bar{\theta} = 0.641$ as the number of opportunities (periods) to receive offers approaches infinity.

Let us now consider a more general case where the offers have a beta distribution so that the density $f(w)$ assumes the form

$$f(w) = \frac{\Gamma(a+b)}{\Gamma(a)\Gamma(b)} w^{a-1}(1-w)^{b-1}, \quad \text{for } 0 \leq w \leq 1$$

where $\Gamma(y)$ is the gamma function defined by $\Gamma(y) = \int_0^{\infty} u^{y-1}e^{-u}\,du$. It is well known that when y is a positive integer, we obtain $\Gamma(y) = (y-1)!$, and that when $a = b = 1$, the beta density reduces to the uniform so that $f(w) = 1$ for $0 \leq w \leq 1$.

The Maple output for $a = 5$ and $b = 2$ follows where for $N = 5$, we compute the threshold levels $\theta_t, t = 0, 1, \ldots, 4$.

```
>   restart: # BetaDP.mws
>   Digits:=3;
```

$$Digits := 3$$

```
>   a:=5; b:=2;
```

$$a := 5$$
$$b := 2$$

```
>   Mean:=a/(a+b);
```

$$Mean := \frac{5}{7}$$

```
>   f:=GAMMA(a+b)/(GAMMA(a)*GAMMA(b))*
    w^(a-1)*(1-w)^(b-1);
```

$$f := 30\,w^4\,(1 - w)$$

```
>   N:=5; theta[N]:=0;
```

$$N := 5$$
$$\theta_5 := 0$$

```
>   for t from N-1 by -1 to 0 do;
>   F[t+1]:=int(f,w=0..theta[t+1]);
>   A[t+1]:=int(w*f,w=theta[t+1]..1);
>   theta[t]:=evalf(theta[t+1]*F[t+1]+A[t+1]);
>   od:
>   seq(theta[t],t=0..N);
```

$$.838, .830, .806, .765, .714, 0$$

We find the limiting value $\bar{\theta} = 0.704$ (with $r = 0.10$) as the unique real valued root of a polynomial of degree 5:

```
>   restart: # BetaDP.mws (Part 2)
>   Digits:=3;
```

$$Digits := 3$$

```
>   r:=0.10;
```

$$r := .10$$

```
>   a:=5;b:=2;
```

$$a := 5$$
$$b := 2$$

```
>   f:=GAMMA(a+b)/(GAMMA(a)*GAMMA(b))
    *w^(a-1)*(1-w)^(b-1);
```

$$f := 30\,w^4\,(1 - w)$$

```
>   F:=int(f,w=0..theta);
```

$$F := -5\,\theta^6 + 6\,\theta^5$$

```
>   A:=int(w* f,w=theta..1);
```

$$A := \frac{5}{7} + \frac{30}{7}\,\theta^7 - 5\,\theta^6$$

```
>   Asy:=(1+r)*theta=theta* F+A;
```

$$Asy := 1.10\,\theta = \theta\,(-5\,\theta^6 + 6\,\theta^5) + \frac{5}{7} + \frac{30}{7}\,\theta^7 - 5\,\theta^6$$

```
>   fsolve(Asy,theta);
```

$$.704$$

Compared to the limiting case $\bar{\theta} = 0.641$ obtained for uniformly distributed offers (with mean 1/2), the beta-distributed offers with a mean of $a/(a+b) = 5/7$ have higher threshold levels. This should be expected since with the higher mean of the latter case, the decision maker can afford to be picky and reject offers that are not high enough.

6.8 Summary

Dynamic programming (DP) is useful in solving different types of sequential decision problems with Bellman's principle of optimality. We started this chapter by introducing a simple discrete-time, discrete-state sequential decision problem known as the stagecoach problem. Sequential decision problems with a quadratic performance criterion and a linear system normally give rise to closed-form solutions that may require an iterative process of solving a sequence of recursive equations. Using Maple's symbolic manipulation capabilities, we were able to develop closed-form solutions for a large class of such problems. A finite horizon workforce planning problem with constraints was solved with the help of Maple's piecewise() function. A gambling problem modeled using stochastic DP was solved that gave rise to a myopic policy. Finally, optimal stopping problems were considered where the stopping boundary was computed by solving a nonlinear difference equation numerically.

6.9 Exercises

1. Consider the following road network where a driver can travel only on the one-way streets as indicated by arrows. The driver starts at the intersection labelled "A" and must reach the final destination at the intersection labelled "H". The travel distances between any two intersections are indicated next to the arrows. Define the states and decisions and find the optimal route for

the driver.

$$
\begin{array}{ccccccc}
A & \longrightarrow 5 & D & \longrightarrow 3 & E & \longrightarrow 7 & H \\
\downarrow 3 & & \uparrow 6 & & \downarrow 1 & & \uparrow 2 \\
B & \longrightarrow 8 & C & \longrightarrow 2 & F & \longrightarrow 3 & G
\end{array}
$$

2. Suppose we are given a positive quantity x. Use dynamic programming to divide x into n parts in such a way that the product of the n parts is maximized.

3. Consider the following optimization problem:

$$
\min J = \sum_{i=0}^{N} (qx_i^2 + ru_i^2) + qx_3^2
$$

subject to the state dynamics

$$
x_{i+1} = ax_i + bu_i, \quad i = 0, \ldots, N-1
$$

where $N = 2$, $q = 2$, $r = 4$, $a = 1$ and $b = 0.5$.

(a) Use dynamic programming to solve this problem.

(b) The same problem appeared as a nonlinear programming problem in Exercise 10 of Chapter 5; see page 191. Comment on the nature of the solution produced by the two approaches.

4. Consider the problem of minimizing $J = \sum_{t=0}^{\infty} [u_t^2 + (x_t - u_t)^2]$ subject to the state dynamics $x_{t+1} = a(x_t - u_t)$ with $0 < a < 1$ and $0 \le u_n \le x_n$, $n = 0, 1, 2, \ldots$. Find the optimal policy and the value function for this infinite-stage problem.

5. Use the Hamilton-Jacobi-Bellman equation to minimize the performance criterion $J = \frac{1}{4}x^2(T) + \int_0^T \frac{1}{4}u^2(t)\,dt$ subject to the state dynamics $x'(t) = x(t) + u(t)$ where $T < \infty$.

6. Suppose we need to purchase a particular commodity within the next N days. Successive prices w_t for this commodity are random and i.i.d. with density $f(w_t)$. Determine an optimal policy to purchase this commodity in order to minimize the expected cost assuming that w_t are distributed uniformly between 0 and 1.

7. Consider a machine whose probability of failure depends on the number of items it has produced. Assume that if the machine fails while producing an item, there is a repair cost of c_f. The machine can be overhauled before it fails at a cost of $c_o < c_f$. If the machine is repaired or overhauled, then it is considered good as new. After the production of each unit we may decide

to overhaul the machine or attempt to continue with the production of the next unit.

Define $V_n(x)$ to be the minimum expected cost of producing n more units when the machine has already produced x parts. Let $p(x)$ be the probability of the machine failing while producing the $(x + 1)$st unit *given* that it already has produced x units. Develop the DP functional equations for $V_n(x)$ and use the following data to compute $V_1(x)$, $V_2(x)$, ..., $V_6(x)$ for $x = 0, 1, 2, \ldots, 5$.

$p(0)$	$p(1)$	$p(2)$	$p(3)$	$p(4)$	$p(5)$	c_o	c_f
0	0.2	0.3	0.5	0.8	1.0	1	2

8. A set of items is called "group-testable" if for any of its subsets it is possible to perform a simultaneous (group) test on the subset with an outcome of "success" or "failure"; see Bar-Lev, Parlar and Perry [14]. The "success" outcome indicates that all the tested units are good, and the "failure" outcome indicates that at least one item in the tested subset is defective without knowing which (or how many) are defective.

In industrial problems that involve, e.g., the identification of good electronic chips in a large batch that contain good and bad ones, chips of 100% quality cost much more than the chips of $100q\%$ quality where q is a positive constant that is usually greater than 0.9 but strictly less than 1.

Let K be the set-up cost for each test, u be the group size decision variable, D be the requirement (demand) for 100% quality items, N be the number of $100q\%$ units purchased. Define the state variables x as the number of units of demand that is yet to be satisfied, y be the number of untested units. Let q be the probability that a given unit in the group is "good" and π be the unit cost of having a shortage of 100% quality items after the testing ends.

Define $V(x, y)$ as the minimum expected cost (until the process terminates) when we are currently at state (x, y) and an optimal policy is followed until termination. Write down the DP functional equation for the value function $V(x, y)$ and solve the problem using the following data:

D	N	K	π	q
3	5	20	100	0.95

7
Stochastic Processes

7.1   Introduction

Operations research models can be categorized broadly as deterministic and prob-abilistic. When the variable of interest—e.g., the number of customers in a service system or the number of units in inventory—randomly evolves in time, a proba-bilistic model provides a more accurate representation of the system under study.

A collection of random variables indexed by a parameter T (such as time) is called a *stochastic process* (or "random" process). When the process is observed at discrete time points $0, 1, 2, \ldots$, the process is usually denoted by $\{X_n,\ n = 0, 1, \ldots\}$. The number of students enrolled in an MBA-level operations research course at the start of each academic year is an example of a discrete-time stochas-tic process. When the index parameter is continuous the process is usually denoted by $\{X(t),\ t \geq 0\}$. An example of a continuous-time stochastic process is the price of a company's stock observed at time t.

Stochastic processes are also characterized by their state space S, i.e., the val-ues taken by X_n (or by $X(t)$). If the set S is countable, then the process has a discrete state space. For example, if $X(t)$ represents the number of customers in the system at time t, then S is countable and $X(t)$ has a discrete state space. If S is a subinterval of the real line (an uncountable set), then the process has continuous state space. For example, if $X(t)$ measures the water content of a dam, then $X(t)$ has a continuous state-space. Thus, a stochastic process can be placed in one of four possible categories depending on the nature of its parameter set T (discrete vs. continuous) and state space S (discrete vs. continuous).

The theory of stochastic processes plays a very important role in operations research. Many realistic problems arising in queueing, inventory, reliability and maintainability and other areas are modeled using stochastic processes as the future values of the variables representing the evolution of these systems are not known in advance. For an excellent coverage of applications of stochastic processes in queueing theory, see Cooper [50], Gross and Harris [76], Medhi [130] and Wolff [196]. One of the classical texts in stochastic inventory theory is by Hadley and Whitin [81]. A recent monograph by Şahin [163] provides a rigorous analysis of regenerative stochastic inventory systems. Porteus [151] gives a very readable account of the recent results in stochastic inventory theory. For a review of the applications of stochastic processes in reliability and maintainability, see Shaked and Shantikumar [170].

Several excellent textbooks provide a general overview of the theory and applications of stochastic processes, e.g., Bhat [33], Çınlar [47], Cox and Miller [53], Heyman and Sobel [88], Kao [101], Karlin and Taylor [102], Medhi [131], Ross [157] and Tijms [186].

We will start our discussion of stochastic processes by considering a very simple continuous-time, discrete state space process known as the Poisson process. Then, we will discuss renewal theory that generalizes the Poisson process. This will be followed by discrete- and continuous-time Markov chains.

7.2   Exponential Distribution and Poisson Process

In this section we first describe the properties of exponential distribution as it is intimately related to the Poisson process.

Suppose T is an exponential random variable with parameter $\lambda > 0$ and density

$$f(t) = \lambda e^{-\lambda t}, \quad t \geq 0.$$

As the following Maple results demonstrate, for the exponential r.v., the cumulative distribution function is $\Pr(T \leq t) \equiv F(t) = \int_0^t \lambda e^{-\lambda x} \, dx = 1 - e^{-\lambda t}$ and the survivor probability is $\Pr(T > t) \equiv \bar{F}(t) = \int_t^\infty \lambda e^{-\lambda x} \, dx = e^{-\lambda t}$. If, for example, T is the lifetime of a component, then $F(t)$ gives the probability that the component will fail at or before t time units and the survival probability $\bar{F}(t)$ gives the probability that the component will fail after t time units.

```
>   restart: # exponential.mws
>   assume(lambda>0);
>   f:=unapply(lambda*exp(-lambda*t),t);
```

$$f := t \to \lambda e^{(-\lambda t)}$$

```
>   F:=unapply(int(f(x),x=0..t),t);
```

$$F := t \to -e^{(-\lambda t)} + 1$$

```
>   FBar:=unapply(1-F(t),t);
```

$$FBar := t \rightarrow e^{(-\lambda t)}$$

7.2.1 Memoryless Property of the Exponential Distribution

We are interested in computing the conditional probability that the component will fail after $s + t$ time units *given* that it has already worked for more than t time units. This can be written as

$$\Pr(T > s + t \mid T > t) = \frac{\Pr(T > s + t \text{ and } T > t)}{\Pr(T > t)} = \frac{\Pr(T > s + t)}{\Pr(T > t)}.$$

Evaluating this expression, Maple gives

```
>   Conditional:=FBar(s+t)/FBar(t);
```

$$Conditional := \frac{e^{(-\lambda(s+t))}}{e^{(-\lambda t)}}$$

```
>   simplify(Conditional);
```

$$e^{(-\lambda s)}$$

Thus, the conditional probability is obtained as

$$\Pr(T > s + t \mid T > s) = e^{-\lambda s}.$$

Comparing this to the probability $\Pr(T > s)$ that the (new) component will fail after s time units, we find

```
>   FBar(s);
```

$$e^{(-\lambda s)}$$

This implies that for the exponential random variable we have the interesting property that

$$\Pr(T > s + t \mid T > t) = \frac{\Pr(T > s + t)}{\Pr(T > t)} = \Pr(T > s)$$

or

$$\Pr(T > s + t) = \Pr(T > t)\Pr(T > s). \tag{7.1}$$

This is a very significant result—known as the "memoryless" property of the exponential—as it states that distribution of the remaining life of the component does *not* depend on how long it has been operating. In other words, given that the component has operated for t time units, the conditional probability that it will operate for at least s time units is *independent* of how long it has been in operation. It is this memoryless property of the exponential distribution that facilitates the analysis of a large number models in queueing, inventory theory, etc. For example, the Markovian models presented in Chapter 9, Queueing Systems, make the assumption that either the interarrival or the service time is memoryless, thereby considerably simplifying the analysis.

Writing (7.1) as $\bar{F}(s + t) = \bar{F}(t)\bar{F}(s)$ gives a functional equation with the unknown function $\bar{F}(\cdot)$. It can be shown that the only solution that satisfies this equation is of the form $\bar{F}(t) = e^{-\lambda t}$; thus the only continuous memoryless distribution is the exponential. For a proof of this result, see Ross [157, pp. 24–25].

7.2.2   Hazard Rate Function

Suppose we now wish to compute the conditional probability that the component will fail in the short time interval $(t, t + dt]$ given that it has already survived t time units. Denoting the lifetime r.v. of the component by T, this probability is written as

$$
\begin{aligned}
\Pr(t < T \leq t + dt \mid T > t) &= \frac{\Pr(t < T \leq t + dt \text{ and } T > t)}{\Pr(T > t)} \\
&= \frac{\Pr(t < T \leq t + dt)}{\Pr(T > t)} \\
&\approx \frac{f(t)\, dt}{\bar{F}(t)} = r(t)\, dt
\end{aligned}
$$

where $r(t) = f(t)/\bar{F}(t)$ is the hazard (failure) rate function of the component. Naturally, if $f(t) = \lambda e^{-\lambda t}$, then we find that $r(t) = \lambda$, a constant. For most mechanical equipment, the hazard rate function would be an increasing function of time (Shaked and Shantikumar [170]). In manpower planning applications, Bartholomew [16] has shown that the hazard rate function for the completed length of service (CLS) distributions estimated for the British Civil Service is a decreasing function of time. This implies that the longer a civil servant works for the government, the smaller is the probability that he will resign his job and move elsewhere.

Clearly, for a given density $f(t)$, the hazard rate function $r(t)$ can be uniquely determined from the equation $r(t) = f(t)/\bar{F}(t)$. For example, suppose T is a Weibull r.v. with distribution F and density f:

```
>    restart: # weibull.mws
>    F:=unapply(1-exp(-lambda*x^a),x);
```

$$
F := x \to 1 - e^{(-\lambda x^a)}
$$

```
>    f:=unapply(diff(F(x),x),x);
```

$$
f := x \to \frac{\lambda x^a\, a\, e^{(-\lambda x^a)}}{x}
$$

Using $r(t) = f(t)/\bar{F}(t)$ the hazard rate function r is found as

```
>    r:=unapply(f(x)/(1-F(x)),x);
```

$$
r := x \to \frac{\lambda x^a\, a}{x}
$$

We now deal with the inverse problem: Does the hazard rate function $r(t)$ uniquely determine the density $f(t)$? The answer to this is in the affirmative and it is obtained by solving a simple ordinary differential equation.

Recall that $\bar{F}(t) = \int_t^\infty f(u)\,du$, which implies $\bar{F}'(t) = -f(t)$. Rewriting $r(t) = f(t)/\bar{F}(t)$ we obtain an ordinary differential equation $\bar{F}'(t) = -r(t)\bar{F}(t)$ with the initial condition $\bar{F}(0) = 1$. Solving this ODE using Maple's dsolve() function, we find the expression $\bar{F}(t)$ in terms of the hazard rate function $r(t)$.

```
>  restart: # rODE.mws
>  ODE:=diff(FBar(t),t)=-r(t)*FBar(t);
```

$$ODE := \tfrac{\partial}{\partial t}\,\mathrm{FBar}(t) = -r(t)\,\mathrm{FBar}(t)$$

```
>  Solution:= dsolve({ODE,FBar(0)=1},FBar(t));
```

$$Solution := \mathrm{FBar}(t) = e^{(-\int_0^t r(u)\,du)}$$

```
>  assign(Solution);
```

Next, differentiating $-\bar{F}'(t)$ gives the unique solution for the density $f(t)$ in terms of the hazard rate function $r(t)$.

```
>  f:=unapply(-diff(FBar(t),t),t);
```

$$f := t \rightarrow r(t)\,e^{(-\int_0^t r(u)\,du)}$$

Thus, given the hazard rate function $r(t)$, the density $f(t)$ is obtained uniquely by evaluating an expression involving an integral.

$$f(t) = r(t)e^{-\int_0^t r(u)\,du}. \tag{7.2}$$

Returning to the Weibull example, we see that with the hazard rate function $r(t) = \lambda x^{a-1}a$, the correct density $f(t)$ of the Weibull is obtained from (7.2), as expected.

```
>  restart: # weibull2.mws
>  F:=unapply(1-exp(-lambda*x^a),x):
>  f:=unapply(diff(F(x),x),x):
>  r:=unapply(f(x)/(1-F(x)),x):
>  assume(a>0,lambda>0);
>  f:=r(t)*exp(-int(r(x),x=0..t));
```

$$f := \frac{\lambda\,t^a\,a\,e^{(-t^a\,\lambda)}}{t}$$

7.2.3 The Erlang Random Variable

Suppose T_1 and T_2 are two independent and identically distributed (i.i.d.) exponential random variables (r.v.) with parameter λ. In a queueing example with two stages, each of these r.v.'s could correspond to the time spent in one stage of the system. The total time spent in the *system* is then $S_2 = T_1 + T_2$.

To find the distribution of the new random variable S_2, we load the int-trans() package and compute the Laplace transform (LT) of the exponential r.v., which we denote by fLT[1] .

```
>   restart: # Erlang.mws
```

```
>   with(inttrans);
```

[*addtable, fourier, fouriercos, fouriersin, hankel, hilbert, invfourier, invhilbert, invlaplace, invmellin, laplace, mellin, savetable*]

```
>   assume(lambda>0);
```

```
>   f:=lambda* exp(-lambda* t);
```

$$f := \lambda e^{(-\lambda t)}$$

```
>   fLT[ 1] :=laplace(f,t,s);
```

$$fLT_1 := \frac{\lambda}{s + \lambda}$$

Since S_2 is the sum of two r.v.'s, the LT of S_2 (denoted by fLT[2]) is obtained as the product of the LT of the individual exponential r.v.'s:

```
>   fLT[ 2] :=fLT[ 1] * fLT[ 1] ;
```

$$fLT_2 := \frac{\lambda^2}{(s + \lambda)^2}$$

Inverting the resulting transform, we obtain

```
>   invlaplace(fLT[ 2] ,s,t);
```

$$\lambda^2 t e^{(-\lambda t)}$$

which is the density of the Erlang r.v. with parameters $(2, \lambda)$. Thus, the r.v. S_2 has the density

$$f_{S_2}(t) = \lambda^2 t e^{-\lambda t}, \quad t \geq 0.$$

Continuing with this approach, we now form the new sum $S_3 = \sum_{i=1}^{3} T_i$ whose LT—denoted by fLT[3] —is found as

```
>   fLT[ 3] :=(fLT[ 1] )^3;
```

$$fLT_3 := \frac{\lambda^3}{(s + \lambda)^3}$$

Inverting this LT, we find

```
>   invlaplace(fLT[ 3] ,s,t);
```

$$\frac{1}{2} \lambda^3 t^2 e^{(-\lambda t)}$$

as the density of Erlang with parameters $(3, \lambda)$. Thus, we see that the r.v. S_2 has density

$$f_{S_3}(t) = \frac{1}{2}\lambda^3 t^2 e^{-\lambda t}, \quad t \geq 0.$$

Generalizing these results to a positive integer n by using induction it can be shown that the finite sum $S_n = \sum_{i=1}^{n} T_i$ has the Erlang density

$$f_{S_n}(t) = \frac{\lambda e^{-\lambda t}(\lambda t)^{n-1}}{(n-1)!}, \quad t \geq 0$$

with parameters (n, λ).

7.2.4 Poisson Process

Suppose we "count" the total number of events $N(t)$—e.g., customer arrivals—that have occurred up to time t. Such a counting process has the properties that (i) it is nonnegative and integer valued, (ii) if $s < t$, then $N(s) \leq N(t)$, and (iii) for $s < t$, the quantity $N(t) - N(s)$ equals the number of events that occurred in the interval $(s, t]$.

A counting process $N(t)$ is said to have *independent increments* if the (random) number of arrivals in two disjoint intervals are independent. The *stationary-increment* property of a counting process says that the distribution of the number of events depends only on the length of the interval, not on the position of the interval.

The Poisson process $\{N(t), t \geq 0\}$ with rate $\lambda > 0$ is a special type of counting process with the properties that (i) $N(0) = 0$, (ii) $N(t)$ has independent and stationary increments, (iii) for a short time interval of length h, $\Pr[N(h) = 1] = \lambda h + o(h)$ and (iv) $\Pr[N(h) \geq 2] = o(h)$, where $o(h)$ is the "little-oh" function that approaches zero faster than h, i.e., $\lim_{h \to 0} o(h)/h = 0$.

Based on these four properties, we now show using Maple's help that the number of events in an interval of length t has the distribution

$$\Pr[N(t) = n] = p_n(t) = \frac{(\lambda t)^n e^{-\lambda t}}{n!}, \quad n = 0, 1, 2, \ldots$$

with mean λt.

We first start with $n = 0$. Using the properties of the Poisson process, we can write an expression for the probability of having no events between $(0, t + h]$:

$$
\begin{aligned}
p_0(t + h) &= \Pr[N(t + h) = 0] \\
&= \Pr[N(t) = 0 \text{ and } N(t + h) - N(t) = 0] \quad \text{(independence)} \\
&= \Pr[N(t) = 0]\Pr[N(t + h) - N(t) = 0] \quad \text{(stationarity)} \\
&= p_0(t)[1 - \lambda h + o(h)].
\end{aligned}
$$

Subtracting $p_0(t)$ from both sides of the last equation, dividing by h and letting $h \to 0$ gives

$$p_0'(t) = -\lambda p_0(t)$$

with the initial condition $p_0(0) = 1$ since we assume that initially the system is empty, i.e., $N(0) = 0$. Solving this ODE with Maple gives

```
>    restart:  # Poisson.mws
>    de.0:=diff(p.0(t),t)=-lambda*p.0(t);
```

$$de0 := \frac{\partial}{\partial t} \text{p0}(t) = -\lambda \, \text{p0}(t)$$

```
>    IC.0:=p.0(0)=1;
```

$$IC0 := \text{p0}(0) = 1$$

```
>    assign(dsolve({ de.0,IC.0} ,{ p.0(t)} ));
>    p.0(t);
```

$$e^{(-\lambda t)}$$

That is, $p_0(t) = e^{-\lambda t}$.

For $n = 1$ we have

$$p_1(t + h) = p_1(t)[1 - \lambda h + o(h)] + p_0(t)[\lambda h + o(h)].$$

Subtracting $p_1(t)$ from both sides of the last equation, dividing by h and letting $h \to 0$ gives a new ODE

$$p_1'(t) = -\lambda p_1(t) + \lambda p_0(t)$$

with the condition $p_1(0) = 1$. Maple solves this ODE as

```
>    de.1:=diff(p.1(t),t)=-lambda*p.1(t)+lambda*p.0(t);
```

$$de1 := \frac{\partial}{\partial t} \text{p1}(t) = -\lambda \, \text{p1}(t) + \lambda \, e^{(-\lambda t)}$$

```
>    IC.1:=p.1(0)=0;
```

$$IC1 := \text{p1}(0) = 0$$

```
>    assign(dsolve({ de.1,IC.1} ,{ p.1(t)} ));
>    p.1(t);
```

$$e^{(-\lambda t)} \lambda t$$

That is, $p_1(t) = e^{-\lambda t} \lambda t$.

In general, for $n = 1, 2, 3, \ldots$, we obtain a set of differential-difference equations as

$$p_n(t + h) = p_n(t)[1 - \lambda h + o(h)] + p_{n-1}(t)[\lambda h + o(h)].$$

With the usual operations of subtracting $p_n(t)$ from both sides, dividing both sides by h and letting $h \to 0$, we find an infinite system of ODEs (i.e., differential-difference equations)

$$p_n'(t) = -\lambda p_n(t) + \lambda p_{n-1}(t), \quad n = 1, 2, 3, \ldots$$

We now again use Maple to solve these ODEs and obtain the solution for a few small values of n:

```
>   for n from 2 to 10 do
>   de.n:=diff(p.n(t),t)=-lambda*p.n(t)
    +lambda*p.(n-1)(t);
>   IC.n:=p.n(0)=0;
>   assign(dsolve({de.n,IC.n},{p.n(t)}));
>   p.n(t);
>   od:
>   p.2(t); p.3(t); p.4(t); p.5(t);
```

$$\frac{1}{2} e^{(-\lambda t)} \lambda^2 t^2$$

$$\frac{1}{6} e^{(-\lambda t)} \lambda^3 t^3$$

$$\frac{1}{24} e^{(-\lambda t)} \lambda^4 t^4$$

$$\frac{1}{120} e^{(-\lambda t)} \lambda^5 t^5$$

To show the general result, we use mathematical induction and assume that the result is true for $n = k$, i.e., that

$$p_k(t) = \frac{e^{-\lambda t} (\lambda t)^k}{k!}$$

is true.

```
>   p.k(t):=exp(-lambda*t)*(lambda*t)^k/k!;
```

$$\mathrm{pk}(t) := \frac{e^{(-\lambda t)} (\lambda t)^k}{k!}$$

With this assumption we attempt to solve the ODE for $n = k + 1$, i.e.,

$$p'_{k+1}(t) = -\lambda p_{k+1}(t) + \lambda p_k(t),$$

and hope to find

$$p_{k+1}(t) = \frac{e^{-\lambda t} (\lambda t)^{k+1}}{(k + 1)!}.$$

(We denote $k + 1$ by `kPlus1` in the Maple worksheet.)

```
>   de.(kPlus1):=diff(p.(kPlus1)(t),t)=
    -lambda*p.(kPlus1)(t)
    +lambda*p.k(t);  IC.(kPlus1):=p.(kPlus1)(0)=0;
```

$$dekPlus1 := \tfrac{\partial}{\partial t} \mathrm{pkPlus1}(t) = -\lambda \, \mathrm{pkPlus1}(t) + \frac{\lambda \, e^{(-\lambda t)} (\lambda t)^k}{k!}$$

$$ICkPlus1 := \text{pkPlus1}(0) = 0$$

```
>   assign(dsolve({ de.(kPlus1),IC.(kPlus1)},
    {p.(kPlus1)(t)} ));
>   p.(kPlus1)(t);
```

$$\frac{e^{(-\lambda t)} (\lambda t)^{(k+1)}}{\Gamma(k+1)(k+1)}$$

Maple solves the ODE with the initial condition $p_{k+1}(0) = 1$ and finds the correct result

$$p_{k+1}(t) = \frac{e^{-\lambda t}(\lambda t)^{k+1}}{\Gamma(k+1)(k+1)} = \frac{e^{-\lambda t}(\lambda t)^{k+1}}{(k+1)!},$$

thus proving the proposition.

7.2.5 Interarrival Times of the Poisson Process

There is a very important property that relates the Poisson process to its inter-arrival times. Suppose we have a Poisson process and T_n is the time between the arrival of the nth and $(n-1)$st event, $n = 1, 2, \ldots$. We now show that T_n, $n = 1, 2, \ldots$ is exponentially distributed.

First we compute the survivor probability $\bar{F}_{T_1}(t) = \Pr(T_1 > t)$ of the arrival time of the first event T_1. Since the events $\{T_1 > t\}$ and $\{N(t) = 0\}$ are equivalent (the first event takes place after time t if and only if there are no arrivals until t), we can write

$$\bar{F}_{T_1}(t) = \Pr(T_1 > t) = \Pr[N(t) = 0] = e^{-\lambda t}\frac{(\lambda t)^0}{0!} = e^{-\lambda t}.$$

This implies that $f_{T_1}(t) = -\bar{F}'_{T_1}(t) = \lambda e^{-\lambda t}$, i.e., T_1 is exponential with parameter λ.

To find the distribution of T_2, we condition on the arrival time of the first event T_1:

$$
\begin{aligned}
\Pr(T_2 > t \mid T_1 = s) &= \Pr\{\text{no events in } (s, s+t] \mid T_1 = s\} \\
&= \Pr\{\text{no events in } (s, s+t]\} \quad \text{(independence)} \\
&= e^{-\lambda t} \quad \text{(stationarity)}
\end{aligned}
$$

Repeating the arguments we see that the nth interarrival time T_n, $n = 1, 2, \ldots$ is distributed exponentially with parameter λ.

7.2.6 Density of the Erlang—An Equivalent Derivation

In Section 7.2.3 we showed using Laplace transforms that the density of the Erlang random variable $S_n = \sum_{i=1}^{n} T_i$ with parameters (n, λ) was

$$f_{S_n}(t) = \frac{\lambda e^{-\lambda t}(\lambda t)^{n-1}}{(n-1)!}, \quad t \geq 0.$$

We now show the same results using a much simpler argument.

It is easy to see that the two events $\{S_n \leq t\}$ and $\{N(t) \geq n\}$ are equivalent since the nth event takes place at or before time t if and only if the number of events until t is at least equal to n. Thus,

$$F_{S_n}(t) = \Pr(S_n \leq t) = \Pr[N(t) \geq n] = \sum_{j=n}^{\infty} e^{-\lambda t} \frac{(\lambda t)^j}{j!}.$$

Differentiating $F_{S_n}(t)$ w.r.t. t and simplifying we find the density $f_{S_n}(t) = F'_{S_n}(t)$ as

```
>    restart: # Erlang2.mws
>    FSn:=Sum(exp(-lambda*t)*(lambda*t)^j/j!,
     j=n..infinity);
```

$$FSn := \sum_{j=n}^{\infty} \frac{e^{(-\lambda t)} (\lambda t)^j}{j!}$$

```
>    fSn:=simplify(diff(value(FSn),t));
```

$$fSn := \frac{e^{(-\lambda t)} (\lambda t)^n}{t \, \Gamma(n)}$$

Rewriting the result in a more familiar format, we obtain the density $f_{S_n}(t)$ of the Erlang r.v. with parameters (n, λ).

7.2.7 Nonhomogeneous Poisson Process

Very often, the arrival rate of a Poisson process may be time dependent. For example, the number of vehicles arriving at a bridge or the number of customers arriving at a restaurant may be different at different times of the day. Thus, it may be more accurate to model such phenomena by assuming that the arrival rate $\lambda(t)$ is a function of time t.

A *nonhomogeneous Poisson (NHP) process* $N(t)$ with arrival rate $\lambda(t)$ is a process with the following properties: (i) $N(0) = 0$, (ii) $N(t)$ had independent increments,[1] (iii) $\Pr[N(t + h) - N(t) = 1] = \lambda(t)h + o(h)$ and (iv) $\Pr[N(t + h) - N(t) \geq 2] = o(h)$.

The integrated arrival rate function for the NHP is defined as

$$M(t) = \int_0^t \lambda(u) \, du.$$

This quantity is the mean value function of the process that gives the expected number of arrivals until t. For an ordinary Poisson process it reduces to $M(t) = \lambda t$.

---

[1] Note that an ordinary (homogeneous) Poisson process with constant λ has stationary increments.

For the NHP, the probability distribution of the number of arrivals in the interval $(t, t+s]$ is still Poisson but with parameter $M(t+s) - M(t)$, i.e.,

$$\Pr[N(t+s) - N(t) = n] = e^{-[M(t+s)-M(t)]} \frac{[M(t+s) - M(t)]^n}{n!}, \quad n = 0, 1, 2, \ldots$$

As in ordinary Poisson processes, we show this result using induction.

For a *fixed* t, define $p_n(s) = \Pr[N(t+s) - N(t) = n]$. For $n = 0$ we have

$$\begin{aligned} p_0(s+h) &= \Pr[N(t+s) - N(t) = 0] \\ &= p_0(s)[1 - \lambda(t+s)h + o(h)] \end{aligned}$$

with $p_0(0) = 1$. The usual operations lead to an ordinary differential equation with a variable coefficient given by

$$p_0'(s) = -\lambda(t+s)p_0(s).$$

We now solve this ODE using Maple and show that $p_0(s) = e^{-[M(t+s)-M(t)]}$.

```
>    restart: # NHP.mws
>    de.0:=diff(p.0(s),s)=-lambda(t+s)*p.0(s);
```

$$de0 := \frac{\partial}{\partial s} p0(s) = -\lambda(t+s)\, p0(s)$$

```
>    IC.0:=p.0(0)=1;
```

$$IC0 := p0(0) = 1$$

```
>    assign(dsolve({ de.0,IC.0} ,{ p.0(s)} ));
>    p.0(s);
```

$$e^{\left(- \int_0^s \lambda(t+u)\,du\right)}$$

Thus, we find that $p_0(s) = e^{-\int_0^s \lambda(t+u)\,du}$. It is not immediately obvious how the function $M(t)$ appears in this result, so we attempt to convert it to a different form using a change of variable that requires the `student` package.

```
>    with(student);
```

[D, *Diff, Doubleint, Int, Limit, Lineint, Product, Sum, Tripleint, changevar, combine, completesquare, distance, equate, extrema, integrand, intercept, intparts, isolate, leftbox, leftsum, makeproc, maximize, middlebox, middlesum, midpoint, minimize, powsubs, rightbox, rightsum, showtangent, simpson, slope, summand, trapezoid, value*
]

```
>    Integral:=int(lambda(t+u),u=0..s);
```

$$Integral := \int_0^s \lambda(t+u)\,du$$

```
>    changevar(t+u=z,Integral,z);
```

$$\int_{t}^{t+s} \lambda(z)\,dz$$

Since $\int_{t}^{t+s} \lambda(z)\,dz = M(t+s) - M(t)$, the result follows.

For $n = 1, 2, \ldots$ the differential-difference equations are obtained as

$$p_n'(s) = -\lambda(t+s)p_n(s) + \lambda(t+s)p_{n-1}(s)$$

which can be solved in a similar manner. Using induction we obtain the required result.

Example 72 *Time-dependent arrival rate in a restaurant.* Consider a restaurant where customers arrive in different rates at different times of the day. The time-dependent arrival rate $\lambda(t)$ is conveniently specified using Maple's piecewise() function and displayed with the plot() function in Figure 7.1.

```
>   restart: #NHPExample.mws
>   lambda:=t->piecewise(t<=3,4+2*t,t<=5,10,
    t<=9,10-(t-3));
```

$$\lambda := t \rightarrow \text{piecewise}(t \le 3,\ 4+2t,\ t \le 5,\ 10,\ t \le 9,\ 13-t)$$

```
>   plot(lambda(t),t=0..9,0..10,discont=true);
```

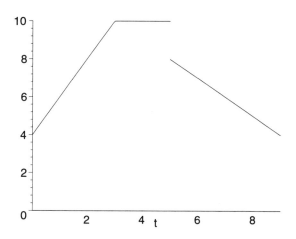

FIGURE 7.1. Time-dependent arrival rate $\lambda(t)$.

In this example, we assume that the restaurant opens at 8 a.m. and stays open for 9 hours. Thus, $t = 0$ corresponds to 8 a.m. and $t = 9$ corresponds to 5 p.m.

Maple easily computes the integrated arrival rate function $M(t)$ (denoted by MInt) as follows:

```
>   MInt:=int(lambda(u),u=0..t);
```

$$MInt := -\frac{1}{2}t^2 \, \text{Heaviside}(t-5) - \frac{5}{2}\text{Heaviside}(t-5) - t^2 \, \text{Heaviside}(t-3)$$

$$- 9\,\text{Heaviside}(t-3) + 4t + t^2 + 3t\,\text{Heaviside}(t-5)$$

$$+ \frac{153}{2}\text{Heaviside}(t-9) - 13t\,\text{Heaviside}(t-9)$$

$$+ \frac{1}{2}t^2\,\text{Heaviside}(t-9) + 6t\,\text{Heaviside}(t-3)$$

Since the result is in terms of Heaviside functions defined as

$$\text{Heaviside}(t) = \begin{cases} 0 & \text{if } t < 0 \\ \text{undefined} & \text{if } t = 0 \\ 1 & \text{if } t > 0, \end{cases}$$

we convert it to the piecewise form and use the `unapply()` function so that $M(t)$ can be evaluated for different values of t.

```
>   M:=unapply(convert(MInt,piecewise,t),t);
```

$$M := t \to \text{piecewise}(t \le 3, \, 4t + t^2, \, t \le 5, \, -9 + 10t, \, t \le 9, \, -\frac{1}{2}t^2 - \frac{23}{2} + 13t,$$

$$9 < t, \, 65)$$

Thus, we find that

$$M(t) = \begin{cases} t^2 + 4t, & 0 \le t \le 3 \\ 10t - 9, & 3 \le t \le 5 \\ -\frac{1}{2}t^2 + 13t - \frac{23}{2}, & 5 \le t \le 9 \end{cases},$$

which is plotted in Figure 7.2.

```
>   plot(M(t),t=0..9,color=black);
```

Next we define the probability $\Pr[N(t+s) - N(t) = n]$ as `PrNHP:=(t,s,n)` `->...` which is a function of three variables (t, s, n). We use the arrow notation so that this probability can be computed by entering the three inputs (t, s, n). For a specific set of values $(2, 2, 15)$ we find that the probability that 15 customers will enter the restaurant between the times $t = 2$ and $t + s = 4$ (i.e., between 8 a.m. and 12 noon) is found to be approximately 0.065:

```
>   PrNHP:=(t,s,n)->exp(-(M(t+s)-M(t)))
    * (M(t+s)-M(t))^n/n!;
```

$$PrNHP := (t, s, n) \to \frac{e^{(-M(t+s)+M(t))} \, (M(t+s) - M(t))^n}{n!}$$

```
>   evalf(PrNHP(2,2,15));
```

$$.06504430041$$

Suppose we are not aware of the above results and attempt to find the probability that 15 customers will arrive between the times $t = 2$ and $t + s = 4$ by using the formula for the ordinary Poisson process. In this case Poisson's arrival rate λ

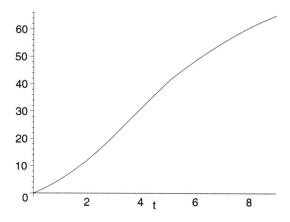

FIGURE 7.2. The integrated arrival rate function $M(t) = \int_0^t \lambda(u)\, du$.

could be computed using the *average* arrival rate of $\lambda_{\text{Avg}} = \frac{1}{t}\int_0^t \lambda(\tau)\, d\tau = 7.22$ customers. The Poisson probability formula, which would provide only an approximation to the true probability, gives 0.101.

>   `lambda[ Avg] :=evalf(int(lambda(t),t=0..9)/9);`

$$\lambda_{Avg} := 7.222222222$$

>   `Poisson:=(t,n)->exp(-lambda[ Avg] * t)`
>   `* (lambda[ Avg] * t)^n/n!;`

$$\text{Poisson} := (t,\, n) \rightarrow \frac{e^{(-\lambda_{Avg}\, t)}\, (\lambda_{Avg}\, t)^n}{n!}$$

>   `Poisson(2,15);`

$$.1013609134$$

>   `Error:=evalf((Poisson(2,15)-PrNHP(2,2,15))`
>   `/PrNHP(2,2,15));`

$$\text{Error} := .5583365914$$

This results in an error of about 55.8%, indicating that the Poisson model provides a poor approximation to the nonhomogeneous Poisson process.

7.2.8   Compound Poisson Process

Consider the following generalization of the ordinary Poisson process $N(t)$ with rate λ. Let $\{Y_i\}$ be discrete-valued i.i.d. random variables with mean $E(Y)$, variance $\text{Var}(Y)$ and the p.g.f. $G_Y(z) = E(z^Y) = \sum_{n=0}^{\infty} a_n z^n$ where $a_n = \Pr(Y = n)$. We assume that the Poisson process $N(t)$ and the sequence $\{Y_i\}$ are independent.

By combining the Poisson process $N(t)$ and the sequence $\{Y_i\}$ in a particular way, we obtain a new process

$$X(t) = \sum_{i=0}^{N(t)} Y_i. \tag{7.3}$$

This new process $X(t)$ is called the *compound Poisson process*.

Such a generalization may be useful in applications where each customer arriving in a store spends a random amount of Y dollars. In this case, $X(t)$ would correspond to the total amount spent by all customers until time t. In a maintenance-type application Y may correspond to the number of hours of work done on a particular vehicle in which case $X(t)$ would be the total time spent on all vehicles until t.

We will now compute the p.g.f. $G_{X(t)}(z) = E[z^{X(t)}]$ of the compound Poisson process $X(t)$ and find its mean and variance assuming that Y is geometric. Conditioning on the number of Poisson events that take place until time t, we can write

$$
\begin{aligned}
G_{X(t)}(z) &= E[z^{X(t)}] \\
&= \sum_{i=0}^{\infty} E\left[z^{X(t)} \mid N(t) = i\right] \Pr[N(t) = i] \\
&= \sum_{i=0}^{\infty} E\left[z^{Y_1+Y_2+\cdots+Y_i}\right] \Pr[N(t) = i] \\
&= \sum_{i=0}^{\infty} [E(z^Y)]^i \Pr[N(t) = i] \\
&= \sum_{i=0}^{\infty} [G_Y(z)]^i e^{-\lambda t} \frac{(\lambda t)^i}{i!}.
\end{aligned}
$$

This infinite sum is easily evaluated by Maple to give

```
>    restart: # CompoundP.mws
>    G[X] :=unapply('sum((G[Y](z))^i
     *exp(-lambda*t)*(lambda*t)^i/i!,
     i=0..infinity)',z);
```

$$G_X := z \to \sum_{i=0}^{\infty} \frac{G_Y(z)^i\, e^{(-\lambda t)}\, (\lambda t)^i}{i!}$$

```
>    G[X](z);
```

$$e^{(-\lambda t)}\, e^{(G_Y(z)\lambda t)}$$

Hence, we see that the p.g.f. of the compound Poisson process $X(t)$ is obtained as $G_{X(t)}(z) = e^{-\lambda t} e^{G_Y(z)\lambda t}$.

Now consider the case where Y follows the geometric distribution $a_n = \Pr(Y = n) = (1-p)p^{n-1}$, $n = 1, 2, \ldots$:

```
>    restart: # Stutter.mws
```

```
>    a[ n] :=(1-p)*p^(n-1);
```

$$a_n := (1 - p)\, p^{(n-1)}$$

The p.g.f. of Y is easily found from $G_Y(z) = \sum_{n=1}^{\infty} a_n z^n$ as

```
>    G[ Y] :=unapply(sum(a[ n] * z^n,n=1..infinity),z);
```

$$G_Y := z \rightarrow -\frac{(1 - p)\, z}{p\, z - 1}$$

```
>    G[ Y] (z);
```

$$-\frac{(1 - p)\, z}{p\, z - 1}$$

Next, evaluating the expression for the p.g.f. $G_{X(t)}(z) = e^{-\lambda t} e^{G_Y(z)\lambda t}$ of $X(t)$, we find

```
>    G[ X] :=unapply(sum(((G[ Y] (z))^n
     * exp(-lambda* t)* (lambda* t)^n/n!,
     n=0..infinity),z);
```

$$G_X := z \rightarrow e^{(-\lambda t)}\, e^{(\frac{-z\lambda t + z\lambda t p}{p z - 1})}$$

Expected Value of $X(t)$

The first moment $E[X(t)]$ of the $X(t)$ process is equal to $G'_{X(t)}(1)$ (denoted by G[X, 1]). To compute the first moment, we first ask Maple to differentiate $G_{X(t)}(z)$ w.r.t. z and find

```
>    G[ X,1] :=unapply(diff(G[ X] (z),z),z);
```

$$G_{X,1} := z \rightarrow e^{(-\lambda t)}\, \left(\frac{-\lambda t + \lambda t\, p}{p\, z - 1} - \frac{(-z\lambda t + z\lambda t\, p)\, p}{(p\, z - 1)^2}\right) e^{(\frac{-z\lambda t + z\lambda t p}{p z - 1})}$$

Substituting $z = 1$ and simplifying, we obtain a simple expression for $E[X(t)]$ as

```
>    EX:=simplify(G[ X,1] (1));
```

$$EX := -\frac{\lambda t}{-1 + p}$$

Variance of $X(t)$

To compute the variance of $X(t)$ we need the second factorial moment of $X(t)$, which is found by evaluating $G''_{X(t)}(1)$. The second derivative $G''_{X(t)}(z)$ (denoted by G[X, 2]) is a very complicated expression and it would be rather difficult to compute it manually. But using Maple, this is achieved relatively easily. Computing $G''_{X(t)}(z)$, substituting $z = 1$ and simplifying, we find

```
>    G[ X,2] :=unapply(diff(G[ X] (z),z$2),z);
```

$$G_{X,2} := z \to e^{(-\lambda t)} \left(-2\,\frac{(-\lambda t + \lambda t\, p)\, p}{(p z - 1)^2} + 2\,\frac{(-z\, \lambda t + z\, \lambda t\, p)\, p^2}{(p z - 1)^3}\right) e^{\left(\frac{-z\lambda t + z\lambda t p}{p z - 1}\right)}$$

$$+ e^{(-\lambda t)} \left(\frac{-\lambda t + \lambda t\, p}{p z - 1} - \frac{(-z\, \lambda t + z\, \lambda t\, p)\, p}{(p z - 1)^2}\right)^2 e^{\left(\frac{-z\lambda t + z\lambda t p}{p z - 1}\right)}$$

```
>   simplify(G[ X, 2] (1));
```

$$\frac{\lambda t\, (2\, p + \lambda t)}{(-1 + p)^2}$$

Since the second moment of $X(t)$ is $E\{[X(t)]^2\} = G''_{X(t)}(1) + G'_{X(t)}(1)$,

```
>   EX2:=simplify(G[ X, 2] (1)+G[ X, 1] (1));
```

$$EX2 := \frac{\lambda t\, (p + \lambda t + 1)}{(-1 + p)^2}$$

we obtain the variance $\mathrm{Var}[X(t)] = E\{[X(t)]^2\} - \{E[X(t)]\}^2$ of the compound Poisson process $X(t)$ simply as

```
>   VarX:=simplify(EX2-EX^2);
```

$$VarX := \frac{\lambda t\, (p + 1)}{(-1 + p)^2}$$

Distribution of $X(t)$

The results found in the preceding paragraphs provide the exact expressions for the first few moments of the compound Poisson process for geometrically distributed Y. When Y is a nonnegative integer random variable, it is also possible to compute symbolically the exact distribution of the $X(t)$ process using a recursive scheme due to Adelson [5]. Also see Tijms [187, pp. 29–30] who provides a faster method based on the fast Fourier transform.

As before, let $a_n = \Pr(Y = n)$, $n = 1, 2, \ldots$ be the density of Y. Defining $r_n(t) = \Pr[X(t) = n]$, $n = 0, 1, \ldots$ as the probability density of $X(t)$, Adelson's method computes these probabilities as

$$r_0(t) = e^{-\lambda t(1-a_0)},$$

$$r_n(t) = \frac{\lambda t}{n} \sum_{j=0}^{n-1}(n - j)a_{n-j}r_j, \quad n = 1, 2, \ldots$$

Note that if, for example, Y corresponds to the number of units of a product purchased by customers who arrive according to a Poisson process with rate λ, then $X(t)$ is the total number of units purchased until time t and $r_n(t)$ is the probability that exactly n units will be purchased by time t.

The recursive equations for, say, $n = 0, 1, 2, 3$ are easily formed and computed with Maple.

```
>   restart: # Adelson.mws
>   Total:= 3;
```

$$Total := 3$$

```
>   r[ 0] :=exp(-lambda* t* (1-a[ 0] ));
```

$$r_0 := e^{(-\lambda t (1-a_0))}$$

```
>   for n from 1 to Total do
    r[ n] :=(lambda* t/n)* sum((n-j)* a[ n-j] * r[ j] ,
    j=0..n-1) od:
```

The expressions for $r_1(t)$, $r_2(t)$ and $r_3(t)$ are found in terms of the as yet unspecified values of a_n, $n = 0, 1, 2, 3$.

```
>   r[ 1] ;  r[ 2] ;  r[ 3] ;
```

$$\lambda t\, a_1\, e^{(-\lambda t (1-a_0))}$$

$$\frac{1}{2} \lambda t \left(2\, a_2\, e^{(-\lambda t (1-a_0))} + a_1^2\, \lambda t\, e^{(-\lambda t (1-a_0))}\right)$$

$$\frac{1}{3} \lambda t \left(3\, a_3\, \%1 + 2\, a_2\, \lambda t\, a_1\, \%1 + \frac{1}{2}\, a_1\, \lambda t \left(2\, a_2\, \%1 + a_1^2\, \lambda t\, \%1\right)\right)$$

$$\%1 := e^{(-\lambda t (1-a_0))}$$

If we assume that Y is geometric, i.e., that $a_n = \Pr(Y = n) = (1 - p)p^{n-1}$, $n = 1, 2, \ldots$, we can obtain the exact forms of the probabilities $r_n(t)$, $n = 1, 2, 3$.

```
>   a[ 0] :=0;
```

$$a_0 := 0$$

```
>   for n from 1 to Total do a[ n] :=(1-p)* p^ (n-1) od;
```

$$a_1 := 1 - p$$

$$a_2 := (1 - p)\, p$$

$$a_3 := (1 - p)\, p^2$$

```
>   r[ 1] ;  r[ 2] ;  r[ 3] ;
```

$$\lambda t\, (1 - p)\, e^{(-\lambda t)}$$

$$\frac{1}{2} \lambda t \left(2\, (1 - p)\, p\, e^{(-\lambda t)} + (1 - p)^2\, \lambda t\, e^{(-\lambda t)}\right)$$

$$\frac{1}{3} \lambda t \left(3\, (1 - p)\, p^2\, e^{(-\lambda t)} + 2\, (1 - p)^2\, p\, \lambda t\, e^{(-\lambda t)}\right.$$
$$\left. + \frac{1}{2}\, (1 - p)\, \lambda t \left(2\, (1 - p)\, p\, e^{(-\lambda t)} + (1 - p)^2\, \lambda t\, e^{(-\lambda t)}\right)\right)$$

We now let $\lambda = 1$ and $p = 0.6$ and plot the probability functions for $r_n(t)$, $n = 0, 1, 2, 3$. These functions are displayed in Figure 7.3.

```
>   lambda:=1; p:=0.6;
```

$$\lambda := 1$$

$$p := .6$$

```
>   plot([ r[ 0] ,seq(r[ n] ,n=1..3)] ,t=0..4,color=black);
```

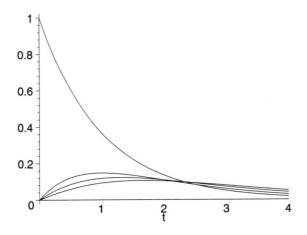

FIGURE 7.3. Probability functions $r_n(t) = \Pr[X(t) = n]$ for $n = 0, 1, 2, 3$ and $0 \le t \le 4$ where $r_0(0) = 1$.

7.3   Renewal Theory

Renewal theory is concerned with extending the results found for the exponential/Poisson duality. As we showed in Section 7.2.5, in a Poisson process the distribution of the interarrival times are i.i.d. exponential random variables. When the interarrival time distribution is non exponential, the resulting counting process becomes a *renewal process*.

Because of its generality, renewal theory has found wide applicability in different areas of operations research. Many problems in inventory and queueing theory have been solved using tools from renewal theory. For excellent coverage of the inventory applications of renewal theory, see Tijms [187, Chapter 1]. Şahin's monograph [163] provides a rigorous renewal-theoretic analysis of a class of (s, S) inventory systems. A book by Nelson [137] describes the use of renewal theory in queueing and performance modeling. Bartholomew [16] uses renewal theory to model manpower systems and other social processes. Crane and Lemoine [56] use the regenerative method of renewal theory to perform simulation analyses more efficiently. As these examples indicate, renewal theory and renewal theoretic arguments have been successfully used to solve a large number of important problems in operations research.

7.3.1 Preliminaries

We consider a sequence of i.i.d. random variables $\{X_n, n = 1, 2, \ldots\}$ with common distribution F, density f, mean $\mu = E(X_n)$ and variance σ^2. The r.v. X_n denotes the interarrival time between the nth and $(n-1)$st events (arrivals/renewals). We assume $F(0) = \Pr(X_n = 0) < 1$, so the mean time between events $\mu > 0$. The quantity $1/\mu$ is the arrival rate of the process.

We define $S_n = \sum_{i=1}^{n} X_i, n = 1, 2, \ldots$ as the arrival time of the nth event with $S_0 = 0$ indicating that there is an event at time $t = 0$. As before, we define $N(t)$ as the number of renewals in the interval $(0, t]$ and write $N(t) = \max\{n : S_n \le t\}$. In Section 7.2.6 we argued that the two events $\{S_n \le t\}$ and $\{N(t) \ge n\}$ were equivalent since the nth event takes place at or before time t if and only if the number of events until t is at least equal to n. These arguments are still valid even if the interarrival times are non exponential. Thus to compute the distribution of the process $N(t)$, we can write

$$
\begin{aligned}
p_n(t) &= \Pr[N(t) = n] \\
&= \Pr[N(t) \ge n] - \Pr[N(t) \ge n + 1] \\
&= \Pr(S_n \le t) - \Pr(S_{n+1} \le t) \\
&= F_n(t) - F_{n+1}(t) \qquad\qquad (7.4)
\end{aligned}
$$

where $F_n(t) = F_1(t) \circledast F_{n-1}(t) = \int_0^\infty F_1(t - u) f_{n-1}(u)\, du$ is the n-fold convolution of F with itself.

7.3.2 Distribution of $N(t)$

We now show using Laplace transforms (LT) that it is relatively easy to develop explicit expressions for the probability $p_n(t)$ provided that the LT of the interarrival density is not too complicated.

Let $\tilde{p}(s) = \mathcal{L}\{p_n(t)\}$, $\tilde{F}_n(s) = \mathcal{L}\{F_n(t)\}$ and $\tilde{f}_n(s) = \mathcal{L}\{f_n(t)\}$ be the LTs of $p_n(t)$, $F_n(t)$ and $f_n(t)$, respectively. Since

$$
\tilde{F}_n(s) = \frac{1}{s}\tilde{f}_n(s)
$$

and $\tilde{f}_n(s) = [\tilde{f}(s)]^n$, taking the LT of both sides of (7.4), we obtain

$$
\tilde{p}_n(s) = \frac{1}{s}[\tilde{f}(s)]^n[1 - \tilde{f}(s)].
$$

We now describe an example where we can invert $\tilde{p}_n(s)$ to obtain closed-form expressions for the probability $p_n(t)$.

Example 73 *Distribution of $N(t)$ with Erlang($2, \lambda$) interarrivals.* In this example, we assume that the interarrivals are distributed as Erlang with parameters $(2, \lambda)$, i.e.,

$$
f(t) = \lambda^2 t e^{-\lambda t}.
$$

This Erlang r.v. has mean $\mu = 2/\lambda$ and variance $\sigma^2 = 2/\lambda^2$.

```
>    restart: # DistN.mws
>    k:=2;
```

$$k := 2$$

```
>    f:=lambda*(lambda*t)^(k-1)*exp(-lambda*t)/(k-1)!;
```

$$f := \lambda^2 t\, e^{(-\lambda t)}$$

We load the integral transforms package inttrans() in order to use the inverse transformation facilities of Maple.

```
>    with(inttrans);
```

[*addtable, fourier, fouriercos, fouriersin, hankel, hilbert, invfourier, invhilbert, invlaplace, invmellin, laplace, mellin, savetable*]

The LT of Erlang(2, λ) is easily found as

```
>    fLT:=laplace(f,t,s);
```

$$fLT := \frac{\lambda^2}{(s+\lambda)^2}$$

For $n = 1$, we attempt to find the probability $p_n(t)$. We form the LT of $p_1(t)$ using $\tilde{p}_1(s) = \frac{1}{s}[\tilde{f}(s)]^1[1 - \tilde{f}(s)]$.

```
>    pLT[ 1] :=(1/s)*fLT*(1-fLT);
```

$$pLT_1 := \frac{\lambda^2\left(1 - \dfrac{\lambda^2}{(s+\lambda)^2}\right)}{s\,(s+\lambda)^2}$$

Inverting $\tilde{p}_1(s)$ gives the probability $p_1(t) = \lambda^2\left(\frac{1}{6}\lambda t^3 e^{-\lambda t} + \frac{1}{2}t^2 e^{-\lambda t}\right)$:

```
>    p[ 1] :=invlaplace(pLT[ 1] ,s,t);
```

$$p_1 := \lambda^2\left(\frac{1}{6}\lambda t^3 e^{(-\lambda t)} + \frac{1}{2}t^2 e^{(-\lambda t)}\right)$$

For $n = 2$ and $n = 3$, similar but more complicated expressions for the Laplace transforms $\tilde{p}_n(t)$ are obtained. These are easily inverted to give

```
>    pLT[ 2] :=(1/s)*fLT^2*(1-fLT);
```

$$pLT_2 := \frac{\lambda^4\left(1 - \dfrac{\lambda^2}{(s+\lambda)^2}\right)}{s\,(s+\lambda)^4}$$

```
>    p[ 2] :=invlaplace(pLT[ 2] ,s,t);
```

$$p_2 := \lambda^4\left(\frac{1}{120}\lambda t^5 e^{(-\lambda t)} + \frac{1}{24}t^4 e^{(-\lambda t)}\right)$$

```
>    pLT[ 3] :=(1/s)*fLT^3*(1-fLT);
```

$$pLT_3 := \frac{\lambda^6 \left(1 - \dfrac{\lambda^2}{(s+\lambda)^2}\right)}{s(s+\lambda)^6}$$

> p[3] :=invlaplace(pLT[3] ,s,t);

$$p_3 := \lambda^6 \left(\frac{1}{5040} \lambda t^7 e^{(-\lambda t)} + \frac{1}{720} t^6 e^{(-\lambda t)}\right)$$

It is worth noting that if the interarrival density were $f(t) = \lambda e^{-\lambda t}$, i.e., exponential, then the distribution of the counting process $N(t)$ would be Poisson, as expected. The reader is invited to check this result by setting $k = 1$ on the second line of the Maple worksheet and executing the worksheet with \Edit \Execute \Worksheet.

7.3.3 Renewal Function and Renewal Density

As we observe, if the interarrival distribution is not too complicated, then it may be possible to find explicit expressions for the distribution of the counting process $N(t)$. However, in some cases this may not be possible or we may need only to compute the mean of the renewal process, i.e., $M(t) = E[N(t)]$. This deterministic function is called the *renewal function* and it can often be computed relatively easily using Laplace transforms.

It is easy to see that

$$
\begin{aligned}
M(t) &= E[N(t)] = \sum_{n=1}^{\infty} n \Pr[N(t) = n] \\
&= \sum_{n=1}^{\infty} n[F_n(t) - F_{n+1}(t)] \\
&= \sum_{n=1}^{\infty} F_n(t)
\end{aligned}
$$

where the last equality is obtained using telescopic cancellations.

It is usually easier to work with *renewal density* $m(t)$, which is the first derivative of the renewal function, i.e., $m(t) = M'(t) = \sum_{n=1}^{\infty} f_n(t)$. The renewal density has an interesting and useful interpretation: Since $f_n(t)\,dt$ is approximately the probability that the nth renewal occurs in $(t, t + dt)$, the sum $m(t)\,dt = \sum_{n=1}^{\infty} f_n(t)\,dt$ is the probability that one renewal (the 1st or the 2nd or 3rd and so on) occurs in $(t, t + dt)$.

Defining $\tilde{m}(s) = \mathcal{L}\{m(t)\}$ as the LT of $m(t)$, and performing the usual operations on $m(t) = \sum_{n=1}^{\infty} f_n(t)$, we find

$$\tilde{m}(s) = \sum_{n=1}^{\infty} \tilde{f}_n(s) = \sum_{n=1}^{\infty} [\tilde{f}(s)]^n$$

$$= \frac{\tilde{f}(s)}{1 - \tilde{f}(s)}. \qquad (7.5)$$

Note that the infinite sum $\sum_{n=1}^{\infty} [\tilde{f}(s)]^n$ converges since

$$0 < \tilde{f}(s) = \int_0^\infty e^{-st} f(t)\, dt < \int_0^\infty f(t)\, dt = 1.$$

Example 74 *Renewal function $M(t)$ for Erlang(3,1) interarrivals.* In this example, the interarrival r.v. is Erlang with parameters $k = 3$ and $\lambda = 1$ having mean $\mu = k/\lambda = 3$ and variance $\sigma^2 = k/\lambda^2 = 3$. Since $k = 3$, the density is given by $f(t) = \frac{1}{2} t^2 e^{-t}$.

```
>    restart:  # RenewalFunc.mws
>    with(inttrans):
>    k:=3;
```

$$k := 3$$

```
>    lambda:=1;
```

$$\lambda := 1$$

```
>    f:=lambda* (lambda*t)^(k-1)*exp(-lambda*t)/(k-1)!;
```

$$f := \frac{1}{2} t^2 e^{(-t)}$$

The LT $\tilde{f}(s)$ of the interarrival density is computed easily as

```
>    fLT:=laplace(f,t,s);
```

$$fLT := \frac{1}{(1+s)^3}$$

Using the result for $\tilde{m}(s)$ in (7.5), we find the LT of the renewal density in terms of $\tilde{f}(s)$:

```
>    mLT:=fLT/(1-fLT);
```

$$mLT := \frac{1}{(1+s)^3 \left(1 - \dfrac{1}{(1+s)^3}\right)}$$

The resulting transform $\tilde{m}(s)$ appears complicated, but Maple easily inverts it and finds the renewal density $m(t)$. Since the renewal function $M(t) = \int_0^t m(u)\, du$ is computed by integrating the renewal density $m(t)$, we also find $M(t)$ as follows.

```
>    m:=unapply(invlaplace(mLT,s,t),t);
```

$$m := t \to \frac{1}{3} - \frac{1}{3} e^{(-3/2\,t)} \cos(\frac{1}{2}\sqrt{3}\,t) - \frac{1}{3} e^{(-3/2\,t)} \sqrt{3} \sin(\frac{1}{2}\sqrt{3}\,t)$$

```
>    M:=int(m(u),u=0..t);
```

$$M := \frac{1}{3} t + \frac{1}{3} e^{(-3/2\,t)} \cos(\frac{1}{2}\sqrt{3}\,t) + \frac{1}{9} e^{(-3/2\,t)} \sqrt{3} \sin(\frac{1}{2}\sqrt{3}\,t) - \frac{1}{3}$$

We note that the $M(t)$ function has a linear asymptotic term ($\frac{1}{3}t$), two transient terms that approach zero as $t \to 0$ and a bias term (the constant $-\frac{1}{3}$). We find that

```
>    limit(M/t,t=infinity);
```

$$\frac{1}{3},$$

i.e.,

$$\lim_{t\to\infty} \frac{M(t)}{t} = \frac{1}{\mu} = \frac{1}{3}.$$

This limiting result holds in general and it is known as the *elementary renewal theorem*.

Now, plotting the renewal function $M(t)$ and its asymptote $\frac{1}{3}t - \frac{1}{3}$ on the same graph, we see that the latter approximates the former *very* accurately for $t \geq 2$. These results are displayed in Figure 7.4.

```
>    #plot({ M,t/3-1/3} ,t=0..4);
```

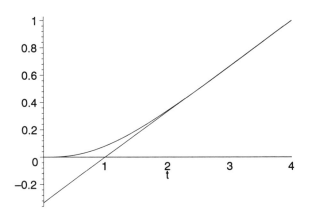

FIGURE 7.4. The renewal function $M(t)$ and its asymptote $\frac{1}{3}t - \frac{1}{3}$ for Erlang(3,1) inter-arrival times.

Finally, a simple check reveals that

$$\lim_{t\to\infty}\left[M(t) - \frac{t}{3}\right] = -\frac{1}{3}.$$

This result is generalized to obtain

$$\lim_{t\to\infty}\left[M(t) - \frac{t}{\mu}\right] = \frac{\sigma^2 - \mu^2}{2\mu^2},$$

which can be seen to be true from the following Maple statements.

```
>    mu:=k/lambda; sigma:=sqrt(k/lambda^2);
```

$$\mu := 3$$

$$\sigma := \sqrt{3}$$

> `limit(M-t/mu,t=infinity);`

$$\frac{-1}{3}$$

> `(sigma^2-mu^2)/(2*mu^2);`

$$\frac{-1}{3}$$

7.3.4 Renewal Equation and the Key Renewal Theorem

In this section we use the powerful "renewal argument" and write an integral equation for the renewal function $M(t)$. We then use Laplace transforms to solve the resulting integral equation. We also provide a brief discussion of the key renewal theorem that is useful in obtaining limiting results as time tends to infinity.

Renewal Equation

Suppose we fix t and condition on the occurrence of the *first* renewal time X_1. Thus

$$M(t) = E[N(t)] = \int_0^\infty E[N(t) \mid X_1 = x] f(x) \, dx.$$

Depending on the relative values of x and t, the conditional expectation assumes different values. If the first renewal occurs after t, then there must have been no renewals until t, and hence $E[N(t) \mid X_1 = x] = 0$. On the other hand, if the first renewal occurs at or before t, then the expected number of renewals until t is 1 (the first renewal) plus $M(t - x)$. These arguments give

$$E[N(t) \mid X_1 = x] = \begin{cases} 0, & \text{for } x > t \\ 1 + M(t - x), & \text{for } x \leq t. \end{cases}$$

Thus, we can write

$$M(t) = \int_t^\infty 0 \cdot f(t) \, dt + \int_0^t [1 + M(t - x)] f(x) \, dx, \qquad (7.6)$$

or

$$M(t) = F(t) + \int_0^t M(t - x) f(x) \, dx.$$

This is known as the *renewal equation* (or the *integral equation of renewal theory*) and it is a computationally more useful result than $M(t) = \sum_{n=1}^\infty F_n(t)$, which is given in terms of the n-fold convolutions of the distribution F.

Now consider the more general integral equation

$$g(t) = h(t) + \int_0^t g(t-x)f(x)\,dx, \quad t \geq 0 \qquad (7.7)$$

where $h(t)$ and $f(t)$ are known functions and $g(t)$ is the unknown function. If we let $h(t) = F(t)$ and $g(t) = M(t)$, then the integral equation for $g(t)$ reduces to the renewal equation in (7.6). We now solve (7.7) using Laplace transforms.

Let $\tilde{g}(s)$ and $\tilde{h}(s)$ be the Laplace transforms of $g(t)$ and $h(t)$, respectively. Taking the LT of equation (7.7), we obtain

$$\tilde{g}(s) = \tilde{h}(s) + \tilde{g}(s)\tilde{f}(s).$$

Solving for $\tilde{g}(s)$ gives

$$\tilde{g}(s) = \frac{\tilde{h}(s)}{1-\tilde{f}(s)} = \tilde{h}(s)\left[1 + \frac{\tilde{f}(s)}{1-\tilde{f}(s)}\right] = \tilde{h}(s) + \tilde{h}(s)\tilde{m}(s) \qquad (7.8)$$

where the last equality is obtained from (7.5). Inversion of (7.8) gives

$$g(t) = h(t) + \int_0^t h(t-x)m(x)\,dx.$$

Specializing this result to the renewal equation, we find

$$M(t) = F(t) + \int_0^t F(t-x)m(x)\,dx$$

as the solution for $M(t)$. Provided that the renewal density is not too complicated, the integral may be evaluated explicitly. We now provide an example of this result using Maple.

Example 75 *Solution of the renewal equation with Erlang(2,2) interarrival times.* In this example the interarrival times are Erlang(k,λ) with $k = 2$ and $\lambda = 2$. This implies that the mean interarrival time is $\mu = k/\lambda = 1$. After defining the density $f(t)$, we easily find the distribution $F(t) = \int_0^t f(u)\,du$ using Maple's int() function.

```
>   restart: # RenewalEqn.mws
>   with(inttrans):
>   k:=2;
```

$$k := 2$$

```
>   lambda:=2;
```

$$\lambda := 2$$

```
>   f:=unapply(lambda* (lambda* t)^(k-1)
    * exp(-lambda* t)/(k-1)!,t);
```

$$f := t \to 4\,t\,e^{(-2t)}$$

```
>  F:=unapply(int(f(u),u=0..t),t);
```

$$F := t \rightarrow -2t\,e^{(-2t)} - e^{(-2t)} + 1$$

We compute the LTs of the probability density $\tilde{f}(s)$ and the renewal density $\tilde{m}(s)$ and invert the latter to find $m(t) = 1 - e^{-4t}$.

```
>  fLT:=laplace(f(t),t,s);
```

$$fLT := 4\,\frac{1}{(s+2)^2}$$

```
>  mLT:=fLT/(1-fLT);
```

$$mLT := 4\,\frac{1}{(s+2)^2\,(1 - 4\,\dfrac{1}{(s+2)^2})}$$

```
>  m:=unapply(invlaplace(mLT,s,t),t);
```

$$m := t \rightarrow 1 - e^{(-4t)}$$

Now using the solution $M(t) = F(t) + \int_0^t F(t-x)m(x)\,dx$, we compute the renewal function $M(t)$.

```
>  M:=unapply(F(t)+int(F(t-x)*m(x),x=0..t),t);
```

$$M := t \rightarrow 1 + \frac{1}{4}e^{(-4t)} + e^{(-4t)}\,t\,e^{(4t)} - e^{(-4t)}\,e^{(4t)} - \frac{1}{4}e^{(-2t)}\,e^{(2t)}$$

```
>  simplify(M(t));
```

$$-\frac{1}{4} + \frac{1}{4}e^{(-4t)} + t$$

Thus, $M(t) = -\frac{1}{4} + \frac{1}{4}e^{-4t} + t$.

As a final check, we find the renewal function using the obvious result $M(t) = \int_0^t m(u)\,du$, which agrees with the solution found.

```
>  MIntm:=int(m(u),u=0..t);
```

$$MIntm := -\frac{1}{4} + \frac{1}{4}e^{(-4t)} + t$$

Key Renewal Theorem

It is worth noting that in this case—and in other cases—it may be much easier to simply integrate the renewal density $m(t)$ and find $M(t)$ instead of using the more complicated solution that involves the integral $\int_0^t F(t-x)m(x)\,dx$. We discuss this solution of the renewal equation since it has other important uses in renewal theory, such as finding the limiting probabilities. These are found using the *key renewal theorem* (KRT), which says that for large t, provided that $h(t)$ is *directly*

Riemann integrable,[2] (DRI), then

$$\lim_{t \to \infty} g(t) = \frac{1}{\mu} \int_0^\infty h(t)\,dt. \tag{7.9}$$

For example, Ross [157, p. 67] discusses an application of this result to find the limiting probability that an alternating renewal process (ARP) will be found in one of its two possible states. An alternating renewal process can be in either ON state (for example, working) or in OFF state (for example, idle). Initially the system is ON and it stays there for a random length of time X_1; after that it stays idle for Y_1. Then it goes to the ON state and stays there for X_2 time units and then switches to OFF state and stays OFF for a length Y_2. This process continues forever. Assuming that the vectors (X_n, Y_n), $n = 1, 2, \ldots$ are i.i.d., Ross shows that if $P(t)$ is defined as the probability that the system is ON at time t, then

$$P(t) = \bar{F}_X(t) + \int_0^t \bar{F}_X(t-z) m_Z(z)\,dz. \tag{7.10}$$

Here, with $F_X(t)$ as the distribution of the ON random variable X, $\bar{F}_X(t) = 1 - F_X(t)$ is the survivor function and $m_Z(t)$ is the renewal density for the cycle time $Z = X + Y$. We now present an example that computes the probability $P(t)$ for finite and infinite t.

Example 76 *The ARP and the Key Renewal Theorem.* We assume that the ON times are Erlang(1,3)—that is, exponential with parameter $\lambda_X = 3$—and the OFF times are Erlang(1,2)—exponential with parameter $\lambda_Y = 2$. This implies that $\mu_X = \frac{1}{3}$, $\mu_Y = \frac{1}{2}$ and $\mu_Z = \mu_X + \mu_Y = \frac{5}{6}$ are the mean times for the ON period X, the OFF period Y and the cycles time Z, respectively.

```
>   restart:  # KRT.mws
>   with(inttrans):
>   k[ X] :=1;
```

$$k_X := 1$$

```
>   lambda[ X] :=3;
```

$$\lambda_X := 3$$

```
>   mu[ X] :=k[ X] /lambda[ X] ;
```

$$\mu_X := \frac{1}{3}$$

```
>   k[ Y] :=1;
```

$$k_Y := 1$$

```
>   lambda[ Y] :=2;
```

---

[2]Sufficient conditions for $h(t)$ to be DRI are as follows: (i) $h(t) \geq 0$, (ii) $h(t)$ is nonincreasing, and (iii) $\int_0^\infty h(t)\,dt < \infty$; see Ross [157, p. 64] and Feller [65, pp. 362–363].

$$\lambda_Y := 2$$

> `mu[ Y] :=k[ Y] /lambda[ Y] ;`

$$\mu_Y := \frac{1}{2}$$

> `mu[ Z] :=mu[ X] +mu[ Y] ;`

$$\mu_Z := \frac{5}{6}$$

For the ON period, we have the density f_X, distribution F_X, survivor function $\bar{F}_X$ and the Laplace transform $\tilde{f}_X$ as follows:

> `f[ X] :=unapply(lambda[ X] * (lambda[ X] * t)^(k[ X] -1)`
> `* exp (-lambda[ X] * t)/(k[ X] -1)!,t);`

$$f_X := t \to 3 e^{(-3t)}$$

> `F[ X] :=unapply(int(f[ X] (u),u=0..t),t);`

$$F_X := t \to -e^{(-3t)} + 1$$

> `FBar[ X] :=t->1-F[ X] (t); FBar[ X] (t);`

$$FBar_X := t \to 1 - F_X(t)$$

$$e^{(-3t)}$$

> `fLT[ X] :=laplace(f[ X] (t),t,s);`

$$fLT_X := 3 \frac{1}{s+3}$$

Similarly, for the OFF period we define the density f_Y and distribution F_Y, and find the survivor function $\bar{F}_Y$ and the Laplace transform $\tilde{f}_Y$:

> `f[ Y] :=unapply(lambda[ Y] * (lambda[ Y] * t)^(k[ Y] -1)`
> `* exp (-lambda[ Y] * t)/(k[ Y] -1)!,t);`

$$f_Y := t \to 2 e^{(-2t)}$$

> `F[ Y] :=unapply(int(f[ Y] (u),u=0..t),t);`

$$F_Y := t \to -e^{(-2t)} + 1$$

> `FBar[ Y] :=t->1-F[ Y] (t); FBar[ Y] (t);`

$$FBar_Y := t \to 1 - F_Y(t)$$

$$e^{(-2t)}$$

> `fLT[ Y] :=laplace(f[ Y] (t),t,s);`

$$fLT_Y := 2 \frac{1}{s+2}$$

Since $Z = X + Y$, with independent X and Y, the LT of Z is the product of the LTs $\tilde{f}_X$ and $\tilde{f}_Y$.

> `fLT[ Z] :=fLT[ X] * fLT[ Y] ;`

$$fLT_Z := 6 \frac{1}{(s+3)(s+2)}$$

Using the expression for the LT of the renewal density $\tilde{m}(s)$ in (7.5) we find the LT of the renewal density for Z:

```
>   mLT[ Z] :=fLT[ Z] /(1-fLT[ Z] );
```

$$mLT_Z := 6 \frac{1}{(s+3)(s+2)(1 - 6\dfrac{1}{(s+3)(s+2)})}$$

Inversion of the LT gives the renewal density $m_Z(t)$ as

```
>   m[ Z] :=unapply(invlaplace(mLT[ Z] ,s,t),t);
```

$$m_Z := t \rightarrow \frac{6}{5} - \frac{6}{5}e^{(-5t)}$$

The probability $P(t)$ is computed using equation (7.10). We see that $P(t) = \frac{2}{5} + \frac{3}{5}e^{-5t}$ and $\lim_{t\to\infty} P(t) = \frac{2}{5}$; i.e., the limiting probability of being in the ON state is 2/5.

```
>   P:=unapply(FBar[ X] (t)+int(FBar[ X] (t-x)*m[ Z] (x),
    x=0..t),t);
```

$$P := t \rightarrow \frac{2}{5}e^{(-5t)}e^{(5t)} + \frac{3}{5}e^{(-5t)}$$

```
>   limit(P(t),t=infinity);
```

$$\frac{2}{5}$$

We also plot the $P(t)$ function and observe that it converges to 2/5 as indicated. This is presented in Figure 7.5.

```
>   #plot(P(t),t=0..10,0..1);
```

From the KRT we know that since $P(t) = \bar{F}_X(t) + \int_0^t \bar{F}_X(t-z)m_Z(z)\,dz$ and since $\bar{F}_X(t)$ is DRI,

$$\lim_{t\to\infty} P(t) = \frac{1}{\mu_Z}\int_0^\infty \bar{F}_X(t)\,dt = \frac{\mu_X}{\mu_X + \mu_Y} = \frac{\frac{1}{3}}{\frac{1}{3}+\frac{1}{2}} = \frac{2}{5}.$$

We obtain the expected results.

```
>   int(FBar[ X] (t),t=0..infinity)/mu[ Z] ;
```

$$\frac{2}{5}$$

```
>   mu[ X] /(mu[ X] +mu[ Y] );
```

$$\frac{2}{5}$$

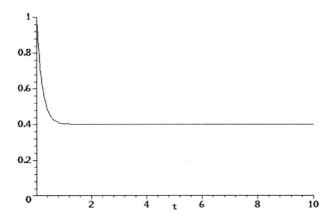

FIGURE 7.5. The probability $P(t)$ converges to $\frac{2}{5}$ as t tends to infinity.

7.3.5   Renewal Reward Process

This process generalizes the compound Poisson process that was discussed in Section 7.2.8. Consider a renewal process $N(t)$ with interarrival times $X_1, X_2, \ldots$ having the common distribution F and suppose that a "reward" R_n is earned at the time of the nth renewal. We assume that the rewards $\{R_n, \; n = 1, 2, \}$ and the vectors $\{(X_n, R_n), \; n = 1, 2, \ldots\}$ are i.i.d.

Similar to the construction of the compound Poisson process, we let

$$R(t) = \sum_{n=1}^{N(t)} R_n$$

where $R(t)$ denotes the total reward earned until t. Computing the limiting values of the average reward is an extremely important problem in stochastic optimization applications found in, for example, inventory theory and queueing processes and it is given by the following theorem; see, Ross [157, pp. 78–79].

Theorem 12 *Renewal Reward Theorem (RRT): If $E(R)$ and $\mu = E(X)$ are both finite, then the long-run average expected reward per unit time is*

$$\lim_{t \to \infty} \frac{R(t)}{t} = \frac{E(R)}{E(X)} \; with \; probability \; 1.$$

Consider a stochastic system that replicates itself probabilistically at renewal points $X_1, X_1 + X_2, X_1 + X_2 + X_3, \ldots$ If we suppose that a renewal signifies the start of a *cycle*, then this theorem states that the average reward per unit time is simply the expected reward per cycle divided by the expected cycle length. This is a very powerful result in optimization of stochastic systems as it reduces an infinite horizon optimization problem to a static (calculus) problem.

We now discuss an example where an application of the RRT greatly simplifies the analysis of the problem.

Example 77 *Age replacement policy for a car.* This example is adapted from an exercise in Ross [158, Problem 7.14]. Consider a car whose lifetime X has distribution F and density f. The replacement policy adopted by the owner is as follows: "Buy a new car as soon as the old car either breaks down or reaches the age of T years." The cost of a new car is c_1 dollars. If the car has to be replaced after it breaks down, then there is an additional cost of c_2 dollars. If the owner sells a T-year old car in working order, then he receives $R(T)$ dollars. The problem is to find the long-run average cost and determine the optimal replacement time T^* that minimizes the owner's average cost.

It is easy to see that every time a new car is purchased, either due to a breakdown or planned replacement after using the car for T years, a new cycle starts. To compute the long-run average cost, we need to determine the expected cycle cost $E(C)$ and expected cycle length $E(L)$.

Conditioning on the occurrence time of the first breakdown, we can write the expected cycle cost as $E(C) = \int_0^\infty E(C \mid X_1 = x) f(x)\,dx$ where

$$E(C \mid X_1 = x) = \begin{cases} c_1 - R(T), & x > T \\ c_1 + c_2, & x \le T. \end{cases}$$

Thus,

$$\begin{aligned} E(C) &= \int_T^\infty [c_1 - R(T)] f(x)\,dx + \int_0^T (c_1 + c_2) f(x)\,dx \\ &= c_1 + c_2 F(T) - R(T)[1 - F(T)]. \end{aligned}$$

Similarly, we can write the expected cycle length as $E(L) = \int_0^\infty E(L \mid X_1 = x) f(x)\,dx$ where

$$E(L \mid X_1 = x) = \begin{cases} T, & x > T \\ x, & x \le T \end{cases}$$

so that

$$\begin{aligned} E(L) &= \int_T^\infty T f(x)\,dx + \int_0^T x f(x)\,dx \\ &= [1 - F(T)]T + \int_0^T x f(x)\,dx. \end{aligned}$$

Now using the RRT, we obtain the long-run average expected cost per unit time as

$$C(T) = \frac{E(C)}{E(L)} = \frac{c_1 + c_2 F(T) - R(T)[1 - F(T)]}{[1 - F(T)]T + \int_0^T x f(x)\,dx}.$$

Optimization of this cost function for $c_1 = 16$ and $c_2 = 1$ with $R(T) = 10 - T$ is performed as follows.

```
>    restart: # CarReplace.mws
>    c[ 1] :=16; c[ 2] :=1;
```

$$c_1 := 16$$

$$c_2 := 1$$

```
>    R:=T->10-T;
```

$$R := T \to 10 - T$$

We assume that the lifetime random variable is uniform with density $f(x) = 1/(b-a)$ where $a = 2$ and $b = 8$.

```
>    a:=2; b:=8;
```

$$a := 2$$

$$b := 8$$

```
>    f:=x->1/(b-a);
```

$$f := x \to \frac{1}{b-a}$$

The distribution $F(x) = \int_0^x f(u)\,du$ is easily computed to give

```
>    F:=unapply(int(f(u),u=2..x),x);
```

$$F := x \to \frac{1}{6}x - \frac{1}{3}$$

```
>    F(x);
```

$$\frac{1}{6}x - \frac{1}{3}$$

The cost function $C(T)$ is obtained as a somewhat complicated nonlinear expression.

```
>    Top:=c[ 1] +c[ 2]*F(T)-R(T)*(1-F(T));
```

$$Top := \frac{47}{3} + \frac{1}{6}T - (10 - T)\left(\frac{4}{3} - \frac{1}{6}T\right)$$

```
>    Bottom:=int(x*f(x),x=0..T)+T*(1-F(T));
```

$$Bottom := \frac{1}{12}T^2 + T\left(\frac{4}{3} - \frac{1}{6}T\right)$$

```
>    C:=unapply(Top/Bottom,T);
```

$$C := T \to \frac{\dfrac{47}{3} + \dfrac{1}{6}T - (10 - T)\left(\dfrac{4}{3} - \dfrac{1}{6}T\right)}{\dfrac{1}{12}T^2 + T\left(\dfrac{4}{3} - \dfrac{1}{6}T\right)}$$

When the cost function is plotted—which we leave to the reader—one observes that it is a strictly convex function with a unique minimum in the interior of the interval $[a, b] = [2, 8]$.

```
>    #plot(C(T),T=a..b);
```

Differentiating the cost function, equating the result to zero and solving—using `fsolve()`—gives the optimal replacement time as $T^* = 5.15$ years.

```
> CT:=normal(diff(C(T),T));
```

$$CT := 2\,\frac{3\,T^2 + 28\,T - 224}{T^2\,(T-16)^2}$$

```
> TStar:=fsolve(CT(T),T,a..b);
```

$$TStar := 5.153946575$$

As a final check we find the second derivative $C''(T)$ of the cost function and evaluate it at $T^* = 5.15$, which results in a positive value implying that the solution we just found is indeed the global minimum. A comparison of the value of the cost function at the end points a and b with T^* also indicates that we have located the true optimal solution.

```
> CTT:=unapply(normal(diff(C(T),T$2)),T);
```

$$CTT := T \to -4\,\frac{3\,T^3 + 42\,T^2 - 672\,T + 3584}{T^3\,(T-16)^3}$$

```
> CTT(TStar);
```

$$.03771347490$$

```
> evalf(C(a)); C(TStar); evalf(C(b));
```

$$3.428571429$$

$$3.054091250$$

$$3.187500000$$

The renewal reward theorem was popularized by Ross [156] and it has been applied successfully in different areas of operations research. For example, Gürler and Parlar [78] and Parlar [143] use the RRT to model a stochastic inventory system with two unreliable suppliers who may be unavailable for random durations. Taylor [184] uses it to model a repair problem for a deteriorating item and Tijms [186, Ch. 1] describes different applications of RRT in queueing theory.

7.4   Discrete-Time Markov Chains

Consider a discrete-time stochastic process $\{X_n,\ n = 0, 1, 2, \ldots\}$ with finite or countably infinite state space S. Suppose that this process has the property that the conditional distribution of X_{n+1} given the past information $X_n, X_{n-1}, \ldots, X_1, X_0$ depends *only* on the value of X_n but not on $X_{n-1}, \ldots, X_1, X_0$, i.e., for any set of values $j, i, i_{n-1}, \ldots, i_0$ belonging to the state space S,

$$\begin{aligned}\Pr(X_{n+1} = j \ &| \ X_n = i,\ X_{n-1} = i_{n-1}, \ldots, X_0 = i_0) \\ &= \ \Pr(X_{n+1} = j \mid X_n = i) = p_{ij}^{(n,n+1)}\end{aligned}$$

for $n = 0, 1, 2, \ldots$ A *time-homogeneous* Markov chain has the property that the transition probabilities $p_{ij}^{(n,n+1)}$ are constant over time; i.e., for all $n = 0, 1, 2, \ldots$

$$\Pr(X_{n+1} = j \mid X_n = i) = p_{ij}, \quad i, j \in S.$$

We assume throughout this chapter that the Markov chains we consider are time homogeneous. The transition probabilities p_{ij} form the one-step transition matrix $\mathbf{P} = [p_{ij}]$ and they satisfy (i) $p_{ij} \geq 0$ for $i, j \in S$ and (ii) $\sum_{j \in S} p_{ij} = 1$ for $i \in S$, thus making $\mathbf{P}$ a "stochastic matrix." The distribution of the initial state X_0 and the transition matrix $\mathbf{P}$ completely determine the distribution of a future state $X_n, n = 1, 2, \ldots$ of the Markov chain.

Example 78 *Random walk.* As a simple example, consider a coin-tossing game where the possible outcomes are (i) heads (H) with probability p and (ii) tails (T) with probability $q = 1 - p$. Suppose that H and T correspond to a gain of \$1 and a loss of \$1, respectively. If we denote the outcome of the nth toss by Z_n, then

$$Z_n = \begin{cases} 1 & \text{if heads with probability } p \\ -1 & \text{if tails with probability } q. \end{cases}$$

Now consider the partial sum $X_n = Z_1 + Z_2 + \cdots + Z_n$. Assuming $X_0 = 0$, the sum X_n may represent the accumulated wealth of a gambler in the first n plays of the coin-tossing game who earns a random amount Z_n in each play. Since we can write $X_{n+1} = X_n + Z_{n+1}$, it is easy to see that the X_n process (known as the "random walk") is a Markov chain as the distribution of X_{n+1} depends only on the value of X_n (and on the r.v. Z_{n+1}), *not* on $X_{n-1}, \ldots, X_1, X_0$.

To compute the transition probabilities $p_{ij} = \Pr(X_{n+1} = j \mid X_n = i), i, j \in S = \{\ldots, -2, -1, 0, 1, 2, \ldots\}$ of this simple Markov chain we proceed formally as follows:

$$\begin{aligned} p_{ij} &= \Pr(X_{n+1} = j \mid X_n = i) \\ &= \Pr(X_n + Z_{n+1} = j \mid X_n = i) \\ &= \Pr(i + Z_{n+1} = j) \\ &= \Pr(Z_{n+1} = j - i). \end{aligned}$$

Since $Z_{n+1} = \pm 1$, we see that $j = i + 1 \Leftrightarrow Z_{n+1} = 1$ and $j = i - 1 \Leftrightarrow Z_{n+1} = -1$. Thus, if the process is in state i, the only possible movements are (i) to $i + 1$ ("up") with probability p and (ii) to $i - 1$ ("down") with probability $q = 1 - p$, i.e., $p_{i,i+1} = p$ and $p_{i,i-1} = q, i \in S$. The infinite dimensional transition matrix

P of this random walk process is then obtained as

$$
\begin{array}{c|ccccccc}
\nearrow & \cdots & -2 & -1 & 0 & 1 & 2 & \cdots \\
\hline
\vdots & \ddots & & & & & & \cdots \\
-2 & & 0 & p & & & & \\
-1 & & q & 0 & p & & & \\
0 & & & q & 0 & p & & \\
1 & & & & q & 0 & p & \\
2 & & & & & q & 0 & \\
\vdots & \vdots & & & & & & \ddots
\end{array}
$$

Transition matrices of Markov chains can be displayed graphically using a "digraph" (or a "transition diagram"). The digraph for the random walk example is displayed in Figure 7.6.

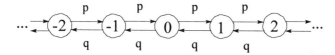

FIGURE 7.6. Digraph for the random walk example.

The one-period dependency exhibited in this example (that X_{n+1} is a function of X_n only) is known as the *Markovian property*[3] of a Markov chain. Despite the simplicity of this property a surprisingly large number of practical problems in inventory theory, queues, telecommunications and so on have been successfully analyzed using Markov chains; see, for example, Tijms [186, Chapter 2] for an excellent discussion of these applications.

Example 79 *A periodic-review (s, S) inventory system.* Consider a single product for which demands Y_n during successive weeks $n = 1, 2, \ldots$ are i.i.d. with the common density $g_Y(k) = \Pr(Y = k)$, $k = 0, 1, 2, \ldots$. We let X_n be the inventory position at the end of week n and assume that excess demand is lost. Management uses an (s, S) ordering policy, that is, after observing the level of X_n, they order

---

[3]It is interesting to note that the Russian mathematician Andrei Andreivich Markov (1856–1922) arrived at the notion of Markov chains when he investigated the alteration of vowels and consonants in Pushkin's poem *Onegin* and discovered that there was indeed some kind of dependency between the occurrence of these vowels and consonants (Takács [183, p. 6]).

Q_n units at the end of week n where

$$Q_n = \begin{cases} S - X_n & \text{if } 0 \le X_n < s \\ 0 & \text{if } s \le X_n \le S. \end{cases}$$

In inventory theory S and s are known as the "order-up-to level" and the "re-order point," respectively. We may assume that orders are placed on a Saturday evening before the weekend closing and they arrive on Monday morning just before the store reopens. Thus, for modeling purposes, deliveries can be assumed to be instantaneous. Since the excess demand is lost and the maximum possible value of X_n is S, the state space of this stochastic process is $S = \{0, 1, 2, \ldots, S\}$. We now show that X_n is a Markov chain.

If, at the end of week n the inventory position X_n is below the reorder point s, i.e., if $X_n < s$, then the management orders $S - X_n$ units, which increases the inventory position at the start of the following week (on Monday morning) to S. Since demand during $n + 1$ is a random quantity represented by Y_{n+1} and since excess demand is lost, we obtain

$$X_{n+1} = \max(S - Y_{n+1}, 0), \quad \text{if } 0 \le X_n < s$$

as the inventory position at the end of the following week. However, if $s \le X_n \le S$, then no action is taken and we have

$$X_{n+1} = \max(X_n - Y_{n+1}, 0), \quad \text{if } s \le X_n \le S.$$

We see that X_{n+1} depends *only* on X_n but not on $X_{n-1}, X_{n-2}, \ldots, X_1, X_0$; thus X_n is a Markov chain. We now find the transition matrix $\mathbf{P} = [p_{ij}]$ of this Markov chain.

Assume that the demand r.v. Y is Poisson with rate λ, i.e.,

$$g_Y(k) = \frac{\lambda^k e^{-\lambda}}{k!}, \quad k = 0, 1, 2, \ldots$$

and define a complementary cumulative-type probability as

$$\hat{G}(k) = \Pr(Y \ge k) = 1 - \sum_{m=0}^{k-1} g_Y(m)$$

For $0 \le i < s$, we have

$$\begin{aligned} p_{ij} &= \Pr(X_{n+1} = j \mid X_n = i) \\ &= \Pr[\max(S - Y_{n+1}, 0) = j] \\ &= \begin{cases} \Pr(Y_{n+1} \ge S) = \hat{G}(S), & \text{if } j = 0 \\ \Pr(Y_{n+1} = S - j) = g(S - j), & \text{if } j = 1, \ldots, S. \end{cases} \end{aligned}$$

Similarly, for $s \le i \le S$, we obtain

$$\begin{aligned} p_{ij} &= \Pr(X_{n+1} = j \mid X_n = i) \\ &= \Pr[\max(i - Y_{n+1}, 0) = j] \\ &= \begin{cases} \Pr(Y_{n+1} \ge i) = \hat{G}(i), & \text{if } j = 0 \\ \Pr(Y_{n+1} = i - j) = g(i - j), & \text{if } j = 1, \ldots, S \end{cases} \end{aligned}$$

with the condition that if $i - j < 0$ then $\Pr(Y_{n+1} = i - j) = g(i - j) = 0$. With $\lambda = 1$ and $(s, S) = (1, 4)$, the transition matrix $\mathbf{P}$ is found as follows.

```
>    restart: # PeriodicsS.mws
>    lambda:=1;
```

$$\lambda := 1$$

```
>    g:=k->'if'(k>=0,lambda^k*exp(-lambda)/k!,0);
```

$$g := k \rightarrow \text{'if'}\left(0 \le k, \frac{\lambda^k e^{(-\lambda)}}{k!}, 0\right)$$

```
>    GHat:=k->1-sum(g(m),m=0..k-1);
```

$$GHat := k \rightarrow 1 - \left(\sum_{m=0}^{k-1} g(m)\right)$$

```
>    s:=1; S:=4;
```

$$s := 1$$

$$S := 4$$

The next two Maple commands generate the probabilities p_{ij} for $0 \le i < s$ and $s \le i \le S$, $j \in S$ using the results obtained.

```
>    for i from 0 to s-1 do
     for j from 0 to S do if j=0 then
     p[i,j]:=GHat(S) else
     p[i,j]:= g(S-j) fi od od;
>    for i from s to S do
     for j from 0 to S do
     if j=0 then p[i,j]:=GHat(i)
     else p[i,j]:= g(i-j) fi od od;
```

A simple check reveals that the probabilities in each row indeed sum to 1, i.e., $\sum_{j \in S} p_{ij} = 1$ for $i \in S$:

```
>    for i from 0 to S do
     evalf(sum(p[0,column],column=0..S))
     od;
```

$$1.$$

$$1.$$

$$1.$$

$$1.$$

$$1.$$

Each row of the $\mathbf{P}$ matrix is generated and placed into an array named PArray whose row and column indices are $0..S$ and $0..S$.

```
>    for i from 0 to S do
     Row[i]:=evalf([seq(p[i,j],j=0..S)],4)
     od;
```

$$Row_0 := [.0188, .06133, .1840, .3679, .3679]$$

$$Row_1 := [.6321, .3679, 0, 0, 0]$$

$$Row_2 := [.2642, .3679, .3679, 0, 0]$$

$$Row_3 := [.0802, .1840, .3679, .3679, 0]$$

$$Row_4 := [.0188, .06133, .1840, .3679, .3679]$$

```
>   PArray:=array(0..S,0..S,[ seq(Row[ i] ,i=0..S)] ):
```

Converting the array to a matrix we obtain the transition matrix **P**:

```
>   P:=convert(PArray,matrix,4);
```

$$P := \begin{bmatrix} .0188 & .06133 & .1840 & .3679 & .3679 \\ .6321 & .3679 & 0 & 0 & 0 \\ .2642 & .3679 & .3679 & 0 & 0 \\ .0802 & .1840 & .3679 & .3679 & 0 \\ .0188 & .06133 & .1840 & .3679 & .3679 \end{bmatrix}$$

Note that if the process is in state 1, i.e., $X_n = 1$, it is impossible to make a transition to states 2, 3 or 4 since no order will be placed at the end of week n. This means that week $n + 1$ will start with 1 unit and end with either 0 units (if there is a demand) or 1 unit (if there is no demand). The other entries in rows 2 and 3 with zero transition probabilities have a similar interpretation.

7.4.1 Chapman-Kolmogorov Equations

In the foregoing discussion, we defined the one-step transition as $p_{ij} = \Pr(X_{n+1} = j \mid X_n = i)$. We now generalize and define the n-step transition as $p_{ij}^{(n)} = \Pr(X_{m+n} = j \mid X_m = i)$ for $n = 0, 1, 2, \ldots$ and $i, j \in S$. Since we assumed that the Markov chain is time homogeneous, i.e., that the probability is invariant in m, we can write the n-step transition as

$$\Pr(X_{m+n} = j \mid X_m = i) = \Pr(X_n = j \mid X_0 = i).$$

To compute $p_{ij}^{(n)}$ we proceed by conditioning on the state visited in the first transition:

$$\begin{aligned} p_{ij}^{(n)} &= \Pr(X_n = j \mid X_0 = i) \\ &= \sum_{k \in S} \Pr(X_n = j \mid X_1 = k, X_0 = i) \Pr(X_1 = k \mid X_0 = i) \\ &= \sum_{k \in S} p_{ik} p_{kj}^{(n-1)}. \end{aligned} \qquad (7.11)$$

If we now define $\mathbf{P}^{(n)} = [p_{ij}^{(n)}]$ as the matrix of n-step transition probabilities, we see that Chapman-Kolmogorov equations in (7.11) can be written conveniently in matrix notation as

$$\mathbf{P}^{(n)} = \mathbf{P} \times \mathbf{P}^{(n-1)}.$$

Since $\mathbf{P}^{(2)} = \mathbf{P} \times \mathbf{P}^{(1)} = \mathbf{P}^2$, $\mathbf{P}^{(3)} = \mathbf{P} \times \mathbf{P}^{(2)} = \mathbf{P}^3$ and so on, we find that the n-step transition matrix $\mathbf{P}^{(n)}$ is simply the nth power $\mathbf{P}^n$ of the one-step transition matrix $\mathbf{P}$, i.e.,

$$\mathbf{P}^{(n)} = \mathbf{P}^n.$$

Example 80 *A periodic-review (s, S) inventory system (continued)*. Using Maple's `evalm()` function we compute the nth power of the transition matrix for different values of n. First, we set `Digits:=4` in order to fit the output on a printed page. (Note that `Digits` is an environment variable that specifies the number of digits carried in floating point calculations with a default of 10.)

```
>    Digits:=4; # PeriodicsS.mws (continued)
```

$$Digits := 4$$

For $n = 2$, 8 and 64 we compute $\mathbf{P}^n$ and note that for $n = 64$ the probabilities seem to have (almost) converged to the same values in each row. This is not a coincidence and as we will discuss later, for a large class of Markov chains this convergence is achieved regardless of the initial state.

```
>    evalm(P^2);
```

$$\begin{bmatrix} .1241 & .1817 & .2743 & .2777 & .1423 \\ .2444 & .1742 & .1163 & .2325 & .2325 \\ .3347 & .2870 & .1840 & .09720 & .09720 \\ .2445 & .2757 & .2856 & .1649 & .02951 \\ .1241 & .1817 & .2743 & .2777 & .1423 \end{bmatrix}$$

```
>    evalm(P^8);
```

$$\begin{bmatrix} .2217 & .2225 & .2220 & .2051 & .1293 \\ .2220 & .2223 & .2209 & .2051 & .1303 \\ .2231 & .2231 & .2212 & .2041 & .1293 \\ .2226 & .2233 & .2220 & .2043 & .1287 \\ .2217 & .2225 & .2220 & .2051 & .1293 \end{bmatrix}$$

```
>    evalm(P^64);
```

$$\begin{bmatrix} .2235 & .2240 & .2228 & .2058 & .1301 \\ .2235 & .2240 & .2228 & .2058 & .1301 \\ .2235 & .2240 & .2228 & .2058 & .1301 \\ .2235 & .2241 & .2229 & .2059 & .1301 \\ .2235 & .2240 & .2228 & .2058 & .1301 \end{bmatrix}$$

$$\%1 := [.2235, .2240, .2228, .2058, .1301]$$

We should point out that since $p_{ij}^{(n)} = \Pr(X_n = j \mid X_0 = i)$, the probabilities we have computed so far are all *conditional* on the knowledge of the initial state. To find the *unconditional* probabilities $\pi_j^{(n)} = \Pr(X_n = j)$ of finding the process in state j after n, we condition on the initial state X_0 and write

$$\pi_j^{(n)} = \sum_{i \in S} \Pr(X_n = j \mid X_0 = i) \Pr(X_0 = i)$$

$$= \sum_{i \in S} p_{ij}^{(n)} \pi_i^{(0)}. \tag{7.12}$$

Now defining the vector of n-step unconditional probabilities as $\Pi^{(n)} = [\pi_0^{(n)}, \pi_1^{(n)}, \ldots]$ with $\Pi^{(0)}$ as the initial distribution vector with $\sum_{i \in S} \pi_i^{(0)} = 1$, we can write (7.12) using matrix/vector notation as

$$\Pi^{(n)} = \Pi^{(0)} \mathbf{P}^{(n)}.$$

Equivalently, if we had conditioned on the position of the process at the $(n-1)$st step as $\pi_j^{(n)} = \sum_{i \in S} \Pr(X_n = j \mid X_{n-1} = i) \Pr(X_{n-1} = i)$, then $\Pi^{(n)}$ would be obtained as

$$\Pi^{(n)} = \Pi^{(n-1)} \mathbf{P}. \tag{7.13}$$

7.4.2 Limiting Probabilities and the Stationary Distribution

It is useful to note that (7.13) can be written as

$$\Pi^{(n)} - \Pi^{(n-1)} = \Pi^{(n-1)} \mathbf{P} - \Pi^{(n-1)}$$

$$= \Pi^{(n-1)} \mathbf{Q}$$

where $\mathbf{I}$ is the identity matrix and $\mathbf{Q} = \mathbf{P} - \mathbf{I}$ is a matrix with rows adding up to 0. If the vectors $\Pi^{(n)}$ converge to a constant vector $\Pi = [\pi_0, \pi_1, \ldots]$ for large n, then $\mathbf{0} = \Pi \mathbf{Q}$, or $\Pi = \Pi \mathbf{P}$. Thus, the limiting probabilities can be found by solving the linear system $\Pi = \Pi \mathbf{P}$ with the normalizing condition $\sum_{i \in S} \pi_i = 1$.

To formalize this argument, we suppose that $\lim_{n \to \infty} p_{ij}^{(n)} = \pi_j$ for all $i \in S = \{0, 1, 2, \ldots\}$; i.e., after a long time the conditional probability

$$\Pr(\text{process in state } j \mid \text{process started in } i)$$

becomes independent of the initial state i. This implies that there exists the limit

$$\lim_{n \to \infty} \mathbf{P}^{(n)} = \begin{array}{c} 0 \\ 1 \\ 2 \\ \vdots \end{array} \left(\begin{array}{cccc} \pi_0 & \pi_1 & \pi_2 & \cdots \\ \pi_0 & \pi_1 & \pi_2 & \cdots \\ \pi_0 & \pi_1 & \pi_2 & \cdots \\ \vdots & \vdots & \vdots & \ddots \end{array} \right)$$

where $\{\pi_j\}$ are the limiting (steady-state) probabilities.

Now consider the n-step unconditional probabilities $\pi_j^{(n)} = \sum_{i \in S} \pi_i^{(0)} p_{ij}^{(n)}$. Taking the limit of both sides, we have

$$\lim_{n \to \infty} \pi_j^{(n)} = \lim_{n \to \infty} \sum_{i \in S} \pi_i^{(0)} p_{ij}^{(n)}$$

$$= \sum_{i \in S} \pi_i^{(0)} \lim_{n \to \infty} p_{ij}^{(n)}$$

$$= \sum_{i \in S} \pi_i^{(0)} \pi_j$$

$$= \pi_j \sum_{i \in S} \pi_i^{(0)}$$

$$= \pi_j.$$

Thus, $\pi_j^{(n)}$ converges to the same limit as $p_{ij}^{(n)}$ and it is independent of the initial probabilities.

When they exist, the unconditional probabilities π_j can be found solving $\Pi = \Pi \mathbf{P}$ with the normalizing condition $\Pi \mathbf{e} = 1$ where $\mathbf{e} = [1, 1, \ldots, 1]'$ is a column vector of 1s. The solution to $\Pi = \Pi \mathbf{P}$ and $\Pi \mathbf{e} = 1$ (when it exists) gives the *stationary* distribution of the Markov chain.

Example 81 *The stationary distribution of the periodic-review* (s, S) *inventory system.* Returning to the periodic-review (s, S) inventory example, we now find the stationary distribution of the Markov chain representing the inventory position X_n.

The next eleven lines are the same as before; therefore they are terminated with a colon (:) in order not to repeat any of the already known results.

```
>    restart: # PeriodicsSLimit.mws

>    lambda:=1:

>    g:=k-> 'if'(k>=0,lambda^k*exp(-lambda)/k!,0):

>    GHat:=k->1-sum(g(m),m=0..k-1):

>    s:=1: S:=4:

>    for i from 0 to s-1 do
     for j from 0 to S do if j=0 then
     p[ i,j] :=GHat(S) else
     p[ i,j] := g(S-j)  fi
     od od;

>    for i from s to S do
     for j from 0 to S do
     if j=0 then p[ i,j] :=GHat(i)
     else p[ i,j] := g(i-j)  fi
     od od;

>    for i from 0 to S do
     evalf(sum(p[ 0,column] ,column=0..S))
     od:

>    for i from 0 to S do
     Row[ i] :=evalf([ seq(p[ i,j] ,j=0..S)] ,4)
     od:

>    PArray:=array(0..S,0..S,[ seq(Row[ i] ,i=0..S)] ):

>    P:=convert(PArray,matrix,4);
```

$$P := \begin{bmatrix} .0188 & .06133 & .1840 & .3679 & .3679 \\ .6321 & .3679 & 0 & 0 & 0 \\ .2642 & .3679 & .3679 & 0 & 0 \\ .0802 & .1840 & .3679 & .3679 & 0 \\ .0188 & .06133 & .1840 & .3679 & .3679 \end{bmatrix}$$

We define the vector of unknown stationary probabilities π_j

```
>   pi:=vector(5,[] );
```

$$\pi := \text{array}(1..5,\ [])$$

and evaluate $\Pi = \Pi P$.

```
>   EqnList:=evalm(pi - pi &* P);
```

$$EqnList := \Big[.9812\,\pi_1 - .6321\,\pi_2 - .2642\,\pi_3 - .0802\,\pi_4 - .0188\,\pi_5,$$
$$.6321\,\pi_2 - .06133\,\pi_1 - .3679\,\pi_3 - .1840\,\pi_4 - .06133\,\pi_5,$$
$$.6321\,\pi_3 - .1840\,\pi_1 - .3679\,\pi_4 - .1840\,\pi_5,$$
$$.6321\,\pi_4 - .3679\,\pi_1 - .3679\,\pi_5,\ .6321\,\pi_5 - .3679\,\pi_1\Big]$$

```
>   EqnSetR:=convert(EqnList,set);
```

$$EqnSetR := \{.6321\,\pi_4 - .3679\,\pi_1 - .3679\,\pi_5,$$
$$.9812\,\pi_1 - .6321\,\pi_2 - .2642\,\pi_3 - .0802\,\pi_4 - .0188\,\pi_5,$$
$$.6321\,\pi_2 - .06133\,\pi_1 - .3679\,\pi_3 - .1840\,\pi_4 - .06133\,\pi_5,$$
$$.6321\,\pi_3 - .1840\,\pi_1 - .3679\,\pi_4 - .1840\,\pi_5,\ .6321\,\pi_5 - .3679\,\pi_1\}$$

Since one of the equations in the system $\Pi = \Pi P$ is always redundant (see Heyman and Sobel [88, p. 243] for a discussion), we delete an arbitrary equation—say, the first one—using the set theoretic minus () function. We also introduce the normalizing condition $\Pi e = 1$ as a new equation.

```
>   EqnSet:=EqnSetR minus { EqnList[ 1]};
```

$$EqnSet := \{.6321\,\pi_4 - .3679\,\pi_1 - .3679\,\pi_5,$$
$$.6321\,\pi_2 - .06133\,\pi_1 - .3679\,\pi_3 - .1840\,\pi_4 - .06133\,\pi_5,$$
$$.6321\,\pi_3 - .1840\,\pi_1 - .3679\,\pi_4 - .1840\,\pi_5,\ .6321\,\pi_5 - .3679\,\pi_1\}$$

```
>   SumProb:={ sum(pi[ k] ,k=1..5)=1};
```

$$SumProb := \{\pi_1 + \pi_2 + \pi_3 + \pi_4 + \pi_5 = 1\}$$

After defining the linear system as the union () of two sets of equations, we solve it and obtain the stationary distribution. As a check, we note that the sum of these probabilities is unity.

```
>   Sys:=EqnSet union SumProb;
```

$Sys := \{.6321\,\pi_4 - .3679\,\pi_1 - .3679\,\pi_5,$
$.6321\,\pi_2 - .06133\,\pi_1 - .3679\,\pi_3 - .1840\,\pi_4 - .06133\,\pi_5,$
$.6321\,\pi_3 - .1840\,\pi_1 - .3679\,\pi_4 - .1840\,\pi_5,\ .6321\,\pi_5 - .3679\,\pi_1,$
$\pi_1 + \pi_2 + \pi_3 + \pi_4 + \pi_5 = 1\}$

```
>   Solution:=solve(Sys);
```

$Solution := \{\pi_5 = .1293182802,\ \pi_2 = .2225162422,\ \pi_1 = .2221856072,$
$\pi_3 = .2213947096,\ \pi_4 = .2045851608\}$

```
>   assign(Solution); sum(pi[ k] ,k=1..5);
```

$$1.000000000$$

We should note that since we had to define Π as a vector with five entries indexed as 1, 2, 3, 4, 5, the indices of the solution must be adjusted by shifting them by -1, i.e., Pr(stationary inventory level is $j - 1$) $= \pi_j, j = 1, 2, \ldots, 5$.

The fact that one of the equations in the linear system $\Pi = \Pi\mathbf{P}$ is redundant can be illustrated as follows.

First consider a Markov chain with two states and the transition matrix given by

$$\mathbf{P} = \begin{matrix} 1 \\ 2 \end{matrix} \begin{pmatrix} 0.7 & 0.3 \\ 0.4 & 0.6 \end{pmatrix}.$$

Expanding the system $\Pi = \Pi\mathbf{P}$ we find

$$\begin{aligned} \pi_1 &= 0.7\pi_1 + 0.4\pi_2 \\ \pi_2 &= 0.3\pi_1 + 0.6\pi_2. \end{aligned}$$

If we express π_2 in the first equation in terms of π_1, we find $\pi_2 = \frac{3}{4}\pi_1$. Similarly, writing π_2 in the second equation in terms of π_1 again gives $\pi_2 = \frac{3}{4}\pi_1$. Thus, these two equations are linearly dependent.

To illustrate the redundancy of one of the equations in a three state Markov chain we use Maple's three-dimensional plotting facilities. Consider the Markov chain with the transition matrix

$$\mathbf{P} = \begin{pmatrix} .3 & .2 & .5 \\ .4 & .3 & .3 \\ .1 & .2 & .7 \end{pmatrix}.$$

```
>   restart: # RedundantEqn.mws

>   with(linalg):

Warning, new definition for norm

Warning, new definition for trace
```

```
>  with(plots):
>  Row[ 1] :=[ .3,.2,.5] ; Row[ 2] :=[ .4,.3,.3] ;
   Row[ 3] :=[ .1,.2,.7] ;
```

$$Row_1 := [.3, .2, .5]$$

$$Row_2 := [.4, .3, .3]$$

$$Row_3 := [.1, .2, .7]$$

```
>  P:=array(1..3,1..3,[ seq(Row[ i] ,i=1..3)] );
```

$$P := \begin{bmatrix} .3 & .2 & .5 \\ .4 & .3 & .3 \\ .1 & .2 & .7 \end{bmatrix}$$

We first define the vector of unknown probabilities as $\Pi = [x, y, z]$.

```
>  pi:=vector(3,[ x,y,z] );
```

$$\pi := [x, y, z]$$

```
>  EqnList:=evalm(pi - pi &* P);
```

$$EqnList := [.7x - .4y - .1z, .7y - .2x - .2z, .3z - .5x - .3y]$$

```
>  solve({ seq(EqnList[ k] ,k=1..3)} );
```

$$\{y = 1.066666667x, z = 2.733333333x, x = x\}$$

When Maple solves the three equations generated by $\Pi = \Pi P$, it finds in-finitely many solutions as $y = 1.0\bar{6}x$ and $z = 2.7\bar{3}x$ for $0 \le x \le 1$. We also use Maple's plot3d() function to plot the three planes corresponding to the linear system $\Pi = \Pi P$. The planes intersecting at infinitely many points in the (x, y, z) space are shown in Figure 7.7.

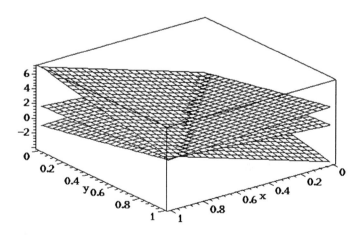

FIGURE 7.7. Three planes generated by $\Pi = \Pi P$ intersect at infinitely many points.

```
>   EqnSetR:=convert(EqnList,set);
```

$$EqnSetR := \{.7x - .4y - .1z, .7y - .2x - .2z, .3z - .5x - .3y\}$$

```
>   for k from 1 to 3 do
    Plane[ k] :=solve(EqnList[ k] ,z) od;
```

$$Plane_1 := 7.x - 4.y$$

$$Plane_2 := 3.500000000\,y - 1.x$$

$$Plane_3 := 1.666666667\,x + y$$

```
>   p[ 1] :=plot3d(Plane[ 1] ,x=0..1,y=0..1):
```

```
>   p[ 2] :=plot3d(Plane[ 2] ,x=0..1,y=0..1):
```

```
>   p[ 3] :=plot3d(Plane[ 3] ,x=0..1,y=0..1):
```

```
>   #display({ p[ 1] ,p[ 2] ,p[ 3]
    } ,axes=boxed,shading=none,orientation=[ 52,47] );
```

After introducing the normalizing condition $\Pi\mathbf{e} = 1$, the new system is solved to give the unique stationary probabilities as $x = .208\bar{3}$, $y = .22\bar{2}$ and $z = .569\bar{4}$.

```
>   EqnSet:=EqnSetR minus { EqnList[ 1]} ;
```

$$EqnSet := \{.7y - .2x - .2z, .3z - .5x - .3y\}$$

```
>   SumProb:={ sum(pi[ j] ,j=1..3)=1} ;
```

$$SumProb := \{x + y + z = 1\}$$

```
>   Sys:=EqnSet union SumProb;
```

$$Sys := \{.7y - .2x - .2z, .3z - .5x - .3y, x + y + z = 1\}$$

```
>   Solution:=solve(Sys);
```

$$Solution := \{z = .5694444444, y = .2222222222, x = .2083333333\}$$

```
>   assign(Solution); sum(pi[' k'] ,' k' =1..3);
```

$$.9999999999$$

Example 82 *Machine maintenance (Heyman and Sobel [88]).* Consider a machine that is examined once a day and classified as either working perfectly (state 0), having minor defects (state 1), or having failed (state 2). Suppose the number of consecutive days a machine is in state 0 (*up*) follows the geometric distribution $u^{k-1}(1 - u)$, $k = 1, 2, \ldots$, and the number of consecutive days a machine is in state 1 (*down*) follows the geometric distribution $d^{k-1}(1 - d)$, $k = 1, 2, \ldots$. When the machine leaves state 0, it makes a transition to state 2 (i.e., it *f*ails) with probability f and when it leaves state 1 (with probability d), it always enters state 2. Assume that *r*epair activity lasts a random number of days and is also geometric with distribution $r^{k-1}(1 - r)$, $r = 1, 2, \ldots$. The repair process is not totally reliable and it results in a machine that is in perfectly *w*orking state 0 with probability w and in minor defects state 1 with probability $1 - w$.

We define X_n to be the state of the machine on day n and show that X_n is a Markov chain by using the memoryless property of the geometric distribution.

First, suppose the machine is in state 0. The length of time the machine is in this perfectly working state is independent of the states before it entered this state. When $X_n = 0$, it does not matter how long the machine has been in state 0; the probability that it will stay in the same state is u, i.e.,

$$p_{00} = \Pr(X_{n+1} = 0 \mid X_n = 0, X_{n-1} = i_{n-1}, \ldots, X_0 = i_0) = u.$$

Similarly, we obtain

$$p_{01} = \Pr(X_{n+1} = 1 \mid X_n = 0, X_{n-1} = i_{n-1}, \ldots, X_0 = i_0) = (1 - u)(1 - f)$$

since the probability of making a transition to 1 from 0 requires that (i) the machine makes a transition out of state 0, and (ii) it moves into state 1 with probability $1 - f$. Using the same argument, the transition from state 0 to state 2 is obtained as

$$p_{02} = \Pr(X_{n+1} = 2 \mid X_n = 0, X_{n-1} = i_{n-1}, \ldots, X_0 = i_0) = (1 - u)f.$$

The other probabilities can be determined using similar arguments and we find the transition matrix as

$$\mathbf{P} = \begin{pmatrix} p_{00} & p_{01} & p_{02} \\ p_{10} & p_{11} & p_{12} \\ p_{20} & p_{21} & p_{22} \end{pmatrix} = \begin{pmatrix} u & (1 - u)(1 - f) & (1 - u)f \\ 0 & d & 1 - d \\ (1 - r)w & (1 - r)(1 - w) & r \end{pmatrix}.$$

It is easy to check that the transition probabilities satisfy $p_{ij} \geq 0$ for $i, j \in S$ and $\sum_{j \in S} p_{ij} = 1$ for $i \in S = \{0, 1, 2\}$.

Although this Markov chain's transition probabilities are given symbolically, Maple easily solves the equations $\Pi = \Pi\mathbf{P}$ and $\Pi\mathbf{e} = 1$ and finds the stationary distribution in terms of the symbolic parameters u, f, d, r and w:

```
>   restart: # Machine.mws
>   Row[ 0] :=[ u, (1-u)* (1-f) , (1-u)* f] ;
```

$$Row_0 := [u, (1 - u)(1 - f), (1 - u)f]$$

```
>   Row[ 1] :=[ 0,d,1-d] ;
```

$$Row_1 := [0, d, 1 - d]$$

```
>   Row[ 2] :=[ (1-r)*w, (1-r)* (1-w) , r] ;
```

$$Row_2 := [(1 - r)w, (1 - r)(1 - w), r]$$

```
>   for i from 0 to 2 do
    simplify(sum(Row[ i][ j] ,j=1..3)) od;
```

$$1$$

$$1$$

$$1$$

```
>   PArray:=array(0..2,0..2,[ seq(Row[ i] ,i=0..2)] ):
>   P:=convert(PArray,matrix,3);
```

$$P := \begin{bmatrix} u & (1-u)(1-f) & (1-u)f \\ 0 & d & 1-d \\ (1-r)w & (1-r)(1-w) & r \end{bmatrix}$$

```
>   pi:=vector(3,[] );
```

$$\pi := \text{array}(1..3, [])$$

```
>   EqnList:=evalm(pi - pi &* P);
```

$$EqnList := \left[\pi_1 - \pi_1 u - \pi_3 (1-r)w, \right.$$
$$\pi_2 - \pi_1 (1-u)(1-f) - \pi_2 d - \pi_3 (1-r)(1-w),$$
$$\left. \pi_3 - \pi_1 (1-u)f - \pi_2 (1-d) - \pi_3 r \right]$$

```
>   EqnSetR:=convert(EqnList,set);
```

$$EqnSetR := \{ \pi_1 - \pi_1 u - \pi_3 (1-r)w,$$
$$\pi_2 - \pi_1 (1-u)(1-f) - \pi_2 d - \pi_3 (1-r)(1-w),$$
$$\pi_3 - \pi_1 (1-u)f - \pi_2 (1-d) - \pi_3 r \}$$

```
>   EqnSet:=EqnSetR minus { EqnList[ 1]} ;
```

$$EqnSet := \{ \pi_2 - \pi_1 (1-u)(1-f) - \pi_2 d - \pi_3 (1-r)(1-w),$$
$$\pi_3 - \pi_1 (1-u)f - \pi_2 (1-d) - \pi_3 r \}$$

```
>   SumProb:={ sum(pi[ k] ,k=1..3)=1} ;
```

$$SumProb := \{ \pi_1 + \pi_2 + \pi_3 = 1 \}$$

```
>   Sys:=EqnSet union SumProb;
```

$$Sys := \{ \pi_2 - \pi_1 (1-u)(1-f) - \pi_2 d - \pi_3 (1-r)(1-w),$$
$$\pi_1 + \pi_2 + \pi_3 = 1, \pi_3 - \pi_1 (1-u)f - \pi_2 (1-d) - \pi_3 r \}$$

```
>   Solution:=solve(Sys,{ seq(pi[ k] ,k=1..3)} );
```

$$Solution := \{ \pi_1 = -\frac{w(-r-d+1+rd)}{\%1},$$
$$\pi_2 = \frac{u-1-rwf-wuf+wf-ur+r+rwuf}{\%1},$$
$$\pi_3 = -\frac{-u-d+ud+1}{\%1} \}$$
$$\%1 := d - rwf - rwd - wuf + wf - ud - ur - w + r + rwuf$$
$$+ rw - 2 + 2u + wd$$

```
>   assign(Solution); simplify(sum(pi[ k] ,k=1..3));
```

1

As in the inventory example, since we had to define Π as a vector with three entries indexed as $1, 2, 3$, the indices of the solution must be adjusted by shifting them by -1, i.e., Pr(machine is in state $j - 1$) $= \pi_j$, $j = 1, 2, 3$.

In some "pathological" cases, although the stationary distribution exists, the limiting (steady-state) probabilities do not exist; i.e., $\lim_{n \to \infty} \mathbf{P}^{(n)}$ does not converge. As a simple example of this unusual case, consider the following problem.

Example 83 *A "pathological" case.* Suppose the transition matrix is given by

$$\mathbf{P} = \begin{pmatrix} 0 & 1 \\ 1 & 0 \end{pmatrix}.$$

When we compute the higher powers of P, we observe an interesting result:

```
>    restart: # Pathological.mws
>    Row[ 1] :=[ 0,1] ;  Row[ 2] :=[ 1,0] ;
```
$$Row_1 := [0, 1]$$
$$Row_2 := [1, 0]$$
```
>    P:=matrix(2,2,[ seq(Row[ k] ,k=1..2)] );
```
$$P := \begin{bmatrix} 0 & 1 \\ 1 & 0 \end{bmatrix}$$
```
>    evalm(P^2);
```
$$\begin{bmatrix} 1 & 0 \\ 0 & 1 \end{bmatrix}$$
```
>    evalm(P^3);
```
$$\begin{bmatrix} 0 & 1 \\ 1 & 0 \end{bmatrix}$$
```
>    evalm(P^4);
```
$$\begin{bmatrix} 1 & 0 \\ 0 & 1 \end{bmatrix}$$

These imply that

$$\mathbf{P}^{(n)} = \begin{cases} \begin{pmatrix} 1 & 0 \\ 0 & 1 \end{pmatrix} & \text{if } n \text{ is even} \\ \begin{pmatrix} 0 & 1 \\ 1 & 0 \end{pmatrix} & \text{if } n \text{ is odd;} \end{cases}$$

thus the limit $\lim_{n \to \infty} \mathbf{P}^{(n)}$ does not exist. Hence, this Markov chain has no limiting probabilities.

However, if we solve the linear system $\Pi = \Pi \mathbf{P}$ with $\Pi \mathbf{e} = 1$, we find $\pi_1 = \pi_2 = \frac{1}{2}$ as the stationary distribution of this pathological Markov chain:

```
>    pi:=vector(2,[ ] );
```

$$\pi := \mathrm{array}(1..2, [])$$

```
> EqnList:=evalm(pi - pi &* P);
```

$$EqnList := [\pi_1 - \pi_2, \pi_2 - \pi_1]$$

```
> EqnSetR:=convert(EqnList,set);
```

$$EqnSetR := \{\pi_1 - \pi_2, \pi_2 - \pi_1\}$$

```
> EqnSet:=EqnSetR minus { EqnList[ 1]};
```

$$EqnSet := \{\pi_2 - \pi_1\}$$

```
> SumProb:={ sum(pi[ k] ,k=1..2)=1};
```

$$SumProb := \{\pi_1 + \pi_2 = 1\}$$

```
> Sys:=EqnSet union SumProb;
```

$$Sys := \{\pi_2 - \pi_1, \pi_1 + \pi_2 = 1\}$$

```
> Solution:=solve(Sys);
```

$$Solution := \{\pi_1 = \frac{1}{2}, \pi_2 = \frac{1}{2}\}$$

What is interesting about this case is that although $\lim_{n\to\infty} \mathbf{P}^{(n)}$ does not exist, the process periodically visits states 1 and 2 in a deterministic fashion. Thus, the process spends 50% of the time in state 1 and 50% of the time in state 2, implying that the *long-run proportion of time the system stays in state* $j = 1, 2$ is $\frac{1}{2}$. It can be shown that, in general, the probabilities π_j found by solving $\Pi = \Pi \mathbf{P}$ and $\Pi \mathbf{e} = 1$ can be interpreted as the long-run proportion of time the (periodic) Markov chain is in state j; see Ross [157, p. 111].

7.4.3   Classification of States

The states of a Markov chain can be classified according to some properties of the system. By classifying the states of a Markov chain, it becomes possible to identify its limiting behavior.

We say that state j is **reachable** (or, **accessible**) from state i if $p_{ij}^{(n)} > 0$ for some $n \geq 0$. This is denoted by $i \to j$. For example, if

$$\mathbf{P}_2 = \begin{array}{c} 0 \\ 1 \end{array} \left(\begin{array}{cc} \frac{1}{2} & \frac{1}{2} \\ 0 & 1 \end{array} \right),$$

then state 1 is reachable from 0 $(0 \to 1)$ but 0 is not reachable from 1 $(1 \nrightarrow 0)$.

States i and j are said to **communicate** (denoted by $i \longleftrightarrow j$) if they are reachable from each other. For example, if

$$\mathbf{P}_3 = \begin{array}{c} 0 \\ 1 \\ 2 \end{array} \left(\begin{array}{ccc} & \times & \times \\ & \times & \\ \times & & \end{array} \right)$$

where × denotes a positive probability, then $2 \to 1$ (since $2 \to 0$ and $0 \to 1$), $0 \longleftrightarrow 2$, but $1 \nrightarrow 2$. Thus, in this Markov chain not all states communicate.

The set of all states that communicate with each other is called an **equivalence class**. For example, if

$$
\mathbf{P}_4 = \begin{array}{c} 0 \\ 1 \\ 2 \\ 3 \\ 4 \end{array} \left(\begin{array}{ccccc} \times & \times & & & \\ \times & \times & & & \\ & & & 1 & \\ & & \times & & \times \\ & & & 1 & \end{array} \right),
$$

then states $\{0, 1\}$ and $\{2, 3, 4\}$ form two distinct equivalence classes. When a Markov chain has a large number of states it may be difficult to identify its equivalence classes by visual inspection. For an efficient algorithm for identifying the equivalence classes, see Kao [101, pp. 172–173].

If all the states of a Markov chain belong to one equivalence class, the chain is said to be **irreducible**. In such a Markov chain all states communicate. For example, the Markov chain representing the (s, S) inventory system is irreducible since all its states communicate. The three Markov chains with transition matrices $\mathbf{P}_2$, $\mathbf{P}_3$ and $\mathbf{P}_4$ are *not* irreducible.

The equivalence classification of states takes into account the *external* (i.e., between "neighbors") relationship between states. The following concepts of periodicity and recurrence identify states with respect to their *internal* nature.

The **period** $d(i)$ of a state i is the greatest common divisor (g.c.d.) of all integers $n = 1, 2, \ldots$ for which $p_{ii}^{(n)} > 0$. When the period of a state is 1, the state is called **aperiodic**. For example, suppose $p_{ii}^{(1)} > 0$, $p_{ii}^{(2)} > 0$, $p_{ii}^{(3)} = 0$, $p_{ii}^{(4)} > 0$, ... Since the g.c.d. of $1, 2, 4, \ldots$ is 1, the state i is aperiodic. In the random walk model in Example 78, for all i we have $p_{ii}^{(2n)} > 0$ and $p_{ii}^{(2n+1)} = 0$ for $n = 0, 1, 2, \ldots$. Thus, $d(i) = 2$ for all i and the chain is periodic.

Now define the **first passage time** probability as

$$
f_{ii}^{(n)} = \Pr(X_n = i,\ X_k \neq i,\ k = 1, 2, \ldots, n-1 \mid X_0 = i).
$$

Thus, $f_{ii}^{(n)}$ is the probability that the process visits state i for the *first* time in n steps after starting in state i. The probability of *ever* visiting state i after starting in the same state is then $f_{ii} = \sum_{n=1}^{\infty} f_{ii}^{(n)}$.

The state i is said to be **recurrent** (persistent) if $f_{ii} = 1$, **transient** if $f_{ii} < 1$. This means that for a recurrent state i, the process eventually returns to i with probability 1. Thus for a recurrent state i we have $\sum_{n=1}^{\infty} f_{ii}^{(n)} = 1$, implying that $\{f_{ii}^{(n)}, n = 1, 2, \ldots\}$ is a p.d.f. It is usually difficult to test f_{ii} for the value it takes, but it can be shown that state i is recurrent if and only if $\sum_{n=1}^{\infty} p_{ii}^{(n)} = \infty$; see Ross [157, p. 105].

Example 84 *Random walk revisited.* As an example of identifying the recurrent and transient states of a Markov chain, consider again the random walk model introduced in Example 78. Here the transition probabilities are given as

$$p_{i,i+1} = p = 1 - p_{i,i-1}, \quad i = 0, \pm 1, \pm 2, \ldots$$

for $0 < p < 1$. In this chain all states communicate, thus all states are either recurrent or transient. We therefore consider state 0 and examine its properties.

Since it is impossible to return to 0 in an odd number of transitions, we have

$$p_{00}^{(2n+1)} = 0, \quad n = 0, 1, 2, \ldots.$$

But, there is a positive probability (binomial) of returning to 0 after an even number of transitions:

$$p_{00}^{(2n)} = \binom{2n}{n} p^n q^n = \frac{(2n)!}{n!n!} p^n q^n, \quad n = 1, 2, \ldots.$$

This expression is simplified using Stirling's approximation of a factorial as $n! \sim n^{n+\frac{1}{2}} e^{-n} \sqrt{2\pi}$ where '$\sim$' is a shorthand notation for "approximated in the ratio."

```
>    restart: # RecurrenceRW.mws
>    Stirling:=n->n^(n+1/2)*exp(-n)*sqrt(2*Pi);
```

$$Stirling := n \rightarrow n^{(n+1/2)} e^{(-n)} \sqrt{2\pi}$$

Substituting and simplifying, Maple finds that the probability $p_{00}^{(2n)}$ of returning to 0 in an even number of transitions is

$$p_{00}^{(2n)} = \frac{(4pq)^n}{\sqrt{n\pi}}.$$

```
>    Even:=(Stirling(2*n)/(Stirling(n)*Stirling(n)))
     *p^n*q^n;
```

$$Even := \frac{1}{2} \frac{(2n)^{(2n+1/2)} e^{(-2n)} \sqrt{2} \, p^n \, q^n}{\sqrt{\pi} \, (n^{(n+1/2)})^2 \, (e^{(-n)})^2}$$

```
>    c:=simplify(Even);
```

$$c := \frac{4^n \, p^n \, q^n}{\sqrt{n} \, \sqrt{\pi}}$$

We now analyze the properties of the infinite sum $\sum_{n=1}^{\infty} p_{ii}^{(2n)}$. If it converges, then we conclude that the state 0 is transient; otherwise it is recurrent.

To simplify the analysis, we write $r = 4pq$ and first assume that $0 < r < 1$, i.e., $p \neq \frac{1}{2}$. Using the ratio test of calculus we compute $\lim_{n \to \infty} |a_{n+1}/a_n|$. Intuitively, when $p \neq \frac{1}{2}$ there is a positive probability that the process, initially at the origin, will drift to plus (or minus) infinity when $p > q$ (or, $p < q$) without returning to the origin.

```
>    assume(0<r,r<1);
>    a:=n->r^n/sqrt(Pi*n);
```

$$a := n \rightarrow \frac{r^n}{\sqrt{\pi\, n}}$$

```
>   limit(abs(a(n+1)/a(n)),n=infinity);
```

$$r$$

Since the limit converges to $r \in (0, 1)$, we conclude that when $p \neq \frac{1}{2}$, the sum $\sum_{n=1}^{\infty} p_{ii}^{(2n)}$ converges and hence the chain is transient.

When $r = 1$ (or $p = \frac{1}{2}$), the integral test of calculus reveals that the integral

$$\int_1^{\infty} \frac{1}{\sqrt{\pi\, n}}\, dn$$

diverges to infinity. Thus, in this case, the chain is recurrent.

```
>   b:=n->1/sqrt(Pi*n);
```

$$b := n \rightarrow \frac{1}{\sqrt{\pi\, n}}$$

```
>   int(b(n),n=1..infinity);
```

$$\infty$$

```
>   evalf(int(b(n),n=1..100000));
```

$$355.6964440$$

Intuitively, when $p = \frac{1}{2}$, the process fluctuates around the origin and even if strays away, it eventually returns to state 0.

For a recurrent state i, **mean recurrence time** μ_{ii} is defined as the expected number of transitions needed to return to i, i.e., $\mu_{ii} = \sum_{n=1}^{\infty} n f_{ii}^{(n)}$. If $\mu_{ii} < \infty$, the state i is said to be **positive recurrent**, and if $\mu_{ii} = \infty$ the state i is **null recurrent**.

Finally, we define a state to be **ergodic** if it is (i) aperiodic and (ii) positive recurrent.

With these definitions, we are now ready to present a very important theorem that identifies the long-run behavior of a Markov chain. For a proof, see Ross [157, p. 109].

Theorem 13 *An irreducible Markov chain belongs to one of the following three classes:*

	Periodicity	
	Aperiodic	**Periodic**
Transient	$p_{ij}^{(n)} \rightarrow 0$ and π_j	
Null-Recurrent	do not exist	
Positive-Recurrent	$p_{ij}^{(n)} \rightarrow \pi_j$ are uniquely found from $\Pi = \Pi\mathbf{P}$ and $\Pi\mathbf{e} = 1$	$p_{ij}^{(n)} \rightarrow ?$ π_j are long-run fraction of time in j

We now discuss some examples where this theorem can be used to identify the limiting properties of the Markov chains.

Example 85 *A recurrent and periodic chain.* Consider the Markov chain with the transition matrix

$$\mathbf{P} = \begin{array}{c} 1 \\ 2 \\ 3 \\ 4 \end{array} \begin{pmatrix} 0 & 1 & 0 & 0 \\ 0 & 0 & 1 & 0 \\ 0 & 0 & 0 & 1 \\ \frac{1}{3} & 0 & \frac{2}{3} & 0 \end{pmatrix}.$$

First, we note that all states of the chain communicate; hence the chain is irreducible and Theorem 13 can be applied.

After finding $\mathbf{P}^n$ and printing the first few values of $p_{11}^{(n)}$ for $n = 2, 3, \ldots, 50$, we see that $p_{11}^{(n)} = 0$ for $n = 1, 2, 3$ and $p_{11}^{(n)} > 0$ for $n = 4, 6, 8, \ldots$. This implies that the chain is periodic with $d(i) = 2$.

```
>   restart: # PeriodicChain.mws
>   Row[ 1] :=[ 0,1,0,0] ; Row[ 2] :=[ 0,0,1,0] ;
    Row[ 3] :=[ 0,0,0,1] ; Row[ 4] :=[ .2,0,.8,0] ;
```

$$Row_1 := [0, 1, 0, 0]$$

$$Row_2 := [0, 0, 1, 0]$$

$$Row_3 := [0, 0, 0, 1]$$

$$Row_4 := [.2, 0, .8, 0]$$

```
>   P:=matrix(4,4,[ seq(Row[ i] ,i=1..4)] );
```

$$P := \begin{bmatrix} 0 & 1 & 0 & 0 \\ 0 & 0 & 1 & 0 \\ 0 & 0 & 0 & 1 \\ .2 & 0 & .8 & 0 \end{bmatrix}$$

```
>   for n from 2 to 50 do
    p[ n] :=evalm(P^n) od: n:=' n' :
>   for n from 2 to 15 do
    print(n,p[ n][ 1,1] ) od; n:=' n' :
```

$$2, 0$$

$$3, 0$$

$$4, .2$$

$$5, 0$$

$$6, .16$$

$$7, 0$$

$$8, .168$$

$$9, 0$$

$$10, .1664$$

$$11, 0$$

$$12, .16672$$

$$13, 0$$

$$14, .166656$$

$$15, 0$$

Next, we define the partial sum $S_k = \sum_{n=1}^{k} p_{11}^{(n)}$ and find that the difference $S_{k+1} - S_k$ approaches the constant $0.16\bar{6}$. This has the implication that the infinite sum $\sum_{n=1}^{\infty} p_{11}^{(n)}$ diverges; hence all states of the chain are recurrent.

```
>    for k from 2 to 50 do
     S[ k] :=evalf(sum(p[ n][ 1,1] ,n=2..k)) od:
```

```
>    for k from 35 to 49 do
     Difference[ k] :=S[ k+1] -S[ k]  od;
```

$$Difference_{35} := .166666667$$

$$Difference_{36} := 0$$

$$Difference_{37} := .166666667$$

$$Difference_{38} := 0$$

$$Difference_{39} := .166666667$$

$$Difference_{40} := 0$$

$$Difference_{41} := .166666667$$

$$Difference_{42} := 0$$

$$Difference_{43} := .166666667$$

$$Difference_{44} := 0$$

$$Difference_{45} := .166666667$$

$$Difference_{46} := 0$$

$$Difference_{47} := .166666667$$

$$Difference_{48} := 0$$

$$Difference_{49} := .166666667$$

At $n = 50$, the process has stabilized but the rows of $\mathbf{P}^{50}$ are *not* identical. Thus, the chain does not possess limiting probabilities.

```
>    evalf(evalm(p[ 50] ),4);
```

$$\begin{bmatrix} .1667 & 0 & .8333 & 0 \\ 0 & .1667 & 0 & .8333 \\ .1667 & 0 & .8333 & 0 \\ 0 & .1667 & 0 & .8333 \end{bmatrix}$$

However, the linear system $\Pi = \Pi P$ with $\Pi e = 1$ does have a solution implying that the stationary distribution exists. We find this distribution with Maple's help as follows.

```
>  pi:=vector(4,[] );
```

$$\pi := \text{array}(1..4, \, [])$$

```
>  EqnList:=evalm(pi - pi &* P);
```

$$EqnList := [\pi_1 - .2\,\pi_4, \ \pi_2 - \pi_1, \ \pi_3 - \pi_2 - .8\,\pi_4, \ \pi_4 - \pi_3]$$

```
>  EqnSetR:=convert(EqnList,set);
```

$$EqnSetR := \{\pi_3 - \pi_2 - .8\,\pi_4, \ \pi_1 - .2\,\pi_4, \ \pi_2 - \pi_1, \ \pi_4 - \pi_3\}$$

```
>  EqnSet:=EqnSetR minus { EqnList[ 1]} ; k:=' k' :
```

$$EqnSet := \{\pi_3 - \pi_2 - .8\,\pi_4, \ \pi_2 - \pi_1, \ \pi_4 - \pi_3\}$$

```
>  SumProb:={ sum(pi[ k] ,k=1..4)=1} ;
```

$$SumProb := \{\pi_1 + \pi_2 + \pi_3 + \pi_4 = 1\}$$

```
>  Sys:=EqnSet union SumProb;
```

$$Sys := \{\pi_1 + \pi_2 + \pi_3 + \pi_4 = 1, \ \pi_3 - \pi_2 - .8\,\pi_4, \ \pi_2 - \pi_1, \ \pi_4 - \pi_3\}$$

In this case $\pi_1 = \pi_2 = 0.08\bar{3}$, $\pi_3 = \pi_4 = 0.416\bar{6}$, indicating that the process spends about 8.3% of its time in states 1 and 2 and about 42% of its time in states 3 or 4.

```
>  Solution:=solve(Sys);
```

$$Solution := \{\pi_1 = .08333333333, \ \pi_4 = .4166666667, \ \pi_3 = .4166666667,$$
$$\pi_2 = .08333333333\}$$

```
>  assign(Solution); sum(pi[ k] ,k=1..4);
```

$$1.000000000$$

7.4.4 Imbedded Markov Chain Technique

Consider the single-server queueing system with Poisson arrivals and general service time S with distribution $B(t)$, density $b(t)$ and mean $E(S) = 1/\mu$.[4] We are interested in the properties of the stochastic process $X(t)$ defined as the number of customers in the system at time t.

---

[4] As we will discuss in more detail in Chapter 9, Queueing Systems, this queue is denoted by the shorthand notation $M/G/1$.

Naturally, this process is not a (discrete-time) Markov *chain* since its time index t is continuous. The process $X(t)$ is not even *Markovian* since the knowledge of the system at time t is not sufficient to determine the distribution of the state at time $t + s$. For example, suppose we wish to determine the probability that there is one customer present in the system at time $t + \Delta t$ given that there was one customer present at time t. This probability can be written as

$$
\begin{aligned}
\Pr[X(t + \Delta t) = 1 \mid X(t) = 1] &= [1 - \lambda \Delta t + o(\Delta t)] \\
&\quad \times \Pr\{\text{no departures in } (t, t + \Delta t]\} \\
&= [1 - \lambda \Delta t + o(\Delta t)][1 - r_B(u) \Delta t + o(\Delta t)]
\end{aligned}
$$

where $r_B(u) \Delta t$ is the probability that a customer who has already spent u time units in service by time t will leave the system within the next Δt time units. Here, $r_B(u)$ is the hazard rate function for the service time random variable S. For an exponentially distributed service time S with parameter μ, the hazard rate would be a constant $r_B(u) = \mu$. But when S is general, $r_B(u)$ requires the knowledge of the past (prior to t), thus making the $X(t)$ process non-Markovian.

Using the method of supplementary variables due to Cox [52], it is possible to expand the state space of this process and transform it to a Markovian system. In this section, we will describe a different technique known as the *imbedded Markov chain* approach that uses the concept of regeneration point. This approach is due to Kendall [105].

In order to make the process Markovian we proceed as follows. Instead of observing the system at all points of time t, we focus our attention on those epochs when a customer leaves the system. We define X_n to be the number of customers left behind by the nth departure, and Y_n the number of customers arriving during the service period of the nth customer ($n \geq 1$). This gives

$$
X_{n+1} = \begin{cases} Y_{n+1} & \text{if } X_n = 0 \\ X_n - 1 + Y_{n+1} & \text{if } X_n > 0, \end{cases}
$$

indicating that the process X_n is now a Markov chain since X_{n+1} depends only on X_n.

To develop the transition probabilities p_{ij} for this Markov chain, we need to find the probability distribution of the number of customers that arrive during an arbitrary customer's service interval. Conditioning on the length of the service time we find

$$
\begin{aligned}
\Pr(Y_n = j) &= \int_0^\infty \Pr(Y_n = j \mid S = s) \, dB(s) \\
&= \int_0^\infty e^{-\lambda s} \frac{(\lambda s)^j}{j!} \, dB(s), \quad j = 0, 1, 2, \dots.
\end{aligned}
$$

For $i = 0$, we have

$$
p_{00} = \Pr(X_{n+1} = 0 \mid X_n = 0) = \Pr(Y_{n+1} = 0) = \int_0^\infty e^{-\lambda s} \, dB(s) = a_0
$$

$$p_{01} = \Pr(X_{n+1} = 1 \mid X_n = 0) = \Pr(Y_{n+1} = 1) = \int_0^\infty e^{-\lambda s} \lambda s \, dB(s) = a_1$$

$$\vdots$$

$$p_{0j} = \Pr(X_{n+1} = j \mid X_n = 0) = \Pr(Y_{n+1} = j) = \int_0^\infty \frac{e^{-\lambda s}(\lambda s)^j}{j!} \, dB(s) = a_j.$$

Proceeding in a similar manner for $i = 1, 2, \ldots$, we obtain the transition matrix **P** of the imbedded Markov chain for the $M/G/1$ queue as

$$\mathbf{P} = \begin{array}{c} 0 \\ 1 \\ 2 \\ 3 \\ \vdots \end{array} \left(\begin{array}{cccccc} a_0 & a_1 & a_2 & a_3 & \cdots \\ a_0 & a_1 & a_2 & a_3 & \cdots \\ 0 & a_0 & a_1 & a_2 & \cdots \\ 0 & 0 & a_0 & a_1 & \cdots \\ \vdots & \vdots & \vdots & \vdots & \ddots \end{array} \right).$$

The digraph for this Markov chain is displayed in Figure 7.8.

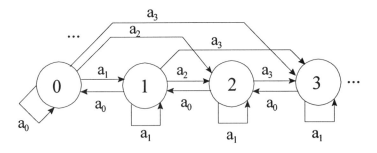

FIGURE 7.8. The digraph for the transition matrix of the embedded Markov chain of the $M/G/1$ queue.

Inspection of the transition matrix and the digraph reveals when $a_j > 0$ ($j = 0, 1, 2, \ldots$) the chain is irreducible and aperiodic. Since the chain has infinitely many states, we need to examine its recurrence and transience properties to determine whether it is ergodic. We define $\rho = E(Y_n)$ to be the mean number of arrivals during service time of an arbitrary customer.[5] Prabhu [152, p. 62] has shown that the chain is positive recurrent, null recurrent or transient according to $\rho < 1$, $\rho = 1$ or $\rho > 1$. We assume that $\rho < 1$ and proceed to find the stationary distribution of the imbedded Markov chain using $\Pi = \Pi \mathbf{P}$ and $\Pi \mathbf{e} = 1$.

---

[5]Note that $\rho = E(Y_n) = \int_0^\infty E(Y_n \mid S = s) \, dB(s) = \int_0^\infty \lambda s \, dB(s) = \lambda E(S) = \lambda/\mu$.

Writing the equations in the system $\Pi = \Pi\mathbf{P}$ explicitly, we obtain

$$
\begin{aligned}
\pi_0 &= \pi_0 a_0 + \pi_1 a_0 \\
\pi_1 &= \pi_0 a_1 + \pi_1 a_1 + \pi_2 a_0 \\
\pi_2 &= \pi_0 a_2 + \pi_1 a_2 + \pi_2 a_1 + \pi_3 a_0 \\
&\ \ \vdots
\end{aligned}
$$

We let $G(z) = \sum_{n=0}^{\infty} \pi_n z^n$ and $A(z) = \sum_{n=0}^{\infty} a_n z^n$ be the probability-generating function (p.g.f.) of $\{\pi_n\}$ and $\{a_n\}$, respectively. Multiplying both sides of the jth equation in the system $\Pi = \Pi\mathbf{P}$ by z^j, summing and collecting the terms gives

$$
G(z) = \frac{\pi_0 A(z)(z-1)}{z - A(z)} \tag{7.14}
$$

where π_0 is the stationary probability of the empty system.

Given the p.g.f. $A(z)$, the only undetermined quantity in (7.14) is π_0. We now find an expression for π_0 with Maple's help.

We define $A(z)$ and form the function $G(z)$. Since we know that $G(1) = \sum_{n=0}^{\infty} a_n = 1$, we attempt to solve for π_0 using this property. But this attempt fails since we find that a direct substitution of $z = 1$ in $G(z)$ results in $G(1) = 0/0$.

```
>    restart: # ImbeddedMC.mws
>    A:=z->Sum(a[ n] * z^n, n=0..infinity);
```

$$
A := z \to \sum_{n=0}^{\infty} a_n z^n
$$

```
>    G:=z-> (pi[ 0] * A(z)* (z-1))/(z-A(z));
```

$$
G := z \to \frac{\pi_0 A(z)(z-1)}{z - A(z)}
$$

```
>    subs(z=1,numer(G(z)));
```

$$
0
$$

```
>    subs(z=1,denom(G(z)));
```

$$
1 - \left(\sum_{n=0}^{\infty} a_n \right)
$$

Thus, with the hope of finding the limit of $G(z)$ as $z \to 1$, we use l'Hôpital's rule and compute the following:

$$
1 = \lim_{z \to 1} G(z) = \lim_{z \to 1} \frac{\{\pi_0 [A(z) + (z-1)A'(z)]\}}{[1 - A'(1)]}
$$

```
>    diff(numer(G(z)),z);
     Top:=subs(z=1,diff(numer(G(z)),z));
```

$$
\pi_0 \left(\sum_{n=0}^{\infty} \frac{a_n z^n n}{z} \right)(z-1) + \pi_0 \left(\sum_{n=0}^{\infty} a_n z^n \right)
$$

$$Top := \pi_0 \left(\sum_{n=0}^{\infty} a_n \right)$$

```
>   diff(denom(G(z)),z);
    Bottom:=subs(z=1,diff(denom(G(z)),z));
```

$$1 - \left(\sum_{n=0}^{\infty} \frac{a_n z^n n}{z} \right)$$

$$Bottom := 1 - \left(\sum_{n=0}^{\infty} a_n n \right)$$

Thus, $1 = \pi_0/(1 - \rho)$ since $\sum_{n=0}^{\infty} a_n = 1$ and $\sum_{n=0}^{\infty} na_n = E(Y) = \rho$ implying that $\pi_0 = 1 - \rho$.

```
>   solve(1=Top/Bottom,pi[0]);
```

$$-\frac{-1 + \left(\sum_{n=0}^{\infty} a_n n \right)}{\sum_{n=0}^{\infty} a_n}$$

In conclusion, the p.g.f. $G(z)$ of the unknown probabilities $\{\pi_j\}$ is obtained as

$$G(z) = \frac{(1 - \rho)(z - 1)A(z)}{z - A(z)}.$$

We can now compute the probabilities π_n using

$$\pi_n = \frac{1}{n!} \left(\frac{d^n G(z)}{dz^n} \right)_{z=0}. \tag{7.15}$$

It is also possible to compute the expected number of customers in the system by evaluating

$$E(X) = G'(1) = \sum_{n=0}^{\infty} n\pi_n. \tag{7.16}$$

Before we compute these quantities, we will present a simplification of the p.g.f. $A(z)$ and write it in terms of the Laplace transform (LT) of the service time S. Since

$$A(z) = \sum_{n=0}^{\infty} a_n z^n = \sum_{n=0}^{\infty} z^n \left[\int_0^{\infty} e^{-\lambda t} \frac{(\lambda t)^n}{n!} dB(t) \right],$$

Maple evaluates this infinite sum as

```
>   restart: # pgfA.mws
>   A:=z->Int(exp(-lambda*t)*b(t),t=0..infinity)
    *Sum((lambda*t*z)^n/n!,n=0..infinity);
```

$$A := z \to \int_0^\infty e^{(-\lambda t)} \, b(t) \, dt \left(\sum_{n=0}^\infty \frac{(\lambda t z)^n}{n!} \right)$$

```
> value(A(z));
```

$$\int_0^\infty e^{(-\lambda t)} \, b(t) \, dt \, e^{(\lambda t z)}$$

Rewriting this result as $A(z) = \int_0^\infty e^{-(\lambda - \lambda z)t} \, dB(t) = \tilde{b}(\lambda - \lambda z)$, we see that the p.g.f. $A(z)$ can be expressed as the LT $\tilde{b}(\cdot)$ of the service time S evaluated at $\lambda - \lambda z$. Hence, the p.g.f. of the distribution $\{\pi_n\}$ is given by

$$G(z) = \frac{(1 - \rho)(z - 1)\tilde{b}(\lambda - \lambda z)}{z - \tilde{b}(\lambda - \lambda z)},$$

which is known as the *Pollaczek-Khinchine (P-K) formula*. We now present some examples where the P-K formula is used to find the stationary distribution $\{\pi_n\}$.

Example 86 *The $M/M/1$ model.* In this model the service time S is assumed exponential with rate μ. With this assumption, Maple finds the p.g.f. $G(z)$ as follows where bLT is the Laplace transform of S.

```
> restart: # PKMM1.mws
> rho:=lambda/mu;
```

$$\rho := \frac{\lambda}{\mu}$$

```
> G:=z->((1-rho)*bLT(lambda-lambda*z)*(z-1))
  /(z-bLT(lambda-lambda*z)));
```

$$G := z \to \frac{(1 - \rho) \, \mathrm{bLT}(\lambda - \lambda z)(z - 1)}{z - \mathrm{bLT}(\lambda - \lambda z)}$$

```
> with(inttrans):
> b:=mu*exp(-mu*t);
```

$$b := \mu \, e^{(-\mu t)}$$

```
> bLT:=unapply(laplace(b,t,z),z);
```

$$\mathrm{bLT} := z \to \frac{\mu}{z + \mu}$$

```
> normal(G(z));
```

$$\frac{-\mu + \lambda}{\lambda z - \mu}$$

After loading the genfunc() package, we invert the resulting p.g.f. explicitly and obtain the well-known result for the stationary probabilities of the $M/M/1$ queue as $\pi_n = (1 - \rho)\rho^n$, $n = 1, 2, \ldots$.

```
> with(genfunc):
```

```
>    rgf_expand(G(z),z,n);
```

$$-\frac{(-\mu + \lambda)\,(\frac{\lambda}{\mu})^n}{\mu}$$

The exact values of π_n for $n = 1, 2, 3$ and the expected number of customers in the system $E(X) = \lambda/(\mu - \lambda)$ are also easily computed using (7.15) and (7.16), respectively.

```
>    for k from 1 to 3 do
     pi[k]:=subs(z=0,normal(diff(G(z),z$k)))/k!
     od;
```

$$\pi_1 := -\frac{\lambda\,(-\mu + \lambda)}{\mu^2}$$

$$\pi_2 := -\frac{\lambda^2\,(-\mu + \lambda)}{\mu^3}$$

$$\pi_3 := -\frac{\lambda^3\,(-\mu + \lambda)}{\mu^4}$$

```
>    EX:=normal(limit(diff(G(z),z),z=1));
```

$$EX := -\frac{\lambda}{-\mu + \lambda}$$

The next example deals with a more complicated case where the service time S is Erlang with k stages (with mean $1/\mu$).

Example 87 *The $M/E_k/1$ model.* Here we have essentially the same set of commands as in the previous example except that the density $b(t)$ of k-stage Erlang r.v. S is given by[6]

$$b(t) = \frac{(\mu k)^k t^{k-1} e^{-k\mu t}}{(k-1)!}, \quad t > 0.$$

We set $k = 2$ and again find that Maple successfully inverts the resulting p.g.f. $G(z)$ and finds the stationary probabilities exactly.

```
>    restart: # PKErlang.mws
```
```
>    rho:=lambda/mu;
```

$$\rho := \frac{\lambda}{\mu}$$

```
>    G:=z->((1-rho)*bLT(lambda-lambda*z)*(z-1))
     /(z-bLT(lambda-lambda*z));
```

---

[6]This is a slightly different but an equivalent way of writing the density of the Erlang r.v. with k stages. Here, each stage is exponential with rate $k\mu$ and the mean duration of the Erlang is $k[1/(k\mu)] = 1/\mu$; see Kleinrock [110, pp. 119–126].

$$G := z \rightarrow \frac{(1-\rho)\,\mathrm{bLT}(\lambda - \lambda z)\,(z-1)}{z - \mathrm{bLT}(\lambda - \lambda z)}$$

```
>  with(inttrans):
>  k:=2;
```

$$k := 2$$

```
>  b:=(mu*k)^k*t^(k-1)*exp(-k*mu*t)/(k-1)!;
```

$$b := 4\,\mu^2\, t\, e^{(-2\mu t)}$$

```
>  bLT:=unapply(laplace(b,t,z),z);
```

$$bLT := z \rightarrow 4\,\frac{\mu^2}{(z+2\mu)^2}$$

```
>  normal(G(z));
```

$$4\,\frac{\mu\,(\mu - \lambda)}{\lambda^2 z^2 - 4\lambda z \mu - \lambda^2 z + 4\mu^2}$$

```
>  with(genfunc):
>  pi:=unapply(rgf_expand(G(z),z,n),n);
```

$$\pi := n \rightarrow -8\,\frac{\sqrt{8\lambda\mu + \lambda^2}\,\mu\,(\mu - \lambda)\,(2\,\dfrac{\lambda}{4\mu + \lambda + \sqrt{8\lambda\mu + \lambda^2}})^n}{\lambda\,(8\mu + \lambda)\,(4\mu + \lambda + \sqrt{8\lambda\mu + \lambda^2})}$$
$$-\,8\,\frac{\sqrt{8\lambda\mu + \lambda^2}\,\mu\,(\mu - \lambda)\,(-2\,\dfrac{\lambda}{-4\mu - \lambda + \sqrt{8\lambda\mu + \lambda^2}})^n}{\lambda\,(8\mu + \lambda)\,(-4\mu - \lambda + \sqrt{8\lambda\mu + \lambda^2})}$$

To check the results we let $\lambda = 5$ and $\mu = 7$ and compute the first 1000 terms of π_n, $n = 0, 1, 2, \ldots, 1000$ which gives $\sum_{n=0}^{1000} \pi_n = 1.0$, as expected.

```
>  lambda:=5; mu:=7;
```

$$\lambda := 5$$

$$\mu := 7$$

```
>  Sum(pi(n),n=0..1000); evalf(%);
```

$$\sum_{n=0}^{1000}\left(-\frac{112}{305}\,\frac{\sqrt{305}\,(10\,\dfrac{1}{33+\sqrt{305}})^n}{33+\sqrt{305}} - \frac{112}{305}\,\frac{\sqrt{305}\,(-10\,\dfrac{1}{-33+\sqrt{305}})^n}{-33+\sqrt{305}}\right)$$

$$1.000000000$$

```
>  lambda:='lambda'; mu:='mu';
```

$$\lambda := \lambda$$

$$\mu := \mu$$

The first few terms in $\{\pi_n\}$ and the expected number of customers $E(X)$ are also found exactly as follows.

```
>    for k from 1 to 3 do
     pi[ k] :=subs(z=0,normal(diff(G(z),z$k)))/k!
     od;
```

$$\pi_1 := \frac{1}{4} \frac{(\mu - \lambda)\,\lambda\,(4\,\mu + \lambda)}{\mu^3}$$

$$\pi_2 := \frac{1}{16} \frac{(\mu - \lambda)\,\lambda^2\,(12\,\mu^2 + 8\,\lambda\,\mu + \lambda^2)}{\mu^5}$$

$$\pi_3 := \frac{1}{64} \frac{(\mu - \lambda)\,\lambda^3\,(\lambda^3 + 40\,\mu^2\,\lambda + 32\,\mu^3 + 12\,\lambda^2\,\mu)}{\mu^7}$$

```
>    EX:=normal(limit(diff(G(z),z),z=1));
```

$$EX := \frac{1}{4} \frac{(-\lambda + 4\,\mu)\,\lambda}{\mu\,(\mu - \lambda)}$$

7.4.5   Transient Behavior of Markov Chains

As we discussed in Section 7.4.1, the n-step transition matrix $\mathbf{P}^{(n)}$ can be computed by finding $\mathbf{P}^n$. However, as n increases, the computation of $\mathbf{P}^n$ may become tedious and may even give rise to numerical inaccuracies for some ill-conditioned transition matrices. We now present an *exact* approach for computing the transient probabilities $\mathbf{P}^{(n)}$ that is based on generating functions. This approach is a particularly attractive one since Maple has powerful facilities for inverting generating functions.

First recall that

$$\begin{aligned} \mathbf{P}^{(n)} &= \mathbf{P}^{(n-1)}\mathbf{P} \\ &= \mathbf{P}^{(0)}\mathbf{P}^n, \quad n = 1, 2, \ldots \end{aligned} \tag{7.17}$$

where $\mathbf{P}^{(0)}$ is defined as the identity matrix $\mathbf{I}$. We define

$$\mathbf{G}(z) = \sum_{n=0}^{\infty} \mathbf{P}^{(n)} z^n$$

as the matrix-generating function of $\mathbf{P}^{(n)}$. Multiplying both sides of (7.17) and summing over $n = 1, 2, \ldots$ gives

$$\sum_{n=1}^{\infty} \mathbf{P}^{(n)} z^n = \sum_{n=1}^{\infty} \mathbf{P}^{(n-1)} \mathbf{P} z^n.$$

Adding and subtracting $\mathbf{P}^{(0)} z^0$, we get

$$\sum_{n=1}^{\infty} \mathbf{P}^{(n)} z^n + \mathbf{P}^{(0)} z^0 - \mathbf{P}^{(0)} z^0 = z \sum_{n=1}^{\infty} \mathbf{P}^{(n-1)} \mathbf{P} z^{n-1}$$

$$\mathbf{G}(z) - \mathbf{P}^{(0)}z^0 \;=\; z\mathbf{G}(z)\mathbf{P}.$$

Solving for $\mathbf{G}(z)$, we find

$$\mathbf{G}(z) = \mathbf{P}^{(0)}(\mathbf{I} - z\mathbf{P})^{-1}.$$

Since $\mathbf{P}^{(n)} = \mathbf{P}^{(0)}\mathbf{P}^n$ and $\mathbf{G}(z)$ is the generating function of $\mathbf{P}^{(n)}$, the matrix expression $(\mathbf{I} - z\mathbf{P})^{-1}$ must be the generating function of the nth power $\mathbf{P}^n$ as the following diagram indicates:

$$
\begin{array}{ccccc}
\mathbf{G}(z) & = & \mathbf{P}^{(0)} & (\mathbf{I} - z\mathbf{P})^{-1} \\
\updownarrow & & \updownarrow & \updownarrow \\
\mathbf{P}^{(n)} & = & \mathbf{P}^{(0)} & \mathbf{P}^n
\end{array}
$$

Thus, to compute the n-step transient probability matrix $\mathbf{P}^n$, one only needs to invert the matrix-generating function $(\mathbf{I} - z\mathbf{P})^{-1}$. We now present an example of this method.

Example 88 *Transient solution of a three-state Markov chain.* Consider a Markov chain with the transition matrix given as

$$
\mathbf{P} = \begin{pmatrix} 0 & 1 & 0 \\ 1/2 & 0 & 1/2 \\ 3/10 & 1/2 & 1/5 \end{pmatrix}.
$$

After loading Maple's `linalg()` and `genfunc()` packages, we find the inverse of the matrix $(I - zP)^{-1}$ whose entries are functions of z.

```
>    restart: # TransientMC.mws
>    with(linalg):
```

Warning, new definition for norm

Warning, new definition for trace

```
>    with(genfunc);
```

$[rgf\_charseq,\ rgf\_encode,\ rgf\_expand,\ rgf\_findrecur,\ rgf\_hybrid,$
$rgf\_norm,\ rgf\_pfrac,\ rgf\_relate,\ rgf\_sequence,\ rgf\_simp,\ rgf\_term,$
$termscale]$

```
>    Row[ 1] :=[ 0,1,0] ;
```

$$Row_1 := [0,\ 1,\ 0]$$

```
>    Row[ 2] :=[ 1/2,0,1/2] ;
```

$$Row_2 := [\frac{1}{2},\ 0,\ \frac{1}{2}]$$

```
>    Row[ 3] :=[ 3/10,5/10,2/10] ;
```

$$Row_3 := [\frac{3}{10},\ \frac{1}{2},\ \frac{1}{5}]$$

```
>   P:=matrix(3,3,[ seq(Row[k],k=1..3)] );
```

$$P := \begin{bmatrix} 0 & 1 & 0 \\ \dfrac{1}{2} & 0 & \dfrac{1}{2} \\ \dfrac{3}{10} & \dfrac{1}{2} & \dfrac{1}{5} \end{bmatrix}$$

```
>   ID:=array(identity,1..3,1..3);
```

$$ID := \mathrm{array}(identity,\ 1..3,\ 1..3,\ [])$$

```
>   IDminuszP:=evalm(ID-z* P);
```

$$IDminuszP := \begin{bmatrix} 1 & -z & 0 \\ -\dfrac{1}{2}z & 1 & -\dfrac{1}{2}z \\ -\dfrac{3}{10}z & -\dfrac{1}{2}z & 1-\dfrac{1}{5}z \end{bmatrix}$$

```
>   inv:=inverse(IDminuszP);
```

$$inv := \begin{bmatrix} \dfrac{-20+4z+5z^2}{\%1} & 4\dfrac{z(-5+z)}{\%1} & -10\dfrac{z^2}{\%1} \\ -\dfrac{z(10+z)}{\%1} & 4\dfrac{-5+z}{\%1} & -10\dfrac{z}{\%1} \\ -\dfrac{z(5z+6)}{\%1} & -2\dfrac{z(5+3z)}{\%1} & 10\dfrac{-2+z^2}{\%1} \end{bmatrix}$$

$$\%1 := -20+4z+15z^2+z^3$$

By applying $\mathrm{rgf\_expand}()$ to each element of the matrix $(\mathbf{I}-z\mathbf{P})^{-1}$, Maple inverts the generating functions. This operation generates the exact expressions for each of the nine entries of the $\mathbf{P}^{(n)}$ matrix denoted by Pn:

```
>   Pn:=map(rgf_expand,inv,z,n);
```

$Pn :=$

$$\left[\frac{11}{37} + \frac{(\frac{98}{37} - \frac{262}{407}\sqrt{11})\%2}{8 - 2\sqrt{11}} + \frac{(\frac{262}{407}\sqrt{11} + \frac{98}{37})\%1}{8 + 2\sqrt{11}}\right.,$$

$$\frac{16}{37} + \frac{(\frac{82}{37} - \frac{408}{407}\sqrt{11})\%2}{8 - 2\sqrt{11}} + \frac{(\frac{408}{407}\sqrt{11} + \frac{82}{37})\%1}{8 + 2\sqrt{11}},$$

$$\left.\frac{10}{37} + \frac{(\frac{670}{407}\sqrt{11} - \frac{180}{37})\%2}{8 - 2\sqrt{11}} + \frac{(-\frac{670}{407}\sqrt{11} - \frac{180}{37})\%1}{8 + 2\sqrt{11}}\right]$$

$$\left[\frac{11}{37} + \frac{(-\frac{13}{37} - \frac{3}{407}\sqrt{11})\%2}{8 - 2\sqrt{11}} + \frac{(\frac{3}{407}\sqrt{11} - \frac{13}{37})\%1}{8 + 2\sqrt{11}}\right.,$$

$$\frac{16}{37} + \frac{(\frac{8}{37} + \frac{73}{407}\sqrt{11})\%2}{8 - 2\sqrt{11}} + \frac{(-\frac{73}{407}\sqrt{11} + \frac{8}{37})\%1}{8 + 2\sqrt{11}},$$

$$\left.\frac{10}{37} + \frac{(\frac{5}{37} - \frac{70}{407}\sqrt{11})\%2}{8 - 2\sqrt{11}} + \frac{(\frac{70}{407}\sqrt{11} + \frac{5}{37})\%1}{8 + 2\sqrt{11}}\right]$$

$$\left[\frac{11}{37} + \frac{(-\frac{87}{37} + \frac{293}{407}\sqrt{11})\%2}{8 - 2\sqrt{11}} + \frac{(-\frac{293}{407}\sqrt{11} - \frac{87}{37})\%1}{8 + 2\sqrt{11}}\right.,$$

$$\frac{16}{37} + \frac{(-\frac{103}{37} + \frac{332}{407}\sqrt{11})\%2}{8 - 2\sqrt{11}} + \frac{(-\frac{332}{407}\sqrt{11} - \frac{103}{37})\%1}{8 + 2\sqrt{11}},$$

$$\left.\frac{10}{37} + \frac{(-\frac{625}{407}\sqrt{11} + \frac{190}{37})\%2}{8 - 2\sqrt{11}} + \frac{(\frac{625}{407}\sqrt{11} + \frac{190}{37})\%1}{8 + 2\sqrt{11}}\right]$$

$$\%1 := (-\frac{1}{8 + 2\sqrt{11}})^n$$

$$\%2 := (-\frac{1}{8 - 2\sqrt{11}})^n$$

To evaluate one of these entries of the $\mathbf{P}^{(n)}$ matrix, say, $p_{11}^{(n)}$ at $n = 5$, we use the unapply() function and find that $p_{11}^{(5)} = 0.2185$.

```
>    P[ 1,1] :=unapply(Pn[ 1,1] ,n);
```

$P_{1,1} := n \rightarrow$

$$\frac{11}{37} + \frac{(\frac{98}{37} - \frac{262}{407}\sqrt{11})\,(-\dfrac{1}{8-2\sqrt{11}})^n}{8-2\sqrt{11}} + \frac{(\frac{262}{407}\sqrt{11} + \frac{98}{37})\,(-\dfrac{1}{8+2\sqrt{11}})^n}{8+2\sqrt{11}}$$

```
>   evalf(P[1,1](5));
```

$$.2185000001$$

Taking the limit as $n \rightarrow \infty$ of each entry gives the limiting probabilities of the Markov chain. The reader would appreciate that performing these inversions manually without the help of Maple would be very difficult, if not impossible.

```
>   map(limit,Pn,n=infinity);
```

$$\begin{bmatrix} -\dfrac{55}{37}\%1 & -\dfrac{80}{37}\%1 & -\dfrac{50}{37}\%1 \\[2mm] -\dfrac{55}{37}\%1 & -\dfrac{80}{37}\%1 & -\dfrac{50}{37}\%1 \\[2mm] -\dfrac{55}{37}\%1 & -\dfrac{80}{37}\%1 & -\dfrac{50}{37}\%1 \end{bmatrix}$$

$$\%1 := \frac{1}{(-4+\sqrt{11})\,(4+\sqrt{11})}$$

```
>   evalf(%);
```

$$\begin{bmatrix} .2972972971 & .4324324322 & .2702702701 \\ .2972972971 & .4324324322 & .2702702701 \\ .2972972971 & .4324324322 & .2702702701 \end{bmatrix}$$

Thus, the long-run probability of finding the process in state $j = 1, 2, 3$ is independent of the initial state.

7.5 Continuous-Time Markov Chains

As we discussed in Section 7.4, if a stochastic process $\{X_n, n = 0, 1, \ldots\}$ is a discrete-time Markov chain with state space S, it spends exactly one time unit in some state $i \in S$ before making a transition. With probability p_{ii} the process returns to the same state and with probability $1 - p_{ii}$ it makes a transition to another state $j \neq i$. Hence, the probability distribution of time spent in state i is *geometric* with parameter $1 - p_{ii}$. That is, if we define N_i as the (random) number of time units the process spends in i, then its density is given by $\Pr(N_i = k) = p_{ii}^{k-1}(1 - p_{ii})$, $k = 1, 2, \ldots$ with mean duration $E(N_i) = 1/(1 - p_{ii})$. We now present a continuous-time analogue of this process.

Consider a continuous-time stochastic process $\{X(t), t \geq 0\}$ that spends an *exponential* length of time, T_i, with parameter v_i in state $i \in S$; that is, the density of T_i is $f_{T_i}(t) = v_i e^{-v_i t}, t \geq 0$ with mean duration $E(T_i) = 1/v_i$. The parameter v_i of the exponential distribution is the transition rate out of state i per unit time. When the process leaves state i, it enters state j with probability p_{ij} where $p_{ii} = 0$ and $\sum_{j \neq i} p_{ij} = 1$ for all i. Hence, the process $\{X(t), t \geq 0\}$ goes from state to state according to a Markov chain with transition matrix $\mathbf{P} = [p_{ij}]$, but each time it visits a state, it stays there a random length (exponential) of time that is independent of the past behavior. Now consider two states i and j such that $i \neq j$. Since the process leaves state i with rate v_i and goes into state j with probability p_{ij}, we define $q_{ij} = v_i p_{ij}$ as the *transition rate* from i to j. Note that as $\sum_{j \neq i} p_{ij} = 1$ we have $\sum_{j \neq i} q_{ij} = v_i$. The q_{ij} values constitute the entries of the *infinitesimal generator* matrix $\mathbf{Q} = [q_{ij}]$ of the CTMC.

Such a process $\{X(t), t \geq 0\}$ is a continuous-time Markov chain (CTMC) with state space S. The process has the property that for $i, j \in S$,

$$\Pr[X(t+s) = j \quad | \quad X(s) = i, X(u) = x(u), 0 \leq u < s]$$
$$= \quad \Pr[X(t+s) = j \mid X(s) = i]$$

for all $s, t \geq 0$ and $x(u), 0 \leq u \leq s$.

Analogous to a discrete-time Markov chain, in a CTMC the conditional distribution of $X(t+s)$ given the past history over $0 \leq u \leq s$ depends only on the current state $X(s)$ at time s. We say that $\{X(t), t \geq 0\}$ is a *time-homogeneous* CTMC if the conditional probability $\Pr[X(t+s) = j \mid X(s) = i]$ is independent of s. In this case we write

$$\Pr[X(t+s) = j \mid X(s) = i] = p_{ij}(t)$$

where $p_{ij}(t)$ is called the *transition function* from state i to state j. This quantity is analogous to the transition probability $p_{ij}^{(n)}$ of a discrete-time Markov chain.

Many realistic problems arising in manufacturing and communication systems are commonly modeled using continuous time Markov chains. The Markovian queueing models that will be presented in Chapter 9, Queueing Systems, are all CTMCs since, in those models, one assumes that interarrival and service times are exponential, thus making the process Markovian.

We now discuss some examples of continuous-time Markov chains.

Example 89 *Birth and death processes.* Consider a CTMC that can make transitions from state i only to its neighboring states $i-1$ or $i+1$ for $i \in S = \{0, 1, \ldots\}$. When the process is in state i, the transition rates are given by $q_{i,i+1} = \lambda_i$ for $i = 0, 1, \ldots$ and $q_{i,i-1} = \mu_i$ for $i = 1, 2, \ldots$. Usually, the state $X(t)$ of the process represents the size of some population (e.g., customers in a queueing system). In the case of a queueing system with a single server this corresponds to the following: When there are i customers in the system, the customers arrive according to a Poisson process with rate λ_i and the service time is exponential with rate μ_i, meaning that the arrival and service rates can be, in general, state dependent. The parameters λ_i and μ_i are called the *birth rates* and the *death rates*, respectively. Note that since $\sum_{j \neq i} q_{ij} = v_i$, the transition rate out of state i is

$$v_i = \begin{cases} \lambda_0, & \text{for } i = 0 \\ \lambda_i + \mu_i, & \text{for } i = 1, 2, \ldots \end{cases}$$

and the infinitesimal generator $\mathbf{Q}$ is given by

$$\mathbf{Q} = \begin{matrix} 0 \\ 1 \\ 2 \\ \vdots \end{matrix} \begin{pmatrix} ? & \lambda_0 & & & \\ \mu_1 & ? & \lambda_1 & & \\ & \mu_2 & ? & \lambda_2 & \\ \vdots & \vdots & \vdots & \vdots & \ddots \end{pmatrix}.$$

The entries with question marks (?) in the infinitesimal generator matrix $\mathbf{Q}$ will be identified shortly.

To compute the transition probabilities p_{ij} we first define X_i as the time until the next birth (exponential with rate λ_i) and Y_i as the time until the next death (also exponential but with rate μ_i). Thus, $p_{01} = $ Pr(a birth at some time in the future) = Pr($X_0 < \infty$) = 1. To compute $p_{12} = $ Pr(a birth before a death) = Pr($X_1 < Y_1$), we proceed by conditioning on the occurrence of the time of death, Y_1, which gives

$$\begin{aligned} p_{12} &= \text{Pr}(X_1 < Y_1) = \int_0^\infty \text{Pr}(X_1 < Y_1 \mid Y_1 = t) \mu_1 e^{-\mu_1 t} \, dt \\ &= \int_0^\infty (1 - e^{-\lambda_1 t}) \mu_1 e^{-\mu_1 t} \, dt. \end{aligned}$$

We compute this integral with Maple as follows.

```
>    restart: # p12.mws
>    assume(lambda[ 1] >0,mu[ 1] >0);
```

```
Error, (in assume) indexed names cannot be assumed
```

Since Maple is unable to use the command assume() with the indexed variables, we temporarily remove the indices from λ_1 and μ_1 to compute the integral.

```
>    assume(lambda>0,mu>0);
>    p[ 1,2] :=Int((1-exp(-lambda*t))*mu*exp(-mu*t),
     t=0..infinity);
```

$$p_{1,2} := \int_0^\infty (1 - e^{(-\lambda t)}) \, \mu \, e^{(-\mu t)} \, dt$$

```
>    value(%);
```

$$\frac{\lambda}{\mu + \lambda}$$

Thus, in general, we obtain

$$p_{i,i+1} = \frac{\lambda_i}{\lambda_i + \mu_i} = 1 - p_{i,i-1}$$

as the transition probabilities of the birth-death processes.

Example 90 *Pure birth (Poisson) process.* It is interesting to note that if no deaths are possible (i.e., when $\mu_i = 0$ for all $i = 1, 2, \ldots$) and if the arrival rate is a constant (i.e., $v_i = \lambda_i = \lambda$ for $i = 0, 1, \ldots$) the birth-death process reduces to the ordinary Poisson process with state space $S = \{0, 1, 2, \ldots\}$. For this process, the transition probabilities are $p_{i,i+1} = 1$ for $i = 0, 1, \ldots$

Example 91 *A self-service car wash and vacuum facility.* Consider a small self-service car wash and vacuum facility where the service on a car is performed in two stages, (i) wash and (ii) vacuum. We assume that there is only space for one car at the facility, i.e., that queues are not permitted. Potential customers arrive according to a Poisson process with rate λ. If the facility is empty, the customer enters, washes his car (time required distributed exponentially with rate μ_1), vacuums it (time required distributed exponentially with rate μ_2) and then departs. If the facility is busy, the arriving customer goes elsewhere since there is no space for waiting.

The stochastic process $X(t)$ representing the status of the facility at time t is a CTMC with state space $S = \{0, 1, 2\}$ where state 0 corresponds to the empty system and states 1 and 2 correspond to car being washed and being vacuumed, respectively. Thus, the transition rates are $v_0 = \lambda$, $v_1 = \mu_1$ and $v_2 = \mu_2$. Since the operations are performed sequentially, the states are visited in the order $0 \rightarrow 1 \rightarrow 2$, implying that the transition matrix $P = [p_{ij}]$ for this CTMC is

$$
P = \begin{matrix} 0 \\ 1 \\ 2 \end{matrix} \begin{pmatrix} 0 & 1 & 0 \\ 0 & 0 & 1 \\ 1 & 0 & 0 \end{pmatrix}.
$$

The infinitesimal generator $Q = [q_{ij}]$ is then

$$
Q = \begin{matrix} 0 \\ 1 \\ 2 \end{matrix} \begin{pmatrix} ? & \lambda & 0 \\ 0 & ? & \mu_1 \\ \mu_2 & 0 & ? \end{pmatrix}.
$$

7.5.1 Kolmogorov Differential Equations

In discrete-time Markov chains the n-step transition probability matrix $\mathbf{P}^{(n)} = [p_{ij}^{(n)}]$ was computed using $\mathbf{P}^{(n)} = \mathbf{P}^{(n-1)}\mathbf{P}$ where $p_{ij}^{(n)} = \Pr(X_n = j \mid X_0 = i)$. Using the Chapman-Kolmogorov equations, it is easy to show that $p_{ij}^{(m+n)} = \Pr(X_{m+n} = j \mid X_0 = i) = \sum_{k \in S} p_{ik}^{(m)} p_{kj}^{(n)}$, implying that $\mathbf{P}^{(m+n)} = \mathbf{P}^{(m)}\mathbf{P}^{(n)}$. We now consider the continuous-time analogues of these quantities.

As we indicated at the start of this section, for a time homogeneous CTMC, we have

$$
\Pr[X(t+s) = j \mid X(s) = i] = \Pr[X(t) = j \mid X(0) = i] = p_{ij}(t)
$$

where $p_{ij}(t)$ is the probability that the process that is in state i will be in state j after an additional time t. It follows that the continuous time version of the

Chapman-Kolmogorov equations is given by

$$\Pr[X(t+s)=j \mid X(0)=i] = p_{ij}(t+s) = \sum_{k\in S} p_{ik}(t)p_{kj}(s), \qquad (7.18)$$

which follows from the Markovian property of the CTMC. Unfortunately, in its present form, this result is not very useful as it cannot be used to solve for the unknown functions $p_{ij}(t)$.

Before we develop differential equations whose solution would, in principle, provide us with the probability functions $p_{ij}(t)$, we need to discuss a lemma.

Lemma 1 *For state i in S, we have*

$$\lim_{h\to 0} \frac{1-p_{ii}(h)}{h} = v_i,$$

i.e., the limit is equal to the transition rate v_i out of state i. Additionally, for any two states i and j in S,

$$\lim_{h\to 0} \frac{p_{ij}(h)}{h} = q_{ij},$$

i.e., the limit is equal to the transition rate q_{ij} from i to j.

Proof. Note that since $p_{ii}(h) = \Pr[X(h)=i \mid X(0)=i]$, we find

$$
\begin{aligned}
1-p_{ii}(h) &= \Pr[X(h) \text{ is anywhere but } i \mid X(0)=i]\\
&= \Pr[\text{a transition in } h \mid X(0)=i] = v_i h + o(h).
\end{aligned}
$$

Dividing by h and letting $h \to 0$ gives the first result. Similarly, since $p_{ij}(h) = \Pr[X(h)=j \mid X(0)=i] = v_i h p_{ij} + o(h)$, dividing by h and letting $h \to 0$ gives the second result. ∎

Kolmogorov's Backward Differential Equations

Using the Chapman-Kolmogorov equations (7.18), we can write $p_{ij}(h+t) = \sum_{k=0}^{\infty} p_{ik}(h)p_{kj}(t)$. This is the conditional probability that the process will be in state j at time $h+t$ given that it was in state i at time 0 and it is obtained by conditioning on the state *back* at time h. Subtracting $p_{ij}(t)$ from both sides, removing the term corresponding to $k=i$ and collecting the terms we have,

$$
\begin{aligned}
p_{ij}(h+t) - p_{ij}(t) &= \sum_{k=0}^{\infty} p_{ik}(h)p_{kj}(t) - p_{ij}(t)\\
&= \sum_{\substack{k=0\\k\neq i}}^{\infty} p_{ik}(h)p_{kj}(t) - [1-p_{ii}(h)]p_{ij}(t).
\end{aligned}
$$

Dividing by h and letting $h \to 0$ gives

$$
p'_{ij}(t) = v_i \sum_{\substack{k\neq i}}^{\infty} p_{ik}p_{kj}(t) - v_i p_{ij}(t)
$$

$$= \sum_{k \neq i}^{\infty} q_{ik} p_{kj}(t) - v_i p_{ij}(t) \qquad (7.19)$$

as Kolmogorov's *backward* differential equations with the initial conditions $p_{ij}(0) = \delta_{ij}$, for $i, j \in S$ where δ_{ij} is the Kronecker delta. The solution of this infinite system of differential equations, if available, would give the transient probabilities $p_{ij}(t)$ for $t \geq 0$.[7]

Suppose we now define $q_{ii} = -v_i$. Rewriting (7.19), we get

$$p_{ij}'(t) = \sum_{k=0}^{\infty} q_{ij} P_{kj}(t),$$

or, using matrix notation,

$$\mathbf{P}'(t) = \mathbf{Q}\mathbf{P}(t), \quad \text{with } \mathbf{P}(0) = \mathbf{I}$$

where $\mathbf{I}$ is the identity matrix. We see that with this definition, the diagonal elements of the ith row of the infinitesimal generator $\mathbf{Q}$ matrix is the negative of the sum of all other elements in row i. For example, for the birth-death process of Example 89, the $\mathbf{Q}$ matrix is

$$\mathbf{Q} = \begin{pmatrix} -\lambda_0 & \lambda_0 & & & \\ \mu_1 & -(\mu_1 + \lambda_1) & \lambda_1 & & \\ & \mu_2 & -(\mu_1 + \lambda_1) & \lambda_2 & \\ \vdots & \vdots & \vdots & \vdots & \ddots \end{pmatrix}.$$

Similarly, for the car wash model in Example 91, we have

$$\mathbf{Q} = \begin{pmatrix} -\lambda & \lambda & 0 \\ 0 & -\mu_1 & \mu_1 \\ \mu_2 & 0 & -\mu_2 \end{pmatrix}.$$

We first discuss a simple example with $N = 2$ states.

Example 92 *A Markovian queue with no waiting space.* Consider a machine that works (stays ON in state 1) for a random length of time distributed exponentially with rate μ. When the machine breaks down and moves to the OFF state 0, it takes a random length of time to repair it that is distributed exponentially with rate λ. This two state CTMC ($N = 2$) is thus a Markovian queue with a single server and no waiting space. It is clear that for this model the transition rates from 0 and 1 are $v_0 = \lambda$ and $v_1 = \mu$, respectively. It then follows that $q_{01} = \lambda$ and $q_{10} = \mu$.

For $i = j = 0$ the Kolmogorov differential equation is written as

$$\begin{aligned} p_{00}'(t) &= q_{01} p_{10}(t) - v_0 p_{00}(t) \\ &= \lambda p_{10}(t) - \lambda p_{00}(t) \end{aligned}$$

---

[7]For a CTMC with N states, this system has N^2 ordinary differential equations. Its solution, if available, gives the transient probabilities $p_{ij}(t)$ for $t \geq 0$ and $i, j \in \{1, 2, \ldots, N\}$.

with $p_{00}(0) = 1$. The other three differential equations are obtained in a similar manner, which gives

$$\begin{pmatrix} p'_{00}(t) & p'_{01}(t) \\ p'_{10}(t) & p'_{11}(t) \end{pmatrix} = \begin{pmatrix} -\lambda & \lambda \\ \mu & -\mu \end{pmatrix} \begin{pmatrix} p_{00}(t) & p_{01}(t) \\ p_{10}(t) & p_{11}(t) \end{pmatrix}.$$

This system of $N \times N = 4$ equations can be written as $\mathbf{P}'(t) = \mathbf{Q}\mathbf{P}(t)$, with $\mathbf{P}(0) = \mathbf{I}$ where $\mathbf{P}(t) = [p_{ij}(t)]$ and $\mathbf{P}'_{ij}(t) = [p'_{ij}(t)]$.

We now use Maple to solve these differential equations.[8]

Maple has a `matrixDE()` command that can solve a system of time-dependent, linear ODEs of the form $\mathbf{X}'(t) = \mathbf{A}(t)\mathbf{X}(t) + \mathbf{B}(t)$. This command returns a pair of matrices $[\mathbf{K}(t), \mathbf{L}(t)]$ of dimensions $n \times n$ and $n \times 1$, respectively. If $\mathbf{B}(t) = \mathbf{0}$, then we also have $\mathbf{L}(t) = \mathbf{0}$. In our case since $\mathbf{P}'(t) = \mathbf{Q}\mathbf{P}(t)$, the term $\mathbf{B}(t)$ is not present and thus $\mathbf{L}(t) = \mathbf{0}$.

A particular solution of the system is then written in the form $\mathbf{P}(t) = \mathbf{K}(t)\mathbf{C}_0 + \mathbf{L}(t)$ where $\mathbf{C}_0$ is an $n \times 1$ vector to be determined. Using the condition at $t = 0$ that $\mathbf{P}(0) = \mathbf{I}$, we find $\mathbf{C}_0 = [\mathbf{K}(0)]^{-1}\mathbf{P}(0) = [\mathbf{K}(0)]^{-1}$.

```
>   restart: # MatrixDE.mws

>   with(DEtools): with(linalg):

Warning, new definition for adjoint

Warning, new definition for norm

Warning, new definition for trace

>   Q:=matrix(2,2,[ -lambda,lambda,mu,-mu] );
```

$$Q := \begin{bmatrix} -\lambda & \lambda \\ \mu & -\mu \end{bmatrix}$$

```
>   Sol:=matrixDE(Q,t);
```

$$Sol := \begin{bmatrix} \begin{bmatrix} 1 & e^{(-(\lambda+\mu)t)} \\ 1 & -\dfrac{e^{(-(\lambda+\mu)t)}\mu}{\lambda} \end{bmatrix}, [0, 0] \end{bmatrix}$$

```
>   K:=Sol[ 1] ;
```

---

[8]Since $p_{00}(t) + p_{01}(t) = 1$ and $p_{10}(t) + p_{11}(t) = 1$, we could have considerably simplified the solution of this system by solving the lower dimensional system with the unknown functions $p_{00}(t)$ and $p_{10}(t)$ as

$$\begin{aligned} p'_{00}(t) &= -\lambda p_{00}(t) + \lambda p_{10}(t) \\ p'_{10}(t) &= \mu p_{00}(t) - \mu p_{10}(t). \end{aligned}$$

The present Maple worksheet is more straightforward since it solves the system $\mathbf{P}'(t) = \mathbf{Q}\mathbf{P}(t)$ directly.

$$K := \begin{bmatrix} 1 & e^{(-(\lambda+\mu)t)} \\ 1 & -\dfrac{e^{(-(\lambda+\mu)t)}\,\mu}{\lambda} \end{bmatrix}$$

```
>   C[ 0] :=simplify(subs(t=0,inverse(K)));
```

$$C_0 := \begin{bmatrix} \dfrac{\mu}{\lambda+\mu} & \dfrac{\lambda}{\lambda+\mu} \\ \dfrac{\lambda}{\lambda+\mu} & -\dfrac{\lambda}{\lambda+\mu} \end{bmatrix}$$

```
>   P:=evalm(K &* C[ 0] );
```

$$P := \begin{bmatrix} \dfrac{\mu}{\lambda+\mu}+\dfrac{\%1\,\lambda}{\lambda+\mu} & \dfrac{\lambda}{\lambda+\mu}-\dfrac{\%1\,\lambda}{\lambda+\mu} \\ \dfrac{\mu}{\lambda+\mu}-\dfrac{\%1\,\mu}{\lambda+\mu} & \dfrac{\lambda}{\lambda+\mu}+\dfrac{\%1\,\mu}{\lambda+\mu} \end{bmatrix}$$

$$\%1 := e^{(-(\lambda+\mu)t)}$$

To summarize, the explicit solution $\mathbf{P}(t)$ for the linear system of differential equations $\mathbf{P}'(t) = \mathbf{Q}\mathbf{P}(t)$, $\mathbf{P}(0) = \mathbf{I}$ for this two state CTMC is obtained as follows:

$$\mathbf{P}(t) = \begin{matrix} 0 \\ 1 \end{matrix}\left(\begin{matrix} \frac{\mu}{\lambda+\mu}+\frac{\lambda}{\lambda+\mu}e^{-(\lambda+\mu)t} & \frac{\lambda}{\lambda+\mu}-\frac{\lambda}{\lambda+\mu}e^{-(\lambda+\mu)t} \\ \frac{\mu}{\lambda+\mu}-\frac{\mu}{\lambda+\mu}e^{-(\lambda+\mu)t} & \frac{\lambda}{\lambda+\mu}+\frac{\mu}{\lambda+\mu}e^{-(\lambda+\mu)t} \end{matrix} \right).$$

As a check, we see that the solution found satisfies the original systems of ODEs.

```
>   RHS:=evalm(Q &* P);
```

$$RHS :=$$
$$\left[-\lambda\left(\frac{\mu}{\lambda+\mu}+\frac{\%1\,\lambda}{\lambda+\mu}\right)+\lambda\left(\frac{\mu}{\lambda+\mu}-\frac{\%1\,\mu}{\lambda+\mu}\right), \right.$$
$$\left. -\lambda\left(\frac{\lambda}{\lambda+\mu}-\frac{\%1\,\lambda}{\lambda+\mu}\right)+\lambda\left(\frac{\lambda}{\lambda+\mu}+\frac{\%1\,\mu}{\lambda+\mu}\right) \right]$$
$$\left[\mu\left(\frac{\mu}{\lambda+\mu}+\frac{\%1\,\lambda}{\lambda+\mu}\right)-\mu\left(\frac{\mu}{\lambda+\mu}-\frac{\%1\,\mu}{\lambda+\mu}\right), \right.$$
$$\left. \mu\left(\frac{\lambda}{\lambda+\mu}-\frac{\%1\,\lambda}{\lambda+\mu}\right)-\mu\left(\frac{\lambda}{\lambda+\mu}+\frac{\%1\,\mu}{\lambda+\mu}\right) \right]$$

$$\%1 := e^{(-(\lambda+\mu)t)}$$

```
>   seq(seq(normal(diff(P[ i,j] ,t)-RHS[ i,j] ),
    i=1..2),j=1..2);
```

$$0, 0, 0, 0$$

Kolmogorov's Forward Differential Equations

Again using the Chapman-Kolmogorov equations (7.18), we can write $p_{ij}(t + h) = \sum_{k=0}^{\infty} p_{ik}(t)p_{kj}(h)$. This is the conditional probability that the process will be in state j at time $t + h$ given that it was in state i at time 0. This equation is obtained by conditioning on the state at time t. Following steps similar to the ones used in developing the backward equations, we find

$$
\begin{aligned}
p'_{ij}(t) &= \sum_{k \neq j}^{\infty} p_{ik}(t)v_k p_{kj} - v_j p_{ij}(t) \\
&= \sum_{k \neq j}^{\infty} p_{ik}(t)q_{kj} - v_j p_{ij}(t)
\end{aligned}
$$

as Kolmogorov's *forward* differential equations with the initial conditions $p_{ij}(0) = \delta_{ij}$ for $i, j \in S$ where δ_{ij} is the Kronecker delta. Defining $q_{jj} = -v_j$, these equations can also be written in matrix form as $\mathbf{P}'(t) = \mathbf{P}(t)\mathbf{Q}$ with $\mathbf{P}(0) = \mathbf{I}$.[9]

The Exponential Matrix

To find the transition functions $\mathbf{P}(t)$ for the simple Markovian queue in Example 92 we had to solve a system of DEs that required us to type several lines of Maple commands. We now describe a method based on the spectral representation of matrices that considerably simplifies the process.

First, note that if we had a *scalar* differential equation given by $p'(t) = qp(t)$, with $p(0) = 1$, the solution would simply be $p(t) = e^{qt} = \sum_{n=0}^{\infty} (qt)^n/n!$. It can be shown that for the matrix DE system $\mathbf{P}'(t) = \mathbf{Q}\mathbf{P}(t)$, $\mathbf{P}(0) = \mathbf{I}$, the solution assumes a similar form as $\mathbf{P}(t) = e^{\mathbf{Q}t}$, where $e^{\mathbf{Q}}$ is the exponential matrix defined by $e^{\mathbf{Q}t} = \sum_{n=0}^{\infty} (\mathbf{Q}t)^n/n!$. The computation of the exponential matrix requires the use of *spectral representation theorem* of linear algebra that is discussed in Karlin and Taylor [102, pp. 539–541] and Medhi [131, pp. 196–197]. Essentially, if the generator matrix $\mathbf{Q}$ of a CTMC has distinct eigenvalues, then $\mathbf{Q}$ can be expressed as $\mathbf{Q} = \mathbf{A}\mathbf{D}\mathbf{A}^{-1}$ where $\mathbf{A}$ is a non-singular matrix (formed with the right eigenvectors of $\mathbf{Q}$) and $\mathbf{D}$ is a diagonal matrix having its diagonal elements as the eigenvalues of $\mathbf{Q}$. Then the transient solution of the CTMC is obtained as $\mathbf{P}(t) = \mathbf{A}e^{\mathbf{D}t}\mathbf{A}^{-1}$ where $e^{\mathbf{D}t}$ is a diagonal matrix constructed with the eigenvalues of $\mathbf{Q}$.

Maple can compute the exponential matrix $\mathbf{P}(t) = e^{\mathbf{Q}t}$ with the `exponential(Q,t)` command as we will now show by re-solving the Markovian queue problem of Example 92:

```
>    restart: # QueueExpQt.mws

>    with(linalg):
```

---

[9]For the Markovian queue example with no waiting space, the forward equations look different but their solution is the same as the backward equations.

```
Warning, new definition for norm

Warning, new definition for trace

>   Q:=matrix(2,2,[ -lambda,lambda,mu,-mu] );
```

$$Q := \begin{bmatrix} -\lambda & \lambda \\ \mu & -\mu \end{bmatrix}$$

```
>   P:=exponential(Q,t);
```

$$P := \begin{bmatrix} \dfrac{\%1\,\lambda + \mu}{\lambda + \mu} & -\dfrac{\lambda\,(-1 + \%1)}{\lambda + \mu} \\[2ex] -\dfrac{\mu\,(-1 + \%1)}{\lambda + \mu} & \dfrac{\mu\,\%1 + \lambda}{\lambda + \mu} \end{bmatrix}$$

$$\%1 := e^{(-(\lambda + \mu)\,t)}$$

By specifying that both λ and μ are positive, we also show that the probabilities converge (as in a discrete time Markov chain) to the same values independent of the initial state.

```
>   assume(lambda>0,mu>0);
>   PLimit:=map(limit,P,t=infinity);
```

$$PLimit := \begin{bmatrix} \dfrac{\mu}{\lambda + \mu} & \dfrac{\lambda}{\lambda + \mu} \\[2ex] \dfrac{\mu}{\lambda + \mu} & \dfrac{\lambda}{\lambda + \mu} \end{bmatrix}$$

As another—and more impressive—example of the use of exponential(), consider the car wash facility of Example 91 represented by a CTMC with three states. In this case, following the same steps we compute the transient probability matrix $\mathbf{P}(t)$ as follows.

```
>   restart: # CarWashExpQt.mws
>   with(linalg): with(DEtools):

Warning, new definition for norm

Warning, new definition for trace

Warning, new definition for adjoint

>   Q:=matrix(3,3,[ -lambda,lambda,0,
    0,-mu[ 1] ,mu[ 1] ,
    mu[ 2] ,0,-mu[ 2] ] );
```

$$Q := \begin{bmatrix} -\lambda & \lambda & 0 \\ 0 & -\mu_1 & \mu_1 \\ \mu_2 & 0 & -\mu_2 \end{bmatrix}$$

In this problem the solution $\mathbf{P}(t)$ takes up several pages of output. In order to obtain a view of the general structure of the solution, we use the `sprint()` facility in the `\share` folder. This utility allows the user to look at the "top-level" structure of an expression to obtain a bird's-eye view of the expression.[10]

```
>    P:=exponential(Q,t):
```

```
>    with(share): with(sprint):
```

```
See ?share and ?share,contents for information about the
share library
```

```
Share Library:    sprint
```

```
Author: Monagan, Michael.
```

```
Description:  Utility routine 'short print' for displaying
large expressions, allowing the user to look at the
structure (top levels) of a large expression.
```

```
>    sprint(P[ 1,1] );  sprint(P[ 1,2] );  sprint(P[ 1,3] );
```

$$\frac{<< +16 >>}{(\mu_1 + \mu_2 + \lambda + \sqrt{<< +6 >>})(\mu_1 + \mu_2 + \lambda - \sqrt{<< +6 >>})\sqrt{<< +6 >>}}$$

$$-2\frac{\lambda << +11 >>}{(\mu_1 + \mu_2 + \lambda + \sqrt{<< +6 >>})(\mu_1 + \mu_2 + \lambda - \sqrt{<< +6 >>})\sqrt{<< +6 >>}}$$

$$2\frac{\lambda\,\mu_1 << +9 >>}{(\mu_1 + \mu_2 + \lambda + \sqrt{<< +6 >>})(\mu_1 + \mu_2 + \lambda - \sqrt{<< +6 >>})\sqrt{<< +6 >>}}$$

The exact solution for the transient probabilities $p_{0j}(t)$, $j = 0, 1, 2$[11] are very large expressions. For example, the first term P[1, 1] corresponding to $p_{00}(t)$ contains a sum in its numerator with 16 terms indicated by $<< +16 >>$. The probabilities in the first row add to 1.

```
>    Sum(P[ 1,j] ,j=1..3);
```

$$\sum_{j=1}^{3} P_{1,j}$$

```
>    is(value(%)=1);
```

true

If we specify the parameter values by letting, say, $\lambda = 10$, $\mu_1 = 5$ and $\mu_2 = 30$, the conditional probability that the system is empty at time t *given* that it was

---

[10]Note that the share library must be installed before this utility [and other `share()` utilities] can be used. Once installed, entering `with(share);` followed by `?share,contents` gives a complete list of all available share utilities.

[11]Note that the states 0, 1, 2 in the CTMC correspond to the rows 1, 2 and 3 of the Q matrix.

empty at time 0 is obtained as a complicated (but exact) expression involving exponential functions. For $t = 0.12$, the transient probability is found as 0.396.

>   lambda:=10; mu[1] :=5; mu[2] :=30;

$$\lambda := 10$$
$$\mu_1 := 5$$
$$\mu_2 := 30$$

>   P[1,1] ;

$$\frac{1}{25}((-11500\,e^{(-1/2\,(45+\sqrt{25})\,t)} + 11500\,e^{(-1/2\,(45-\sqrt{25})\,t)}$$
$$+ 700\,\sqrt{25}\,e^{(-1/2\,(45-\sqrt{25})\,t)} + 600\,\sqrt{25} + 700\,\sqrt{25}\,e^{(-1/2\,(45+\sqrt{25})\,t)})$$
$$\sqrt{25})\,/((45 + \sqrt{25})\,(45 - \sqrt{25}))$$

>   evalf(subs(t=0.12,P[1,1]));

$$.3962472752$$

Finally, as $t \to \infty$, the limiting probabilities are found as $[0.3, 0.6, 0.1]$ for each row.

>   PLimit:=evalf(map(limit,P,t=infinity));

$$PLimit := \begin{bmatrix} .3000000000 & .6000000000 & .1000000000 \\ .3000000000 & .6000000000 & .1000000000 \\ .3000000000 & .6000000000 & .1000000000 \end{bmatrix}$$

Unconditional State Probabilities

Analogous to the unconditional probabilities $\pi_j^{(n)}$ of a discrete-time Markov chain, we now define $\pi_j(t) = \Pr[X(t) = j]$ to be the unconditional probability that the CTMC will be found in state j at time t. First, conditioning on the state at time t we have $\pi_j(t + h) = \sum_{k=0}^{\infty} \pi_k(t) p_{kj}(h)$. Next, using arguments similar to those employed in the development of the forward DEs, we find that

$$\pi_j'(t) = \sum_{k \neq j}^{\infty} \pi_k(t) q_{kj} - v_j \pi_j(t)$$

is the system of differential equations representing the unconditional probabilities. Note that for a CTMC with a finite number of states, say N, this system has only N differential equations. Again defining $q_{jj} = -v_j$, these differential equations can be written using vector/matrix notation as

$$\Pi'(t) = \Pi(t)\mathbf{Q}$$

with $\Pi(0)$ as the vector of initial probabilities.

Here is the Maple solution of the unconditional probabilities for the Markovian queue problem of Example 92 where $\Pi(0) = [c_0, c_1]^T$ is not specified but $c_0 + c_1 = 1$. We solve this system of ODEs using the `dsolve` command.

```
>    restart:# TwoStateUncond.mws
>    Q:=matrix(2,2,[ -lambda,lambda,mu,-mu] );
```

$$Q := \begin{bmatrix} -\lambda & \lambda \\ \mu & -\mu \end{bmatrix}$$

```
>    pi:=vector(2,[ seq(pi.j(t),j=0..1)] );
```

$$\pi := [\pi 0(t),\ \pi 1(t)]$$

```
>    RHS:=evalm(pi&*Q);
```

$$RHS := [-\pi 0(t)\,\lambda + \pi 1(t)\,\mu,\ \pi 0(t)\,\lambda - \pi 1(t)\,\mu]$$

```
>    for j from 0 to 1 do
     DE.j:=diff(pi.j(t),t)=RHS[ j+1]  od;
```

$$DE0 := \tfrac{\partial}{\partial t}\,\pi 0(t) = -\pi 0(t)\,\lambda + \pi 1(t)\,\mu$$

$$DE1 := \tfrac{\partial}{\partial t}\,\pi 1(t) = \pi 0(t)\,\lambda - \pi 1(t)\,\mu$$

```
>    DESys:={ seq(DE.j,j=0..1)};
```

$$DESys := \{\tfrac{\partial}{\partial t}\,\pi 1(t) = \pi 0(t)\,\lambda - \pi 1(t)\,\mu,\ \tfrac{\partial}{\partial t}\,\pi 0(t) = -\pi 0(t)\,\lambda + \pi 1(t)\,\mu\}$$

```
>    for j from 0 to 1 do IC.j:=pi.j(0)=c[ j]  od;
```

$$IC0 := \pi 0(0) = c_0$$

$$IC1 := \pi 1(0) = c_1$$

```
>    ICSys:={ seq(IC.j,j=0..1)};
```

$$ICSys := \{\pi 0(0) = c_0,\ \pi 1(0) = c_1\}$$

```
>    dsolve(DESys union
     ICSys,{ seq(pi.j(t),j=0..1)} );
```

$$\left\{ \pi 0(t) = \frac{\mu\,(c_1 + c_0)}{\lambda + \mu} + \frac{(\lambda\,c_0 - c_1\,\mu)\,e^{(-t\lambda - t\mu)}}{\lambda + \mu}, \right.$$

$$\left. \pi 1(t) = -\frac{-\dfrac{\lambda\,\mu\,(c_1 + c_0)}{\lambda + \mu} + \dfrac{(\lambda\,c_0 - c_1\,\mu)\,e^{(-t\lambda - t\mu)}\,\mu}{\lambda + \mu}}{\mu} \right\}$$

```
>    simplify(%);
```

$$\{\pi 1(t) = -\frac{-\lambda\,c_1 - \lambda\,c_0 + \%1\,\lambda\,c_0 - \%1\,c_1\,\mu}{\lambda + \mu},$$

$$\pi 0(t) = \frac{c_1\,\mu + c_0\,\mu + \%1\,\lambda\,c_0 - \%1\,c_1\,\mu}{\lambda + \mu}\}$$

$$\%1 := e^{(-(\lambda + \mu)t)}$$

```
>   expand(%);
```

$$\{\pi 1(t) = \frac{\lambda c_1}{\lambda + \mu} + \frac{\lambda c_0}{\lambda + \mu} - \frac{\lambda c_0}{(\lambda + \mu) e^{(t\,\lambda)} e^{(t\,\mu)}} + \frac{c_1 \mu}{(\lambda + \mu) e^{(t\,\lambda)} e^{(t\,\mu)}},$$

$$\pi 0(t) = \frac{c_1 \mu}{\lambda + \mu} + \frac{c_0 \mu}{\lambda + \mu} + \frac{\lambda c_0}{(\lambda + \mu) e^{(t\,\lambda)} e^{(t\,\mu)}} - \frac{c_1 \mu}{(\lambda + \mu) e^{(t\,\lambda)} e^{(t\,\mu)}}\}$$

To summarize, the explicit solution $\Pi(t)$ for the linear system of DEs $\Pi'(t) = \Pi Q$ is obtained as

$$\Pi(t) = \begin{matrix} 0 \\ 1 \end{matrix} \left(\begin{matrix} \frac{\mu}{\lambda+\mu} + c_0 \frac{\lambda}{\lambda+\mu} e^{-(\lambda+\mu)t} - c_1 \frac{\mu}{\lambda+\mu} e^{-(\lambda+\mu)t} \\ \frac{\lambda}{\lambda+\mu} - c_0 \frac{\lambda}{\lambda+\mu} e^{-(\lambda+\mu)t} + c_1 \frac{\mu}{\lambda+\mu} e^{-(\lambda+\mu)t} \end{matrix} \right),$$

which, as $t \to \infty$, becomes

$$\lim_{t\to\infty} \Pi(t) = \begin{matrix} 0 \\ 1 \end{matrix} \left(\begin{matrix} \frac{\mu}{\lambda+\mu} \\ \frac{\lambda}{\lambda+\mu} \end{matrix} \right),$$

as expected.

Although the three-state car wash model of Example 91 is more complicated, Maple can still solve explicitly for the unconditional probabilities $\Pi(t)$. However, the solution is very long and does not reveal any insights. Thus, we leave the analysis of this example to the reader who can easily solve the model by replacing the definition of the Q matrix in the Maple worksheet `TwoStateUncond.mws` to `(3,3,[ -lambda,lambda,0,0,-mu1,mu1,mu2,0,-mu2] );` and by changing all the limits on the index j from `0..1` to `0..2`.

7.5.2   Limiting Probabilities

As we discussed in Section 7.4.2, the limiting probabilities $\pi_j = \lim_{n\to\infty} p_{ij}^{(n)}$ of a discrete-time Markov chain can be found—when they exist—by solving the linear system $\Pi = \Pi P$ with $\Pi e = 1$. One approach that can be used to find the limiting probabilities of a CTMC is the Kolmogorov differential equations. Suppose we consider the forward equations

$$p_{ij}'(t) = \sum_{k\neq j} p_{ik}(t)q_{kj} - v_j p_{ij}(t), \quad p_{ij}(0) = \delta_{ij}.$$

If we assume that the limiting probabilities $\pi_j = \lim_{t\to\infty} p_{ij}(t)$ exist,[12] then the derivative $p_{ij}'(t)$ must converge to 0. This gives

$$v_j \pi_j = \sum_{k\neq j} \pi_j q_{kj}, \quad j \in \mathcal{S} \qquad (7.20)$$

---

[12] They will exist provided that the transition matrix of the embedded Markov chain for the CTMC is irreducible and positive recurrent.

$$\sum_{j \in \mathcal{S}} \pi_j \;=\; 1.$$

The "balance" equations (7.20) have a very useful interpretation. Since v_j is the rate of leaving state j and π_j is the limiting probability of being in state j, the quantity $v_j \pi_j$ is the steady-state *output* rate from state j. Similarly, q_{kj} is the transition rate from state k into state j for $k \neq j$. Thus, the sum $\sum_{k \neq j} \pi_j q_{kj}$ is the steady-state *input* rate into state j. In summary, we have, for any state $j \in \mathcal{S}$,

Output rate from $j =$ **Input** rate to j.

Before using the balance equation approach to find the limiting probabilities for some of the models discussed previously, we note the following.

As before, if we define $q_{jj} = -v_j$, the system of equations in (7.20) can be re-written more compactly as $\mathbf{0} = \Pi \mathbf{Q}$ where $\mathbf{Q}$ is the infinitesimal generator matrix of the CTMC.

It should be interesting to note that for discrete-time Markov chains, the limiting probabilities are also found using a similar expression, i.e., by writing $\Pi = \Pi \mathbf{P}$ as $\mathbf{0} = \Pi \mathbf{Q}$ where $\mathbf{Q} = \mathbf{P} - \mathbf{I}$ is a matrix with rows adding up to 0 (as in the infinitesimal generator matrix $\mathbf{Q}$ of the CTMC).

Example 93 *A Markovian queue with no waiting space revisited.* We know from our previous discussion that the limiting probabilities for this model are

$$[\pi_0, \pi_1] = \left[\frac{\mu}{\lambda + \mu}, \frac{\lambda}{\lambda + \mu} \right].$$

Using the balance equation approach we have the following linear equations for each state $j = 0, 1$:

	Rate Out	$=$	Rate In
State 0:	$\lambda \pi_0$	$=$	$\mu \pi_1$
State 1:	$\mu \pi_1$	$=$	$\lambda \pi_0.$

Since the probabilities must sum to unity, we also include $\pi_0 + \pi_1 = 1$ as another linear equation.

As in discrete-time Markov chains, one of the balance equations is always redundant. Hence, we remove, say, the first equation before solving for the two unknowns π_0 and π_1.

```
>   restart: # TwoStateCTMCLimit.mws
>   eq[ 0] :=lambda*pi[ 0] =mu*pi[ 1] ;
```

$$eq_0 := \lambda \pi_0 = \mu \pi_1$$

```
>   eq[ 1] :=mu*pi[ 1] =lambda*pi[ 0] ;
```

$$eq_1 := \mu \pi_1 = \lambda \pi_0$$

```
>   Unity:=sum(pi[ j] ,j=0..1)=1;
```

$$Unity := \pi_0 + \pi_1 = 1$$

```
>    Eqns:={ seq(eq[ k] ,k=1)}  union { Unity} ;
```

$$Eqns := \{\mu \pi_1 = \lambda \pi_0, \pi_0 + \pi_1 = 1\}$$

```
>    Vars:={ seq(pi[ j] ,j=0..1)} ;
```

$$Vars := \{\pi_0, \pi_1\}$$

```
>    with(linalg):
```

Warning, new definition for norm

Warning, new definition for trace

```
>    solve(Eqns,Vars); assign(%);
```

$$\{\pi_0 = \frac{\mu}{\mu + \lambda}, \pi_1 = \frac{\lambda}{\mu + \lambda}\}$$

Naturally, the results agree with what we had found before using different methods.

Example 94 *The self-service car wash and vacuum facility revisited.* In this problem there are three states and the balance equations are obtained as follows:

	Rate Out	=	Rate In
State 0:	$\lambda \pi_0$	=	$\mu_2 \pi_2$
State 1:	$\mu_1 \pi_1$	=	$\lambda \pi_0$
State 2:	$\mu_2 \pi_2$	=	$\mu_1 \pi_1$

Noting, as usual, that $\sum_{j=0}^{2} \pi_j = 1$ and following the steps used in the Markovian queue example, Maple finds that limiting probabilities are given by

$$[\pi_0, \pi_1, \pi_2] = \left[\frac{\mu_2 \mu_1}{\mu_2 \mu_1 + \mu_2 \lambda + \mu_1 \lambda}, \frac{\mu_2 \lambda}{\mu_2 \mu_1 + \mu_2 \lambda + \mu_1 \lambda}, \frac{\lambda \mu_1}{\mu_2 \mu_1 + \mu_2 \lambda + \mu_1 \lambda}\right]:$$

```
>    restart: # CarWashCTMCLimit.mws
>    eq[ 0] :=lambda* pi[ 0] =mu[ 2] * pi[ 2] ;
```

$$eq_0 := \lambda \pi_0 = \mu_2 \pi_2$$

```
>    eq[ 1] :=mu[ 1] * pi[ 1] =lambda* pi[ 0] ;
```

$$eq_1 := \mu_1 \pi_1 = \lambda \pi_0$$

```
>    eq[ 2] :=mu[ 2] * pi[ 2] =mu[ 1] * pi[ 1] ;
```

$$eq_2 := \mu_2 \pi_2 = \mu_1 \pi_1$$

```
>    Unity:=sum(pi[ j] ,j=0..2)=1;
```

$$Unity := \pi_0 + \pi_1 + \pi_2 = 1$$

```
>    Eqns:={ seq(eq[ k] ,k=1..2)}  union { Unity} ;
```

$$Eqns := \{\mu_1 \pi_1 = \lambda \pi_0, \mu_2 \pi_2 = \mu_1 \pi_1, \pi_0 + \pi_1 + \pi_2 = 1\}$$

```
>   Vars:={ seq(pi[ j] ,j=0..2)} ;
```

$$Vars := \{\pi_0, \pi_2, \pi_1\}$$

```
>   with(linalg):
```

Warning, new definition for norm

Warning, new definition for trace

```
>   solve(Eqns,Vars); assign(%);
```

$$\{\pi_2 = \frac{\lambda \mu_1}{\%1}, \ \pi_0 = \frac{\mu_2 \mu_1}{\%1}, \ \pi_1 = \frac{\mu_2 \lambda}{\%1}\}$$
$$\%1 := \mu_2 \mu_1 + \mu_2 \lambda + \mu_1 \lambda$$

```
>   pi[ 0] ; pi[ 1] ; pi[ 2] ;
```

$$\frac{\mu_2 \mu_1}{\mu_2 \mu_1 + \mu_2 \lambda + \mu_1 \lambda}$$

$$\frac{\mu_2 \lambda}{\mu_2 \mu_1 + \mu_2 \lambda + \mu_1 \lambda}$$

$$\frac{\lambda \mu_1}{\mu_2 \mu_1 + \mu_2 \lambda + \mu_1 \lambda}$$

```
>   normal(pi[ 0] +pi[ 1] +pi[ 2] );
```

$$1$$

In this example, we entered each balance equation explicitly and then solved the resulting equations after including $\sum_{j=0}^{2} \pi_j = 1$ to illustrate the use of the "Rate Out = Rate In" approach. Some readers may feel that since the infinitesimal generator $\mathbf{Q}$ of this CTMC is already available, it may be simpler to generate the equation using the system $\mathbf{0} = \Pi\mathbf{Q}$. This is true, and the reader is invited to try this approach to solve the problem with Maple's help.

Example 95 *The self-service car wash and vacuum facility with extra capacity.* In order to increase the profitability of the car wash facility, the owner is considering increasing the capacity of the facility to two cars at a time—one at the washing location, the other at the vacuuming location.[13] As before, there is no possibility of forming a queue and the cars are first washed and then vacuumed.

With the additional capacity, a customer arriving when the wash space is free can enter the system. After the customer finishes washing his car, he must wait until the vacuuming space is free. This, of course, blocks the access to the wash space for other potential customers. Once the vacuuming space is freed up, the customer waiting in the wash space moves to the vacuum space and completes the

---

[13]Currently, there is only one space at the facility.

service. The arrival and service rates λ, μ_1 and μ_2 have the same interpretation as before.[14]

In this more complicated problem, the states must be defined as a vector since a knowledge of the number of customers in the facility, $X(t)$, is not sufficient to describe the system. Hence we define the system as $(X_1(t), X_2(t))$ where $X_1(t)$ is the state of the wash space (idle, busy or blocked) and $X_2(t)$ is the state of the vacuuming position (idle or busy). With this definition of the state vector, we obtain five possible values describing the system:

STATE	DESCRIPTION
(0,0)	Empty system
(1,0)	Busy car wash, idle vacuum
(0,1)	Idle car wash, busy vacuum
(1,1)	Busy car wash, busy vacuum
(b, 1)	Idle car wash blocked by busy vacuum

With this description of state space we can now generate the balance equations by equating the "out" rate to the "in" rate.

	Rate Out	=	Rate In
State (0,0):	$\lambda \pi_{00}$	=	$\mu_2 \pi_{01}$
State (0,1):	$(\mu_2 + \lambda)\pi_{01}$	=	$\mu_1 \pi_{10} + \mu_2 \pi_{b1}$
State (1,0):	$\mu_1 \pi_{10}$	=	$\lambda \pi_{00} + \mu_2 \pi_{11}$
State (1,1):	$(\mu_1 + \mu_2)\pi_{11}$	=	$\lambda \pi_{01}$
State (b, 1):	$\mu_2 \pi_{b1}$	=	$\mu_1 \pi_{11}$

Maple finds explicitly the solution of the system [minus the equation for State (0,0)] with the equation $\pi_{00} + \pi_{01} + \pi_{10} + \pi_{11} + \pi_{b1} = 1$:

```
>   restart: # CarWashExtraCTMCLimit.mws
>   eq[ 0] :=lambda*pi[ 0,0] =mu[ 2] *pi[ 0,1] ;
```

$$eq_0 := \lambda \pi_{0,0} = \mu_2 \pi_{0,1}$$

```
>   eq[ 1] := (mu[ 2] +lambda)*pi[ 0,1] =mu[ 1] *pi[ 1,0]
    +mu[ 2] *pi[ b,1] ;
```

$$eq_1 := (\mu_2 + \lambda) \pi_{0,1} = \mu_1 \pi_{1,0} + \mu_2 \pi_{b,1}$$

```
>   eq[ 2] :=mu[ 1] *pi[ 1,0] =lambda*pi[ 0,0]
    +mu[ 2] *pi[ 1,1] ;
```

$$eq_2 := \mu_1 \pi_{1,0} = \lambda \pi_{0,0} + \mu_2 \pi_{1,1}$$

```
>   eq[ 3] := (mu[ 1] +mu[ 2] )*pi[ 1,1] =lambda*pi[ 0,1] ;
```

---

[14]This could also be the description of a shoeshine shop or a barbershop with two chairs. In the shoeshine shop, the first chair is used to polish shoes and the second chair used to buff them. In the barbershop, the first chair is used to wash the customers' hair and the second chair is used to cut it; see, Goodman [74, Section 11.4] and Ross [158, Section 3.3].

$$eq_3 := (\mu_1 + \mu_2)\,\pi_{1,1} = \lambda\,\pi_{0,1}$$

```
>    eq[ 4] :=mu[ 2] *pi[ b, 1] =mu[ 1] *pi[ 1, 1] ;
```

$$eq_4 := \mu_2\,\pi_{b,1} = \mu_1\,\pi_{1,1}$$

```
>    Unity:=pi[ 0, 0] +pi[ 0, 1] +pi[ 1, 0] +pi[ 1, 1] +pi[ b, 1] =1;
```

$$Unity := \pi_{0,0} + \pi_{0,1} + \pi_{1,0} + \pi_{1,1} + \pi_{b,1} = 1$$

```
>    Eqns:={ seq(eq[ k] ,k=1..4)} union { Unity} ;
```

$$Eqns := \{(\mu_2 + \lambda)\,\pi_{0,1} = \mu_1\,\pi_{1,0} + \mu_2\,\pi_{b,1},\ \mu_1\,\pi_{1,0} = \lambda\,\pi_{0,0} + \mu_2\,\pi_{1,1},$$
$$(\mu_1 + \mu_2)\,\pi_{1,1} = \lambda\,\pi_{0,1},\ \mu_2\,\pi_{b,1} = \mu_1\,\pi_{1,1},$$
$$\pi_{0,0} + \pi_{0,1} + \pi_{1,0} + \pi_{1,1} + \pi_{b,1} = 1\}$$

```
>    Vars:={ pi[ 0, 0] ,pi[ 0, 1] ,pi[ 1, 0] ,pi[ 1, 1] ,pi[ b, 1] } ;
```

$$Vars := \{\pi_{0,0},\ \pi_{0,1},\ \pi_{b,1},\ \pi_{1,1},\ \pi_{1,0}\}$$

```
>    with(linalg) :
```

Warning, new definition for norm

Warning, new definition for trace

```
>    solve(Eqns,Vars); assign(%);
```

$$\{\pi_{0,0} = \frac{\mu_2{}^2\,\mu_1\,(\mu_1 + \mu_2)}{\%1},\ \pi_{1,0} = \frac{\mu_2{}^2\,\lambda\,(\mu_1 + \mu_2 + \lambda)}{\%1},$$

$$\pi_{0,1} = \frac{(\mu_1 + \mu_2)\,\lambda\,\mu_2\,\mu_1}{\%1},\ \pi_{b,1} = \frac{\mu_1{}^2\,\lambda^2}{\%1},\ \pi_{1,1} = \frac{\lambda^2\,\mu_2\,\mu_1}{\%1}\}$$

$$\%1 := \mu_2{}^2\,\mu_1{}^2 + \mu_2{}^3\,\mu_1 + 2\,\mu_2{}^2\,\lambda\,\mu_1 + \mu_2{}^3\,\lambda + \mu_2{}^2\,\lambda^2 + \mu_2\,\mu_1{}^2\,\lambda$$
$$+ \lambda^2\,\mu_2\,\mu_1 + \lambda^2\,\mu_1{}^2$$

```
>    normal(pi[ 0, 0] +pi[ 0, 1] +pi[ 1, 0] +pi[ 1, 1] +pi[ b, 1] );
```

$$1$$

Since Maple collects many of the terms in the solution, we print the π_{00} value in a slightly more detailed form.

```
>    pi[ 0, 0] ;
```

$$\mu_2{}^2\,\mu_1\,(\mu_1 + \mu_2)\,\Big/\,(\mu_2{}^2\,\mu_1{}^2 + \mu_2{}^3\,\mu_1 + 2\,\mu_2{}^2\,\lambda\,\mu_1 + \mu_2{}^3\,\lambda + \mu_2{}^2\,\lambda^2$$
$$+ \mu_2\,\mu_1{}^2\,\lambda + \lambda^2\,\mu_2\,\mu_1 + \lambda^2\,\mu_1{}^2)$$

Example 96 *Response areas for emergency units.* Consider two emergency units $i = 1, 2$ (e.g., ambulances) that serve two areas $j = A, B$, in a city. Normally, units 1 and 2 respond to emergencies in their respective areas A and B. However, if a call arrives from either area when only one unit is available the call is served by the available unit. When both units are busy, the arriving calls are lost (i.e., alternative measures are taken to deal with the emergency).

We assume that arrivals of calls from area j follow a Poisson process with rate λ_j. Service times in the areas are independent random variables where the time required to serve a call from area $j = A, B$ by unit $i = 1, 2$ is exponential with rate μ_{ij}.

As in Example 95, this model requires the state to be represented by a vector $(X_1(t), X_2(t))$ where

$$X_i(t) = \begin{cases} 0 & \text{if unit } i \text{ is idle} \\ 1 & \text{if unit } i \text{ is serving area 1} \\ 2 & \text{if unit } i \text{ is serving area 2.} \end{cases}$$

Thus the vector process $(X_1(t), X_2(t))$ is a CTMC with state space $S = \{(0, 0), (0, 1), (0, 2), (1, 0), (1, 1), (1, 2), (2, 0), (2, 1), (2, 2)\}$ that consists of nine states.

Let $\pi_{ij}, i, j = 0, 1, 2$ be the limiting probabilities of this CTMC. Using the balance equation approach we find the nine equations for this process and enter them in the Maple worksheet as follows:[15]

```
>    restart: # Ambulances.mws
>    eq[ 1] :=(lambda[ 1] +lambda[ 2] )*pi[ 0,0]
     =mu[ 2,1] *pi[ 0,1] +mu[ 2,2] *pi[ 0,2] +
     mu[ 1,1] *pi[ 1,0] +mu[ 1,2] *pi[ 2,0] ;
```

$$eq_1 := (\lambda_1 + \lambda_2)\,\pi_{0,0} = \mu_{2,1}\,\pi_{0,1} + \mu_{2,2}\,\pi_{0,2} + \mu_{1,1}\,\pi_{1,0} + \mu_{1,2}\,\pi_{2,0}$$

```
>    eq[ 2] :=(lambda[ 1] +lambda[ 2] +mu[ 2,1] )*pi[ 0,1]
     =mu[ 1,1] *pi[ 1,1] +mu[ 1,2] *pi[ 2,1] ;
```

$$eq_2 := (\lambda_1 + \lambda_2 + \mu_{2,1})\,\pi_{0,1} = \mu_{1,1}\,\pi_{1,1} + \mu_{1,2}\,\pi_{2,1}$$

```
>    eq[ 3] :=(lambda[ 1] +lambda[ 2] +mu[ 2,2] )*pi[ 0,2]
     =lambda[ 2] *pi[ 0,0] +mu[ 1,1] *pi[ 1,2]
     +mu[ 1,2] *pi[ 2,2] ;
```

$$eq_3 := (\lambda_1 + \lambda_2 + \mu_{2,2})\,\pi_{0,2} = \lambda_2\,\pi_{0,0} + \mu_{1,1}\,\pi_{1,2} + \mu_{1,2}\,\pi_{2,2}$$

```
>    eq[ 4] :=(lambda[ 1] +lambda[ 2] +mu[ 1,1] )*pi[ 1,0]
     =lambda[ 1] *pi[ 0,0] +mu[ 2,1] *pi[ 1,1]
     +mu[ 2,2] *pi[ 1,2] ;
```

$$eq_4 := (\lambda_1 + \lambda_2 + \mu_{1,1})\,\pi_{1,0} = \lambda_1\,\pi_{0,0} + \mu_{2,1}\,\pi_{1,1} + \mu_{2,2}\,\pi_{1,2}$$

```
>    eq[ 5] :=(mu[ 1,1] +mu[ 2,1] )*pi[ 1,1]
     =lambda[ 1] *pi[ 0,1] +lambda[ 1] *pi[ 1,0] ;
```

$$eq_5 := (\mu_{1,1} + \mu_{2,1})\,\pi_{1,1} = \lambda_1\,\pi_{0,1} + \lambda_1\,\pi_{1,0}$$

```
>    eq[ 6] :=(mu[ 1,1] +mu[ 2,2] )*pi[ 1,2]
     =lambda[ 1] *pi[ 0,2] +lambda[ 2] *pi[ 1,0] ;
```

$$eq_6 := (\mu_{1,1} + \mu_{2,2})\,\pi_{1,2} = \lambda_1\,\pi_{0,2} + \lambda_2\,\pi_{1,0}$$

```
>    eq[ 7] :=(lambda[ 1] +lambda[ 2] +mu[ 1,2] )*pi[ 2,0]
     =mu[ 2,1] *pi[ 2,1] +mu[ 2,2] *pi[ 2,2] ;
```

---

[15] For the infinitesimal generator matrix $\mathbf{Q}$ of this process, see Kao [101, p. 245]

$$eq_7 := (\lambda_1 + \lambda_2 + \mu_{1,2})\,\pi_{2,0} = \mu_{2,1}\,\pi_{2,1} + \mu_{2,2}\,\pi_{2,2}$$

```
>   eq[ 8] :=(mu[ 1,2] +mu[ 2,1] )*pi[ 2,1]
    =lambda[ 2] *pi[ 0,1] +lambda[ 1] *pi[ 2,0] ;
```

$$eq_8 := (\mu_{1,2} + \mu_{2,1})\,\pi_{2,1} = \lambda_2\,\pi_{0,1} + \lambda_1\,\pi_{2,0}$$

```
>   eq[ 9] :=(mu[ 1,2] +mu[ 2,2] )*pi[ 2,2]
    =lambda[ 2] *pi[ 0,2] +lambda[ 2] *pi[ 2,0] ;
```

$$eq_9 := (\mu_{1,2} + \mu_{2,2})\,\pi_{2,2} = \lambda_2\,\pi_{0,2} + \lambda_2\,\pi_{2,0}$$

Next, we introduce the equation $\sum_{i=0}^{2}\sum_{j=0}^{2}\pi_{ij} = 1$, define the nine equations of the system and solve the resulting system using `solve()`.

```
>   Unity:=sum(sum(pi[ i,j] ,j=0..2),i=0..2)=1;
```

$Unity := \pi_{0,0} + \pi_{0,1} + \pi_{0,2} + \pi_{1,0} + \pi_{1,1} + \pi_{1,2} + \pi_{2,0} + \pi_{2,1} + \pi_{2,2} = 1$

```
>   Eqns:={ seq(eq[ k] ,k=1..8)}  union { Unity} ;
```

$Eqns := \{(\lambda_1 + \lambda_2)\,\pi_{0,0} = \mu_{2,1}\,\pi_{0,1} + \mu_{2,2}\,\pi_{0,2} + \mu_{1,1}\,\pi_{1,0} + \mu_{1,2}\,\pi_{2,0},$
$(\lambda_1 + \lambda_2 + \mu_{2,1})\,\pi_{0,1} = \mu_{1,1}\,\pi_{1,1} + \mu_{1,2}\,\pi_{2,1},$
$(\lambda_1 + \lambda_2 + \mu_{1,2})\,\pi_{2,0} = \mu_{2,1}\,\pi_{2,1} + \mu_{2,2}\,\pi_{2,2},$
$(\mu_{1,2} + \mu_{2,1})\,\pi_{2,1} = \lambda_2\,\pi_{0,1} + \lambda_1\,\pi_{2,0},$
$(\lambda_1 + \lambda_2 + \mu_{2,2})\,\pi_{0,2} = \lambda_2\,\pi_{0,0} + \mu_{1,1}\,\pi_{1,2} + \mu_{1,2}\,\pi_{2,2},$
$(\lambda_1 + \lambda_2 + \mu_{1,1})\,\pi_{1,0} = \lambda_1\,\pi_{0,0} + \mu_{2,1}\,\pi_{1,1} + \mu_{2,2}\,\pi_{1,2},$
$(\mu_{1,1} + \mu_{2,1})\,\pi_{1,1} = \lambda_1\,\pi_{0,1} + \lambda_1\,\pi_{1,0},$
$(\mu_{1,1} + \mu_{2,2})\,\pi_{1,2} = \lambda_1\,\pi_{0,2} + \lambda_2\,\pi_{1,0},$
$\pi_{0,0} + \pi_{0,1} + \pi_{0,2} + \pi_{1,0} + \pi_{1,1} + \pi_{1,2} + \pi_{2,0} + \pi_{2,1} + \pi_{2,2} = 1\}$

```
>   Vars:={ seq(seq(pi[ i,j] ,j=0..2),i=0..2)} ;
```

$Vars := \{\pi_{0,0},\ \pi_{0,1},\ \pi_{0,2},\ \pi_{1,0},\ \pi_{2,0},\ \pi_{2,1},\ \pi_{1,1},\ \pi_{1,2},\ \pi_{2,2}\}$

```
>   with(linalg) :
```

`Warning, new definition for norm`

`Warning, new definition for trace`

```
>   solve(Eqns,Vars): assign(%):
```

This model is originally due to Carter, Chaiken and Ignall [40], but it is also discussed in Tijms [186]. After presenting this model, Tijms [186, p. 116] states (somewhat pessimistically): "These linear equations must be solved *numerically*." But Maple succeeds in solving the linear system *symbolically* (i.e., exactly)! However, the solution for each unknown π_{00}, π_{01} and so on takes up several pages of output. To appreciate the nature of the solution, the reader is invited to replace the colon (:) in front of the `pi[ 0,0]` by a semi-colon (;).

```
>   pi[ 0,0] :
```

A check reveals that probabilities add up to 1.

```
>   simplify(sum(sum(pi[ i,j] ,j=0..2),i=0..2));
```

$$1$$

We again use share utility `sprint()` in Maple's `\share` folder.

```
>   with(share); with(sprint);
```

```
See ?share and ?share,contents for information about the
share library
```

$$[]$$

```
Share Library:   sprint
```

```
Author: Monagan, Michael.
```

```
Description:  Utility routine 'short print' for displaying
large expressions, allowing the user to look at the
structure (top levels) of a large expression.
```

$$[\text{``sprint''}]$$

```
>   sprint(pi[ 0,0] );
```

$$\frac{\mu_{2,2}\,\mu_{1,2}\,\mu_{2,1}\,\mu_{1,1} << +79 >>}{<< +275 >>}$$

Thus, we see that π_{00} is a very large symbolic expression—a fraction—that has a sum of 79 terms in its numerator and a sum of 275 terms in its denominator!

We can, of course, perform numerical analysis once we find the symbolic solution for the unknowns. For example, setting the values of λ_1, λ_2, μ_{11}, μ_{12}, μ_{21}, μ_{22} as

```
>   lambda[ 1] :=0.1;  lambda[ 2] :=0.125;
```

$$\lambda_1 := .1$$

$$\lambda_2 := .125$$

```
>   mu[ 1,1] :=0.5;  mu[ 1,2] :=0.25;
    mu[ 2,1] :=0.2;  mu[ 2,2] :=0.4;
```

$$\mu_{1,1} := .5$$

$$\mu_{1,2} := .25$$

$$\mu_{2,1} := .2$$

$$\mu_{2,2} := .4$$

the numerical value of the probability of the idle system is computed as $\pi_{00} = 0.569$.

```
>   pi[ 0,0] ;
```

$$.5692678429$$

Finally, note that the sum $\sum_{i=1}^{2} \sum_{j=1}^{2} \pi_{ij}$ is the probability that both units are busy and thus the incoming call is lost to the system. This probability can now be computed easily as 10.48%.

```
>   Sum(Sum(pi[ i,j] ,j=1..2),i=1..2); value(%);
```

$$\sum_{i=1}^{2} \left(\sum_{j=1}^{2} \pi_{i,j} \right)$$

$$.1048372876$$

Using the model presented in this example, one can analyze decision problems such as the optimal number of units to assign in each location in order to minimize the probability of lost calls or the optimal number of designated areas to serve within the city boundaries.

7.6 Summary

This chapter started with a detailed description of the exponential/Poisson duality. Building on this duality, we presented renewal theory where the renewal function was computed explicitly using Maple's `invlaplace()` function that inverts a large class of Laplace transforms. Next, discrete-time Markov chains were discussed and their limiting and transient behavior was analyzed using Maple's `solver` and generating function inverter `rgf_expand()`. We completed the chapter with an analysis of continuous-time Markov chains. Limiting probabilities were again computed analytically using the `solve()` function. To compute the transient probabilities, we used Maple's `matrixDE()`, `dsolve()` and `exponential()` functions. We can now start our analysis of specific applications in inventory management and queueing processes.

7.7 Exercises

1. Provide a reasonable definition of the memoryless property for a nonnegative discrete random variable N. Based on your definition show that $P\{N = n\} = (1 - p)p^{n-1}$, $n = 1, 2, \ldots$ is the unique discrete memoryless distribution.

2. Let X and Y be independent exponential random variables with means $E(X) = 1/\lambda$ and $E(Y) = 1/\mu$.

   (a) Find the distribution of $Z = \min(X, Y)$.

   (b) Now, compute $\Pr(X < Y)$.

3. Show that the probability generating function $G_{N(t)}(z) = E\left[z^{N(t)}\right]$ of a nonhomogeneous Poisson process $\{N(t), t \geq 0\}$ with rate $\lambda(t)$ is $G_{N(t)}(z) = e^{(z-1)M(t)}$ where $M(t) = \int_0^t \lambda(u)\, du$.

4. Suppose we have a counting process $\{N(t); t \geq 0\}$ where $N(t)$ denotes the number of arrivals in the interval $(0, t]$. Consider the following infinite system of differential-difference equations for $P_n(t) = \Pr[N(t) = n]$:

$$
\begin{aligned}
P_0'(t) &= -\lambda P_0(t), \quad P_0(0) = 1 \\
P_n'(t) &= -\lambda P_n(t) + \lambda P_{n-1}(t), \quad n = 1, 2, \ldots
\end{aligned}
$$

Solve this system using generating functions.

HINT: For fixed t, define the probability generating function $\tilde{P}(z, t) = \sum_{n=0}^{\infty} z^n P_n(t)$, $|z| < 1$. Next find a solution for $\tilde{P}(z, t)$ and extract its coefficients.

5. Let $N(t)$ be a Poisson process with rate λ. Assume that T is a random variable with density $g(t) = \mu e^{-\mu t}$ and define N as the number of renewals in a time interval of random length $(0, T]$. Find the probability $\Pr(N = n)$, for $n = 0, 1, 2, \ldots$

6. In a renewal process let $A(t) = t - S_{N(t)}$ and $B(t) = S_{N(t)+1} - t$.

   (a) Find $\Pr\{A(t) \leq a\}$ and $\Pr\{B(t) \leq b\}$ for finite t.

   (b) Assume that interarrival times are exponential with $\mu = E(X)$. Find the above quantities. Then let $t \to \infty$. Do the results agree with your intuition?

7. Consider a renewal process where the interarrival times are i.i.d. with density $f(t) = a_1 \lambda_1 e^{-\lambda_1 t} + a_2 \lambda_2 e^{-\lambda_2 t}$ for $t \geq 0$ with $a_1 + a_2 = 1$. Assume $a_1 = 0.3$, $\lambda_1 = 1$ and $\lambda_2 = 2$ and find the renewal function $M(t)$.

8. Consider the following transition probability matrix for a Markov chain.

$$
\mathbf{P} = \begin{bmatrix} 0 & 3/4 & 1/4 \\ 1/4 & 0 & 3/4 \\ 1/4 & 1/4 & 1/2 \end{bmatrix}.
$$

Using Maple's genfunc() function, find the transient probabilities for this chain, i.e., find $\mathbf{P}^{(n)}$ for finite n. Next, find the stationary distribution $\{\pi_j, j = 1, 2, 3\}$ and compare it to above when $n \to \infty$.

9. Let $\{X_n, n = 0, 1, \ldots\}$ be a Markov chain with state space $S = \{1, 2, 3\}$. The transition probability matrix for this chain is given as

$$
\mathbf{P} = \begin{bmatrix} p & q & 0 \\ \frac{1}{2}p & \frac{1}{2} & \frac{1}{2}q \\ 0 & p & q \end{bmatrix}
$$

with $p + q = 1$ and $0 < p < 1$. Find the transient probabilities for this chain. Med, 103

10. Let $\{X_n, n = 1, 2, \ldots\}$ be an irreducible and aperiodic MC with state space $S = \{1, 2, \ldots, m\}$ and $P_{ij} = P_{ji}$, $i, j \in S$. Find $\lim_{n \to \infty} \Pr(X_n = k)$. (This is an example of a doubly-stochastic matrix.)

11. Consider a CTMC with three states 0, 1 and 2. This CTMC has the infinitesimal generator

$$\mathbf{Q} = \begin{matrix} 0 \\ 1 \\ 2 \end{matrix} \begin{pmatrix} * & a_1\lambda & (1 - a_1)\lambda \\ 2a_1\mu & * & 0 \\ 2\mu(1 - a_1) & 0 & * \end{pmatrix}$$

where the values assumed by the diagonal entries denoted by $*$ are such that the row sums are zero. Find the long-run probability that the system is in state 0.

12. The infinitesimal generator $\mathbf{Q}$ of a three-state CTMC assumes the form

$$\mathbf{Q} = \begin{bmatrix} -2\lambda & 2\lambda & 0 \\ 0 & -2\lambda & 2\lambda \\ \mu & 0 & -\mu \end{bmatrix}.$$

Let $\lambda = 1$ and $\mu = 8$ and determine the transient probabilities of this CTMC. Next, let $t \to \infty$ and find the limiting probabilities.

8
Inventory Models

8.1 Introduction

According to recent estimates the total value of inventories in the United States is in excess of $1 trillion [135, p. 184]. Almost 80% of this investment in inventories is in manufacturing, wholesale and retail sectors. Many of the mathematical models that we will discuss in this chapter have proved very useful in managing inventories and their applications have resulted in substantial savings for the companies employing them.

The term "inventory" can be defined as a stock of commodity that will be used to satisfy some future demand for commodity. Due to the random nature of customer arrivals and the amount required by the customers, future demand is usually not known in advance. Inventory processes are influenced by the customers generating the demand, the management making the inventory order decisions and the supplier(s) providing the required amount of the commodity to the management.

In order to satisfy future demands, management has to answer two basic questions relating to timing and magnitude: (1) *When* should an inventory replenishment order be placed? (2) *How much* should be ordered? In the last 50 years, thousands of inventory models have been developed to answer these questions under different assumptions on the demand process, cost structure and the characteristics of the inventory system. For recent reviews, see Barancsi et al. [15], Porteus [151] and Lee and Nahmias [120]. The objective in almost all of these models is to minimize the total costs of operating the system. This is achieved by finding the optimal values of timing and magnitude decision variables that minimize the objective function. (When the model involves random components such as random demand or random yield, the objective is to minimize the expected cost.)

There are several reasons why an organization may want to keep inventories. First, it may be either impossible or economically unsound to have goods arrive in a system exactly when the demands occur. Although in manufacturing organizations the just-in-time system has made inventories almost obsolete, the retail sector still needs goods on the shelves to attract customers. Even in a family unit, we see a type of inventory management when the weekly food shopping excursions result in purchases of large quantities of bread, milk, etc., that may last the family more than a few days.[1] This is the transaction motive for ordering large lots and keeping inventories in order to reduce the fixed cost of ordering.

A second reason for keeping inventories is the precautionary motive, which is based on the fact that it is usually very difficult to predict the random demand, and hence inventory is a hedge against uncertainty [120]. The randomness of demand and lead time are the most common factors of uncertainty, but as we saw in the 1970s during the OPEC oil embargoes, in some cases even the supply can be random, thus adding to the uncertainty in the inventory system.

Another reason for keeping inventories is the speculative motive. When the price of a commodity is expected to increase, it may be more profitable to purchase large quantities of the commodity and keep it for future use when the price will be higher. There are other motives — such as long transportation times and logistics reasons — for keeping inventories. We refer the reader to an early paper by Arrow [9] and the recent book by Nahmias [135, Section 4.2] for a discussion of these and other motives for keeping inventories.

8.2   Classification of Inventory Models

In the last 50 years thousands of inventory models have been developed by operations researchers. These models have ranged from simple extensions of the classical economic order quantity (EOQ) model [81] to complex multilevel control systems [168]. In recent years some researchers have started investigating the possibility of classifying the inventory models. Review papers by Aggarwal [6], Nahmias [134] and Silver [174] and a book by Barancsi et al. [15] provide classifications of inventory models that emphasize the wide variety of problem situations. There were also some attempts to incorporate artificial intelligence technology to automate the selection of the correct inventory model in a given situation. The first paper dealing with such a problem appears to be by Parlar [142], who developed a knowledge-based expert system (EXPIM) that could identify and recommend up to 30 production/inventory models.

---

[1] While in North America a weekly grocery list may include several loaves of bread that may last the family many meals, in many other countries bread is still purchased daily from the local bakery.

Among the classifications published in recent years, the one by Silver [174] is probably the most detailed. In his scheme, Silver lists a number of factors that can each take on two or more values. For example, inventory models are classified according to the number of items involved (single vs. multiple), nature of demand (deterministic vs. probabilistic), length of decision horizon (single period vs. multiperiod), time variability of the demand process (stationary vs. time varying), nature of the supply process (deterministic or random lead time), procurement cost structure (discount possibility), backorders vs. lost sales when there is a shortage, shelf-life considerations (obsolescence or perishability) and single vs. multiple stocking points.

One can define a set of "standard problems" that have mathematically developed optimal solutions. These problems have found wide applicability or have potential applications with minor modifications. Among these problems, EOQ-type models are classified as single item with deterministic demand and stationary conditions. In multi-items problems, deterministic and stationary demand under budget or space constraints are normally solved using the Lagrange multiplier technique. The single-item problem with deterministic but time-varying parameters (also known as the dynamic lot size problem) was solved by Wagner and Whitin [191] using dynamic programming.

In continuous-review models it is assumed that the inventory level is known at all times. In many supermarkets that have scanning devices at the checkout counters, it is possible to keep track of the inventory level of any item. Hence, in these systems transaction information is easily available that makes the application of continuous-review models possible. In periodic-review models the inventory level is measured only in discrete time points when inventory is physically counted.

The continuous-review (Q, r) policy where r is the fixed order point and Q is the fixed order quantity is used to control the single-item problem with stationary, probabilistic demand with known distribution. Variants of this policy such as the (R, S) periodic replenishment system where R is the review interval and S is the order-up-to level, and the (R, r, S) policy are also used, for which efficient solution techniques have been developed. Decision rules are also available for multi-item probabilistic stationary problem under a budget constraint. The single-item, single-product problem (i.e., the classical newsvendor model) is the simplest of the probabilistic inventory models and its variants with multi-items with budget constraints are solved using the Lagrangian technique. The coordinated control of items under deterministic or probabilistic stationary demand with or without discounts and multiechelon stationary situations are more complicated models for which decision rules are available.

For a description of the solution of these and other inventory models, we refer the reader to the books by Banks and Fabrycky [13], Hax and Candea [84], Silver and Peterson [175], and Tersine [185].

8.3   Costs Associated with Inventory Models

A majority of inventory models are built with the assumption that the objective is to minimize the total (expected) cost of procurement, holding (carrying) cost and penalty cost for shortages.

8.3.1   Procurement Cost

This cost corresponds to the dollar amount spent to procure (or produce) the commodity. If we denote by $c(y)$ the cost of procuring y units, then

$$c(y) = \begin{cases} K + cy & \text{if } y > 0 \\ 0 & \text{if } y = 0 \end{cases} \tag{8.1}$$

is a procurement cost function that has a fixed setup cost component K that is incurred when an order is placed. When demand is probabilistic and a periodic-review system is in place, such a cost structure results in the optimality of the well-known (s, S) policy, as proved by Scarf [166].[2]

8.3.2   Holding (or Carrying) Cost

This cost is the sum of all costs that are proportional to the level of inventory on hand at any time. These costs include mainly the opportunity cost of alternative investments (i.e., the loss in revenue that results in having cash tied up in inventory), taxes and insurance and cost of providing the physical space to store the items. If $x(t)$ is the on-hand inventory at any time t and h is the holding cost per unit per time, then the total holding cost incurred from time 0 to time T is $h \int_0^T x(t)\,dt$.

8.3.3   Penalty Cost

An inventory system is operated in order to satisfy the customers' demands, and thus if the demands are not met there must be some economic loss. In inventory literature the penalty cost is assumed to be a nondecreasing function of time t, $\pi(t)$, to represent the cost of each unit short for which the backorder remain on the books. Usually (as in Hadley and Whitin [81, Section 1-11]) the function $\pi(t)$ is specialized to a linear form as $\pi(t) = \pi_1 t + \pi_2$, where π_1 is the time

---

[2]In inventory literature the symbol s appears under different disguises. In Hadley and Whitin's [81, Chapter 4] continuous-review (Q, r) model where Q is the lotsize and r is the reorder point, s is used to denote the *safety stock*, which is defined as the expected value of the inventory level when a procurement arrives. In Scarf's [166] periodic-review (s, S) model, s denotes the reorder point. In this chapter, in order to present a development parallel to those in Hadley and Whitin and Scarf, we use s as the safety stock when we discuss the continuous-review models. But when a periodic-review model is under consideration, s corresponds to the reorder point.

dependent per unit backorder cost with dimension \$/unit/time and π_2 is the per unit backorder cost with dimension \$/unit. [Here π_1 plays a role similar to the holding cost h (\$/unit/time) that is used in the computation of the holding cost component.]

Let $b(t)$ be the number of units on backorder at time t during a shortage interval of $[0, T]$. Then $\int_0^T \pi_1 b(t)\, dt + \pi_2 b(T)$ is the total penalty cost incurred during $[0, T]$. In practice both π_1 and π_2 are difficult to estimate since they usually represent intangible factors such as the loss of goodwill resulting from a delay in meeting demands. In a recent paper, Çetinkaya and Parlar [42] consider the two backorder costs indirectly by introducing two safety-stock type constraints. Thus in their form the problem becomes one of minimizing the ordering and holding costs subject to two appropriate constraints.

8.4 Deterministic Inventory Models

We will begin our presentation by discussing some inventory models where demand and every other problem parameter are deterministic and stationary. The first example of such a model dates back to the early 20th century when Ford Harris [83] and R. H. Wilson [195] discovered the EOQ formula. The assumptions of deterministic and stationary demands may seem very restrictive but as we will see, the solution to the EOQ model serves as a good starting point even for more complex and realistic problems.

Before we develop the optimization models, it is worthwhile to briefly describe an important theorem from renewal theory. The renewal reward theorem (RRT)—as discussed in Section 7.3.5 of Chapter 7, Stochastic Processes—is a powerful tool used in the optimization of stochastic systems. Once a regenerative cycle of a controlled stochastic process is identified, one forms an average cost objective function simply as the ratio of the *expected* cycle cost to the *expected* cycle time. The resulting objective function is then optimized with respect to the decision variables of the model. For example, in an inventory problem, these decision variables may be the quantity ordered Q and the reorder point r (the level of inventory that triggers an order). The usefulness of the RRT is in reducing an infinite horizon optimization problem to a static nonlinear program that involves the computation of only the *first moments* of the cycle-related random variables.

It is fortunate that even if we have a deterministic problem, the renewal reward theorem may still be useful in forming the objective function provided that we identify the regenerative cycles of the process in a correct manner. Once the cycles are identified, we can then develop the average cost per unit time as the total cycle cost divided by the total cycle length.

8.4.1   The Basic EOQ Model

In this model we make the following assumptions:

1. The demand rate is known and constant (stationary) at λ units per unit time.[3]

2. Shortages are not permitted.

3. Order leadtime is zero, i.e., the delivery is instantaneous.

4. There are three types of costs.

   (a) Purchase cost is c dollars per unit and setup (order) cost is K dollars per order.

   (b) Holding cost is h dollars per unit per unit time. This cost is usually defined as $h = ic$ where i is the inventory carrying charge, the cost in dollars of carrying \$1 of inventory per unit time.

   (c) Since shortages are not allowed, the shortage cost is infinitely high.

It should be clear that since leadtime is zero and demand is known with certainty, it is optimal to order only when the inventory level $I(t)$ at time t reaches zero.[4] As each order must be of the same size, we assume that the order quantity is always Q units.

Referring to Figure 8.1, we see that a regenerative cycle starts when the inventory level is increased to Q units. Since demand is λ units per time, it takes $T = Q/\lambda$ time units for the cycle to end. Thus the deterministic cycle length is simply

$$E(\text{Cycle length}) = \text{Cycle length} = Q/\lambda.$$

The cycle cost computation is slightly more complicated as we need to account for the total cost of ordering and holding in a given cycle. At any time $t \in [0, T]$, the inventory level is $I(t) = Q - \lambda t$. Thus, during each cycle the total holding cost is

$$h \int_0^T (Q - \lambda t)\, dt = h(QT - \tfrac{1}{2}\lambda T^2) = \tfrac{1}{2}hTQ = \tfrac{1}{2}hQ^2/\lambda.$$

The total cycle cost is found by adding the order cost K to the holding cost:[5]

$$E(\text{Cycle cost}) = \text{Cycle cost} = K + \tfrac{1}{2}hQ^2/\lambda.$$

Now we let Maple take over and perform the necessary computations to find the optimal order quantity.

---

[3] In this chapter the "unit time" will normally correspond to one year.

[4] For a rigorous proof of this result, see Arrow, Harris and Marschak [11].

[5] Since the demands must be met ultimately (and there is no quantity discount for larger orders), we do not include the procurement cost per cycle cQ in the cycle cost. If this cost were included, then dividing it by the cycle length Q/λ would give $c\lambda$, a constant that would not affect the optimal decision.

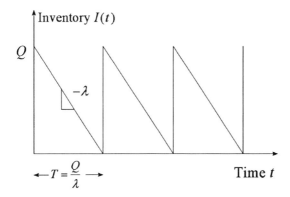

FIGURE 8.1. Inventory level fluctuations in the basic EOQ model.

```
>    restart: # EOQ.mws
```

We use the cycle cost and the cycle length denoted by EC and EL, respectively, to write the expression for the average cost $G(Q) = EC/EL = \lambda K/Q + \frac{1}{2}hQ$.

```
>    EC:=K+(1/2)*h*Q^2/lambda;
```

$$EC := K + \frac{1}{2}\frac{h\,Q^2}{\lambda}$$

```
>    EL:=Q/lambda;
```

$$EL := \frac{Q}{\lambda}$$

```
>    G:=expand(EC/EL);
```

$$G := \frac{\lambda\,K}{Q} + \frac{1}{2}\,Q\,h$$

Differentiating G once and then twice to find G_Q and G_{QQ}, we see that the unique positive solution of the necessary condition will be the global optimal solution since the second derivative is always positive for $Q > 0$.

```
>    GQ:=diff(G,Q);
```

$$GQ := -\frac{\lambda\,K}{Q^2} + \frac{1}{2}\,h$$

```
>    GQQ:=normal(diff(G,Q$2));
```

$$GQQ := 2\frac{\lambda\,K}{Q^3}$$

Solving the necessary condition and recalling that every parameter is positive, we pick $+\sqrt{2K\lambda/h}$ as the optimal solution.

```
>  sol:=solve(GQ,Q);
```

$$sol := \frac{\sqrt{2}\sqrt{h\lambda K}}{h}, \ -\frac{\sqrt{2}\sqrt{h\lambda K}}{h}$$

```
>  QOpt:=simplify(sol[ 1] ,power,symbolic);
```

$$QOpt := \frac{\sqrt{2}\sqrt{\lambda}\sqrt{K}}{\sqrt{h}}$$

The minimum value of the average cost denoted by $GOpt$ is also obtained as a square root expression.

```
>  GOpt:=subs(Q=QOpt,G);
```

$$GOpt := \sqrt{h}\sqrt{2}\sqrt{\lambda}\sqrt{K}$$

To show that the optimal cost is *not* very sensitive to small deviations from the true optimal value, we substitute $Q = \delta \times QOpt$ in the objective function G where δ is a positive constant. Simplifying gives

```
>  GNonOpt:=normal(subs(Q=delta*QOpt,G)/GOpt);
```

$$GNonOpt := \frac{1}{2}\frac{1+\delta^2}{\delta}$$

Here, $(GNonOpt - 1) \times 100$ is the percentage deviation from the minimum cost if the order quantity is chosen as $\delta \times QOpt$. Note that even if we deviate from the optimal solution by 100%, i.e., when $\delta = 2$, we find

```
>  evalf(subs(delta=2,GNonOpt));
```

$$1.250000000$$

implying that the cost increases by only 25%. When δ is very, very large, an additional 100% increase in δ (that is, $\delta_{New} = \delta_{Old} + 1$) increases the cost by only 50%. This is shown by differentiating $GNonOpt$ with respect to δ and letting $\delta \to \infty$, which gives $1/2$.

```
>  limit(diff(GNonOpt,delta),delta=infinity);
```

$$\frac{1}{2}$$

8.4.2   The EOQ Model with Planned Backorders

We now consider an extension of the basic EOQ model and assume that it may be permissible for the inventory system to be out of stock when a demand occurs. Many businesses such as capital-goods firms that deal with expensive products and some service industries that cannot store their services operate with substantial backlogs. Thus planned backorders may be an acceptable policy to follow in some cases.

In this model, we retain every assumption we made while discussing the basic EOQ model except the requirement that there must be no shortages. Here we assume that the demands occurring when the system is out of stock are backordered until a new shipment arrives. At that time all backorders are satisfied before the shipment can be used to meet other demands.

We let $s \geq 0$ be the number of backorders on the books when an order is placed and a shipment of Q units arrives immediately.[6] After satisfying the backorders, the cycle will start with $Q - s$ units, which will be depleted to zero after $T_1 = (Q - s)/\lambda$ time units. If T is the cycle length, then $T_2 = T - T_1$ is the length of a subcycle during which backorders will be incurred. The inventory process for this model is depicted in Figure 8.2.

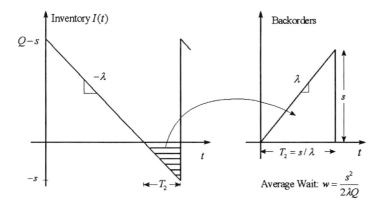

FIGURE 8.2. Inventory and backorder levels.

The inventory holding cost per cycle is

$$h \int_0^{T_1} (Q - s - \lambda t)\, dt = \frac{h}{2\lambda}(Q - s)^2$$

and the backorder cost per cycle is

$$\pi_1 \int_0^{T_2} \lambda t\, dt + \pi_2 s = \frac{\pi_1 s^2}{2\lambda} + \pi_2 s$$

---

[6]This is essentially a continuous-review model with zero leadtime and deterministic demand. Hence the reorder point s in this model is also the safety stock.

since $T_2 = s/\lambda$. Noting that the procurement cost can again be ignored (since the demands will ultimately be satisfied), the average cost is found as

$$G = \frac{\lambda K}{Q} + \frac{1}{2Q}h(Q-s)^2 + \frac{1}{Q}\left(\tfrac{1}{2}\pi_1 s^2 + \pi_2 \lambda s\right)$$

These expressions are entered in Maple and the optimal solution is computed as follows:

```
>   restart: # EOQBackOrder.mws
>   T1:=(Q-s)/lambda;
```

$$T1 := \frac{Q-s}{\lambda}$$

```
>   HoldingC:=normal(h* int(Q-s-lambda*t,t=0..T1));
```

$$HoldingC := \frac{1}{2}\frac{h(-Q+s)^2}{\lambda}$$

```
>   T:=Q/lambda;
```

$$T := \frac{Q}{\lambda}$$

```
>   T2:=normal(T-T1);
```

$$T2 := \frac{s}{\lambda}$$

```
>   BackorderC:=pi[1] * int(lambda*t,t=0..T2)+pi[2]*s;
```

$$BackorderC := \frac{1}{2}\frac{\pi_1 s^2}{\lambda} + \pi_2 s$$

```
>   EC:=K+HoldingC+BackorderC;
```

$$EC := K + \frac{1}{2}\frac{h(-Q+s)^2}{\lambda} + \frac{1}{2}\frac{\pi_1 s^2}{\lambda} + \pi_2 s$$

```
>   EL:=T;
```

$$EL := \frac{Q}{\lambda}$$

```
>   G:=expand(EC/EL);
```

$$G := \frac{\lambda K}{Q} + \frac{1}{2}Qh - hs + \frac{1}{2}\frac{hs^2}{Q} + \frac{1}{2}\frac{\pi_1 s^2}{Q} + \frac{\lambda \pi_2 s}{Q}$$

For ease of exposition, we consider the simpler case of this problem with $\pi_2 = 0$.

```
>   G:=subs(pi[2]=0,G);
```

$$G := \frac{\lambda K}{Q} + \frac{1}{2}Qh - hs + \frac{1}{2}\frac{hs^2}{Q} + \frac{1}{2}\frac{\pi_1 s^2}{Q}$$

The necessary conditions are found by differentiating the objective function.

```
>   GQ:=diff(G,Q); Gs:=diff(G,s);
```

$$GQ := -\frac{\lambda K}{Q^2} + \frac{1}{2}h - \frac{1}{2}\frac{h s^2}{Q^2} - \frac{1}{2}\frac{\pi_1 s^2}{Q^2}$$

$$Gs := -h + \frac{h s}{Q} + \frac{\pi_1 s}{Q}$$

Here we check to see if the objective function is strictly convex in the decision variables (Q, s). First we load the linalg() package and compute the Hessian matrix of G.

```
>   with(linalg):
```

Warning, new definition for norm

Warning, new definition for trace

```
>   H:=hessian(G,[ Q,s] );
```

$$H := \begin{bmatrix} 2\dfrac{\lambda K}{Q^3} + \dfrac{h s^2}{Q^3} + \dfrac{\pi_1 s^2}{Q^3} & -\dfrac{h s}{Q^2} - \dfrac{\pi_1 s}{Q^2} \\ -\dfrac{h s}{Q^2} - \dfrac{\pi_1 s}{Q^2} & \dfrac{h}{Q} + \dfrac{\pi_1}{Q} \end{bmatrix}$$

Maple gives two conditions in order for G to be positive definite.

```
>   definite(H,positive_def);
```

$$-2\frac{\lambda K}{Q^3} - \frac{h s^2}{Q^3} - \frac{\pi_1 s^2}{Q^3} < 0 \text{ and}$$

$$-(2\frac{\lambda K}{Q^3} + \frac{h s^2}{Q^3} + \frac{\pi_1 s^2}{Q^3})(\frac{h}{Q} + \frac{\pi_1}{Q}) + (-\frac{h s}{Q^2} - \frac{\pi_1 s}{Q^2})^2 < 0$$

A further simplification shows that the second term is also negative, indicating that G is positive definite.[7]

```
>   Condition2:=-(2*lambda*K/(Q^3)+h*s^2/(Q^3)
    +pi[ 1] *s^2/(Q^3))*(h/Q+pi[ 1] /Q)
    +(-h*s/(Q^2)-pi[ 1] *s/(Q^2))^2;
```

$$Condition2 := -(2\frac{\lambda K}{Q^3} + \frac{h s^2}{Q^3} + \frac{\pi_1 s^2}{Q^3})(\frac{h}{Q} + \frac{\pi_1}{Q}) + (-\frac{h s}{Q^2} - \frac{\pi_1 s}{Q^2})^2$$

```
>   simplify(Condition2);
```

$$-2\frac{(h + \pi_1)\lambda K}{Q^4}$$

Solution of the necessary conditions gives the values of Q and s.

---

[7]We should note that we have *not* reentered the second term manually but simply copied it to the clipboard using the mouse and then pasted it to the next line.

```
>  sol:=solve({ GQ, Gs} ,{ Q, s} );
```

$$sol := \{s = \text{RootOf}((h\,\pi_1 + \pi_1{}^2)\_Z^2 - 2\,\lambda\,K\,h),$$
$$Q = \frac{\text{RootOf}((h\,\pi_1 + \pi_1{}^2)\_Z^2 - 2\,\lambda\,K\,h)\,(h + \pi_1)}{h}\}$$

However, this result is not very readable, so we turn it into a more familiar form using the `allvalues()` function.

```
>  OptSol:=allvalues(sol);
```

$$OptSol := \{Q = \frac{\sqrt{2}\,\%2\,(h + \pi_1)}{\%1\,h},\ s = \frac{\sqrt{2}\,\%2}{\%1}\},$$
$$\{Q = -\frac{\sqrt{2}\,\%2\,(h + \pi_1)}{\%1\,h},\ s = -\frac{\sqrt{2}\,\%2}{\%1}\}$$
$$\%1 := h\,\pi_1 + \pi_1{}^2$$
$$\%2 := \sqrt{\pi_1\,(h + \pi_1)\,\lambda\,K\,h}$$

```
>  OptSol[ 1] ;
```

$$\left\{Q = \frac{\sqrt{2}\,\sqrt{\pi_1\,(h + \pi_1)\,\lambda\,K\,h}\,(h + \pi_1)}{(h\,\pi_1 + \pi_1{}^2)\,h},\ s = \frac{\sqrt{2}\,\sqrt{\pi_1\,(h + \pi_1)\,\lambda\,K\,h}}{h\,\pi_1 + \pi_1{}^2}\right\}$$

```
>  assign(OptSol[ 1] );
>  simplify(Q,power,symbolic);
```

$$\frac{\sqrt{2}\,\sqrt{h + \pi_1}\,\sqrt{\lambda}\,\sqrt{K}}{\sqrt{\pi_1}\,\sqrt{h}}$$

```
>  simplify(s,power,symbolic);
```

$$\frac{\sqrt{2}\,\sqrt{\lambda}\,\sqrt{K}\,\sqrt{h}}{\sqrt{\pi_1}\,\sqrt{h + \pi_1}}$$

As the backorder cost π_1 approaches infinity, the solution should converge to the one we found for the basic EOQ model, which is indeed the case.

```
>  limit(Q,pi[ 1] =infinity);
```

$$\frac{\sqrt{\lambda\,K\,h}\,\sqrt{2}}{h}$$

```
>  limit(s,pi[ 1] =infinity);
```

$$0$$

Note that since $\lim_{\pi_1 \to \infty} s = 0$ since for very large backorder costs, it would not be desirable to have any backorders.

As a simple example consider the case with a demand of $\lambda = 2000$ units per year, order cost of $K = \$25$ per order, holding cost of $h = \$10$ per unit per year and a backorder cost of $\pi_1 = \$30$ per unit per year. Using the foregoing solution, Maple computes the optimal values of the decision variables Q and s as approximately 115 and 29, respectively. The minimum cost is computed as $866.02.

>   `lambda:=2000; K:=25; pi[ 1] :=30; h:=10;`

$$\lambda := 2000$$

$$K := 25$$

$$\pi_1 := 30$$

$$h := 10$$

>   `Q:=evalf(Q); s:=evalf(s);`

$$Q := 115.4700538$$

$$s := 28.86751346$$

>   `G;`

$$866.0254037$$

8.4.3 Analysis of Implied Backorder Costs

In practical problems it is usually very difficult to accurately estimate the value of the backorder cost. In this section we redefine the simpler problem with $\pi_2 = 0$ presented in Section 8.4.2. The new formulation is a nonlinear programming problem with a service measure constraint defined as the average time a demand must wait until it is filled. Such a constraint may be easier to specify than the backorder cost π_1. With this different formulation of our new problem, we obtain results that give implicit *estimates* of the backorder cost π_1.

Since management would normally be averse to increased backorder delays, we introduce a service measure w as the average time a demand must wait until it is filled. With this definition we refer to Figure 8.2 and find

$$w = \frac{\text{total waiting time per cycle}}{\text{demand per cycle}} = \frac{1}{Q} \int_0^{T_2} \lambda t \, dt = \frac{s^2}{2\lambda Q}.$$

Now suppose that management specifies an upper limit $W > 0$ on w. Our problem then reduces to

$$\min \quad \frac{K\lambda}{Q} + \frac{h(Q-s)^2}{2Q}$$

$$\text{s.t.} \quad \frac{s^2}{2\lambda Q} \leq W.$$

Note that in this formulation the backorder cost π_1 does not appear.

It is clear that W must be positive; otherwise the problem reduces to the basic EOQ model with optimal $s = 0$ and optimal $Q = \sqrt{2K\lambda/h}$.

The Lagrangian of this nonlinear program is written as

$$\mathcal{L} = \frac{K\lambda}{Q} + \frac{h(Q-s)^2}{2Q} + \mu\left(\frac{s^2}{2Q} - \lambda W\right).$$

Since the objective function and the feasible set is convex, the Kuhn-Tucker (KT) conditions are necessary and sufficient for optimality.

The KT conditions are $\partial\mathcal{L}/\partial s = 0$, $\partial\mathcal{L}/\partial Q = 0$, $\partial\mathcal{L}/\partial\mu \leq 0$, $\mu\cdot(\partial\mathcal{L}/\partial\mu) = 0$ and $\mu \geq 0$. The first three of these are easily obtained by Maple as follows.

```
>    restart: # ImpliedBO.mws
>    L:=K* lambda/Q+(1/2)*h*(Q-s)^2/Q
     +mu*((1/2)*s^2/Q-lambda*W);
```

$$L := \frac{K\lambda}{Q} + \frac{1}{2}\frac{h(Q-s)^2}{Q} + \mu\,(\frac{1}{2}\frac{s^2}{Q} - \lambda\,W)$$

```
>    Ls:=diff(L,s);
```

$$Ls := -\frac{h(Q-s)}{Q} + \frac{\mu s}{Q}$$

```
>    LQ:=diff(L,Q);
```

$$LQ := -\frac{K\lambda}{Q^2} + \frac{h(Q-s)}{Q} - \frac{1}{2}\frac{h(Q-s)^2}{Q^2} - \frac{1}{2}\frac{\mu s^2}{Q^2}$$

```
>    Lmu:=diff(L,mu);
```

$$Lmu := \frac{1}{2}\frac{s^2}{Q} - \lambda\,W$$

Solving for the first two equations we find the optimal values of the decision variables (s, Q). Comparing this solution with the one we found in Section 8.4.2, we see that they are the same with $\pi_1 = \mu$ and that $s/Q = h/(h + \mu)$.

```
>    sol:=solve({ Ls, LQ },{ s,Q });
```

$$sol := \{s = \text{RootOf}((h\,\mu + \mu^2)\,\_Z^2 - 2\,K\,\lambda\,h),$$
$$Q = \frac{\text{RootOf}((h\,\mu + \mu^2)\,\_Z^2 - 2\,K\,\lambda\,h)\,(h + \mu)}{h}\}$$

```
>    SolOpt:=allvalues(sol);
```

$$SolOpt := \{Q = \frac{\sqrt{2}\%1\,(h + \mu)}{(h\,\mu + \mu^2)\,h}, \; s = \frac{\sqrt{2}\%1}{h\,\mu + \mu^2}\},$$
$$\{Q = -\frac{\sqrt{2}\%1\,(h + \mu)}{(h\,\mu + \mu^2)\,h}, \; s = -\frac{\sqrt{2}\%1}{h\,\mu + \mu^2}\}$$
$$\%1 := \sqrt{\mu\,(h + \mu)\,K\,\lambda\,h}$$

```
>    SolOpt[ 1] ;
```

$$\{Q = \frac{\sqrt{2}\,\sqrt{\mu\,(h+\mu)\,K\,\lambda\,h}\,(h+\mu)}{(h\,\mu + \mu^2)\,h}, \; s = \frac{\sqrt{2}\,\sqrt{\mu\,(h+\mu)\,K\,\lambda\,h}}{h\,\mu + \mu^2}\}$$

```
>   assign(SolOpt[ 1] );

>   s/Q;
```

$$\frac{h}{h+\mu}$$

```
>   eq:=simplify(subs(SolOpt[ 1] ,s^2/(2*Q)=lambda*W));
```

$$eq := \frac{1}{2}\,\frac{K\,\lambda\,h^2\,\sqrt{2}}{(h+\mu)\,\sqrt{\mu\,(h+\mu)\,K\,\lambda\,h}} = \lambda\,W$$

We note that if $\mu = 0$, no solution can exist for this problem since the first KT condition $\partial\mathcal{L}/\partial s = 0$ implies $s = Q$, thus reducing the second KT condition to $-K\lambda/Q^2 = 0$, which is impossible. Hence $\mu > 0$ and the constraint $s^2/(2\lambda Q) \leq W$ must be satisfied as an equality.

Substituting the optimal values of s and Q into $s^2/(2\lambda Q) = W$ gives a nonlinear equation (a fourth-order polynomial) in μ. This equation has a unique solution since the term in its left-hand side is a strictly decreasing function of μ that reduces from infinity to zero.

```
>   LHSeq:=lhs(eq);
```

$$LHSeq := \frac{1}{2}\,\frac{K\,\lambda\,h^2\,\sqrt{2}}{(h+\mu)\,\sqrt{\mu\,(h+\mu)\,K\,\lambda\,h}}$$

```
>   normal(diff(LHSeq,mu));
```

$$-\frac{1}{4}\,\frac{(4\,\mu + h)\,\sqrt{2}\,h^3\,\lambda^2\,K^2}{(h+\mu)\,(\mu\,(h+\mu)\,K\,\lambda\,h)^{(3/2)}}$$

```
>   assume(h>0, K>0, lambda>0);

>   limit(LHSeq,mu=0,right);
```

$$\infty$$

```
>   limit(LHSeq,mu=infinity);
```

$$0$$

In summary, if the specification of the backorder cost π_1 is difficult, then an alternate approach can be used where an upper limit W on w is specified. The new problem has a solution with exactly the same structure as the original backorder model. Given the value of W, the implied value μ of the backorder cost π_1 is computed after finding the unique solution of a fourth-order polynomial.

A more general version of this approach is presented in Çetinkaya and Parlar [42] where implied backorder costs are computed for the two types of backorder costs π_1 and π_2.

8.4.4 Quantity Discounts

In the models presented so far we made the assumption that the unit purchase cost c is a constant regardless of the number of units procured. This meant that the total procurement cost per cycle was cQ. Since the cycle length was Q/λ, the average procurement cost per unit time was found as $cQ/(Q/\lambda) = c\lambda$ which is independent of the order quantity decision variable Q.

In many real-world applications the unit cost of an item is not always independent of the quantity purchased. In some cases, larger purchase quantities of a product may lead to reductions in the unit cost. A common example of such a quantity discount is the case of the "baker's dozen" where some bakeries give their customers 13 units of an item (such as bagels) for the price of 12. Other examples of lower unit prices for bulk purchases include computer diskettes purchased in a box of 10, office stationary purchased in bulk and soft drink cans purchased in a case of 24.

In the general quantity discount model we will analyze in this section, we will define $b_1, b_2, \ldots, b_{n-1}$ as the order quantity points (break points) where a price change occurs. The order-quantity-dependent unit purchase price $c(Q)$ is described as

$$c(Q) = \begin{cases} p_1, & \text{if } b_0 = 0 \le Q < b_1 \\ p_k, & \text{if } b_{k-1} \le Q < b_k \text{ for } k = 2, \ldots, n-1 \\ p_n, & \text{if } b_{n-1} \le Q < b_n = \infty. \end{cases} \qquad (8.2)$$

Since larger order quantities are associated with lower prices, we have $p_1 > p_2 > \cdots > p_n$. Recalling the definition of the holding cost $h = ic$, we see that since c is a function of Q, the order-quantity-dependent order cost would be $h(Q) = ic(Q)$. With this definition, and retaining every other assumption we made when we introduced the basic EOQ model, the new objective function $G(Q)$ is obtained as

$$G(Q) = \frac{\lambda K}{Q} + \frac{1}{2}h(Q)Q + \lambda c(Q). \qquad (8.3)$$

The objective is to compute the optimal Q that will minimize the average cost function $G(Q)$ in (8.3). Since the holding and procurement costs are both discontinuous functions of Q, the objective function $G(Q)$ is also discontinuous — see Figure 8.3 — and its minimum value has to be obtained using a procedure that is more complicated than the one we employed for the basic EOQ model.

Essentially, the procedure for computing the optimal order quantity involves three steps.

1. For each region $b_{k-1} \le Q < b_k, k = 1, \ldots, n$ defined in (8.2), compute an order quantity Q using the EOQ formula with the unit cost associated for that category.

2. It is possible that some of the Q's found in Step 1 may be too small to qualify for the related discount price; i.e., they may be infeasible. In that

case adjust the order quantity upward to the nearest order quantity point at which the unit price will be reduced.

3. For each of the order quantities found in Steps 1 and 2, compute the average cost $G(Q)$ in (8.3) using the unit price that is appropriate for that quantity. The order quantity yielding the lowest cost is the optimal one.

When there is a large number of break points, this procedure outlined above may be cumbersome to implement manually. However, using Maple's graphical and numerical capabilities, the optimal solution to the quantity discount problem can be found very easily.

As an example, consider a problem with an annual demand of $\lambda = 1000$ units, inventory carrying charge of $i = 0.10$ and order cost of $K = \$100$ per order.

```
>    restart: # QuanDisc.mws
>    lambda:=1000; i:=0.10; K:=100;
```

$$\lambda := 1000$$

$$i := .10$$

$$K := 100$$

There are two break points at $b_1 = 100$ and $b_2 = 300$ with prices $p_1 = 50$, $p_2 = 49$ and $p_3 = 48.50$. Thus, the procurement cost function is

$$c(Q) = \begin{cases} 50, & \text{if } b_0 = 0 \leq Q < 100 \\ 49, & \text{if } 100 \leq Q < 300 \\ 48.50, & \text{if } 300 \leq Q < \infty. \end{cases}$$

```
>    b[ 1] :=100; b[ 2] :=300;
     p[ 1] :=50; p[ 2] :=49; p[ 3] :=48.50;
```

$$b_1 := 100$$

$$b_2 := 300$$

$$p_1 := 50$$

$$p_2 := 49$$

$$p_3 := 48.50$$

We enter this information in Maple using the `piecewise()` function and form the objective function $G(Q)$.

```
>    c:=Q->piecewise (Q<b[ 1] ,p[ 1] ,Q<b[ 2] ,p[ 2] ,  p[ 3] );
```

$$c := Q \rightarrow \text{piecewise}(Q < b_1, p_1, Q < b_2, p_2, p_3)$$

```
>    c (Q);
```

$$\begin{cases} 50 & Q < 100 \\ 49 & Q < 300 \\ 48.50 & \text{otherwise} \end{cases}$$

```
>    h:=Q->i*c (Q);
```

$$h := Q \to i\,c(Q)$$

> G:=Q->K* lambda/Q+(1/2)*h(Q)*Q+lambda*c(Q);

$$G := Q \to \frac{K\lambda}{Q} + \frac{1}{2}h(Q)\,Q + \lambda\,c(Q)$$

A visual inspection of the plot of $G(Q)$ in Figure 8.3 reveals that the optimal solution is at the last break point $b_2 = 300$.

> #plot(G(Q),Q=50..500,discont=true);

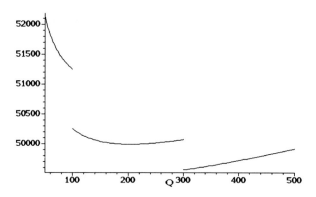

FIGURE 8.3. The cost function $G(Q)$ under quantity discounts.

Maple's ability to compute the derivatives of piecewise functions — such as $G(Q)$ — permits us to simplify the first two steps of the procedure given on page 346. We differentiate $G(Q)$, equate the result to zero and solve to find the only feasible EOQ in the interval $[b_1, b_2) = [100, 300)$ as $q = 202.03$.

> q:=solve(diff(G(Q),Q),Q);

$$q := -200., 202.0305089$$

> q[2];

$$202.0305089$$

At this EOQ point, the cost function assumes the value $G(202.03) = \$49989.94$.

> G(q[2]);

$$49989.94949$$

However, the order quantity at the next break point $b_2 = 300$ gives a lower cost as $G(300) = \$49560.83$, indicating that the optimal order quantity must be $Q = b_2 = 300$ units.

> G(b[2]);

$$49560.83333$$

8.5   Probabilistic Inventory Models

A common assumption in all the models presented so far was the deterministic nature of every relevant factor such as demand, cost data and supplier availability. In reality, many of these factors are not deterministic but random. For example, due to the uncertainty present in consumer preferences the actual number of units of a product a retailer hopes to sell in the coming month is usually not known in advance. Future levels of interest rates, which play a crucial role in determining of the holding cost h, are subject to random fluctuations. Due to unexpected failures in production processes or unplanned work stoppages caused by labor unrest, a supplier may sometimes be unable to deliver the product ordered by the retailer.

In this section we will develop several inventory models that will involve at least one random factor (such as demand or supplier availability). We will assume that although the actual outcome of the random variable of interest cannot be predicted in advance, we have a probability distribution representing the possible outcomes.

We will start by describing a continuous-review inventory model with random demand where backorders are allowed. First, we will present an heuristic treatment of the approximate formulation of this model. We will then briefly discuss an exact version of the model for the case of Poisson demands and constant lead-time. Periodic-review models with random demand (including the single-period newsvendor model) will also be introduced. We will also discuss some models where yield is random and where the supplier status is a random process.

8.5.1   The Continuous-Review Model: Approximate Formulation

Consider an inventory system that is controlled by selecting an order quantity (or lot size) Q and a reorder point r. This policy calls for ordering Q units when the on-hand inventory drops to the level r. As in the basic EOQ model, we make some assumptions that will facilitate the development of the model.

1. Since the system is continuous review, the on-hand inventory level is known at all times.

2. Demand is random but stationary with a constant *expected* value of λ units per year.

3. There is never more than one order outstanding. This means that when the reorder point r is reached, there are no outstanding orders so that the inventory position (IP)[8] is equal to the inventory level (IL), implying that order quantity Q will always be the same.

---

[8]Inventory position is defined as the amount on hand plus the amount on order minus the number of backorders. Inventory level is the amount on hand minus backorders. Some authors (e.g., Hadley and Whitin [81, p. 163]) use the term "net inventory" instead of "inventory level."

4. The reorder point r is positive.

5. There is a constant positive leadtime of L time units.[9] The random demand during leadtime (DDLT) is denoted by X with probability density $f(x)$ and the expected value $\mu = E(X) = \lambda L$.

6. There are three types of costs:

   (a) Purchase cost is c dollars per unit and order (setup) cost is K dollars per order.

   (b) Holding cost is h dollars per unit per year.

   (c) There is a shortage cost of $\pi (= \pi_1)$ per unit of unsatisfied demand. The time-dependent part of the shortage cost is $\pi_2 = 0$.

In deterministic inventory models such as the basic EOQ model, the inventory level at the time of an order arrival can be determined with certainty. In stochastic inventory models this is not possible since the demand during leadtime is also a random quantity. The expected value of the inventory level at the time of an order arrival is defined as the *safety stock* and it is denoted by s. Since the expected demand during leadtime is μ, we can write $s = r - \mu$.

In this model, the regenerative cycles are of random length and they commence when the inventory position is raised to $Q + r$. Figure 8.4 depicts a sample realization of the expected (average) behavior of the inventory position and inventory level processes. The sample paths of these processes are drawn as lines since we had assume that the demand is stationary with a constant expected value of λ units per year. By using the renewal reward theorem we can construct the average cost per year as follows.

Cycle Length

Since the IP fluctuates between r and $Q + r$, and Q units are used at an average rate of λ units per year, the expected cycle length is $E(\text{length}) = Q/\lambda$.

Cycle Cost

We examine the cycle cost in three parts.

- *Order Cost:* Since each cycle starts with an order, the order cost in a cycle is K dollars.

- *Holding Cost:* To compute this cost, we make an important simplifying approximation and assume that backorders occur only a small fraction of time. This has the implication that the on-hand inventory can now be approximated by the inventory level IL. (As Hadley and Whitin [81] indicate,

---

[9]As we indicated previously, in this chapter the unit time is normally taken as one year.

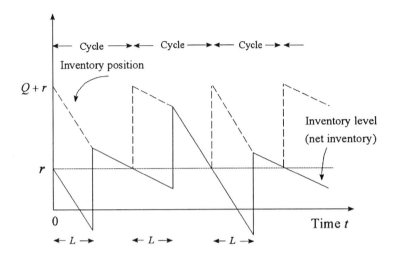

FIGURE 8.4. A sample realization of the expected (average) behavior of inventory position and inventory level processes in the continuous-review model with backorders.

this is a valid approximation when the backorder cost π is large relative to the holding cost h, which is usually the case in practical applications.) Since the average demand rate was assumed to be a constant λ, the expected inventory level varies linearly between s and $Q + s$ and the average inventory during each cycle is approximately $\frac{1}{2}[(Q + s) + s] = \frac{1}{2}Q + s$ where $s = r - \lambda L$. Recalling that each cycle lasts Q/λ years, the expected holding cost per cycle is $E(\text{holding cost}) = h(\frac{1}{2}Q + r - \lambda L)Q/\lambda$.

- *Backorder Penalty Cost:* Defining $\eta(X, r)$ as the random number of backorders per cycle, we have

$$\eta(x, r) = \begin{cases} 0, & \text{if } x < r \\ x - r, & \text{if } x \geq r. \end{cases}$$

Thus, the expected number of backorders in a cycle is $E(\text{backorder cost}) = \pi \int_r^\infty (x - r) f(x)\, dx$.

Combining the three types of costs described and using the renewal reward theorem, the expected average cost $G(Q, r)$ per year is

$$G(Q, r) = \frac{K + h(\frac{1}{2}Q + r - \mu)Q/\lambda + \pi \int_r^\infty (x - r) f(x)\, dx}{Q/\lambda}$$

$$= \frac{\lambda K}{Q} + h(\frac{1}{2}Q + r - \mu) + \frac{\pi \lambda}{Q} \int_r^\infty (x - r) f(x)\, dx.$$

The third term $B(Q, r) = (\pi \lambda / Q) \int_r^\infty (x - r) f(x) \, dx$ in the objective function is the expected backorder cost per year. As we showed in Chapter 5, Nonlinear Programming, despite the claim made by Hadley and Whitin [81, p. 221, problem 4-6] this function is not, in general, jointly convex in Q and r. But, as we also indicated, if ξ is a number such that $f(x)$ is *nonincreasing* for $x \geq \xi$, then $B(Q, r)$ is convex in the region $\{(Q, r) : 0 < Q < \infty \text{ and } \xi \leq r < \infty\}$. Thus, if r is restricted to be suitably large, then $B(Q, r)$ — and the objective function $G(Q, r)$ — is convex within such a region. For example, if $f(x)$ is normal, then $B(Q, r)$ is convex when r exceeds the expected demand during lead time. When $f(x)$ is a strictly decreasing density such as the exponential, then $B(Q, r)$ is convex everywhere.

We now use Maple to find the necessary conditions minimizing $G(Q, r)$ and present an example using an exponential demand density $f(x) = \delta e^{-\delta x}, x \geq 0$.

```
>   restart: # Qr.mws
```

The objective function $G(Q, r)$ and its derivatives G_Q and G_r are entered and evaluated to give the necessary conditions.

```
>   G:=lambda* K/Q+h* (Q/2+r-mu)
    +(pi* lambda/Q)* int ((x-r)* f(x),x=r..infinity
    );
```

$$G := \frac{\lambda K}{Q} + h\left(\frac{1}{2}Q + r - \mu\right) + \frac{\pi \lambda \int_r^\infty (x - r) f(x) \, dx}{Q}$$

```
>   GQ:=diff(G,Q);
```

$$GQ := -\frac{\lambda K}{Q^2} + \frac{1}{2}h - \frac{\pi \lambda \int_r^\infty (x - r) f(x) \, dx}{Q^2}$$

```
>   Gr:=diff(G,r);
```

$$Gr := h + \frac{\pi \lambda \int_r^\infty - f(x) \, dx}{Q}$$

An attempt to solve for the unknowns Q and r directly from the necessary conditions fails. This is understandable since it is impossible to find a closed-form solution for the unknowns from expressions involving integrals whose lower limit includes r. Thus the necessary conditions must be solved numerically once the density $f(x)$ is specified.

```
>   solve ({ GQ,Gr} ,{ Q,r} );
```

```
Error, (in solve) cannot solve expressions with,
int(-(-x+r)* f(x),x =r .. infinity), for, r
```

```
>   Qsol:=solve(GQ,Q);
```

$$Qsol := \frac{\sqrt{2}\sqrt{h\,\lambda\,(K + \int_r^\infty f(x)\,x\,dx\,\pi - \int_r^\infty f(x)\,r\,dx\,\pi)}}{h},$$

$$-\frac{\sqrt{2}\sqrt{h\,\lambda\,(K + \int_r^\infty f(x)\,x\,dx\,\pi - \int_r^\infty f(x)\,r\,dx\,\pi)}}{h}$$

Although we cannot solve the necessary conditions for the Q and r values explicitly, we can at least express Q in terms of r (denoted in the Maple worksheet by Qr) and r in terms of Q (denoted by rQ) easily as follows:

>   Qr:=Qsol[1] ;

$$Qr := \frac{\sqrt{2}\sqrt{h\,\lambda\,(K + \int_r^\infty f(x)\,x\,dx\,\pi - \int_r^\infty f(x)\,r\,dx\,\pi)}}{h}$$

>   readlib(isolate);

proc(*expr, x, n*) ... **end**

>   rQ:=isolate(Gr,int(-f(x),x=r..infinity));

$$rQ := \int_r^\infty - f(x)\,dx = -\frac{h\,Q}{\pi\,\lambda}$$

Following is an example illustrating the solution of the approximate (Q, r) model with an exponential density for the demand during lead time.

>   K:=100; h:=.15; pi:=1; lambda:=10000;
     delta:=1/1000; mu:=1/delta;
     f(x):=delta*exp(-delta*x);

$$K := 100$$
$$h := .15$$
$$\pi := 1$$
$$\lambda := 10000$$
$$\delta := \frac{1}{1000}$$
$$\mu := 1000$$
$$f(x) := \frac{1}{1000}e^{(-1/1000\,x)}$$

The cost function G and its partials now assume specific forms.

>   G;

$$1000000\,\frac{1}{Q} + .07500000000\,Q + .15\,r - 150.00 + 10000000\,\frac{e^{(-1/1000\,r)}}{Q}$$

> `GQ;`

$$-1000000\,\frac{1}{Q^2} + .07500000000 - 10000000\,\frac{e^{(-1/1000\,r)}}{Q^2}$$

> `Gr;`

$$.15 - 10000\,\frac{e^{(-1/1000\,r)}}{Q}$$

Since the data and the density $f(x)$ are specified, Maple easily finds the optimal Q and r values as 4785 and 2634, respectively. The minimum value of the objective function is obtained as \$962.99.

> `sol:=solve({ GQ, Gr} ,{ Q, r} );`

$$sol := \{Q = -2785.938897,\ r = 3175.120135 - 3141.592654\ I\},$$
$$\{r = 2634.022855,\ Q = 4785.938897\}$$

> `sol[ 2] ; GOpt:=evalf(subs(sol[ 2] ,G));`

$$\{r = 2634.022855,\ Q = 4785.938897\}$$

$$GOpt := 962.9942630$$

8.5.2   The Continuous-Review Model: Exact Formulation

In the exact formulation of the (Q, r) model it is assumed that unit demands are generated by a Poisson process with rate λ. Thus, all variables of interest such Q, r and the inventory levels are discrete integers. There is a constant lead-time of L years and, of course, the shortages are backordered. In this model, the two types of backorder costs π_1 and π_2 assume positive values.

Development of the objective function of this model requires the stationary distribution of the inventory position. This is achieved using tools from stochastic processes (more specifically, continuous-time Markov chains); see Hadley and Whitin [81, Chapter 4]. The average cost $G(Q, r)$ is the sum of expected order cost $\lambda K/Q$, expected holding cost $hS(Q, r)$ and two types of backorder costs $[\pi_1 b(Q, r) + \pi_2 B(Q, r)]$, i.e.,

$$G(Q, r) = \frac{\lambda K}{Q} + hS(Q, r) + \pi_1 b(Q, r) + \pi_2 B(Q, r)$$

where

$$S(Q, r) \;=\; \frac{1}{2}(Q + 1) + r - \lambda L + B(Q, r)$$

$$B(Q, r) \;=\; \frac{1}{Q}[\beta(r) - \beta(r + Q)]$$

$$b(Q, r) \;=\; \frac{\lambda}{Q}[\alpha(r) - \alpha(r + Q)]$$

$$\beta(v) = \frac{1}{2}(\lambda L)^2 \bar{F}(v-1; \lambda L) - \lambda L v \bar{F}(v; \lambda L) + \frac{1}{2}v(v+1)\bar{F}(v+1; \lambda L)$$

$$\alpha(v) = \lambda L \bar{F}(v; \lambda L) - v \bar{F}(v+1; \lambda L).$$

In these expressions,

$$\bar{F}(u; \lambda L) = \sum_{j=u}^{\infty} f(j; \lambda L) = 1 - \sum_{j=0}^{u-1} f(j; \lambda L)$$

is the complementary cumulative distribution of the Poisson process with density

$$f(x; \lambda L) = \frac{e^{-\lambda L}(\lambda L)^x}{x!}, \quad x = 0, 1, \ldots$$

Optimization of $G(Q, r)$ over the nonnegative integers appears to be a very complicated task. However, as usual, Maple's ability to plot graphs and perform symbolic and numeric computations simplifies this problem considerably as we demonstrate below.

```
>    restart: # QrExact.mws
```

The Poisson density $f(x; \lambda L)$ and its complementary cumulative distribution $\bar{F}(x; \lambda L)$ are entered using the unapply() function.

```
>    f:=unapply(exp(-lambda*L)*(lambda*L)^x/x!,x);
```

$$f := x \to \frac{e^{(-\lambda L)}(\lambda L)^x}{x!}$$

```
>    FBar:=unapply(1-Sum(f(j),j=0..(u-1)),u);
```

$$FBar := u \to 1 - \left(\sum_{j=0}^{u-1} \frac{e^{(-\lambda L)}(\lambda L)^j}{j!} \right)$$

The expressions for $\alpha(v)$, $\beta(v)$, $b(Q, r)$, $B(Q, r)$ and $S(Q, r)$ are also defined using the unapply() function so that they may be evaluated when we compute the $G(Q, r)$ function.

```
>    alpha:=unapply(' lambda*L*FBar(v)-v*FBar(v+1)' ,v);
```

$$\alpha := v \to \lambda L\, FBar(v) - v\, FBar(v+1)$$

```
>    beta:=unapply(' ((lambda*L)^2/2)*FBar(v-1)
     -lambda*L*v*FBar(v)
     +(1/2)*v*(v+1)*FBar(v+1)' ,v);
```

$$\beta := v \to \frac{1}{2}\lambda^2 L^2\, FBar(v-1) - \lambda L v\, FBar(v) + \frac{1}{2}v(v+1)\, FBar(v+1)$$

```
>    b:=unapply(' (lambda/Q)*(alpha(r)-alpha(Q+r))' ,
     Q,r);
```

$$b := (Q, r) \to \frac{\lambda(\alpha(r) - \alpha(Q+r))}{Q}$$

```
>    B:=unapply(' (1/Q)*(beta(r)-beta(Q+r))' ,Q,r);
```

$$B := (Q, r) \to \frac{\beta(r) - \beta(Q+r)}{Q}$$

```
>   S:=unapply(' (1/2)*(Q+1)+r-lambda*L+B(Q,r)',Q,r);
```

$$S := (Q, r) \to \frac{1}{2}Q + \frac{1}{2} + r - \lambda L + B(Q, r)$$

Next we define the objective function $G(Q, r)$ in terms of the expressions introduced earlier and enter the data values for λ, L, K, h, π_1 and π_2. Although the variables Q and r are discrete, we plot the continuous contour curves of G in order to obtain some insights for its convexity. We see in Figure 8.5 that G appears to be a function jointly convex in (Q, r).

```
>   G:=unapply(' (lambda/Q)*K+h*S(Q,r)
    +pi[ 1]*b(Q,r)+pi[ 2]*B(Q,r)',Q,r);
```

$$G := (Q, r) \to \frac{\lambda K}{Q} + h\,S(Q, r) + \pi_1\,b(Q, r) + \pi_2\,B(Q, r)$$

```
>   lambda:=400; L:=0.25; K:=0.16; h:=2; pi[ 1] :=0.10;
pi[ 2] :=0.30;
```

$$\lambda := 400$$

$$L := .25$$

$$K := .16$$

$$h := 2$$

$$\pi_1 := .10$$

$$\pi_2 := .30$$

```
>   with(plots):

>   contourplot(G(Q,r),Q=5..35,r=5..120,color=black);
```

Based on the information extracted from the graph of the contour curves, the optimal solution appears to be in the region $[15, 35] \times [80, 120]$. Ideally one should implement a discrete optimization method to locate the optimal solution for this problem. But in this case we will use an exhaustive search to compute the optimal value that minimizes $G(Q, r)$. In this way, we will have an opportunity to introduce the Maple time() function that sets the current clock time and that can be used to compute the CPU time of an operation.

```
>   settime:=time():
>   GL:=[ seq(seq([ Q,r,evalf(G(Q,r))],
    Q=15..35),r=80..120)] :

>   nops(GL);
```

$$861$$

On a Pentium-based desktop computer with a clock speed of 200MHz, Maple took slightly over two minutes of CPU time to evaluate the objective function at

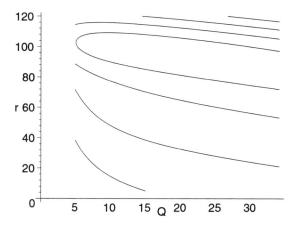

FIGURE 8.5. Contour curves of the $G(Q, r)$ function in the exact formulation of the continuous-review model. Note that although $G(Q, r)$ is a discrete function of discrete variables Q and r, the graph is drawn using continuous variables.

all 861 integer points in the $[15, 35] \times [80, 120]$ region. The optimal solution is found as $Q = 18, r = 96$ and $G = \$32.89$.

>   `cpu_time:=(time()-settime)*seconds;`

$$cpu\_time := 143.474\ seconds$$

The list GL consists of lists of triples with values of $[Q, r, G]$. The next set of statements extracts the triple with the smallest third element G.

```
>   for i from 1 to nops(GL) do if
    GL[ i][ 3] =min(seq(GL[ i][ 3] ,i=1..nops(GL)))
    then print(GL[ i] ) fi od;
```

$$[18,\ 96,\ 32.8953311]$$

>   `evalf(G(19,96));`

$$32.9419609$$

We should note in closing that the same problem was also solved by Hadley and Whitin [81, Section 4-10], who obtained a slightly different—and incorrect—"optimal" solution as $(Q^0, r^0, G(Q^0, r^0)) = (19, 96, \$31.75)$. Although this is an incorrect result[10] due to the flatness of the objective function, using Hadley and Whitin's solution $(19, 96)$ would give $G(19, 96) = \$32.94$ — an increase of only 4 cents, or 0.15%.

---

[10] A Quick BASIC program written to check the results found that Maple's solution was correct.

8.5.3   One-Period (Newsvendor) Model with Random Demand

In this section we treat the one-period inventory model in which only a single in-ventory replenishment decision is made in anticipation of the random demand for the product. This is the simplest probabilistic inventory model but it is important for two reasons. First, many real-life inventory decision problems (such as the order decisions for newspapers, Christmas trees and bakery products) can be rep-resented using the one-period model. Second, insights gained from the analysis of the one-period model provide a useful starting point for the analysis of more complicated — and more realistic — multiperiod dynamic models.

To motivate the discussion let us assume that a newsvendor selling the local newspaper *Hamilton Spectator* in Hamilton, Ontario, must choose the order quan-tity Q of newspapers he will order from the publisher. Each unit he orders will cost him c cents per unit and he sells it for s cents per unit with $c < s$. If he overes-timates the demand and ends up with a surplus at the end of the day, he can return the unsold units for a "salvage" value of v cents per unit where $v < c$. If, on the other hand, he underestimates the demand and finds that more customers wanted to purchase the newspapers than he had available, then he would incur a penalty cost of p cents per unit. Such a penalty cost is usually difficult to measure but it may correspond to a cost of lost "goodwill" or the marginal cost of satisfying the demand through, say, an expedited order.

If the daily demand is represented by a random variable X with density $f(x)$, and if we define Π as the random profit per period, then the realized value π of profit can be written as

$$\pi = \begin{cases} sx + v(Q-x) - cQ & \text{if } x \leq Q \text{ (surplus)} \\ sQ - p(x-Q) - cQ & \text{if } x > Q \text{ (shortage).} \end{cases}$$

Thus, the expected profit $E(\Pi) = P(Q)$ is obtained as

$$\begin{aligned} P(Q) &= \int_0^Q [sx + v(Q-x) - cQ] f(x)\, dx \\ &+ \int_Q^\infty [sQ - p(x-Q) - cQ]\, dx \\ &= (s-v)\mu - (c-v)Q - (s+p-v) \int_Q^\infty (x-Q) f(x)\, dx \end{aligned}$$

where $\mu = E(X)$ is the mean demand.

Using Maple (which in turn uses Leibniz's rule of differentiation under the integral sign), we find the first and second derivatives of $P(Q)$ as follows:

```
>    restart: # Newsvendor.mws
>    P:=(s-v)*mu-(c-v)*Q
     -(s+p-v)*int((x-Q)*f(x),x=Q..infinity);
```

$$P := (s-v)\,\mu - (c-v)\,Q - (s+p-v) \int_Q^\infty (x-Q)\, f(x)\, dx$$

Since $P_{QQ} < 0$, the expected profit function $P(Q)$ is a strictly concave. Thus the solution of the necessary condition $P_Q = 0$ would give the unique maximizing value.

> PQ:=diff(P,Q);

$$PQ := -c + v - (s + p - v) \int_Q^\infty -f(x)\,dx$$

> PQQ:=diff(P,Q$2);

$$PQQ := -(s + p - v)\,f(Q)$$

In order to isolate the term containing the integral term, we use the iso-late() function.

> i:=op(3,op(3,PQ));

$$i := \int_Q^\infty -f(x)\,dx$$

> readlib(isolate);

$$\textbf{proc}(\textit{expr},\ x,\ n)\ \dots\ \textbf{end}$$

> isolate(PQ,i);

$$\int_Q^\infty -f(x)\,dx = \frac{c - v}{-s - p + v}$$

The necessary condition is thus obtained as

$$\bar{F}(Q) = \int_Q^\infty f(x)\,dx = \frac{c - v}{s + p - v}. \tag{8.4}$$

Since $v < c$ and $c < p$, the right-hand side of this expression is always less than unity.

We now solve a numerical example assuming that the demand is a two-stage Erlang random variable with a mean of 100 units.

> c:=15; s:=55; v:=10; p:=20;
  lambda:=1/100; mu:=1/lambda:
  f:=x->2* lambda* (2* lambda)* x* exp (-2* lambda* x);

$$c := 15$$

$$s := 55$$

$$v := 10$$

$$p := 20$$

$$\lambda := \frac{1}{100}$$

$$f := x \rightarrow 4\,\lambda^2\,x\,e^{(-2\lambda x)}$$

> PQ;

$$-5 + \frac{13}{10} Q\, e^{(-1/50\, Q)} + 65\, e^{(-1/50\, Q)}$$

Solving the necessary condition gives two solutions, but the correct one is the positive solution that is equal to 210.82. The maximum value of the expected profit is found as \$3,147.86.

```
>   sol:=evalf(solve(PQ,Q));
```

$$sol := -48.54324892,\ 210.8420750$$

```
>   evalf(subs(Q=sol[2],P));
```

$$3147.867908$$

It is useful to note that the same problem can be restated in terms of expected cost minimization as follows. Define $c_o = c - v$ to be the cost of *overage*, i.e., the loss per unit on any remaining units. Similarly, let $c_u = (s - c) + p$ as the cost of *underage*, i.e., cost per unit incurred if a demand occurs when the system is out of stock. Here c_u is the sum of the potential profit lost $s - c$, and the cost of lost "goodwill" p. With these definitions, the condition in (8.4) can be rewritten as

$$\bar{F}(Q) = \frac{c - v}{s + p - v} = \frac{c - v}{[(s - c) + p] + (c - v)} = \frac{c_o}{c_u + c_o}. \qquad (8.5)$$

This suggests that minimizing a relevant expected cost function (of overage plus underage) is equivalent to maximizing the expected profit function. This follows since the realized value of cost is

$$\text{cost} = \begin{cases} c_o(Q - x) & \text{if } x \le Q \text{ (overage)} \\ c_u(x - Q) & \text{if } x > Q \text{ (underage)} \end{cases}$$

and hence

$$E(\text{Cost}) = G(Q) = \int_0^Q c_o(Q - x)f(x)\,dx + \int_Q^\infty c_u(x - Q)f(x)\,dx.$$

Differentiating $G(Q)$ and equating the result to zero gives $C'(Q) = c_o - c_o \bar{F}(Q) - c_u \bar{F}(Q) = 0$ so that

$$\bar{F}(Q) = \frac{c_o}{c_u + c_o}.$$

This is the same result as in (8.5).

A review by Khouja [107] lists nearly 100 papers that deal with applications and ramifications of the single-period inventory model. Some of the recent work on coordination in supply chain management also involves the use of the single-period model; see, Parlar and Weng [147] and Weng and Parlar [193].

8.5.4   Dynamic Inventory Models

In the previous section we considered the single-period inventory model with stochastic demand and showed that the optimal order quantity can be obtained in terms of the cumulative distribution of the demand random variable. In this section we generalize the model and assume that the inventory system is to be managed over a finite horizon of, say, N periods. Our exposition here follows closely that of Bertsekas [31, Section 4.2].

At the start of each period, the inventory level is observed and a decision is made on the order quantity. If, at the end of a period, there are some unsold items, the system incurs a cost of h dollars per unit held in inventory. If, on the other hand, the period ends with a shortage and the shortages are backordered, then a cost of b dollars per unit is incurred. Clearly, in this model h is the same as the cost of overage c_o, and b is the same as the cost of underage c_u introduced in the previous section where we argued that cost minimization and profit maximization are equivalent. The objective is to minimize the total expected cost of operating the inventory system over N periods.

The dynamic inventory problem is formulated naturally as a dynamic programming (DP) problem. Thus, in this section we adapt the state-equation approach introduced in the Chapter 6, Dynamic Programming, and define state x_k as the inventory available at the start of period k. Similarly, we define control u_k as the quantity ordered (and immediately delivered) at the start of period k. The random demand in each period is defined as w_k with a given density function $f_k(w_k)$. We assume that $w_0, w_1, \ldots, w_{N-1}$ are i.i.d. and that the initial inventory x_0 is a known quantity. The inventory level process is represented by

$$x_{k+1} = x_k + u_k - w_k, \quad k = 0, 1, \ldots, N - 1. \tag{8.6}$$

Although the purchase cost function $c(u)$ can be defined as in (8.1), in this section we will assume that this cost is linear, i.e., $c(u_k) = cu_k$. With this definition, the expected cost that must be minimized is

$$E\left\{ \sum_{k=0}^{N-1} \left[cu_k + h \int_0^{y_k} (y_k - w_k) f_k(w_k)\, dw_k + b \int_{y_k}^\infty (w_k - y_k) f_k(w_k)\, dw_k \right] \right\}$$

where $y_k = x_k + u_k$ is the beginning inventory immediately after receiving an order of $u_k \geq 0$ units. This functional must be minimized with respect to the state dynamics given by (8.6).

In order to apply the DP algorithm, we define $V_k(x_k)$ as the minimum expected cost for periods $k, k + 1, \ldots, N - 1$ when starting at period k and with an initial inventory of x_k units. This gives

$$V_N(x_N) = 0$$

$$
\begin{aligned}
V_k(x_k) &= \min_{u_k \geq 0} \{ cu_k + L(x_k + u_k) + E[V_{k+1}(x_k + u_k - w_k)] \} \\
&= \min_{u_k \geq 0} \left\{ cu_k + L(x_k + u_k) + \int_0^\infty V_{k+1}(x_k + u_k - w_k) f_k(w_k)\, dw_k \right\}
\end{aligned}
$$

where the function $L(y)$ is defined as

$$
\begin{aligned}
L(y) &= hE[\max(0, y - w)] + bE[\max(0, w - y)] \\
&= h \int_0^y (y - w) f(w) \, dw + b \int_y^\infty (w - y) f(w) \, dw.
\end{aligned}
$$

The $L(y)$ function can be interpreted as the expected loss per period due to possible overages or underages. It can be shown that $L(y)$ is strictly convex since $L''(y) = (h + b) f(y) > 0$.

We now state an important theorem concerning the structure of the optimal policy for this problem.

Theorem 14 *Assuming $c > 0$, $h \geq 0$, $b > c$ and linear order cost $c(u) = cu$, the optimal policy $\mu(x)$ is of the form*

$$
\mu_k(x_k) = \begin{cases} S_k - x_k & \text{if } x_k < S_k \\ 0 & \text{if } x_k \geq S_k. \end{cases}
$$

Such a policy is said to be "base stock"—if the starting inventory level in period k is below S_k, a sufficient number of units are ordered to increase it to S_k; otherwise no action is taken.

Since this is one of the central results in stochastic inventory theory, we will provide a brief sketch of the proof. For details, we refer the reader to Bertsekas [30, Section 2.2].

We first define

$$
G_k(y_k) = cy_k + L(y_k) + E[V_{k+1}(y_k - w_k)]
$$

for each period $k = 0, 1, \ldots, N - 1$.

For the terminal period N, we have $V_N(x_N) = 0$, which is convex in x_N. For period $N - 1$, the value function takes the form

$$
\begin{aligned}
V_{N-1}(x_{N-1}) &= \min_{y_{N-1} \geq x_{N-1}} [cy_{N-1} + L(y_{N-1})] - cx_{N-1} \\
&= \min_{y_{N-1} \geq x_{N-1}} G_{N-1}(y_{N-1}) - cx_{N-1}
\end{aligned}
$$

since $E[V_N(y_{N-1} - w_{N-1})] = 0$. This is a single-period problem and it can be shown that the value minimizing the convex[11] function $G_{N-1}(y_{N-1})$ is unique and is found by solving

$$
\bar{F}(y_{N-1}) = \frac{c + h}{b + h}
$$

where $\bar{F}(y)$ is the cumulative distribution function of $f(w)$ evaluated at y. We denote the minimizing value by S_{N-1}.

To determine the form of the policy, we must examine two cases:

---

[11] Convexity of $G(y) = cy + L(y)$ follows from the fact that cy is linear and $L(y)$ is convex.

- If the initial inventory $x_{N-1} < S_{N-1}$, then $y_{N-1} = x_{N-1} + u_{N-1} = S_{N-1}$; i.e., we must order up to S_{N-1}. This implies that the optimal order quantity must be $u_{N-1} = S_{N-1} - x_{N-1}$. Note that in this case $G_{N-1}(S_{N-1}) \leq G(y)$ for all $y \geq x_{N-1}$.

- If the initial inventory $x_{N-1} \geq S_{N-1}$, then $y_{N-1} = x_{N-1} + u_{N-1} = x_{N-1}$; i.e., $u_{N-1} = 0$, thus no units should ne ordered.

In summary, for period $N - 1$, the optimal policy $\mu_{N-1}(x_{N-1})$ is state dependent and it assumes the form

$$\mu_{N-1}(x_{N-1}) = \begin{cases} S_{N-1} - x_{N-1} & \text{if } x_{N-1} < S_{N-1} \\ 0 & \text{if } x_{N-1} \geq S_{N-1}. \end{cases}$$

The value function now becomes

$$V_{N-1}(x_{N-1}) = \begin{cases} c(S_{N-1} - x_{N-1}) + L(S_{N-1}) & \text{if } x_{N-1} < S_{N-1} \\ L(x_{N-1}) & \text{if } x_{N-1} \geq S_{N-1}. \end{cases}$$

It is easy to see that $V_{N-1}(x)$ is a convex function since L is convex. Thus, we have shown that given the convexity of V_N, the value function V_{N-1} is also convex.

Stepping back one more period, we arrive at $N - 2$ for which the value function assumes the form

$$V_{N-2}(x_{N-2}) = \min_{y_{N-2} \geq x_{N-2}} G_{N-2}(y_{N-2}) - c x_{N-2}$$

where $G_{N-2}(y) = cy + L(y) + E[V_{N-1}(y - w)]$. We recall that since V_{N-1} was convex, so is $E[V_{N-1}(y - w)]$, implying that $G_{N-2}(y)$ is also convex.[12] Thus, there exists an S_{N-2} that minimizes $G_{N-2}(y)$ that results in the same policy, i.e.,

$$\mu_{N-2}(x_{N-2}) = \begin{cases} S_{N-2} - x_{N-2} & \text{if } x_{N-2} < S_{N-2} \\ 0 & \text{if } x_{N-2} \geq S_{N-2}. \end{cases}$$

Hence,

$$V_{N-2}(x_{N-2}) = \begin{cases} c(S_{N-2} - x_{N-2}) + L(S_{N-2}) + E[V_{N-1}(S_{N-2} - w_{N-2})], \\ \qquad\qquad\qquad\qquad\qquad\qquad \text{if } x_{N-2} < S_{N-2} \\ L(x_{N-2}) + E[V_{N-1}(x_{N-2} - w_{N-2})], \quad \text{if } x_{N-2} \geq S_{N-2} \end{cases}$$

Continuing with induction we prove that for all $k = N - 1, N - 2, \ldots, 1, 0$, the order-up-to (or base-stock) policy is optimal for the stochastic dynamic inventory problem with the properties stated above.

We should note that although the structure of the optimal policy is available from Theorem 14, we still do not know how to compute the optimal base-stock

---

[12]It is easy to see that if, say, $g(y)$ is convex, then $\phi(y) = E[g(y - w)]$ is also convex since $\phi''(y) = \int_0^\infty g''(y - w) \, dF(w) \geq 0$.

levels $S_0, S_1, \ldots, S_{N-1}$ for each period. The exact computation of these values is usually very difficult since they require the analytic solution of the dynamic programming functional equations for each period. We now describe an example where Maple is used to compute the exact values of the base-stock levels.

We assume that $N = 2$ and the demand in each period is uniformly distributed with density $f(w) = 1/15$. The relevant cost parameters are $c = 4$, $h = 7$ and $b = 10$.

We start by defining the input data.

```
>   restart: # BaseStock.mws
>   c:=4; h:=7; b:=10;
    Low:=0; Hi:=15; f:=w->1/(Hi-Low);
```

$$c := 4$$

$$h := 7$$

$$b := 10$$

$$Low := 0$$

$$Hi := 15$$

$$f := w \rightarrow \frac{1}{Hi - Low}$$

The value function for the terminal period $N = 2$ is $V_2(x_2) = 0$. The function $L(y)$ is defined in the usual manner for each period.

```
>   EV[ 2] :=unapply(0,x);
```

$$EV_2 := 0$$

```
>   L:=unapply(h* int((y-w)* f(x),w=0..y)
    +b* int((w-y)* f(x),w=y..10),y);
```

$$L := y \rightarrow -\frac{1}{10} y^2 + \frac{100}{3} - \frac{2}{3} y (10 - y)$$

For period $k = 1$, we compute symbolically the $G_1(y)$ function and find the optimal base-stock level as $S_1 = 40/17 \approx 2.35$. Using this information, the value function $V_1(x_1)$ for the current period is defined with the piecewise() function.

```
>   G[ 1] :=unapply(c* y+L(y)+EV[ 2] (y),y);
```

$$G_1 := y \rightarrow 4y - \frac{1}{10} y^2 + \frac{100}{3} - \frac{2}{3} y (10 - y)$$

```
>   S[ 1] :=solve(diff(G[ 1] (y),y),y); evalf(%);
```

$$S_1 := \frac{40}{17}$$

$$2.352941176$$

```
>   V[ 1] :=unapply(piecewise(x<S[ 1] ,c* (S[ 1] -x)+L(S[ 1] ),
    simplify(L(x))),x);
```

$$V_1 := x \to \text{piecewise}(x < \frac{40}{17}, \frac{1540}{51} - 4x, \frac{17}{30}x^2 + \frac{100}{3} - \frac{20}{3}x)$$

```
>  V[ 1] (x);
```

$$\begin{cases} \dfrac{1540}{51} - 4x & x < \dfrac{40}{17} \\[2mm] \dfrac{17}{30}x^2 + \dfrac{100}{3} - \dfrac{20}{3}x & \text{otherwise} \end{cases}$$

```
>  Vyw:=V[ 1] (y-w);
```

$$Vyw := \begin{cases} \dfrac{1540}{51} - 4y + 4w & y - w < \dfrac{40}{17} \\[2mm] \dfrac{17}{30}(y-w)^2 + \dfrac{100}{3} - \dfrac{20}{3}y + \dfrac{20}{3}w & \text{otherwise} \end{cases}$$

```
>  op(1,Vyw); op(2,Vyw); op(3,Vyw);
```

$$y - w < \frac{40}{17}$$

$$\frac{1540}{51} - 4y + 4w$$

$$\frac{17}{30}(y-w)^2 + \frac{100}{3} - \frac{20}{3}y + \frac{20}{3}w$$

The expectation of the value function $E[V_1(y - w)]$ is now evaluated as a sum of two integrals which results in a cubic equation.

```
>  EV[ 1] :=unapply(int(op(3,Vyw)* f(w),w=0..y-S[ 1] )
   +int(op(2,Vyw)* f(w),w=y-S[ 1] ..10),y);
```

$$EV_1 := y \to \frac{17}{1350}(y - \frac{40}{17})^3 + \frac{1}{2}(-\frac{17}{225}y + \frac{4}{9})(y - \frac{40}{17})^2 + \frac{17}{450}y^2(y - \frac{40}{17})$$
$$+ \frac{32}{153}y + \frac{85760}{2601} - \frac{4}{9}y(y - \frac{40}{17}) - \frac{4}{15}y(\frac{210}{17} - y) - \frac{2}{15}(y - \frac{40}{17})^2$$

```
>  simplify(EV[ 1] (y));
```

$$\frac{17}{1350}y^3 - \frac{4}{45}y^2 - \frac{376}{153}y + \frac{259840}{7803}$$

For period $k = 0$, the $G_0(y)$ function is computed explicitly. Minimizing this function gives the optimal base-stock level as $S_0 = 4.54$.

```
>  G[ 0] :=unapply(c* y+L(y) +EV[ 1] (y),y);
```

$$G_0 := y \to \frac{644}{153}y - \frac{1}{10}y^2 + \frac{172460}{2601} - \frac{2}{3}y(10 - y) + \frac{17}{1350}(y - \frac{40}{17})^3$$
$$+ \frac{1}{2}(-\frac{17}{225}y + \frac{4}{9})(y - \frac{40}{17})^2 + \frac{17}{450}y^2(y - \frac{40}{17}) - \frac{4}{9}y(y - \frac{40}{17})$$
$$- \frac{4}{15}y(\frac{210}{17} - y) - \frac{2}{15}(y - \frac{40}{17})^2$$

```
>    simplify(G[ 0] (y));
```

$$-\frac{784}{153}y + \frac{43}{90}y^2 + \frac{519940}{7803} + \frac{17}{1350}y^3$$

```
>    evalf(solve(diff(G[ 0] (y),y),y)); S[ 0] :=%[ 1] ;
```

$$4.54562088, \; -29.83973852$$

$$S_0 := 4.54562088$$

To summarize, the optimal base-stock levels are obtained as $S_0 = 4.54$ and $S_1 = 2.35$. The result that $S_0 > S_1$ should be intuitively clear: When there are two periods left we can be adventurous and afford to start with a larger initial inventory. But when there is only one period left, the optimal starting inventory must be somewhat smaller since the unsold items have no terminal value, i.e., $V_2(x_2) = 0$. As in this example, it can be shown that, in general, $S_0 \geq S_1 \geq \cdots \geq S_{N-1}$.

When the order cost function $c(u)$ has a fixed cost component K as in (8.1), the problem becomes much more difficult to solve. However, Scarf [166] has shown using the concept of K-convexity that in this case the optimal policy is of the (s, S)-type, i.e.,

$$\mu_k(x_k) = \begin{cases} S_k - x_k & \text{if } x_k < s_k \\ 0 & \text{if } x_k \geq s_k. \end{cases}$$

For an up-to-date discussion of K-convexity and an enlightening proof of the optimality of the (s, S) policy, see Bertsekas [30, Section 2.2].

Although the optimal policy assumes this simple form, actual computation of the parameters (s_0, S_0), (s_1, S_1), ..., (s_{N-1}, S_{N-1}) for each period is a nontrivial task. However, assuming stationarity of the cost and demand data, one can reduce the problem to the computation of a single pair of numbers (s, S), which simplifies the problem considerably. For a discussion of these issues, we refer the reader to Porteus [150] and references therein.

8.5.5   Diversification Under Yield Randomness

In many problems what is received from the supplier may not be exactly equal to what is ordered by the inventory manager. This discrepancy may be due to shipments that perish during transportation or to clerical errors. Such randomness in the "yield" (the actual amount received) also arises in other contexts including electronic fabrication and assembly, agriculture and chemical processes.

The importance of incorporating the yield randomness into production and inventory models is now well established and in recent years several papers dealing with this issue have been published. Authors such as Silver [173], Shih [172] and Lee and Yano [119] have assumed that when a batch is produced by a single supplier, a random fraction may be defective. Gerchak and Parlar [69] have generalized these models by assuming that yield randomness can be optimally selected jointly with lot sizing decisions. Recently Yano and Lee [197] have published a survey in which they review these and other random yield models.

In this section we consider the issue of source (supplier) diversification under yield randomness and examine the conditions for choosing two suppliers with different yield distributions over a single source. Our exposition parallels the discussion of the second model presented in Gerchak and Parlar [69].

As in the basic EOQ model, we let h be the holding cost per unit per time and λ be the known and constant demand per time. When Q units are ordered, a random number of units are received, denoted by Y_Q. It is assumed that $E(Y_Q) = \mu Q$ and $\text{Var}(Y_Q) = \sigma^2 Q^2$ for $Q \geq 0$. In production-related applications, Y_Q usually represents the number of usable units in a lot of Q units. However, for the sake of generality, we assume that μ does not have to be less than unity and we let $\mu \geq 0$.

First let us suppose that two independent sources are available with K as the cost of setup (ordering) when both sources are used simultaneously. Hence, if $Q_i > 0$ units are ordered and Y_{Q_i} are received from source $i = 1, 2$, we have

$$\text{Cost per cycle} = K + \frac{(Y_{Q_1} + Y_{Q_2})^2}{2\lambda}$$

and

$$\text{Length of cycle} = \frac{Y_{Q_1} + Y_{Q_2}}{\lambda}.$$

If the moments are specified as μ_i, σ_i^2, $i = 1, 2$, then using the renewal reward theorem, we obtain the average cost $G(Q_1, Q_2) = E(\text{Cost per Cycle})/E(\text{Length of Cycle})$ as

$$G(Q_1, Q_2) = \frac{2K\lambda + h[Q_1^2(\sigma_1^2 + \mu_1^2) + 2Q_1 Q_2 \mu_1 \mu_2 + Q_2^2(\sigma_2^2 + \mu_2^2)]}{2(\mu_1 Q_1 + \mu_2 Q_2)}.$$

We demonstrate the minimization of $G(Q_1, Q_2)$ using Maple.

```
>   restart: # Diversify.mws
```

The objective function G and its partial derivatives G_{Q_1}, G_{Q_2} are as follows.

```
>   G:=(2*K*lambda+h*(Q[1]^2*(sigma[1]^2
    +mu[1]^2)+2*Q[1]*Q[2]*mu[1]*mu[2]
    +Q[2]^2*(sigma[2]^2+mu[2]^2)))
    /(2*(mu[1]*Q[1]+mu[2]*Q[2]));
```

$$G := (2\,K\,\lambda +$$
$$h\,({Q_1}^2\,({\sigma_1}^2 + {\mu_1}^2) + 2\,Q_1\,Q_2\,\mu_1\,\mu_2 + {Q_2}^2\,({\sigma_2}^2 + {\mu_2}^2)))$$
$$\Big/(2\,\mu_1\,Q_1 + 2\,\mu_2\,Q_2)$$

```
>   GQ1:=normal(diff(G,Q[1]));
```

$$GQ1 := \frac{1}{2}(h\,Q_1{}^2\,\sigma_1{}^2\,\mu_1 + 2h\,Q_1\,\sigma_1{}^2\,\mu_2\,Q_2 + h\,Q_1{}^2\,\mu_1{}^3$$
$$+ 2h\,Q_1\,\mu_1{}^2\,\mu_2\,Q_2 + h\,Q_2{}^2\,\mu_1\,\mu_2{}^2 - 2\,\mu_1\,K\,\lambda$$
$$- \mu_1\,h\,Q_2{}^2\,\sigma_2{}^2)\Big/(\mu_1\,Q_1 + \mu_2\,Q_2)^2$$

```
>  GQ2:=normal(diff(G,Q[2]));
```

$$GQ2 := -\frac{1}{2}(-h\,Q_1{}^2\,\mu_1{}^2\,\mu_2 - 2h\,Q_1\,\mu_1\,\mu_2{}^2\,Q_2$$
$$- 2h\,Q_2\,\sigma_2{}^2\,\mu_1\,Q_1 - h\,Q_2{}^2\,\sigma_2{}^2\,\mu_2 - h\,Q_2{}^2\,\mu_2{}^3$$
$$+ 2\,\mu_2\,K\,\lambda + \mu_2\,h\,Q_1{}^2\,\sigma_1{}^2)\Big/(\mu_1\,Q_1 + \mu_2\,Q_2)^2$$

Solving the necessary conditions results in a complicated-looking expression which we simplify using the allvalues() function.

```
>  Sol:=solve({GQ1,GQ2},{Q[1],Q[2]});
```

$$Sol := \{Q_1 = 2K\,\lambda\,\mu_1\,\sigma_2{}^2\Big/((\mu_2{}^4\,\sigma_1{}^4 + \sigma_2{}^2\,\sigma_1{}^4\,\mu_2{}^2$$
$$+ 2\,\mu_1{}^2\,\sigma_2{}^2\,\mu_2{}^2\,\sigma_1{}^2 + \mu_1{}^2\,\sigma_2{}^4\,\sigma_1{}^2 + \sigma_2{}^4\,\mu_1{}^4)\text{RootOf}((}$$
$$h\,\mu_2{}^4\,\sigma_1{}^4 + h\,\sigma_2{}^2\,\sigma_1{}^4\,\mu_2{}^2 + 2h\,\mu_1{}^2\,\sigma_2{}^2\,\mu_2{}^2\,\sigma_1{}^2$$
$$+ h\,\mu_1{}^2\,\sigma_2{}^4\,\sigma_1{}^2 + h\,\sigma_2{}^4\,\mu_1{}^4)\_Z^2 - 2\,K\,\lambda)h), Q_2 =$$
$$\text{RootOf}((h\,\mu_2{}^4\,\sigma_1{}^4 + h\,\sigma_2{}^2\,\sigma_1{}^4\,\mu_2{}^2$$
$$+ 2h\,\mu_1{}^2\,\sigma_2{}^2\,\mu_2{}^2\,\sigma_1{}^2 + h\,\mu_1{}^2\,\sigma_2{}^4\,\sigma_1{}^2 + h\,\sigma_2{}^4\,\mu_1{}^4)\_Z^2$$
$$- 2\,K\,\lambda)\sigma_1{}^2\,\mu_2\}$$

```
>  Sol:=allvalues(Sol);
```

$$Sol := \{Q_2 = \frac{\sqrt{2}\,\sqrt{h\,\%2\,K\,\lambda}\,\sigma_1{}^2\,\mu_2}{\%1}, \; Q_1 = \frac{K\,\lambda\,\mu_1\,\sigma_2{}^2\,\%1\,\sqrt{2}}{\%2\,\sqrt{h\,\%2\,K\,\lambda}\,h}$$
$$\}, \{Q_1 = -\frac{K\,\lambda\,\mu_1\,\sigma_2{}^2\,\%1\,\sqrt{2}}{\%2\,\sqrt{h\,\%2\,K\,\lambda}\,h},$$
$$Q_2 = -\frac{\sqrt{2}\,\sqrt{h\,\%2\,K\,\lambda}\,\sigma_1{}^2\,\mu_2}{\%1}\}$$
$$\%1 := h\,\mu_2{}^4\,\sigma_1{}^4 + h\,\sigma_2{}^2\,\sigma_1{}^4\,\mu_2{}^2 + 2h\,\mu_1{}^2\,\sigma_2{}^2\,\mu_2{}^2\,\sigma_1{}^2$$
$$+ h\,\mu_1{}^2\,\sigma_2{}^4\,\sigma_1{}^2 + h\,\sigma_2{}^4\,\mu_1{}^4$$
$$\%2 := \mu_2{}^4\,\sigma_1{}^4 + \sigma_2{}^2\,\sigma_1{}^4\,\mu_2{}^2 + 2\,\mu_1{}^2\,\sigma_2{}^2\,\mu_2{}^2\,\sigma_1{}^2 + \mu_1{}^2\,\sigma_2{}^4\,\sigma_1{}^2$$
$$+ \sigma_2{}^4\,\mu_1{}^4$$

The results still look very unappealing, but we note that there is a relationship between the complicated terms defined by %1 and %2. Maple informs us that these terms are a constant multiple of each other.[13]

```
>    simplify(%1/%2);
```

$$h$$

The solution for Q_1 and Q_2 are assigned in order to manipulate them.

```
>    Sol[ 1] ;
```

$$\{Q_2 = \sqrt{2}\sqrt{h\,\%1\,K\,\lambda}\,\sigma_1{}^2\,\mu_2 \Big/ (h\,\mu_2{}^4\,\sigma_1{}^4 + h\,\sigma_2{}^2\,\sigma_1{}^4\,\mu_2{}^2$$
$$+ 2\,h\,\mu_1{}^2\,\sigma_2{}^2\,\mu_2{}^2\,\sigma_1{}^2 + h\,\mu_1{}^2\,\sigma_2{}^4\,\sigma_1{}^2 + h\,\sigma_2{}^4\,\mu_1{}^4), Q_1$$
$$= K\,\lambda\,\mu_1\,\sigma_2{}^2 (h\,\mu_2{}^4\,\sigma_1{}^4 + h\,\sigma_2{}^2\,\sigma_1{}^4\,\mu_2{}^2$$
$$+ 2\,h\,\mu_1{}^2\,\sigma_2{}^2\,\mu_2{}^2\,\sigma_1{}^2 + h\,\mu_1{}^2\,\sigma_2{}^4\,\sigma_1{}^2 + h\,\sigma_2{}^4\,\mu_1{}^4)\sqrt{2}$$
$$\Big/ (\%1\,\sqrt{h\,\%1\,K\,\lambda}\,h)\}$$
$$\%1 := \mu_2{}^4\,\sigma_1{}^4 + \sigma_2{}^2\,\sigma_1{}^4\,\mu_2{}^2 + 2\,\mu_1{}^2\,\sigma_2{}^2\,\mu_2{}^2\,\sigma_1{}^2 + \mu_1{}^2\,\sigma_2{}^4\,\sigma_1{}^2$$
$$+ \sigma_2{}^4\,\mu_1{}^4$$

```
>    assign(Sol[ 1] );
>    Q1Sol:=simplify(Q[ 1] );
```

$$Q1Sol := \sqrt{2}\,\sigma_2{}^2\,\mu_1\,K\,\lambda \Big/ \mathrm{sqrt}(h(\mu_2{}^4\,\sigma_1{}^4 + \sigma_2{}^2\,\sigma_1{}^4\,\mu_2{}^2$$
$$+ 2\,\mu_1{}^2\,\sigma_2{}^2\,\mu_2{}^2\,\sigma_1{}^2 + \mu_1{}^2\,\sigma_2{}^4\,\sigma_1{}^2 + \sigma_2{}^4\,\mu_1{}^4)K\,\lambda)$$

```
>    Q2Sol:=simplify(Q[ 2] );
```

$$Q2Sol := \sqrt{2}\mathrm{sqrt}(h(\mu_2{}^4\,\sigma_1{}^4 + \sigma_2{}^2\,\sigma_1{}^4\,\mu_2{}^2 + 2\,\mu_1{}^2\,\sigma_2{}^2\,\mu_2{}^2\,\sigma_1{}^2$$
$$+ \mu_1{}^2\,\sigma_2{}^4\,\sigma_1{}^2 + \sigma_2{}^4\,\mu_1{}^4)K\,\lambda)\sigma_1{}^2\,\mu_2 \Big/ (h(\mu_2{}^4\,\sigma_1{}^4$$
$$+ \sigma_2{}^2\,\sigma_1{}^4\,\mu_2{}^2 + 2\,\mu_1{}^2\,\sigma_2{}^2\,\mu_2{}^2\,\sigma_1{}^2 + \mu_1{}^2\,\sigma_2{}^4\,\sigma_1{}^2$$
$$+ \sigma_2{}^4\,\mu_1{}^4))$$

We find after some manipulations that

$$Q_1 = \sqrt{\frac{2\lambda K\mu_1^2\sigma_2^4}{h[(\mu_1^2\sigma_2^2 + \mu_2^2\sigma_1^2)^2 + \mu_2^2\sigma_1^4\sigma_2^2 + \mu_1^2\sigma_2^4\sigma_1^2]}}$$

However Q_2 appears a little different from Q_1, so we attempt to discover a simple relationship between these two variables and see that Q_2/Q_1 is a constant given by $\sigma_1^2\mu_2/(\sigma_2^2\mu_1)$.

```
>    Ratio:=simplify(Q2Sol/Q1Sol);
```

---

[13]Note that in order to see the output in terms of %1 and %2, one should set \Option \Output Display \Typeset Notation. Otherwise Maple gives an error message.

$$Ratio := \frac{\sigma_1{}^2 \, \mu_2}{\sigma_2{}^2 \, \mu_1}$$

This discovery helps in rewriting the expression for Q_2 in a form that is a mirror image of Q_1.

```
> Q2Sol:=simplify(Ratio*Q1Sol);
```

$$Q2Sol := \sigma_1{}^2 \, \mu_2 \, \sqrt{2} \, K \, \lambda \Big/ \text{sqrt}(h(\mu_2{}^4 \, \sigma_1{}^4 + \sigma_2{}^2 \, \sigma_1{}^4 \, \mu_2{}^2$$
$$+ 2\mu_1{}^2 \, \sigma_2{}^2 \, \mu_2{}^2 \, \sigma_1{}^2 + \mu_1{}^2 \, \sigma_2{}^4 \, \sigma_1{}^2 + \sigma_2{}^4 \, \mu_1{}^4)K\,\lambda)$$

i.e.,

$$Q_2 = \sqrt{\frac{2\lambda K \mu_2^2 \sigma_1^4}{h[(\mu_2^2\sigma_1^2 + \mu_1^2\sigma_2^2)^2 + \mu_1^2\sigma_2^4\sigma_1^2 + \mu_2^2\sigma_1^4\sigma_2^2]}}$$

The optimal value of the objective function is also easily calculated using the `simplify()` function.

```
> GOpt:=simplify(G,power,symbolic);
```

$$GOpt := \sqrt{K} \, \sqrt{\lambda} \, \sqrt{2} \, \sqrt{h}\text{sqrt}(\mu_2{}^4 \, \sigma_1{}^4 + \sigma_2{}^2 \, \sigma_1{}^4 \, \mu_2{}^2$$
$$+ 2\mu_1{}^2 \, \sigma_2{}^2 \, \mu_2{}^2 \, \sigma_1{}^2 + \mu_1{}^2 \, \sigma_2{}^4 \, \sigma_1{}^2 + \sigma_2{}^4 \, \mu_1{}^4) \Big/ ($$
$$\mu_2{}^2 \, \sigma_1{}^2 + \mu_1{}^2 \, \sigma_2{}^2)$$

To further simplify this expression, we dissect it using the `nops()` function. Each of the six operands are listed and we observe, after some manipulations of the fifth and sixth operands, that the optimal cost can be written as

$$G^* = \sqrt{2\lambda K h \left(1 + \frac{\sigma_1^2 \sigma_2^2}{\mu_1^2 \sigma_2^2 + \mu_2^2 \sigma_1^2} \right)}. \qquad (8.7)$$

```
> nops(GOpt);
```

$$6$$

```
> seq(op(i,GOpt),i=1..nops(GOpt));
```

$$\sqrt{K}, \ \sqrt{\lambda}, \ \sqrt{2}, \ \sqrt{h}, \text{sqrt}(\mu_2{}^4 \, \sigma_1{}^4 + \sigma_2{}^2 \, \sigma_1{}^4 \, \mu_2{}^2$$
$$+ 2\mu_1{}^2 \, \sigma_2{}^2 \, \mu_2{}^2 \, \sigma_1{}^2 + \mu_1{}^2 \, \sigma_2{}^4 \, \sigma_1{}^2 + \sigma_2{}^4 \, \mu_1{}^4),$$
$$\frac{1}{\mu_2{}^2 \, \sigma_1{}^2 + \mu_1{}^2 \, \sigma_2{}^2}$$

```
> InsideSqrt:=simplify(op(5,GOpt)^2* op(6,GOpt)^2);
```

$$InsideSqrt := \frac{\sigma_1{}^2 \, \sigma_2{}^2 + \mu_2{}^2 \, \sigma_1{}^2 + \mu_1{}^2 \, \sigma_2{}^2}{\mu_2{}^2 \, \sigma_1{}^2 + \mu_1{}^2 \, \sigma_2{}^2}$$

Gerchak and Parlar [69] show that if a single source were used — where the setup cost would be lower — the corresponding optimal costs for each source $i = 1, 2$ would be

$$G_i^* = \sqrt{2\lambda K_i h(1 + \sigma_i^2/\mu_i^2)}. \tag{8.8}$$

Thus, depending on the relative magnitudes of G^*, G_1^* and G_2^*, it would be optimal to use either both sources (if G^* is the smallest), or any of the single sources (if G_1^* or G_2^* is the smallest).

8.6 Summary

Inventory management is one of the oldest and most thoroughly studied areas of operations research. In this relatively short chapter we covered deterministic and probabilistic inventory models. These models ranged from the simplest deterministic demand economic order-quantity model to dynamic inventory models with random demand. Maple's ability to manipulate expressions symbolically simplified the analysis of many nontrivial inventory problems such as the optimal base-stock levels that are obtained by solving a dynamic programming problem. Maple's numerical analysis facilities also helped in computing the optimal decisions for the exact formulation of the continuous-review model with backorders. Maple's graphics were also useful in visualizing some of the objective functions including the discontinuous cost function of the quantity discount problem.

8.7 Exercises

1. Consider the basic EOQ model presented in Section 8.4. Now assume that the demand is an integer quantity; i.e., the units are demanded one at a time and the time between demands τ is deterministic implying that the demand rate is $\lambda = 1/\tau$. Show that the optimal integer order quantity is the largest positive integer Q such that

$$Q(Q-1) < \frac{2\lambda K}{h}.$$

2. In the EOQ models we have discussed, it was implicitly assumed that the supplier was always available, i.e., we would always receive our order (after a possibly positive lead-time). In the classical EOQ model now assume that the supplier is not always available and he can be ON (available) or OFF (unavailable) for random durations.

   (a) Without loss of generality, let the demand rate λ be equal to 1 unit/time. Let us assume that the exponential random variables X and Y correspond to the lengths of the ON and OFF periods, with parameters α

and β, respectively. When an order is placed and the state is ON, an order cost of $\$K$/order is incurred. The holding cost is $\$h$/unit/time and the shortage (backorder) cost is $\$\pi$/unit when the state is OFF and the demand cannot be met. For the sake of generality, also assume that the time dependent part of the backorder cost is $\$\hat{\pi}$/unit/time.

When the inventory level reaches the reorder point r and if the supplier is ON an order for Q units is placed. This brings the inventory up to $Q+r$. If the supplier is OFF when the inventory reaches r, then the decision maker has to wait until the supplier becomes available (ON) before an order can be placed which increases the inventory to $Q+r$. Note that, with Q as the order quantity (which is received immediately if the state is ON), inventory will reach the reorder point r after exactly Q time units since demand rate $\lambda = 1$ as we had assumed.

Develop the objective function for this model. Find the optimal solution for the problem with data given as $K = 10$, $h = 5$, $\pi = 250$, $\hat{\pi} = 25$, $\alpha = 0.25$, $\beta = 2.50$. Compare the results you would obtain for the classical EOQ model where the supplier is always available.

HINT: To compute the expected cycle length and expected cycle cost, condition on the state of the system when inventory reaches the level r.

(b) For the case of two suppliers suggest ways of dealing with the problem.

3. This question considers a generalized version of the single period newsvendor problem. Assume that there are two products which may be substitutable if one of them is out of stock during the period. For simplicity, assume that the beginning inventory for both products is zero, and there is no salvage value for unsold units nor a penalty cost for unsatisfied demand. Use the following notation in your answer:

- u : order quantity of item 1 (decision variable),
- v : order quantity of item 2 (decision variable),
- X : random demand for item 1 with density $f(x)$,
- Y : random demand for item 2 with density $g(y)$,
- s_i : sale price/unit of item i ($i = 1, 2$),
- c_i : purchase cost/unit of item i ($i = 1, 2$),
- a : fraction of customers who will accept a unit of item 2 when item 1 is out of stock
- b : fraction of customers who will accept a unit of item 1 when item 2 is out of stock
- Π : profit (random).

Write down the expressions for expected profit per period, i.e., $E(\Pi)$.

HINT: For given values of u and v, and depending on the values taken by the random variables $X = x$, $Y = y$, one has to consider *four* different possibilities.

4. Consider a stochastic inventory system where unit demand for items arrive in a renewal process at time points T_0, T_1, ... Let $T_n - T_{n-1}$ ($n = 1, 2, ...$) be i.i.d. random variables with mean $1/\lambda$, i.e., the demand rate is λ. Assume that we use an (s, S) policy, that is, when the inventory level falls to s, we order $S - s$ units which are received immediately. Order cost is $\$K$ per order and the purchase cost is $\$c$ per unit. The inventory holding cost is $\$h$ per unit per time.

   Find the best values of s and Q so that the average cost is minimized.

5. Consider the last period of the dynamic inventory problem discussed in Section 8.5.4. Now assume that the cost of purchasing u units is

$$c(u) = \begin{cases} K + cu, & \text{if } u > 0 \\ 0, & \text{if } u = 0. \end{cases}$$

   Show that the optimal policy is now of the (s, S)-type, i.e., it is given in the form

$$\mu(x) = \begin{cases} S - x, & \text{if } x < s \\ 0, & \text{if } x \geq s. \end{cases}$$

6. Consider the optimization problem

$$\min_{u_k \geq 0} E \sum_{k=0}^{N-1} c u_k + h \max(0, x_k + u_k - w_k) + p \max(0, w_k - x_k - u_k)$$

   where x_k is the inventory at the start of period k, u_k is the order quantity at the start of period k (lead-time is assumed zero) and w_k is the random demand with given c.d.f. $F(w_k)$ for $k = 0, 1, \ldots, N - 1$. The parameters c, h and p are the unit costs of purchase, holding and shortage penalty, respectively.

   (a) Find the optimal solution for this problem with $f(w) = F'(w) = 1$, $0 \leq w \leq 1, c = 1, h = 1, p = 2, N = 1$.

   (b) Find the optimal solution for the problem

$$\min_{u_k \geq 0} \lim_{N \to \infty} E \sum_{k=0}^{N-1} \alpha^k [c u_k + h \max(0, x_k + u_k - w_k)$$
$$+ p \max(0, w_k - x_k - u_k)]$$

   with $f(w) = F'(w) = 1, 0 \leq w \leq 1; c = 1, h = 1, p = 2$ and $\alpha = 0.9$.

9

Queueing Systems

9.1 Introduction

The English poet Milton once wrote, "They also serve who only stand and wait," [178, Sonnet 19] but he never elaborated on the average number of people who would wait and the average time spent waiting for service. Of course, as a poet he had no interest in such technicalities and even if he did, the mathematical tools available to him and his contemporaries in the 17th century would not have been of much help in analyzing the queueing processes arising in his day.[1]

Modern queueing theory had its origins in the early 1900s when the Danish engineer A. K. Erlang published his seminal work on congestion in telephone traffic. Since Erlang's time, the number of published (and unpublished) papers and books on queueing theory has shown a tremendous growth. In 1987, Disney and Kiessler [62] estimated at least 5000; given the high growth rate we would estimate the present number around 10,000.

When does a customer requiring service have to "stand and wait"? Queues form when current *demand* for a service exceeds the current *capacity* to provide the service. As an example let us consider a simple (and naïve) deterministic queueing system, e.g., an assembly line operated by a robot. If we assume that

---

[1]Milton, of course, was *not* referring to a physical queueing system when he wrote his Sonnet 19 that mentions "[those] ... who ... wait." Instead, he was comparing angels who serve God and devout people on earth who *wait* with patience for the fulfillment of God's purposes. We have included the reference to Milton since this famous quotation is a favorite of many queueing theorists; see Cox and Smith [54, p. vii].

the time between arrivals of items requiring processing is, say, $a = 3$ minutes and the service (i.e., assembly) time is $s = 2$ minutes, then there will never be any queue in this system and for $s \leq a$, the robot will be idle for $(a - s)/a = 1 - s/a$ fraction of the time. However, if $s > a$, then due to demand exceeding the capacity, arriving items will pile up and the queue size will grow to infinity.[2]

If the arrival and service processes are stochastic, then even when the *average* time $\bar{s}$ to serve a customer is shorter than the *average* time $\bar{a}$ between the arrivals of customers, due to the *variability* in the system, a queue may form. Such congestion phenomena are commonly observed in daily life where customers have to wait for service at a bank, at a restaurant or at the airport checkout counters or jobs arriving at a job shop have to wait for processing or telephone calls arriving at a switchboard may have to wait until the busy signal is cleared.

A shorthand notation has been developed by Kendall [106] to describe a queueing process. The notation involves a series of symbols and slashes such as $A/B/X/Y/Z$. Here, A indicates the interarrival distribution [e.g., M for Memoryless (or exponential or Markovian), D for deterministic]; B indicates the service-time distribution (e.g., G for general, E_k for Erlang with k stages); X indicates the number of (parallel) servers which could be any integer between 1 and ∞; Y indicates the system capacity and Z indicates the queue discipline [e.g., first come, first served ($FCFS$), priority (PR)]. For example, $M/M/1/\infty/FCFS$ is the simplest queueing system with exponential interarrival times, exponential service times, with a single server, no restrictions on the system capacity and a $FCFS$ service discipline. As another example, $M/E_2/3/15/PR$ may represent the emergency department of a hospital with three doctors and a capacity of 15 where the patients' interarrival times is exponential and they have to go through two stages (Erlang with $k = 2$) before service is completed. Normally, if $Y = \infty$ and $Z = FCFS$, these symbols are dropped from the notation and only the first three are used. Thus, $M/M/1$ would be used to represent the single-server Markovian system $M/M/1/\infty/FCFS$.

Queueing processes are studied in order to gain an understanding of their operating characteristics such as the average number of customers in the system or the average time a customer must spend in the system. Once these descriptive analyses are complete, a normative analysis of the system may also be conducted in order to analyze such questions as "How many servers should be available?" "How fast should the servers be?" or "How should the system be designed?" Queueing theory as it has been developed in the last several decades attempts to answer such questions using detailed mathematical analyses. In this chapter we will provide a description of some types of queueing models and demonstrate the usefulness of Maple in solving several queueing problems.

---

[2] The reader who is familiar with the *I Love Lucy* TV shows of the 1950s will recognize that the 'Chocolate Factory' episode corresponds to the case where $s > a$.

9.2   Markovian Queueing Systems

In this section we present queueing models where the arrival and service processes are both exponential and where only the transitions to neighboring states are permissible; i.e., they are based on birth and death processes.

9.2.1   Birth and Death Processes

Before we present the simple $M/M/1$ queueing system, it would be useful to develop expressions representing the evolution of the time-dependent probabilities of the birth and death processes.

We assume that when there are n customers in the system, the arrival rate of the customers is λ_n and the service rate is μ_n; i.e., the arrival and service rates are, in general, state dependent. Thus, when the system is in state n, the probability of a birth (arrival) in a short time interval of length Δt is $\lambda_n \Delta t + o(\Delta t)$, $n \geq 0$, and the probability of a death (departure) is $\mu_n \Delta t + o(\Delta t)$, $n \geq 1$. The queue forms if the server is busy and a new customer arrives and must wait for service. When the service of a customer is completed, he leaves and the customer who has been in the system longest enters the service facility.

Due to the memorylessness of the interarrival and service times, the stochastic process $\{N(t); t \geq 0\}$ where $N(t)$ is the number of customers in the system at time t is a continuous-time Markov chain.

If we define $p_n(t) = \Pr\{N(t) = n\}$ as the probability that there will be n customers in the system at time t, an important problem is that of computing the transient solution of the probability $p_n(t)$ for finite t. In principle, using the birth-death processes that arise from the memorylessness of the arrival and service processes, one can develop an infinite system of first-order linear differential equations (Kolmogorov equations) whose solution would provide us with the transient probabilities $p_n(t)$.

Since the state n at time $t + \Delta t$ can be reached either from state n (if there is neither a birth nor a death) or from state $n - 1$ (if there is a birth and no deaths) or from $n + 1$ (if there is a death and no births), we can write[3] for $n \geq 1$

$$p_n(t + \Delta t) = p_n(t)(1 - \lambda_n \Delta t)(1 - \mu_n \Delta t) + p_{n+1}(t)\mu_{n+1}\Delta t(1 - \lambda_{n+1}\Delta t)$$
$$+ p_{n-1}(t)\lambda_{n-1}\Delta t(1 - \mu_{n-1}\Delta t) + o(\Delta t). \qquad (9.1)$$

Similarly, for $n = 0$, we get

$$p_0(t + \Delta t) = p_0(t)(1 - \lambda_0 \Delta t) + p_1(t)(\mu_1 \Delta t)(1 - \lambda_1 \Delta t) + o(\Delta t). \quad (9.2)$$

Following the standard steps of simplification, carrying the term involving $p_n(t)$ to the left of the equality, dividing both sides by Δt and taking the limit as $\Delta t \to 0$

---

[3]We could also have included the probability of 1 birth and 1 death in the time interval Δt as $p_n(t)(\lambda_n \Delta t)(\mu_n \Delta t)$. But the result is of the order $o(\Delta t)$, and thus it can be safely ignored in the subsequent development.

gives

$$p_n'(t) = -(\lambda_n + \mu_n)p_n(t) + \mu_{n+1}p_{n+1}(t) + \lambda_{n-1}p_{n-1}(t), \quad n \geq 1 \quad (9.3)$$
$$p_0'(t) = -\lambda_0 p_0(t) + \mu_1 p_1(t), \quad n = 0. \quad (9.4)$$

The explicit analytical solution of this infinite system of differential equations (9.3)–(9.4) for finite values of t is impossible to obtain. However, if the limiting solution $p_n = \lim_{t \to \infty} \Pr\{N(t) = n\}$ is assumed to exist, then $\lim_{t \to \infty} p_n'(t) = 0$ and the system can be written as an infinite system of difference equations

$$0 = -(\lambda_n + \mu_n)p_n + \mu_{n+1}p_{n+1} + \lambda_{n-1}p_{n-1}, \quad n \geq 1 \quad (9.5)$$
$$0 = -\lambda_0 p_0 + \mu_1 p_1, \quad n = 0 \quad (9.6)$$

with the condition $\sum_{n=0}^{\infty} p_n = 1$. We will postpone the general solution of this system to Section 9.2.4.

9.2.2 M/M/1 Queueing System

We now use Maple to solve the system of difference (recurrence) equations (9.5)–(9.6) for the *special case* of $\lambda_n = \lambda$ and $\mu_n = \mu$, which reduces the birth and death processes to the $M/M/1$ queue. Note that before using the rsolve() command, we inform Maple of the initial condition that for $n = 1$, $p_1 = (\lambda/\mu)p_0$ (since at $n = 0$, we have $0 = -\lambda p_0 + \mu p_1$).

We first input the difference equations (9.5)–(9.6) using Maple's syntax:

```
>    restart:  # MM1.MWS
>    mm1:={ 0=-(lambda+mu)*p(n)
      +lambda*p(n-1)+mu*p(n+1),p(n=0..1)
      =(lambda/mu)^n*p[ 0] } ;
```

$$mm1 := \{0 = -(\lambda + \mu)\,p(n) + \lambda\,p(n-1) + \mu\,p(n+1),\ p(n = 0..1) = (\frac{\lambda}{\mu})^n\,p_0$$
$$\}$$

Solving the system for the unknown p_n gives

```
>    sol:=rsolve(mm1,p(n));
```

$$sol := (\frac{\lambda}{\mu})^n\,p_0$$

```
>    p:=unapply(sol,n);
```

$$p := n \to (\frac{\lambda}{\mu})^n\,p_0$$

Thus, p_n is obtained as a function of n and the model parameters λ and μ and the (as yet) unknown p_0. Let us now check the solution for the first few values of n:

```
>    p(0); p(1); p(2);
```

$$p_0$$

$$\frac{\lambda\, p_0}{\mu}$$

$$\frac{\lambda^2\, p_0}{\mu^2}$$

Before finding the value of the unknown p_0 in terms of the model parameters, we develop formulas for some of the operating characteristics of the $M/M/1$ queue, e.g., the expected number of customers in the system, $L = \sum_{n=0}^{\infty} n p_n$, and the expected queue size $L_q = \sum_{n=1}^{\infty}(n-1)p_n$. Since the initial state probability p_0 is not yet computed, the results for L and L_q are obtained in terms of this unknown value.

```
>   L:=sum(n*p(n),n=0..infinity);
    Lq:=sum((n-1)*p(n),n=1..infinity);
```

$$L := \frac{p_0\, \mu\, \lambda}{(-\lambda + \mu)^2}$$

$$Lq := \frac{p_0\, \lambda^2}{(-\lambda + \mu)^2}$$

Now, using the fact that $\sum_{n=0}^{\infty} p_n = 1$, we can finally compute the unknown p_0 as

```
>   p[0]:=solve(1=sum(p(n),n=0..infinity),p[0]);
```

$$p_0 := \frac{-\lambda + \mu}{\mu}$$

The expressions for L and L_q can now be obtained in terms of the parameters λ and μ.

```
>   L; Lq;
```

$$\frac{\lambda}{-\lambda + \mu}$$

$$\frac{\lambda^2}{(-\lambda + \mu)\,\mu}$$

Since the quantity λ/μ appears frequently in many queueing formulas, we introduce a new symbol ρ, which is defined as λ/μ. This dimensionless quantity is known as the "traffic intensity" or the "utilization factor" of the service facility.[4] Note that for the $M/M/1$ queue, the expected system size L and the expected queue size L_q are related through the utilization factor ρ, i.e., $L = \rho + L_q$.

---

[4]In multiple-server queueing systems the utilization factor is defined as $\rho = \lambda/(c\mu)$ where c is the number of servers present in a service facility and μ is the service rate of *each* server.

Having developed the expressions for the expected *number of customers* present in the system and in the queue, we now present the formulas for the distributions and the means of the *waiting time* random variables of the $M/M/1$ model.

Let T_q be the random time a customer spends waiting in the *queue* and $W_q(t)$ its distribution function. To develop $W_q(t)$, we argue as follows: It is possible that a customer may *not* have to wait in the queue at all if the system is empty at the time of his arrival. Thus, for $t = 0$,

$$
\begin{aligned}
W_q(0) &= \Pr(T_q = 0) \\
&= \Pr(\text{system empty at time of arrival}) \\
&= p_0 = 1 - \rho.
\end{aligned}
$$

For $t > 0$, we condition on the number of customers found in the system at the time of the arrival and write

$$
\begin{aligned}
W_q(t) &= \Pr(T_q \leq t) \\
&= \sum_{n=1}^{\infty} \Pr(n \text{ service completions in } \leq t \mid \text{arrival finds } n \\
&\quad \text{customers in system}) \times p_n + W_q(0)
\end{aligned}
$$

Since each service time is exponential, the distribution of the completion of n services is Erlang with n stages. Hence, as $p_n = (1 - \rho)\rho^n$, the distribution $W_q(t)$ can be written as follows:

$$
\begin{aligned}
W_q(t) &= \sum_{n=1}^{\infty} \left[\int_0^t \frac{\mu(\mu x)^{n-1} e^{-\mu x}}{(n-1)!} \, dx \right] \times (1 - \rho)\rho^n + (1 - \rho) \\
&= (1 - \rho)\rho \int_0^t \mu e^{-\mu x} \sum_{n=1}^{\infty} \frac{(\mu x \rho)^{n-1}}{(n-1)!} \, dx + (1 - \rho).
\end{aligned}
$$

Now, using Maple's help we compute the foregoing expression.

```
>    restart: # wqt.mws
>    cdfTq:=(1-rho)*rho* Int (mu* exp (-mu* x)
     * Sum( (mu* x* rho) ^ (n-1) / (n-1) !, n=1..infinity),
     x=0..t) + (1-rho);
```

$$
cdfTq := (1 - \rho) \rho \int_0^t \mu \, e^{(-\mu x)} \left(\sum_{n=1}^{\infty} \frac{(\mu x \rho)^{(n-1)}}{(n-1)!} \right) dx + 1 - \rho
$$

```
>    Wqt:=normal (value (cdfTq) );
```

$$
Wqt := -\rho \, e^{(-\mu t + \mu t \rho)} + 1
$$

Thus, the distribution function of T_q is obtained as

$$
W_q(t) = \begin{cases} 1 - \rho, & t = 0 \\ 1 - \rho e^{-\mu(1-\rho)t}, & t > 0. \end{cases}
$$

The probability density function $w_q(t)$ of T_q is found by differentiating $W_q(t)$ with respect to t as

> `wqt:=diff(Wqt,t);`

$$wqt := -\rho\,(-\mu + \mu\,\rho)\,e^{(-\mu\,t + \mu\,t\,\rho)}$$

Using this result (and informing Maple that $0 < \rho < 1$ and $\mu > 0$) we can compute the expected value $E(T_q) = \int_0^\infty t\,w_q(t)\,dt$.

> `assume(rho<1,rho>0,mu>0);`

> `ETq:=0*(1-rho)+Int(t*wqt,t=0..infinity);`

$$ETq := \int_0^\infty -t\,\rho\,(-\mu + \mu\,\rho)\,e^{(-\mu\,t + \mu\,t\,\rho)}\,dt$$

> `simplify(value(ETq));`

$$-\frac{\rho}{\mu\,(-1 + \rho)}$$

Finally, substituting the definition of $\rho = \lambda/\mu$, the expected time spent waiting in the queue is obtained in terms of the model parameters.

> `Wq:=normal(subs(rho=lambda/mu,%));`

$$Wq := \frac{\lambda}{\mu\,(\mu - \lambda)}$$

So far we have found expressions for L, L_q and W_q as the expected number of customers in the system, the expected number of customers in the queue and the expected time spent in the queue, respectively. It should now be easy to see that the expected time spent in the system is $W = W_q + 1/\mu = 1/(\mu - \lambda)$ since W_q is the expected time spent in the queue and $1/\mu$ is the expected time spent in the service. (Recall that the service time distribution was assumed exponential with parameter μ; hence its mean is $1/\mu$.)

By this time the observant reader may have noticed that there is a certain relationship among the expressions for L, L_q, W_q and W. In fact, it turns out that, as we see in the following Maple output, $L = \lambda W$ (and $L_q = \lambda W_q$).

> `W:=normal(Wq+1/mu);`

$$W := \frac{1}{\mu - \lambda}$$

> `L:=lambda/(mu-lambda);`
> `Lq:=lambda^2/(mu*(mu-lambda));`

$$L := \frac{\lambda}{\mu - \lambda}$$

$$Lq := \frac{\lambda^2}{\mu\,(\mu - \lambda)}$$

> `is(L=lambda*W);`

$$true$$

```
>   is(Lq=lambda*Wq);
```

$$true$$

This is known as Little's formula [125] and it can be shown that the formula is not specific to the $M/M/1$ queue, but under very mild assumptions, it is true in general for any queue; see [180] and Wolff [196, p. 236].

9.2.3   Finite Capacity Markovian Queue: $M/M/1/K$

In this section we assume that the queueing system is identical to the $M/M/1$ system of Section 9.2.2 except that when there are K customers present, the new arrival(s) turn away and do not enter the system. Such finite capacity queues are encountered frequently, e.g., at barber shops, drive-ins, and telephone switchboards.

With the limit of K imposed on the system capacity, the infinite system of difference equations given by (9.5)–(9.6) reduces to the following finite system of linear equations, which can be solved easily by Maple.

$$
\begin{aligned}
0 &= -(\lambda_n + \mu_n)p_n + \mu_{n+1}p_{n+1} + \lambda_{n-1}p_{n-1}, \quad n = 1, \dots, K \\
0 &= -\lambda_0 p_0 + \mu_1 p_1, \quad n = 0
\end{aligned}
$$

Assuming again that $\lambda_n = \lambda$ and $\mu_n = \mu$ and using the same set of commands but keeping in mind the fact that the system capacity is K, we solve this problem and compute the expected number of customers L as follows.

```
>   restart: # mm1cap.mws
>   MM1cap:={ 0=-(lambda+mu)*p(n)+lambda*p(n-1)
    +mu*p(n+1),p(n=0..1)=
    (lambda/mu)^n*p[ 0]};
```

$$MM1cap := \{$$

$$0 = -(\lambda + \mu)\,p(n) + \lambda\,p(n-1) + \mu\,p(n+1),\; p(n=0..1) = (\frac{\lambda}{\mu})^n\,p_0\}$$

```
>   sol:=rsolve(MM1cap,p(n));
```

$$sol := (\frac{\lambda}{\mu})^n\,p_0$$

```
>   p:=unapply(sol,n);
```

$$p := n \rightarrow (\frac{\lambda}{\mu})^n\,p_0$$

```
>   L:=sum(n*p(n),n=0..K);
```

$$L := -\frac{p_0\,(\frac{\lambda}{\mu})^{(K+1)}\,\mu\,(-(K+1)\lambda + (K+1)\mu + \lambda)}{(-\lambda + \mu)^2} + \frac{p_0\,\mu\,\lambda}{(-\lambda + \mu)^2}$$

```
>   p[ 0] :=solve(1=sum(p(n),n=0..K),p[ 0] );
```

$$p_0 := -\frac{-\lambda + \mu}{\mu\,((\frac{\lambda}{\mu})^{(K+1)} - 1)}$$

```
>   L:=factor(L);
```

$$L := \frac{-(\frac{\lambda}{\mu})^{(K+1)}\,\lambda\,K + (\frac{\lambda}{\mu})^{(K+1)}\,\mu\,K + (\frac{\lambda}{\mu})^{(K+1)}\,\mu - \lambda}{(-\lambda + \mu)\,((\frac{\lambda}{\mu})^{(K+1)} - 1)}$$

As a final check of the results, we assume that $\lambda < \mu$ and take the limit of L to obtain the original result found in the $M/M/1$ system with infinite capacity.

```
>   assume(lambda<mu,lambda>0);
```

```
>   limit(L,K=infinity);
```

$$\frac{\lambda}{-\lambda + \mu}$$

It should be noted that since the system capacity is finite (i.e., K) and the queue size cannot grow without bound, the results hold not only for $\lambda < \mu$ but also for $\lambda > \mu$. When $\lambda = \mu$, it is easy to show that $p_n = 1/(K+1)$, $n = 0, \ldots, K$ and $L = \frac{1}{2}K$.

9.2.4   Multiserver Queue $M/M/c$

So far we have assumed that the queueing system consists of a single server. In this section we generalize the model to allow for the possibility of multiple servers. To do this we return to the general birth-death process equations (9.5)–(9.6) and attempt to solve them for the steady-state probabilities p_n, $n \geq 0$.

In this case, Maple's `rsolve()` procedure fails to find a solution since the parameters λ_n and μ_n are functions of the state n. We thus proceed in a different manner and solve the system for the first few values of n in order to see a possible pattern in the solution.

```
>   restart: # mmc.mws
```

```
>   eq[ 0] :=-lambda[ 0] *p[ 0] +mu[ 1] *p[ 1] ;
```

$$eq_0 := -\lambda_0\,p_0 + \mu_1\,p_1$$

```
>   assign(solve({ eq[ 0]} ,{ p[ 1]} ));
```

```
>   p[ 1] ;
```

$$\frac{\lambda_0\,p_0}{\mu_1}$$

```
>   eq[ 1] :=-(lambda[ 1] +mu[ 1] )*p[ 1] +mu[ 2] *p[ 2]
    +lambda[ 0] *p[ 0] ;
```

$$eq_1 := -\frac{(\lambda_1 + \mu_1)\,\lambda_0\,p_0}{\mu_1} + \mu_2\,p_2 + \lambda_0\,p_0$$

```
> assign(solve({ eq[ 1]} ,{ p[ 2]} ));

> p[ 2] ;
```

$$\frac{\lambda_0\,p_0\,\lambda_1}{\mu_1\,\mu_2}$$

```
> eq[ 2] :=-(lambda[ 2] +mu[ 2] )*p[ 2] +mu[ 3] *p[ 3]
  +lambda[ 1] *p[ 1] ;
```

$$eq_2 := -\frac{(\lambda_2 + \mu_2)\,\lambda_0\,p_0\,\lambda_1}{\mu_1\,\mu_2} + \mu_3\,p_3 + \frac{\lambda_0\,p_0\,\lambda_1}{\mu_1}$$

```
> assign(solve({ eq[ 2]} ,{ p[ 3]} ));

> p[ 3] ;
```

$$\frac{\lambda_0\,p_0\,\lambda_1\,\lambda_2}{\mu_1\,\mu_2\,\mu_3}$$

For the first few values of n, there does appear to be a structure to the solution of this system. Next, we automate this solution procedure to compute the p_n for higher values of n.

```
> for n from 3 to 10 do

> eq[ n] :=-(lambda[ n] +mu[ n] )*p[ n] +mu[ n+1] *p[ n+1]
  +lambda[ n-1] *p[ n-1] ;

> assign(solve({ eq[ n]} ,{ p[ n+1]} ));

> p[ n] ;

> od:

> p[ 2] ; p[ 3] ; p[ 4] ; p[ 5] ; p[ 6] ;
```

$$\frac{\lambda_0\,p_0\,\lambda_1}{\mu_1\,\mu_2}$$

$$\frac{\lambda_0\,p_0\,\lambda_1\,\lambda_2}{\mu_1\,\mu_2\,\mu_3}$$

$$\frac{\lambda_0\,p_0\,\lambda_1\,\lambda_2\,\lambda_3}{\mu_1\,\mu_2\,\mu_3\,\mu_4}$$

$$\frac{\lambda_0\,p_0\,\lambda_1\,\lambda_2\,\lambda_3\,\lambda_4}{\mu_1\,\mu_2\,\mu_3\,\mu_4\,\mu_5}$$

$$\frac{\lambda_0\,p_0\,\lambda_1\,\lambda_2\,\lambda_3\,\lambda_4\,\lambda_5}{\mu_1\,\mu_2\,\mu_3\,\mu_4\,\mu_5\,\mu_6}$$

Again, we note that the interesting pattern that has emerged so far suggests that the probabilities p_n are in the form

$$p_n = p_0 \prod_{i=1}^{n} \frac{\lambda_{i-1}}{\mu_i}, \quad n \geq 1. \tag{9.7}$$

Naturally, this result must be proved using mathematical induction (for which we refer the reader to Gross and Harris [76, p. 85]). Since $\sum_{n=1}^{\infty} p_n = 1$, i.e.,

$$1 = p_0 \left(1 + \sum_{n=1}^{\infty} \prod_{i=1}^{n} \frac{\lambda_{i-1}}{\mu_i} \right)$$

the existence of the steady-state solution requires the convergence of the infinite series $\sum_{n=1}^{\infty} \prod_{i=1}^{n} \lambda_{i-1}/\mu_i$. For example, when $\lambda_n = \lambda$ and $\mu_n = \mu$, the model reduces to the simple $M/M/1$ queue and we obtain $p_n = (\lambda/\mu)^n (1 - \lambda/\mu)$, $n \geq 0$ provided that $\lambda < \mu$. When $\lambda > \mu$, customers arrive faster than they can be served and thus the queue size diverges to infinity. When $\lambda = \mu$, the queueing system behaves like a symmetric random walk that is null recurrent and has no limiting probabilities [157, p. 106].

Returning to the multiple-server queue $M/M/c$ with c servers, we assume that each server's service time is i.i.d. exponential and the arrival process is Poisson with rate λ. Hence $\lambda_n = \lambda$, and

$$\mu_n = \begin{cases} n\mu, & n = 1, \ldots, c \\ c\mu, & n = c, c+1, \ldots. \end{cases}$$

Using λ_n and μ_n in (9.7), one obtains [76, p. 86] the following expressions as the solution for p_n.

$$p_n = \begin{cases} \dfrac{\lambda^n}{n! \mu^n} p_0, & n = 1, \ldots, c \\ \dfrac{\lambda^n}{c^{n-c} c! \mu^n} p_0, & n = c, c+1, \ldots. \end{cases}$$

After some lengthy algebra p_0 is computed as

$$p_0 = \left[\sum_{n=0}^{c-1} \frac{1}{n!} \left(\frac{\lambda}{\mu} \right)^n + \frac{1}{c!} \left(\frac{\lambda}{\mu} \right)^c \left(\frac{c\mu}{c\mu - \lambda} \right) \right]^{-1}, \tag{9.8}$$

thus completing the analysis of the steady-state probabilities of the $M/M/c$ model.[5]

We now compute the expected queue length $L_q = \sum_{n=c}^{\infty} (n - c) p_n$ using Maple's sum() function.

---

[5]It should be noted here that the condition for the existence of p_0 is $\rho = \lambda/(c\mu) < 1$, implying that the mean arrival rate λ must be less than the maximum potential service rate $c\mu$.

```
>    restart: # mmcLq.mws
>    (n-c)*(lambda/mu)^n/(c^(n-c)*c!)*p[ 0] ;
```

$$\frac{(n-c)(\frac{\lambda}{\mu})^n \, p_0}{c^{(n-c)} \, c!}$$

```
>    Lq:=sum(%,n=c..infinity);
```

$$Lq := \frac{(\frac{\lambda}{\mu})^{(1+c)} \, c \, p_0 \, \mu^2}{c! \, (-\lambda + \mu \, c)^2}$$

Since $L = \lambda/\mu + L_q$, the expected number of customers in the system is found as

```
>    L:=lambda/mu+Lq;
```

$$L := \frac{\lambda}{\mu} + \frac{(\frac{\lambda}{\mu})^{(1+c)} \, c \, p_0 \, \mu^2}{c! \, (-\lambda + \mu \, c)^2}$$

where p_0 is found using (9.8).

As an example, we set $c = 5$, $\lambda = 4$ per hour and $\mu = 2$ per hour and compute the values of p_0, L and W using the following Maple commands:

```
>    restart: # mmcp0.mws
>    p0:=(' sum' ((1/n!)*(lambda/mu)^n,n=0..c-1)
     +(1/c!)*(lambda/mu)^c
     * (c*mu/(c*mu-lambda)))^(-1);
```

$$p0 := \frac{1}{\left(\displaystyle\sum_{n=0}^{c-1} \frac{(\frac{\lambda}{\mu})^n}{n!} \right) + \frac{(\frac{\lambda}{\mu})^c \, c \, \mu}{c! \, (c \, \mu - \lambda)}}$$

```
>    c:=5; lambda:=4; mu:=2;
```

$$c := 5$$
$$\lambda := 4$$
$$\mu := 2$$

```
>    evalf(p0);
```

$$.1343283582$$

```
>    Lq:=' (lambda/mu)^(1+c)*c*p0*mu^2
     /(c!*(mu*c-lambda)^2)' ;
```

$$Lq := \frac{(\frac{\lambda}{\mu})^{(1+c)} \, c \, p0 \, \mu^2}{c! \, (c \, \mu - \lambda)^2}$$

```
>  evalf(Lq);
```

$$.03980099502$$

```
>  L:=evalf(Lq+lambda/mu);
```

$$L := 2.039800995$$

```
>  W:=evalf(L/lambda);
```

$$W := .5099502488$$

9.2.5   Markovian Bulk Arrival System: $M^X/M/1$

In many realistic applications, customers arrive at a queueing system in groups rather than singly. Obvious examples are the arrival of a family of, say, four people at a restaurant, groups of travelers at customs or a box of parts at an assembly line.

In this section we consider a single-server queue in which customers arrive in groups according to a Poisson process with rate λ. We will assume that the group size X is a random variable with the density $\Pr(X = k) = a_k$, $k = 1, 2, \ldots$. In other words, the probability that a group of k customers arrives in an infinitesimal interval of length $o(\Delta t)$ is $\lambda a_k \Delta t + o(\Delta t)$. If the service time is exponential with parameter μ, then the time-dependent balance equations for the resulting system can be written as

$$p_n'(t) = -(\lambda + \mu)p_n(t) + \mu p_{n+1}(t) + \sum_{k=1}^{n} \lambda a_k p_{n-k}(t), \quad n \geq k \geq 1$$

$$p_0'(t) = -\lambda p_0(t) + \mu p_1(t).$$

We note here that $\sum_{k=1}^{n} \lambda a_k p_{n-k}(t)\Delta t$ is the probability that the system state is increased to n from $n - k$ with the arrival of k customers ($k = 1, \ldots, n$), which has a probability of a_k.

Assuming that the steady-state solution exists, we have

$$0 = -(\lambda + \mu)p_n + \mu p_{n+1} + \sum_{k=1}^{n} \lambda a_k p_{n-k}, \quad n \geq k \geq 1 \qquad (9.9)$$

$$0 = -\lambda p_0 + \mu p_1. \qquad (9.10)$$

To solve this system, we introduce the generating function $G(z) = \sum_{n=0}^{\infty} p_n z^n$. Multiplying (9.9) by z^n for $n = 1, 2 \ldots$, adding the result to (9.10) and simplifying (see, [129, p. 3] and [130, p. 182]) we obtain

$$G(z) = \frac{\mu(1 - z)p_0}{\mu(1 - z) - \lambda z[1 - A(z)]}$$

where $A(z) = \sum_{n=0}^{\infty} a_n z^n$ is the generating function of the random variable X. Since $G(1) = 1$, it can be shown that $p_0 = 1 - \rho$ where $\rho = \lambda E(X)/\mu < 1$ is the traffic intensity.

We now apply these results to a specific model where it is assumed that the random variable X has the truncated geometric density, i.e., $a_k = \Pr(X = k) = c(1 - c)^{k-1}, 0 < c < 1, k = 1, 2, \ldots$.

> `restart: # bulkgeom.mws`

We define the density of X and check that the probabilities add up to unity.

> `pdfX:=c* (1-c)^(n-1);`

$$pdfX := c\,(1 - c)^{(n-1)}$$

> `sum(pdfX,n=1..infinity);`

$$1$$

The next three commands define the generating function $A(z)$ of X, compute the mean of X using the property that $E(X) = \sum_{k=1}^{\infty} k a_k = A'(z)\big|_{z=1}$, and write the generating function $G(z)$ of the steady-state probabilities p_n.

> `A:=sum(pdfX* z^n,n=1..infinity);`

$$A := \frac{c\,z}{-z + z\,c + 1}$$

> `EX:=normal(subs(z=1,diff(A,z)));`

$$EX := \frac{1}{c}$$

> `G:=mu* p[ 0] * (1-z) / (mu* (1-z)-lambda* z* (1-A));`

$$G := \frac{\mu\,p_0\,(1 - z)}{\mu\,(1 - z) - \lambda z\left(1 - \dfrac{c\,z}{-z + z\,c + 1}\right)}$$

Since $G(z)$ is given in terms of the unknown p_0, we find this value using the property that $p_0 = \lim_{z \to 1} G(z)$.

> `p0:=limit(G,z=1);`

$$p0 := -\frac{c\,p_0\,\mu}{-\mu\,c + \lambda}$$

> `p[ 0] :=normal(solve(1=p0,p[ 0] ));`

$$p_0 := -\frac{-\mu\,c + \lambda}{c\,\mu}$$

Once p_0 is found, the traffic intensity ρ can be computed as

> `rho:=lambda/ (c* mu);`

$$\rho := \frac{\lambda}{\mu\,c}$$

In order to invert the resulting generating functions we load the `genfunc` library and display $G(z)$, which no longer depends on the probability p_0.

> `with(genfunc);`

[*rgf_charseq, rgf_encode, rgf_expand, rgf_findrecur, rgf_hybrid,*
rgf_norm, rgf_pfrac, rgf_relate, rgf_sequence, rgf_simp, rgf_term,
termscale]
> G;

$$-\frac{(-\mu c + \lambda)(1-z)}{c\left(\mu(1-z) - \lambda z \left(1 - \dfrac{cz}{-z+zc+1}\right)\right)}$$

Finally, the steady-state probabilities p_n are computed explicitly for each n.

> pn:=rgf_expand(G,z,n);

$$pn := -\frac{\lambda(-\mu c + \lambda)\left(\dfrac{\mu - \mu c + \lambda}{\mu}\right)^n}{c(\mu - \mu c + \lambda)\mu}$$

As a check of the results found, the next command adds all the probabilities and finds that they sum to one, i.e., $\sum_{n=0}^{\infty} p_n = 1$.

> normal(p[0]+sum(pn,n=1..infinity));

$$1$$

The expected number of customers in this bulk queueing system, L, can now be easily computed using the property that $L = E(\text{customers in system}) = \sum_{n=1}^{\infty} np_n = G'(z)|_{z=1}$.

> L:=limit(diff(G,z),z=1);

$$L := -\frac{\lambda}{(-\mu c + \lambda)c}$$

9.3 Transient Solutions

The results obtained in Section 9.2 assumed that the system was in operation for long enough so that the steady state was reached. This assumption simplified the model as the time-*dependent* differential-difference equation system (9.3)–(9.4) reduced to a time-*independent* system of difference equations.

However, in many realistic problems, one may be interested in the finite-time (transient) solution of the probabilities. Thus, we now describe the transient solution of some special queueing systems.

9.3.1 Finite Capacity $M/M/1/K$ Queue with $K = 1$

In this simple system there is only a single customer, e.g., a machine, that may require service (e.g., repair) at random times. We will let state 1 correspond to the state of the busy server (machine in repair) and state 0 correspond to the state of the idle server (machine operating). Thus, if λ is the arrival rate of the machine (i.e., the rate at which the process departs state 0) and μ is the service rate of the server (i.e., the rate at which the process departs state 1), then $1/\lambda$ is the average time the machine is in operation and $1/\mu$ is the average time it takes to repair the machine. Using the steps for developing differential equations for birth-death processes, it is easy to show that we have two differential equations representing the evolution of the transient probabilities for the finite capacity $M/M/1/K$ queue with $K = 1$.

$$
\begin{aligned}
p_0'(t) &= -\lambda p_0(t) + \mu p_1(t) \\
p_1'(t) &= -\mu p_0(t) + \lambda p_1(t)
\end{aligned}
$$

Maple's deep knowledge of the solution of systems of differential equations makes it almost trivial to find the solution for this problem where we will assume that initially the system is empty, i.e., $p_0(0) = 1$ and $p_1(0) = 0$.

```
>    restart: # MM11-tra.MWS
>    de0:=diff(p.0(t),t)=-lambda*p.0(t)+mu*p.1(t);
```

$$de0 := \tfrac{\partial}{\partial t}\,p0(t) = -\lambda\,p0(t) + \mu\,p1(t)$$

```
>    de1:=diff(p.1(t),t)=-mu*p.1(t)+lambda*p.0(t);
```

$$de1 := \tfrac{\partial}{\partial t}\,p1(t) = -\mu\,p1(t) + \lambda\,p0(t)$$

```
>    in0:=p.0(0)=1;
```

$$in0 := p0(0) = 1$$

```
>    in1:=p.1(0)=0;
```

$$in1 := p1(0) = 0$$

```
>    sol:=dsolve({ de0,de1, in0,in1} ,{ p.0(t),p.1(t)} );
```

$$
sol := \left\{ p0(t) = \frac{\mu}{\lambda+\mu} + \frac{e^{(-t\lambda-t\mu)}\lambda}{\lambda+\mu},\ p1(t) = \frac{\dfrac{\lambda\mu}{\lambda+\mu} - \dfrac{e^{(-t\lambda-t\mu)}\lambda\mu}{\lambda+\mu}}{\mu} \right\}
$$

```
>    assign(sol);
>    p[ 0] :=unapply(simplify(p0(t)),t);
     p[ 1] :=unapply(simplify(p1(t)),t);
```

$$p_0 := t \to \frac{\mu + \lambda e^{(-(\lambda+\mu)t)}}{\lambda+\mu}$$

$$p_1 := t \to -\frac{\lambda\,(-1 + e^{(-(\lambda+\mu)t)})}{\lambda+\mu}$$

The last line of the Maple output gives the transient solution for the $M/M/1/1$ system. To find the steady-state solution, we let $t \to \infty$ and obtain the steady-state probabilities as $p_0 = \mu/(\lambda+\mu)$ and $p_1 = \lambda/(\lambda+\mu)$.

```
>    assume(lambda>0, mu>0); interface(showassumed=2);
```

```
>    limit(p[ 0] (t),t=infinity);
     limit(p[ 1] (t),t=infinity);
```

$$\frac{\mu}{\lambda + \mu}$$

with assumptions on μ and λ

$$\frac{\lambda}{\lambda + \mu}$$

with assumptions on λ and μ

9.3.2   Ample Server System $M/M/\infty$

Such systems are encountered in self-service-type environments where the customers do not have to wait at all for service. The service rate is assumed to be linearly increasing with the number of customers in the system (i.e., each customer is his own server); thus $\mu_n = n\mu$. If we also assume that $\lambda_n = \lambda$, the differential-difference equation system (9.3)–(9.4) takes the form

$$p_n'(t) = -(\lambda + n\mu)p_n(t) + (n+1)\mu p_{n+1}(t) + \lambda p_{n-1}(t), n \geq 1 \quad (9.11)$$
$$p_0'(t) = -\lambda p_0(t) + \mu p_1(t), \quad n = 0. \quad (9.12)$$

To solve these equations, we make use of the generating function of the transient probabilities defined by

$$G(z, t) = \sum_{n=0}^{\infty} p_n(t)z^n.$$

Multiplying both sides of (9.11) by z^n, summing from 1 to ∞ and adding the equation in (9.12) gives

$$\sum_{n=0}^{\infty} p_n'(t)z^n = -\sum_{n=0}^{\infty}(\lambda + n\mu)p_n(t)z^n + \mu \sum_{n=0}^{\infty}(n+1)p_{n+1}(t)z^n + \lambda \sum_{n=1}^{\infty} p_{n-1}(t)z^n.$$

Since $\sum_{n=1}^{\infty} p_n(t)nz^{n-1} = \partial G(z, t)/\partial z$ and $\sum_{n=0}^{\infty} p_n'(t)z^n = \partial G(z, t)/\partial t$, these equations can be rewritten as a partial differential equation (PDE) in terms of the unknown generating function $G(z, t)$; see Gross and Harris [76, Section 2.10.3]:

$$\frac{\partial G(z, t)}{\partial t} = \mu(1 - z)\frac{\partial G(z, t)}{\partial z} - \lambda(1 - z)G(z, t).$$

At this stage we request Maple's help in solving this PDE.

```
>    restart: # ample.mws
>    PDEGenFunc:=diff(G(z,t),t)=mu* (1-z)*diff(G(z,t),z)
     -lambda* (1-z)*G(z,t);
```

$$PDEGenFunc := \frac{\partial}{\partial t} G(z, t) = \mu (1 - z) (\frac{\partial}{\partial z} G(z, t)) - \lambda (1 - z) G(z, t)$$

```
>   pdsolve(PDEGenFunc,G(z,t));
```

$$G(z, t) = \_F1(-\frac{-t \mu + \ln(-1 + z)}{\mu}) e^{(\frac{\lambda z}{\mu})}$$

The solution is obtained in terms of an arbitrary function $\_F1$. To determine the form of this function, we assume that the system is initially empty, i.e., $p_0(0) = 1$ and $p_n(0) = 0$ for $n > 0$. Thus,

$$G(z, 0) = 1 = \_F1 \left(\frac{-\ln(-1 + z)}{\mu} \right) \exp \left(\frac{\lambda z}{\mu} \right).$$

The next four Maple statements use this information and determine the form of the unknown function.

First, we give a name to the term inside $\_F1$.

```
>   inside:=(t*mu-ln(-1+z))/mu;
```

$$inside := \frac{t \mu - \ln(-1 + z)}{\mu}$$

At $t = 0$, this term takes the value

```
>   subs(t=0,inside);
```

$$-\frac{\ln(-1 + z)}{\mu}$$

which, when equated to, say, y and solved for z gives

```
>   solve(%=y,z);
```

$$e^{(-y \mu)} + 1$$

Substituting the expression for z into $inside$, we obtain the functional form of $\_F1$:

```
>   f1:=y->exp(-lambda*(exp(-y*mu)+1)/mu);
```

$$f1 := y \to e^{\left(-\frac{\lambda (e^{(-y \mu)} + 1)}{\mu} \right)}$$

i.e.,

$$\_F1(y) = \exp \left(-\frac{\lambda \exp(-y\mu) + 1}{\mu} \right).$$

Now that the form of the unknown function $\_F1$ is determined, we can explicitly compute the generating function $G(z, t)$ as follows.

```
>   f1(inside)*exp(lambda/mu* z);
```

$$e^{\left(-\frac{\lambda (e^{(-t \mu + \ln(-1 + z))} + 1)}{\mu} \right)} e^{(\frac{\lambda z}{\mu})}$$

```
>   G(z,t):=simplify(%);
```

$$G(z, t) := e^{\left(-\frac{\lambda(-1+z)(e^{(-t\mu)}-1)}{\mu}\right)}$$

With the explicit form of the generating function available, the transient probabilities $\{p_n(t)\}$ can be obtained by expanding $G(z, t)$ in a power series $\sum_{n=0}^{\infty} a_n z^n$ where, of course, the a_n coefficients are the required transient probabilities $p_n(t)$ since the generating function was defined as $G(z, t) = \sum_{n=0}^{\infty} p_n(t) z^n$.

Using the Maclaurin series, i.e.,

$$a_n = \frac{1}{n!} \frac{\partial^n G(z, t)}{\partial z^n}\bigg|_{z=0},$$

the first few coefficients are computed with Maple's help as follows.

```
>    a[ 0] :=(1/0!)*subs(z=0,G(z,t));
```

$$a_0 := e^{\left(\frac{\lambda(e^{(-t\mu)}-1)}{\mu}\right)}$$

```
>    a[ 1] :=(1/1!)*subs(z=0,diff(G(z,t),z$1));
     a[ 2] =(1/2!)*subs(z=0,diff(G(z,t),z$2));
     a[ 3] =(1/3!)*subs(z=0,diff(G(z,t),z$3));
```

$$a_1 := -\frac{\lambda\,(e^{(-t\mu)}-1)\,e^{\left(\frac{\lambda(e^{(-t\mu)}-1)}{\mu}\right)}}{\mu}$$

$$a_2 = \frac{1}{2}\frac{\lambda^2\,(e^{(-t\mu)}-1)^2\,e^{\left(\frac{\lambda(e^{(-t\mu)}-1)}{\mu}\right)}}{\mu^2}$$

$$a_3 = -\frac{1}{6}\frac{\lambda^3\,(e^{(-t\mu)}-1)^3\,e^{\left(\frac{\lambda(e^{(-t\mu)}-1)}{\mu}\right)}}{\mu^3}$$

These results imply—after using induction—that the transient probabilities of the ample server $M/M/\infty$ system are obtained as

$$p_n(t) = \frac{1}{n!}\left[\frac{\lambda}{\mu}(1 - e^{-\mu t})\right]^n \exp\left[-\frac{\lambda}{\mu}(1 - e^{-\mu t})\right], \quad n \geq 0.$$

Finally, as $t \to \infty$, we see that the steady-state probabilities reduce to

$$p_n = \frac{(\lambda/\mu)^n \exp(-\lambda/\mu)}{n!}, \quad n \geq 0,$$

which is, of course, the Poisson density with parameter λ/μ.

9.4   Queueing Networks

Many realistic queueing systems consist of several single queues (nodes) whose output becomes the input to another queue. Manufacturing or assembly line processes where units must be processed in a sequence of workstations (e.g., drill press, paint shop and packaging) and a university registration process where students must stop at several desks (e.g., advisor, cashier) are examples of *serial* queueing networks. In the more general — *nonserial* — case customers may arrive from outside to any node that may have any number of servers and may depart from any node and even return to nodes previously visited (e.g., for rework).

In this section we will examine some models representing such queueing networks. First, we will describe a simple serial queue with blocking and then we will consider a *Jackson network*, which is a queueing network with a special structure.

9.4.1   Serial Queue with Blocking

Consider a very simple serial queueing network with two nodes (stations) and a single server at each node [76, p. 226]. Customers must sequentially visit node 1 and then node 2.[6] It is assumed that no queues are allowed at any node. If station 2 is occupied and service is completed in node 1, the customer in node 1 must wait (i.e., is *blocked*) until the node 2 customer is cleared. When the system is blocked, any arrivals to node 1 are turned away even if node 2 is empty.

The state of the queueing system is a vector (m, n). Here, $(0, 0)$ is the empty system, $(1, 0)$ is the system with a customer in node 1 only, $(0, 1)$ is the system with customer in node 2 only, $(1, 1)$ is the customers in both nodes and $(b, 1)$ is the blocked system.

We define $p_{mn}(t)$ as the probability that the state of the system is (m, n) and assume that arrivals to the system are Poisson with rate λ and service time is exponential with parameters μ_1 and μ_2 in nodes 1 and 2, respectively. The usual procedure gives the following system of five differential equations:

$$
\begin{aligned}
p'_{00}(t) &= -\lambda p_{00}(t) + \mu_2 p_{01}(t) \\
p'_{10}(t) &= -\mu_1 p_{10}(t) + \mu_2 p_{11}(t) + \lambda p_{00}(t) \\
p'_{01}(t) &= -(\lambda + \mu_2) p_{01}(t) + \mu_1 p_{10}(t) + \mu_2 p_{b1}(t) \\
p'_{11}(t) &= -(\mu_1 + \mu_2) p_{11}(t) + \lambda p_{01}(t) \\
p'_{b1}(t) &= -\mu_2 p_{b1}(t) + \mu_1 p_{11}(t)
\end{aligned}
$$

We will assume that the system is initially empty, i.e., $p_{00}(0) = 1$.

This is a *linear* system of differential equations in the form $\mathbf{P}'(t) = \mathbf{P}(t)\mathbf{A}$, and hence, in principle, it can be solved analytically using the *spectral theorem*

---

[6] An example of such a system is a car wash with two stages—in the first stage the car is washed, and in the second stage it is vacuumed. This was first discussed in Example 91 of Chapter 7, Stochastic Processes.

([102, p. 541], [64, p. 113]) which is implemented by Maple's exponential()
procedure. However, the results are several pages long and not very instructive.
For this reason, we choose to present a discussion of the numerical solution of
this system.

The five differential equations are entered in the usual way using Maple's syntax and the system is defined in the following lines. The initial condition that the
system is empty, i.e., that $p_{00}(0) = 1$, is indicated by the ini:= statement.

```
>   restart: # serial.mws
>   de[ 0,0] :=diff(p[ 0,0] (t),t)=
    -lambda*p[ 0,0] (t)+mu[ 2] *p[ 0,1] (t);
```

$$de_{0,0} := \frac{\partial}{\partial t} p_{0,0}(t) = -\lambda\, p_{0,0}(t) + \mu_2\, p_{0,1}(t)$$

```
>   de[ 1,0] :=diff(p[ 1,0] (t),t)=
    -mu[ 1] *p[ 1,0] (t)+mu[ 2] *p[ 1,1] (t)
    +lambda*p[ 0,0] (t);
```

$$de_{1,0} := \frac{\partial}{\partial t} p_{1,0}(t) = -\mu_1\, p_{1,0}(t) + \mu_2\, p_{1,1}(t) + \lambda\, p_{0,0}(t)$$

```
>   de[ 0,1] :=diff(p[ 0,1] (t),t)=
    -(lambda+mu[ 2] )*p[ 0,1] (t)
    +mu[ 1] *p[ 1,0] (t)+mu[ 2] *p[ b,1] (t);
```

$$de_{0,1} := \frac{\partial}{\partial t} p_{0,1}(t) = -(\lambda + \mu_2)\, p_{0,1}(t) + \mu_1\, p_{1,0}(t) + \mu_2\, p_{b,1}(t)$$

```
>   de[ 1,1] :=diff(p[ 1,1] (t),t)=
    -(mu[ 1] +mu[ 2] )*p[ 1,1] (t)
    +lambda*p[ 0,1] (t);
```

$$de_{1,1} := \frac{\partial}{\partial t} p_{1,1}(t) = -(\mu_1 + \mu_2)\, p_{1,1}(t) + \lambda\, p_{0,1}(t)$$

```
>   de[ b,1] :=diff(p[ b,1] (t),t)=
    -mu[ 2] *p[ b,1] (t)+mu[ 1] *p[ 1,1] (t);
```

$$de_{b,1} := \frac{\partial}{\partial t} p_{b,1}(t) = -\mu_2\, p_{b,1}(t) + \mu_1\, p_{1,1}(t)$$

```
>   sys:=de[ 0,0] ,de[ 1,0] ,de[ 0,1] ,de[ 1,1] ,de[ b,1] :
>   ini:=p[ 0,0] (0)=1,p[ 1,0] (0)=0,
    p[ 0,1] (0)=0,p[ 1,1] (0)=0,p[ b,1] (0)=0;
```

$$ini := p_{0,0}(0) = 1,\ p_{1,0}(0) = 0,\ p_{0,1}(0) = 0,\ p_{1,1}(0) = 0,\ p_{b,1}(0) = 0$$

Numerical values for the parameters λ, μ_1 and μ_2 are specified as 5, 2 and 4,
respectively.

```
>   lambda:=5; mu[ 1] :=2; mu[ 2] :=4;
```

$$\lambda := 5$$

$$\mu_1 := 2$$

$$\mu_2 := 4$$

Maple's dsolve() procedure computes the solution of the system numerically using the Runge-Kutta method.

```
>   F:=dsolve({ sys,ini},
    { p[ 0,0] (t),p[ 1,0] (t),p[ 0,1] (t),p[ 1,1] (t),
    p[ b,1] (t)},type=numeric);
```

$$F := \textbf{proc}(rkf45\_x) \ldots \textbf{end}$$

The solution is displayed at $t = 0$, 1, 2.5 and 20. We note that the steady-state levels have been reached as early as $t = 2.5$.

```
>   F(0); F(1); F(2.5); F(20);
```

$$[t = 0, \; p_{b,1}(t) = 0, \; p_{0,1}(t) = 0, \; p_{0,0}(t) = 1., \; p_{1,0}(t) = 0, \; p_{1,1}(t) = 0]$$

$$[t = 1, \; p_{b,1}(t) = .05192669369649984, \; p_{0,1}(t) = .1471821386619817,$$
$$p_{0,0}(t) = .1197361520137099, \; p_{1,0}(t) = .5616488539244483,$$
$$p_{1,1}(t) = .1195061617033598]$$

$$[t = 2.5, \; p_{b,1}(t) = .06197178680874423, \; p_{0,1}(t) = .1488757203993332,$$
$$p_{0,0}(t) = .1190881473988159, \; p_{1,0}(t) = .5460107337664787,$$
$$p_{1,1}(t) = .1240536116266276]$$

$$[t = 20, \; p_{b,1}(t) = .06203473911889895, \; p_{0,1}(t) = .1488833747939046,$$
$$p_{0,0}(t) = .1191067001829406, \; p_{1,0}(t) = .5459057062649429,$$
$$p_{1,1}(t) = .1240694796393126]$$

To see how the transient probabilities behave for, say, the empty system and the blocked system, we assign the results for $p_{00}(t)$ and $p_{b1}(t)$ to Sol[0,0] and Sol[b,1] using Maple's op() procedure.[7]

```
>   Sol[ 0,0] :=t->rhs(op(2,F(t)));
    Sol[ b,1] :=t->rhs(op(6,F(t)));
```

$$Sol_{0,0} := t \rightarrow \text{rhs}(\text{op}(2, F(t)))$$

$$Sol_{b,1} := t \rightarrow \text{rhs}(\text{op}(6, F(t)))$$

The results are plotted in Figure 9.1.

```
>   #plot({ Sol[ 0,0] ,Sol[ b,1]},0..6,0..1);
```

The steady-state probabilities can be easily computed symbolically since as $t \rightarrow \infty$, $p'_{mn}(t) \rightarrow 0$. The resulting system of linear *algebraic* equations is easily solved by Maple as follows:

```
>   restart: # serial.mws (Steady State)

>   eq[ 0,0] :=-lambda*p[ 0,0] +mu[ 2] *p[ 0,1] ;
```

$$eq_{0,0} := -\lambda \, p_{0,0} + \mu_2 \, p_{0,1}$$

---

[7]If nonadaptive plotting—where graphs may not be as smooth as possible—is acceptable, then there is a simpler way to plot the transient probabilities: After generating the solution to F using dsolve as above, load the plots graphics package using with(plots) and apply the command odeplot(F,[[t,p[0,0] (t)] ,[t,p[b,1] (t)]] ,0..6).

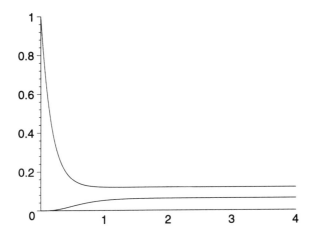

FIGURE 9.1. Transient solution for the empty system and blocked system probabilities $p_{00}(t)$ and $p_{b1}(t)$, respectively, with $p_{00}(0) = 1$ and $p_{b1}(0) = 0$.

```
>   eq[ 1,0] :=-mu[ 1] *p[ 1,0]
    +mu[ 2] *p[ 1,1] +lambda*p[ 0,0] ;
```

$$eq_{1,0} := -\mu_1 p_{1,0} + \mu_2 p_{1,1} + \lambda p_{0,0}$$

```
>   eq[ 0,1] :=-(lambda+mu[ 2] )*p[ 0,1]
    +mu[ 1] *p[ 1,0] +mu[ 2] *p[ b,1] ;
```

$$eq_{0,1} := -(\lambda + \mu_2) p_{0,1} + \mu_1 p_{1,0} + \mu_2 p_{b,1}$$

```
>   eq[ 1,1] :=-(mu[ 1] +mu[ 2] )*p[ 1,1]
    +lambda*p[ 0,1] ;
```

$$eq_{1,1} := -(\mu_1 + \mu_2) p_{1,1} + \lambda p_{0,1}$$

```
>   eq[ b,1] :=-mu[ 2] *p[ b,1]
    +mu[ 1] *p[ 1,1] ;
```

$$eq_{b,1} := -\mu_2 p_{b,1} + \mu_1 p_{1,1}$$

```
>   eq[ unity] :=
    p[ 0,0] +p[ 1,0] +p[ 0,1] +p[ 1,1] +p[ b,1] =1;
```

$$eq_{unity} := p_{0,0} + p_{1,0} + p_{0,1} + p_{1,1} + p_{b,1} = 1$$

```
>   solve ({ eq[ 0,0] ,eq[ 1,0] ,eq[ 0,1] ,
    eq[ 1,1] ,eq[ b,1] ,eq[ unity]} ,
    { p[ 0,0] ,p[ 1,0] ,p[ 0,1] ,p[ 1,1] ,p[ b,1]} );
```

$$\{p_{1,1} = \frac{\lambda^2 \mu_2 \mu_1}{\%1}, \quad p_{1,0} = \frac{\mu_2{}^2 \lambda (\lambda + \mu_1 + \mu_2)}{\%1}, \quad p_{b,1} = \frac{\mu_1{}^2 \lambda^2}{\%1},$$

$$p_{0,0} = \frac{\mu_2{}^2 \mu_1 (\mu_1 + \mu_2)}{\%1}, \quad p_{0,1} = \frac{(\mu_1 + \mu_2) \lambda \mu_2 \mu_1}{\%1}\}$$

$$\%1 := 2 \mu_2{}^2 \lambda \mu_1 + \mu_2{}^2 \mu_1{}^2 + \mu_2{}^3 \mu_1 + \mu_2{}^2 \lambda^2 + \mu_2{}^3 \lambda + \mu_2 \mu_1{}^2 \lambda + \lambda^2 \mu_2 \mu_1 + \lambda^2 \mu_1{}^2$$

When the servers have the same service rate, i.e., when $\mu_1 = \mu_2$, the solution simplifies and reduces to the form that was reported in Gross and Harris [76, p. 225].

```
>   assign(%);
>   mu[ 1] :=mu;  mu[ 2] :=mu;
```

$$\mu_1 := \mu$$

$$\mu_2 := \mu$$

```
>   normal(p[ 0,0] );  normal(p[ 1,0] );
    normal(p[ 0,1] );  normal(p[ 1,1] );
    normal(p[ b,1] );
```

$$2 \frac{\mu^2}{4 \mu \lambda + 2 \mu^2 + 3 \lambda^2}$$

$$\frac{\lambda (\lambda + 2 \mu)}{4 \mu \lambda + 2 \mu^2 + 3 \lambda^2}$$

$$2 \frac{\mu \lambda}{4 \mu \lambda + 2 \mu^2 + 3 \lambda^2}$$

$$\frac{\lambda^2}{4 \mu \lambda + 2 \mu^2 + 3 \lambda^2}$$

$$\frac{\lambda^2}{4 \mu \lambda + 2 \mu^2 + 3 \lambda^2}$$

9.4.2   Jackson Networks

Jackson networks [98] are queueing networks with a special structure that consist of k nodes (queues) where the outside customers arrive at node i according to a Poisson process with mean arrival rate γ_i, $i = 1, \ldots, k$. Each node with an unlimited waiting capacity may have multiple identical servers where each server has an exponential service time with parameter μ_i. When a customer leaves node i, it goes to node j with probability p_{ij} independent of that customer's past history. Thus, one may think of the sequence of nodes visited in a Jackson network as a discrete-time Markov chain with one absorbing state.

The first step in analyzing a Jackson network is the computation of the total input rate λ_i into node i. Before their final departure, the customers may be cycled between different nodes within the network as a result of, say, rework of an item in a manufacturing process. Hence, the total input rate into node i may be different from the arrival rate from outside into node i. Thus, we have

$$\lambda_i = \gamma_i + \sum_j \lambda_j p_{ji}$$

since γ_i is the arrival rate to node i from outside the network and p_{ji} is the probability that the customer leaving node j enters node i. (If node i has c_i servers and if $\rho_i = \lambda_i/(c_i\mu_i) < 1$ for all nodes, then Jackson has shown that node i can be treated as a multiple server $M/M/c_i$ system [98].)

This system may be rewritten in vector/matrix form as $\lambda = \gamma + \lambda P$ and when solved for λ gives

$$\lambda = \gamma(I - P)^{-1}$$

where P is the $k \times k$ (sub-Markov) switching probability matrix.

Let us now consider a Jackson-type manufacturing process where customers arrive at any node of the network with three nodes. The arrival rate vector is $\gamma = (\gamma_1, \gamma_2, \gamma_3)$ and the switching matrix is

$$P = \begin{bmatrix} 0 & p_{12} & p_{13} \\ p_{21} & 0 & p_{23} \\ p_{31} & p_{32} & 0 \end{bmatrix}.$$

We use Maple to perform symbolically the necessary computations and find the total arrival rates at each of the three nodes as a function of model parameters γ and P.

```
>   restart: # JACKSON.MWS
>   with(linalg):
```

Warning, new definition for norm

Warning, new definition for trace

```
>   gammaVector:=array(1..3,
    [ gamma[ 1] ,gamma[ 2] ,gamma[ 3] ] );
```

$$gammaVector := \begin{bmatrix} \gamma_1, & \gamma_2, & \gamma_3 \end{bmatrix}$$

```
>   PMatrix:=matrix(3,3,[ 0,p[ 1,2] ,p[ 1,3] ,
    p[ 2,1] ,0,p[ 2,3] ,p[ 3,1] ,p[ 3,2] ,0] );
```

$$PMatrix := \begin{bmatrix} 0 & p_{1,2} & p_{1,3} \\ p_{2,1} & 0 & p_{2,3} \\ p_{3,1} & p_{3,2} & 0 \end{bmatrix}$$

```
>   IdentityMatrix := array(identity, 1..3,1..3);
```

$$IdentityMatrix := \text{array}(identity, 1..3, 1..3, [])$$

> `IminusP:=evalm(IdentityMatrix-PMatrix);`

$$IminusP := \begin{bmatrix} 1 & -p_{1,2} & -p_{1,3} \\ -p_{2,1} & 1 & -p_{2,3} \\ -p_{3,1} & -p_{3,2} & 1 \end{bmatrix}$$

> `IminusPinv:=inverse(IminusP);`

$IminusPinv :=$

$$\begin{bmatrix} \dfrac{-1+\%3}{\%4} & -\dfrac{p_{1,2}+p_{1,3}\,p_{3,2}}{\%4} & -\dfrac{p_{1,2}\,p_{2,3}+p_{1,3}}{\%4} \\[2ex] -\dfrac{p_{2,1}+p_{2,3}\,p_{3,1}}{\%4} & \dfrac{-1+\%1}{\%4} & -\dfrac{p_{2,3}+p_{1,3}\,p_{2,1}}{\%4} \\[2ex] -\dfrac{p_{2,1}\,p_{3,2}+p_{3,1}}{\%4} & -\dfrac{p_{3,2}+p_{1,2}\,p_{3,1}}{\%4} & \dfrac{-1+\%2}{\%4} \end{bmatrix}$$

$\%1 := p_{3,1}\,p_{1,3}$

$\%2 := p_{2,1}\,p_{1,2}$

$\%3 := p_{2,3}\,p_{3,2}$

$\%4 := -1+\%3+\%2+p_{2,1}\,p_{1,3}\,p_{3,2}+p_{3,1}\,p_{1,2}\,p_{2,3}+\%1$

The total input rate vector λ is found as follows:

> `lambdaVector:=evalm(gammaVector&* IminusPinv);`

$lambdaVector := \Bigg[$

$$\dfrac{\gamma_1\,(-1+\%3)}{\%4} - \dfrac{\gamma_2\,(p_{2,1}+p_{2,3}\,p_{3,1})}{\%4} - \dfrac{\gamma_3\,(p_{2,1}\,p_{3,2}+p_{3,1})}{\%4},$$

$$-\dfrac{\gamma_1\,(p_{1,2}+p_{1,3}\,p_{3,2})}{\%4} + \dfrac{\gamma_2\,(-1+\%1)}{\%4} - \dfrac{\gamma_3\,(p_{3,2}+p_{1,2}\,p_{3,1})}{\%4}$$

,

$$-\dfrac{\gamma_1\,(p_{1,2}\,p_{2,3}+p_{1,3})}{\%4} - \dfrac{\gamma_2\,(p_{2,3}+p_{1,3}\,p_{2,1})}{\%4} + \dfrac{\gamma_3\,(-1+\%2)}{\%4} \Bigg]$$

$\%1 := p_{3,1}\,p_{1,3}$

$\%2 := p_{2,1}\,p_{1,2}$

$\%3 := p_{2,3}\,p_{3,2}$

$\%4 := -1+\%3+\%2+p_{2,1}\,p_{1,3}\,p_{3,2}+p_{3,1}\,p_{1,2}\,p_{2,3}+\%1$

As an example, let us now compute numerically the value of λ assuming that $\gamma = (4, 2, 1)$ and

$$\mathbf{P} = \begin{bmatrix} 0 & 0.55 & 0.25 \\ 0.3 & 0 & 0.65 \\ 0.05 & 0.1 & 0 \end{bmatrix}.$$

Substituting these values into the formulas for λ we developed, Maple gives

```
>    subs ({ gamma[ 1] =4,gamma[ 2] =2,gamma[ 3] =1,
       p[ 1,2] =.55,p[ 1,3] =.25,p[ 2,1] =.3,
       p[ 2,3] =.65,p[ 3,1] =.05,p[ 3,2] =.1} ,
     evalm (lambdaVector));
```

$$[6.126003073, \ 6.013317398, \ 6.440157077]$$

i.e., $\lambda_1 = 6.126$, $\lambda_2 = 6.013$ and $\lambda_3 = 6.440$. These values can now be used in conjunction with the service rate values μ_i to compute the traffic intensities $\rho_i = \lambda_i/(c_i\mu_i)$.

Assuming that $(\mu_1, \mu_2, \mu_3) = (10, 4, 12)$ and $(c_1, c_2, c_3) = (1, 2, 1)$, we obtain $\rho_1 = 0.6126$, $\rho_2 = 0.751$ and $\rho_3 = 0.5366$. Using these values, the expected number of customers in the network can be computed as a sum of the individual L_i values for each node. Since the first and third nodes are $M/M/1$ and the second node is an $M/M/2$ queue, it can be shown that $(L_1, L_2, L_3) = (1.58, 3.45, 1.15)$; thus the expected number of customers in the system is $\sum_{i=1}^{3} L_i = 6.18$.

9.5 Optimization of Queueing Systems

In this chapter we have so far considered the *descriptive* aspects of queues and developed expressions for operating characteristics of different queueing processes. We now turn our attention to an important aspect of queueing theory that deals with *prescriptive* (i.e., normative) issues where the objective is to optimally design and control a queueing system.

Normative queueing models are concerned with finding the optimal system parameters, e.g., optimal number of servers, mean service rate, queue discipline and queue capacity. In general, a design model is *static* in nature in the sense that operating characteristics of a queue are used to superimpose a cost structure in order to compute the optimal value of a model parameter such as c, μ or K. These models are usually optimized using classical optimization techniques such as calculus and nonlinear programming [92, Ch. 17]. As a simple example, suppose it is possible to choose a service rate $\mu > \lambda$ for an $M/M/1$ queue to minimize the cost $C(\mu) = c_1 L + c(\mu)$ where c_1 is the waiting cost per customer per unit time and $c(\mu)$ is the operating cost per unit time as a function μ. If we assume for simplicity that $c(\mu) = c_2\mu$, since $L = \lambda/(\mu - \lambda)$ we obtain $C(\mu) = c_1\lambda/(\mu - \lambda) + c_2\mu$ as the cost function to be minimized. Following the standard steps of optimization we obtain the optimal value of μ as follows.

```
>    restart: # mm1muopt.mws
```

```
>    C:=c[ 1] * lambda/ (mu-lambda) +c[ 2] *mu;
```

$$C := \frac{c_1 \lambda}{\mu - \lambda} + c_2 \mu$$

```
>    diff (C,mu);
```

$$-\frac{c_1 \lambda}{(\mu - \lambda)^2} + c_2$$

```
>    sol:=solve(%,mu);
```

$$sol := \frac{1}{2}\frac{2c_2\lambda + 2\sqrt{c_2 c_1 \lambda}}{c_2}, \frac{1}{2}\frac{2c_2\lambda - 2\sqrt{c_2 c_1 \lambda}}{c_2}$$

```
>    expand(sol[ 1] );
```

$$\lambda + \frac{\sqrt{c_2 c_1 \lambda}}{c_2}$$

Thus, the optimal service rate is obtained as $\mu = \lambda + \sqrt{c_1 \lambda / c_2}$.

Dynamic control models may involve the determination of a decision variable as a function of the number of customers in the system or (as in the transportation queueing example to be discussed) as a function of the customers who are lost to the system. A dynamic control model may also deal with the determination of an optimal policy such as "provide no service (turn the server off) if there are m or fewer customers in the system; but when the number of customers increases to M ($M > m$), turn the server on and continue service until the number in the system drops to m" (Sobel [179]). The first type of dynamic control models may be analyzed using calculus. The second type of dynamic control models are normally analyzed using stochastic dynamic programming techniques.

For a comprehensive bibliography of papers published before 1977 that deal with the static design and dynamic control of queues, see Crabill, Gross and Magazine [55].

9.5.1 *A Transportation Queueing Process*

As an example of a queueing design problem, we consider a transportation queueing process studied by Bhat [32]. It is assumed that the arrival process of taxis at a taxi stand follows a Poisson process and the arrival process of customers follows a renewal process that is independent of the taxi arrivals. This transportation queueing process (an $M/G/1$ queue) is controlled by calling extra taxis whenever the total number of customers lost to the system reaches a certain predetermined number. If w is the predetermined upper limit of the number of lost customers, then in the system described an optimal choice of the number w becomes essential. Bhat finds the transient and steady-state behavior of the process by discovering regeneration points and using renewal theoretic arguments.

If $Q(t)$ is the number of taxis waiting at time t with $Q(0) = i \geq 0$, $A(t)$ is the number of taxis arriving in $(0, t]$ and $D(t)$ is the number of customers arriving in $(0, t]$, then $X(t) = Q(t) + A(t) - D(t)$ where $-X(t)$ denotes the number of customers lost at time t.

For the general $Q(t)$ process Bhat identifies the renewal periods, and using renewal arguments he finds that the limiting expected number of taxis waiting is

$$E(Q) = \frac{3}{w+2} + \frac{\lambda^2 E(V^2)}{2\rho(1-\rho)}$$

where $Q = \lim_{t\to\infty} Q(t)$, λ is the rate of taxi arrivals and V is the interarrival time of consecutive customers with $\rho = \lambda E(V) < 1$. Next, by using the concept of forward recurrence time of renewal theory, he obtains the limiting probability of unsatisfied customers ℓ_k, where

$$\ell_0 = \frac{3}{w+2}, \quad \ell_k = \frac{1}{w+2}, \quad 0 < k \le w-1.$$

The following Maple commands compute the expected number of lost customers in a renewal period (ELost). Using the expression for the expected number of taxis waiting (ETaxis) with c_1 as the cost of making a taxi wait and c_2 as the cost of losing a customer, the total cost C to the system is obtained and optimized to find the optimal value of w:

```
>   restart:  # bhat.mws
>   Sum(k/(w+2),k=1..w-1);
```

$$\sum_{k=1}^{w-1} \frac{k}{w+2}$$

```
>   ELost:=normal(value(%));
```

$$ELost := \frac{1}{2}\frac{w(w-1)}{w+2}$$

```
>   ETaxis:=3/(w+2)+lambda^2*EV2
    /(2*rho*(1-rho));
```

$$ETaxis := 3\frac{1}{w+2} + \frac{1}{2}\frac{\lambda^2 EV2}{\rho(1-\rho)}$$

```
>   C:=c[1]*ETaxis+c[2]*ELost;
```

$$C := c_1\left(3\frac{1}{w+2} + \frac{1}{2}\frac{\lambda^2 EV2}{\rho(1-\rho)}\right) + \frac{1}{2}\frac{c_2 w(w-1)}{w+2}$$

```
>   diff(C,w);
```

$$-3\frac{c_1}{(w+2)^2} + \frac{1}{2}\frac{c_2(w-1)}{w+2} + \frac{1}{2}\frac{c_2 w}{w+2} - \frac{1}{2}\frac{c_2 w(w-1)}{(w+2)^2}$$

```
>   sol:=solve(%,w);
```

$$sol := \frac{1}{2}\frac{-4c_2 + 2\sqrt{6c_2^2 + 6c_2 c_1}}{c_2}, \frac{1}{2}\frac{-4c_2 - 2\sqrt{6c_2^2 + 6c_2 c_1}}{c_2}$$

```
>   wOpt:=collect(normal(sol[1]),c[2]);
```

$$wOpt := -2 + \frac{\sqrt{6c_2{}^2 + 6c_2 c_1}}{c_2}$$

Thus, the optimal solution is $w^* = \sqrt{6(1 + c_1/c_2)} - 2$. Here we note that as the cost of losing a customer approaches infinity ($c_2 \to \infty$), then w^* is obtained as a value close to zero, as expected:

```
>   evalf(limit(wOpt,c[2]=infinity));
```

$$.449489743$$

9.5.2   An $M/M/1$ System with Controlled Arrivals

In dynamic control of some queues, it may be possible to deny admission to entering customers by physically controlling the arrival rate by rejecting every few customers when the system becomes crowded. In [149], Pliska considers such a system for an $M/M/1$ queue with arrival rate λ and service rate μ. It is assumed that arriving customers are turned away with probability $\frac{1}{2}$ whenever k or more customers are present in the system. Thus, when the number of customers in the system is k or more, the actual arrival rate is reduced to $\lambda/2$. In such a case, the queueing system can be modeled as a general birth-death process with a state-dependent arrival rate λ_n such that

$$\lambda_n = \begin{cases} \lambda & \text{for } n < k \\ \lambda/2 & \text{for } n \geq k . \end{cases} \tag{9.13}$$

Using the results presented for the steady-state solution of the birth-death processes in Section 9.2.4 with $c = 1$, λ_n as in (9.13) and $\mu_n = \mu$ for all n and assuming that $\rho = \lambda/(2\mu) < 1$, we obtain the steady-state probabilities p_n as

$$p_n = \begin{cases} (2\rho)^n p_0 & \text{for } n = 0, \ldots, k \\ 2^k \rho^n p_0 & \text{for } n = k, k+1, \ldots . \end{cases}$$

Since $\sum_{n=0}^{\infty} p_n = 1$, the unknown parameter p_0 is obtained as

$$p_0 = \frac{1}{\sum_{n=0}^{k-1}(2\rho)^n + \frac{(2\rho)^k}{1-\rho}} .$$

Now we suppose that each time a customer is served a reward of r is collected, but for each customer in the system a waiting cost of c is incurred. The problem is then to compute the optimal value of k that will maximize the profit per unit time $f(k)$ given as

$$f(k) = \theta r - cL.$$

Here, $\theta = \mu(1 - p_0)$ is the steady-state processing rate that is equal to the rate at which customers depart the system and

$$L = \sum_{n=0}^{\infty} n p_n = p_0 \left[\sum_{n=0}^{k-1} n(2\rho)^n + 2^k \sum_{n=k}^{\infty} n\rho^n \right]$$

$$= p_0 \left[\sum_{n=0}^{k-1} n(2\rho)^n + \frac{2^k}{\rho(1-\rho)^2} - 2^k \sum_{n=0}^{k-1} n\rho^n \right]$$

is the average number of customers in the system which, according to Pliska, "is not a pretty expression but is straightforward to compute."

Since f is not known to be necessarily concave with respect to k, Pliska suggests an exhaustive search procedure to find the optimal value of k. The development of the model and the solution of an example with $\lambda = 16$, $\mu = 14$, $r = 10$, $c = 3$ using an exhaustive search is presented in the following Maple worksheet.

```
>   restart: # pliska.mws
>   p[ 0] :=1/(' sum' ((2* rho)^n,n=0..k-1)
    +(2* rho)^k/(1-rho));
```

$$p_0 := \frac{1}{(\sum_{n=0}^{k-1} (2\rho)^n) + \frac{(2\rho)^k}{1-\rho}}$$

```
>   L:=' p[ 0]'* (' sum' (n* (2* rho)^n,n=0..k-1)
    +2^k* rho/(1-rho)^2
    -2^k*' sum' (n* rho^n,n=0..k-1));
```

$$L := p_0 \left(\left(\sum_{n=0}^{k-1} n(2\rho)^n \right) + \frac{2^k \rho}{(1-\rho)^2} - 2^k \left(\sum_{n=0}^{k-1} n\rho^n \right) \right)$$

```
>   theta:=mu*' (1-p[ 0] )' ;
```

$$\theta := \mu(1 - p_0)$$

```
>   f:=' r* theta-c* L' ;
```

$$f := r\theta - cL$$

```
>   lambda:=16; mu:=14; r:=10;
    c:=3; rho:=(lambda/2)/mu;
```

$$\lambda := 16$$

$$\mu := 14$$

$$r := 10$$

$$c := 3$$

$$\rho := \frac{4}{7}$$

```
>   Digits:=6;
```

$$Digits := 6$$

```
>   for k from 0 to 10 do
    print(k,evalf(f))   od;
```

0, 76.

1, 96.7273

2, 106.495

3, 111.647

4, 114.402

5, 115.743

6, 116.168

7, 115.961

8, 115.294

9, 114.278

10, 112.989

Thus, the optimal solution is $k^* = 6$ with a maximum expected profit of 116.168.

It is worth noting here that if $\rho > \frac{1}{2}$ (as in this example), then $L \to \infty$ as $k \to \infty$ and thus a finite value of $(k^* = 6)$ maximizes the objective function. The reason for this policy is as follows: If the traffic intensity is still high even with controls, it is best to start turning away some of the customers as soon the system size reaches some critical but finite value. However, if $\rho < \frac{1}{2}$, then L is bounded by $2\rho/(1 - 2\rho)$, in which case the optimal value of k may be very large.

9.5.3   Optimal Dynamic Service Rate Control

We now describe an application of deterministic optimal control theory to find the optimal time-dependent service rate $\mu(t)$ in an S-server, finite-capacity (N units), Markovian queue $M(t)/M(t)/S/N$ with nonhomogeneous (i.e., time-dependent) Poisson arrivals. As discussed in Parlar [141], for this process there are $N + 1$ Kolmogorov differential equations representing the time-dependent probabilities $x_n(t)$, $n = 0, \ldots, N$, and the service rate $u(t)$ is the control function that must be found to minimize a cost functional. [In this section we will use a notation that is standard in optimal control theory and denote the states $p_n(t)$ by $x_n(t)$ and the service rate control function $\mu(t)$ by $u(t)$.] The objective functional to be minimized includes the cost of waiting customers plus the cost of service over a specified time interval $[0, T]$ and a final time penalty cost of deviations from an expected queue length.

We assume that the customers arrive according to a Poisson process with time-dependent mean arrival rate $\lambda(t)$ and the service facility has S servers each with an exponential service time of rate $u(t)$ and a maximum of N customers can be in the system. A queueing model fitting this description has been applied to forest fire fighting by Bookbinder and Martell [35], landing clearance of aircraft by Koopman [113] and scheduling police patrol cars in New York City by Kolesar et al. [112]. As described in Gross and Harris [76], the state probabilities for this system are obtained from the Kolmogorov equations (9.1)–(9.2) of the birth-death processes as follows: For $n = 0$,

$$\dot{x}_0(t) = -\lambda(t)x_0(t) + u(t)x_1(t),$$

for $n = 1, \ldots, S - 1$,

$$\dot{x}_n(t) = \lambda(t)x_{n-1}(t) - [\lambda(t) + nu(t)]x_n(t) + (n+1)u(t)x_{n+1}(t),$$

for $n = S, \ldots, N - 1$,

$$\dot{x}_n(t) = \lambda(t)x_{n-1}(t) - [\lambda(t) + Su(t)]x_n(t) + Su(t)x_{n+1}(t),$$

and finally, for $n = N$,

$$\dot{x}_n(t) = \lambda(t)x_{n-1}(t) - Su(t)x_N(t).$$

For this system, the initial condition is $\sum_{n=0}^{N} x_n(0) = 1$. Also, since $x_n(t)$ are probabilities, we also have the point constraints $\sum_{n=0}^{N} x_n(t) = 1$, $x_n(t) \geq 0$, $n = 0, \ldots, N$ for all $t \in [0, T]$. However, if $u(t) \geq 0$ then the point constraints are satisfied automatically; see Klein and Gruver [109]. Defining $P'(t) = [x_0(t), \ldots, x_n(t)]$ where prime denotes transposition, we can write the system more compactly as $\dot{P}(t) = u(t)FP(t) + \lambda(t)GP(t)$ with the constant matrices F and G having dimension $(N + 1) \times (N + 1)$. (For details, see Parlar [141].)

A performance criterion incorporating the cost of waiting customers (for example, fires burning in Bookbinder and Martell's forestry model or aircraft awaiting landing clearance in Koopman's model) and the cost of service rate over $[0, T]$ seems to be a reasonable objective for $M(t)/M(t)/S/N$ queues. Since the state variable $x_n(t)$ is the probability of n customers in the system at time t, and since the system capacity is N units, the expected number of customers $L(t)$ at time t is

$$L(t) = \sum_{n=0}^{N} nx_n(t) = C'P(t)$$

where $C' = (0, 1, 2, \ldots, N)$. We now introduce a service rate cost function $\sigma[u(t)]$ and add a desired (target) expected queue size at the final time T and penalize deviations from this level. Thus, the objective functional to be minimized with a proper choice of $u(t)$ is

$$J = \int_0^T [C'P(t) + \sigma(u)] \, dt + \tfrac{1}{2}b[C'P(T) - k]^2$$

where k is the desired end-of-day expected queue length and b gives the relative cost of deviating from desired k. Although each state variable $x_n(t)$ is restricted to take values between 0 and 1, it is *not* necessary to introduce these constraints explicitly into our formulation since they will be automatically satisfied—as explained above—if $u(t) \geq 0$, for $t \in [0, T]$. Since it may be physically impossible to exceed some upper bound $\bar{u}$ on the control, this implies that we must add to the formulation constraints on the control function in the form $0 \leq u(t) \leq \bar{u}$, $t \in [0, T]$. Now the optimal control problem is given as

$$\min_{u(t)} J = \int_0^T [C'P(t) + \sigma(u)]\,dt + \tfrac{1}{2}b[C'P(T) - k]^2$$

subject to

$$\dot{P}(t) = u(t)FP(t) + \lambda(t)GP(t)$$

$$e'P(0) = 1, \quad e' = (1, 1, \ldots, 1)$$

$$0 \leq u(t) \leq \bar{u}.$$

Introducing a new $(N+2)$nd state variable x_{N+1} as $x_{N+1}(t) = \int_0^t [C'P(\tau) + \sigma(u)]\,d\tau$ so that $\dot{x}_{N+1}(t) = C'P(t) + \sigma(u)$ and $x_{N+1}(0) = 0$, the problem can be reformulated in Mayer form [162, p. 28] as follows:

$$\min_{u(t)} J = \tfrac{1}{2}b[C'P(T) - k]^2 + x_{N+1}(T)$$

subject to

$$\dot{P}(t) = u(t)FP(t) + \lambda(t)GP(t), \quad e'P(0) = 1$$

$$\dot{x}_{N+1}(t) = C'P(t) + \sigma(u), \quad x_{N+1}(0) = 0$$

$$0 \leq u(t) \leq \bar{u}.$$

Using Pontryagin's maximum principle [169], it can be shown that [141] to find the optimal service rate $u(t)$ in an $M(t)/M(t)/S/N$ queue one has to solve the two-point boundary-value problem (TPBVP) given by the following differential and algebraic equations:

$$\dot{W}(t) = -[u(t)F'W(t) + \lambda(t)G'W(t) + C], \quad W(T) = bC[C'P(T) - k] \quad (9.14)$$

$$\dot{P}(t) = u(t)FP(t) + \lambda(t)GP(t), \quad e'P(0) = 1, \quad (9.15)$$

$$\dot{x}_{N+1}(t) = C'P(t) + \sigma(u), \quad x_{N+1}(0) = 0 \quad (9.16)$$

$$u(t) = \begin{cases} 0, & \text{if } g(-P'(t)FW(t)) < 0 \\ g(-P'(t)FW(t)), & \text{if } 0 \leq g(-P'(t)FW(t)) \leq \bar{u} \\ \bar{u}, & \text{if } g(-P'(t)FW(t)) > \bar{u} \end{cases} \quad (9.17)$$

where $g(Z)$ is the inverse function of $d\sigma/du$ evaluated at Z. This TPBVP has $2(N+1)+1$ differential equations where the first $N+1$ for $\dot{W}(t)$ have known final time conditions and the last $N+2$ for $[\dot{P}(t), \dot{x}_{N+1}(t)]$ have known initial time conditions. Also, $\dot{W}(t)$, $\dot{P}(t)$ and $\dot{x}_{N+1}(t)$ all depend on $u(t)$, which in turn depends on the values of $W(t)$ and $P(t)$. Thus, like many optimal control problems arising in engineering applications, this one also requires the application of a numerical technique for its solution.

As an example consider now the case with $N = S = 1$. Letting $\sigma(u) = \frac{1}{2}au^2$, $\bar{u} = 1$ and noting that $x_0(t) + x_1(t) = 1$, the conditions (9.14)–(9.17) reduce to the following TPBVP with three differential equations and one algebraic relation:

$$\dot{w}_1(t) = w_1(t)[\lambda(t) + u(t)] - 1, \quad w_1(T) = b[x_1(T) - k]$$
$$\dot{x}_1(t) = \lambda(t)[1 - x_1(t)] - u(t)x_1(t), \quad x_1(0) = x_{10}$$
$$\dot{x}_2(t) = x_1(t) + \frac{1}{2}a[u(t)]^2, \quad x_2(0) = 0$$

$$u(t) = \begin{cases} 0, & \text{if } w_1(t)x_1(t)/a < 0 \\ w_1(t)x_1(t)/a, & \text{if } 0 \le w_1(t)x_1(t)/a \le 1 \\ 1, & \text{if } w_1(t)x_1(t)/a > 1. \end{cases}$$

For the time-dependent arrival rate we choose $\lambda(t) = 20t^2(1-t)^3$. Note that as $t \to 0$ or $t \to 1$ the arrival rate approaches zero. The values of other parameters are given as $a = 0.4$, $b = 2.5$, $k = 0.1$, $T = 1$, $x_1(0) = 0.8$ and $x_2(0) = 0.0$.

We now solve this TPBVP numerically using Maple's piecewise() function and its knowledge of numerical solution of differential equations.

> restart: # mmsncont.mws

We first define $u(t)$ as a piecewise function in terms of the unknown functions $w_1(t)$ and $x_1(t)$:

> u:=unapply(piecewise(w[1] (t)*x[1] (t)/a<0,
0,w[1] (t)*x[1] (t)/a>=0 and w[1] (t)*x[1] (t)/a<=1,
w[1] (t)*x[1] (t)/a,w[1] (t)*x[1] (t)/a>1,1),t);

$$u := t \to \text{piecewise}(\frac{w_1(t)x_1(t)}{a} < 0,\ 0,$$
$$-\frac{w_1(t)x_1(t)}{a} \le 0 \text{ and } \frac{w_1(t)x_1(t)}{a} - 1 \le 0,\ \frac{w_1(t)x_1(t)}{a},\ 1 < \frac{w_1(t)x_1(t)}{a},$$
1)

Next, the system of differential equations is defined:

> adjoint[1] :=diff(w[1] (t),t)
=w[1] (t)* (lambda(t)+u(t))-1;

$$adjoint_1 := \frac{\partial}{\partial t} w_1(t) = w_1(t) \left(\lambda(t) + \left(\begin{cases} 0 & \%1 < 0 \\ \%1 & -\%1 \le 0 \text{ and } \%1 - 1 \le 0 \\ 1 & 1 < \%1 \end{cases} \right) \right) - 1$$

$$\%1 := \frac{w_1(t)x_1(t)}{a}$$

> state[1] :=diff(x[1] (t),t)
=lambda(t)* (1-x[1] (t))-u(t)*x[1] (t);

$$state_1 := \frac{\partial}{\partial t} x_1(t) = \lambda(t) (1 - x_1(t)) - \left(\begin{cases} 0 & \%1 < 0 \\ \%1 & -\%1 \le 0 \text{ and } \%1 - 1 \le 0 \\ 1 & 1 < \%1 \end{cases} \right) x_1(t)$$

$$\%1 := \frac{w_1(t)x_1(t)}{a}$$

```
>    state[ 2] :=diff(x[ 2] (t),t)
     =x[ 1] (t)+(1/2)*a*u(t)^2;
```

$$state_2 := \frac{\partial}{\partial t} x_2(t) = x_1(t) + \frac{1}{2} a \left(\left\{ \begin{array}{ll} 0 & \%1 < 0 \\ \%1 & -\%1 \le 0 \text{ and } \%1 - 1 \le 0 \\ 1 & 1 < \%1 \end{array} \right. \right)^2$$

$$\%1 := \frac{w_1(t) x_1(t)}{a}$$

```
>    desystem:=adjoint[ 1] ,state[ 1] ,
     state[ 2] ;
```

$$desystem := \frac{\partial}{\partial t} w_1(t) = w_1(t) (\lambda(t) + \%2) - 1,$$

$$\frac{\partial}{\partial t} x_1(t) = \lambda(t) (1 - x_1(t)) - \%2 x_1(t), \quad \frac{\partial}{\partial t} x_2(t) = x_1(t) + \frac{1}{2} a \%2^2$$

$$\%1 := \frac{w_1(t) x_1(t)}{a}$$

$$\%2 := \left\{ \begin{array}{ll} 0 & \%1 < 0 \\ \%1 & -\%1 \le 0 \text{ and } \%1 - 1 \le 0 \\ 1 & 1 < \%1 \end{array} \right.$$

```
>    funcs:={ w[ 1] (t),x[ 1] (t),x[ 2] (t)} ;
```

$$funcs := \{x_2(t), w_1(t), x_1(t)\}$$

Our TPBVP where $w_1(T) = b[x_1(T) - k]$ is normally solved using the "initial value shooting method," which starts with a guess for the value of $w_1(0)$, say w_{10}. With this method, using the initial conditions $w_1(0) = w_{10}, x_1(0) = 0.8$ and $x_2(0) = 0$, the differential equation system is solved numerically. (Maple conveniently incorporates the piecewise algebraic relation for $u(t)$ in the numerical solution of the system.) If at T we observe that $w_1(T) = b[x_1(T) - k]$, then the solution of the TPBVP has been found; otherwise the initial guess is modified and the system is re-solved. For a detailed description of shooting methods see Roberts and Shipman [155]. Sethi and Thompson [169, Ch. 11] provide a summary description of the initial value shooting method for problems that require guessing for a single state variable (which is the case in this example).

In our problem, we manually implemented the initial value shooting method and found the value of $w_1(0)$ as .7475 for which the final time difference is $w_1(T) - b[x_1(T) - k] = .00074165$—an acceptable error. With the definition of $\lambda(t)$, the system is solved numerically using the output=listprocedure option so that the resulting numerical trajectory can be used in plotting the required graphs.

```
>    init:=w[ 1] (0)=.7475, x[ 1] (0)=.8,
     x[ 2] (0)=0;
```

$$init := w_1(0) = .7475, x_1(0) = .8, x_2(0) = 0$$

```
>    lambda:=t->20*t^2* (1-t)^3;
     a:=.4; b:=2.5; k:=.1; T:=1;
```

$$\lambda := t \to 20\,t^2\,(1-t)^3$$

$$a := .4$$

$$b := 2.5$$

$$k := .1$$

$$T := 1$$

```
>    Sol:=dsolve({ desystem,init} ,funcs,
     type=numeric,output=listprocedure);
```

$$Sol := [t = (\mathbf{proc}(t) \ldots \mathbf{end}), x_2(t) = (\mathbf{proc}(t) \ldots \mathbf{end}),$$
$$w_1(t) = (\mathbf{proc}(t) \ldots \mathbf{end}), x_1(t) = (\mathbf{proc}(t) \ldots \mathbf{end})]$$

```
>    wSol[ 1] :=subs(Sol,w[ 1] (t));
     xSol[ 1] :=subs(Sol,x[ 1] (t));
     xSol[ 2] :=subs(Sol,x[ 2] (t));
```

$$wSol_1 := \mathbf{proc}(t) \ldots \mathbf{end}$$

$$xSol_1 := \mathbf{proc}(t) \ldots \mathbf{end}$$

$$xSol_2 := \mathbf{proc}(t) \ldots \mathbf{end}$$

```
>    init; wSol[ 1] (T)-b* (xSol[ 1] (T)-k);
```

$$w_1(0) = .7475,\ x_1(0) = .8,\ x_2(0) = 0$$

$$.0007416587$$

The trajectories for the functions $w_1(t)$ and $x_2(t)$ for $t \in [0, T]$ are presented in Figure 9.2. Note here that $w_1(0) = .7475$ and $x_2(0) = 0$.

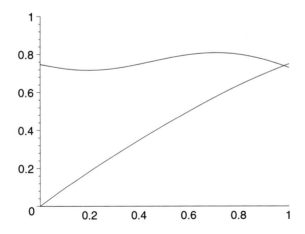

FIGURE 9.2. Optimal trajectories of $w_1(t)$ and $x_2(t)$ for $t \in [0, 1]$ with $w_1(0) = 0.7475$ and $x_2(0) = 0$.

```
>   with(plots):
>   odeplot(Sol,[[ t,w[ 1] (t)],[ t,x[ 2] (t)]] ,
    0..T,view=[ 0..T,0..T] ,color=black);
```

The other trajectories of importance are those that correspond to the control function $u(t)$ and the states $[x_0(t), x_1(t)]$, which are plotted in Figure 9.3 where
$u(0) = 1$, $x_0(0) = 0.2$ and $x_1(0) = 0.8$.

```
>   x[ 0] (t) :=1-x[ 1] (t);
```

$$x_0(t) := 1 - x_1(t)$$

```
>   odeplot(Sol,[[ t,u(t)] ,
    [ t,x[ 0] (t)] ,[ t,x[ 1] (t)]] ,0..T,
    view=[ 0..T,0..T] ,color=black);
```

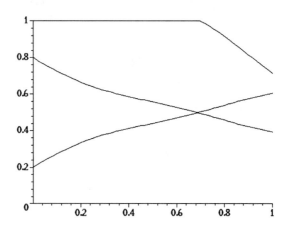

FIGURE 9.3. Optimal trajectories of $u(t)$, $x_0(t)$ and $x_1(t)$ for $t \in [0, 1]$ with $u(0) = 1$,
$x_0(0) = 0.2$ and $x_1(0) = 0.8$.

Finally, since $J = \frac{1}{2}b[x_1(T) - k]^2 + x_2(T)$, this value of the objective functional
is computed as follows:

```
>   J:=(1/2)*b* (xSol[ 1] (T)-k)^2+xSol[ 2] (T);
```

$$J := .8562668471$$

We thus see that an otherwise difficult-to-solve dynamic optimization problem
can be analyzed relatively easily using Maple's dsolve() and piecewise()
functions.[8]

---

[8]For details of an alternative solution method for this control problem, the "Newton-Raphson
boundary condition iteration," see Parlar [141].

9.6   Summary

The main goal of this chapter was to present a survey of Markovian queueing systems.[9] The limiting distribution and the related operating characteristics of different types of Markovian queues were discussed. Using Maple's ordinary and partial differential equation solvers [i.e., `dsolve()` and `pdsolve()`] we were able to find the transient solutions for a finite capacity queue and for the ample server queue. Next, a serial queue and Jackson networks were analyzed. The chapter ended with a detailed discussion of optimization of some types of queueing systems. In particular, Maple's ability to solve differential equations defined by piecewise functions was helpful in determining the optimal control (i.e., service rate) trajectory in a time-dependent queueing system.

9.7   Exercises

1. Yule process is a pure birth process with state-dependent birth rate $\lambda_n = n\lambda$, $n = 0, 1, \ldots$. This process represents the growth of a population where each member gives birth at an exponential rate λ. Let $X(t)$ denote the population size at time t and assume that the population starts with a single member, i.e., $X(0) = 1$. Find the transient solution $p_n(t) = \Pr[X(t) = n]$ for this process.

2. Consider a queueing system where both arrival and service rates are state dependent. Assume that the arrival rate λ_n decreases and service rate μ_n increases as the number n in the system increases, i.e.,

$$\lambda_n = \begin{cases} (N - n)\lambda, & n \leq N \\ 0, & n > N, \end{cases}$$

$$\mu_n = \begin{cases} n\mu, & n \leq N \\ 0, & n > N. \end{cases}$$

What are the differential equations for the probability $p_n(t) = \Pr(n$ in the system at time $t)$?

3. Consider a Markovian queue with two servers where the arrival and service rates are given as

$$\lambda_n = \lambda, \quad n \geq 0$$

$$\mu_n = \begin{cases} \mu, & n = 1 \\ 2\mu, & n \geq 2. \end{cases}$$

Find the steady-state probabilities $p_n = \lim_{t \to \infty} p_n(t)$.

---

[9]An example of a non-Markovian queue ($M/G/1$) was presented in Section 7.4.4 of Chapter 7 where the queue was analyzed using the imbedded Markov chain technique.

4. The $M/M/c/c$ model: Consider a c-server Markovian queue with a capacity of c spaces. When all servers are busy an arriving customer leaves the system without waiting for service. Thus, the arrival and service rates are $\lambda_n = \lambda$, $\mu_n = n\mu$ for $n = 0, 1, 2, \ldots, c - 1$ and $\lambda_n = 0$, $\mu_n = c\mu$ for $n \geq c$. Show that the steady-state probability $p_c(\rho)$ that an arriving customer is lost to the system is given by

$$p_c(\rho) = \frac{\rho^c/c!}{\sum_{k=0}^{c} \rho^k/k!}$$

where $\rho = \lambda/\mu$. Note that $p_c(\rho)$ is a function of both the traffic intensity ρ and the number of servers c.

5. The transient solution $p_n(t)$, $t \geq 0$, $n = 1, 2, \ldots$ for the $M/M/1$ queue with an initially empty system is given as

$$p_n(t) = e^{-(\lambda+\mu)t} \left\{ \rho^{n/2} I_n(at) + \rho^{(n-1)/2} I_{n+1}(at) \right.$$
$$\left. + (1 - \rho)\rho^n \sum_{k=n+2}^{\infty} \rho^{-k/2} I_k(at) \right\}$$

where $I_n(x)$ is the modified Bessel function of the first kind, $\rho = \lambda/\mu$ and $a = 2\sqrt{\lambda\mu}$; see Medhi [130, p. 120] and Saaty [161, p. 340]. Let $\lambda = 1$ and $\mu = 2$ and plot the transient probability trajectory for $0 \leq t \leq 10$ and $n = 1, 2, 3$.

HINT: In Maple the modified Bessel function of the first kind is represented by BesselI.

6. Let W_n, $n \geq 1$ be the waiting time of the nth customer in a $G/G/1$ queue. Show that $W_{n+1} = \max(0, W_n + V_n - U_n)$ where V_n is the service time of the nth customer and U_n is the interarrival time between the nth and $(n + 1)$st customer.

7. Consider a closed queueing network with three nodes and two customers. Assume that a customer chooses the next node which has the fewest number of customers and ties are broken by flipping an unbiased coin. Departure rate from node j is μ_j, $j = 1, 2, 3$. The state space of the stochastic process representing this queue can be described by a vector with three components (n_1, n_2, n_3). Write down the infinitesimal generator for this stochastic process and compute the steady state probabilities.

8. Consider the $M/M/c/c$ model presented in Exercise 4. Since $p_c(\rho)$ is the probability that an arriving customer is lost to the system, the *effective* arrival rate is $\lambda_{\text{eff}} = [1 - p_c(\rho)]\lambda$. If each arriving customer generates an average of r dollars of profit and each server costs $k\mu$ dollars per hour, the average profit rate can be written as $A = r\lambda_{\text{eff}} - ck\mu$. Let $\rho = 1$ (i.e.,

$\lambda = \mu$) and determine whether it is preferable to have two servers instead of one.

HINT: The results should be found in terms of the parameters r and k.

10
Simulation

10.1   Introduction

Simulation[1] is a computer-based probabilistic modeling tool that is used to imitate the behavior of an existing or a proposed system. If an existing system such as a bank branch is experiencing long queues and long customer waiting times, the manager of the branch may use simulation to examine different alternatives for improving service without actually making any physical changes in the system.

For example, the effect of increasing the number of available tellers and/or changing the queue design (e.g., combining parallel queues into a single queue) on the average queue length and the average waiting times can be examined without physically altering the system. Behavior of the proposed system can be observed by running computer simulations of different alternatives over time and measuring the operating characteristics of the system.

Of course, in some cases the system under study may be simple enough so that the queueing models that were presented in Chapter 9 may be used to model it. But, if the random variables representing the arrival and/or service processes are not exponential (thus making the system non-Markovian) and if the proposed design is complex (e.g., the bank branch may be a complicated queueing network) none of the available queueing models may be useful in modeling the problem. In those cases, simulation becomes a necessity in order to model and analyze the system.

---

[1] "Simulation" is derived from the Latin word *simulare* which means "to copy, imitate or represent."

Along with linear programming and statistical methods, simulation has become one of the most important operations research techniques used to solve practical problems.[2] There are now several very powerful and flexible simulation software programs—such as Arena [104] and Visual SLAM [153]—that simplify the modeling and simulation of almost any real-life system that exhibits probabilistic behavior.

The most basic computational ingredient in simulation involves the generation of random variables distributed uniformly between 0 and 1. These are then used to generate other random variables—such as the Weibull or Erlang random variables that may represent a teller's service time—and the simulation is run long enough to obtain an estimate of the operating characteristics of the system, e.g., average waiting time of customers and the percentage of time the facility is idle. Since the result of each run is a random sample, a simulation study is a statistical experiment that must be conducted using statistical tools such as point estimation, confidence intervals, hypothesis testing. For a discussion of these important statistical issues, the reader should consult specialized simulation textbooks such as Law and Kelton [118, Chs. 8–9] and Banks, Carson and Nelson [12, Ch. 12].

Clearly, Maple is not designed to be used as a full-fledged simulation language. However, Maple's ability to generate random variates from nearly 20 continuous and discrete random variables (such as the exponential, gamma, Weibull, normal, binomial and Poisson) makes it a useful teaching tool for presenting many interesting simulation examples.

Example 97 *Port operations.* As a motivating example, consider the day-to-day operations of a port that is used to unload tankers containing imported goods. The tankers arrive very early in the morning and with the available workforce, it is possible to unload a maximum of four tankers in one day. Based on historical data, the number of tankers arriving, X, is binomially distributed with parameters $n = 5$ and $p = 0.7$, i.e., $\Pr(X = k) = \binom{5}{k}(0.7)^k(0.3)^{5-k}$, $k = 0, 1, \ldots, 5$. Since the workers can unload a maximum of four tankers per day, any tankers still unloaded must wait until the next day for unloading. The problem is to find the long-run average number of tankers delayed and propose improvements if necessary.

This appears to be a queueing problem with a random number of tankers arriving with an average arrival rate of $\lambda = np = 3.5$ per day and a maximum service rate of $m = 4$ tankers per day. However, since arrival events take place at discrete time (i.e., during early mornings) and the number of arriving tankers is binomial, none of the queueing models presented in Chapter 9, Queueing Systems, seems to be suitable for this problem. Thus, we decide to simulate this process to find out the long-run average number of tankers delayed.

---

[2]See the paper by Lane, Mansour and Harpell [116] that describes the results of a survey of educators and practitioners who teach and use operations research techniques. The survey found that the three techniques mentioned in this paragraph were consistently rated by the OR educators and practitioners as the most important.

In this case, it is not too difficult to see that if we define X_t as the total number of delayed tankers at the beginning of day t, then we can write

$$X_{t+1} = X_t + A_t - \min(4, X_t + A_t), \quad t = 1, 2, \ldots$$

where $X_1 = x_1$ is the number of tankers waiting when simulation starts at the beginning of day 1 and A_t is the number of tankers arriving on day t. This equation follows since $X_t + A_t$ is the total number of tankers that need to be unloaded on day t, and since the workers cannot unload more than four tankers per day, the number of tankers delayed to day $t+1$ is the difference $X_t + A_t - \min(4, X_t + A_t)$.

In this problem, even though we have a mathematical model of the process, it may not be easy to find analytical expressions for the average number of delayed tankers. The fact that the arrivals A_t are binomially distributed and the state equations involve the $\min(\cdot, \cdot)$ operator makes this a nontrivial problem.[3]

The following Maple worksheet simulates this problem for 25 days, computes the average number of delayed tankers and displays the histogram plot of the distribution of the number delayed.

```
>    restart: # Tankers.mws
>    with(stats): #readlib(randomize): randomize():
```

We first set the values of $m = 4$ and the number of days of simulation $T = 25$.

```
>    m:=4; T:=25;
```

$$m := 4$$

$$T := 25$$

The binomial arrival process has parameters $(n, p) = (5, 0.7)$.

```
>    n:=5; p:=.7;
```

$$n := 5$$

$$p := .7$$

```
>    pb:=(n,k)->binomial(n,k)*p^k*(1-p)^(n-k);
```

$$pb := (n, k) \to binomial(n, k)\, p^k\, (1 - p)^{(n-k)}$$

We compute the binomial probabilities $p(k)$, $k = 1, \ldots, n = 5$ and check to see that $\sum_{k=1}^{n} p(k) = 1$.

```
>    seq([ k,pb(n,k)] ,k=0..n); sum(pb(n,k),k=0..n);
```

$$[0, .00243], [1, .02835], [2, .13230], [3, .30870], [4, .36015], [5, .16807]$$

$$1.$$

How would we go about simulating the binomial random variable X with parameters $(n, p) = (5, 0.7)$? Since the density function $p(k)$ of X is available, we

---

[3]Actually, it is also possible to model this problem as a discrete-time Markov chain. Once the transition probability matrix of the chain is identified, the steady-state distribution of x_t can be computed numerically.

could first compute the cumulative distribution function $P(k) = \sum_{i=0}^{k} p(i)$ of X as given in the following table.

k	0	1	2	3	4	5
$p(k)$	.00243	.02835	.1323	.3087	.3603	.1681
$P(k)$	.00243	.03078	.1631	.4718	.8319	1.0

Next, generating a *uniform* random number U between 0 and 1, we could simulate the binomial random variable X using the formula

$$X = \begin{cases} 0, & \text{if } 0.00000 < U < 0.00243 \\ 1, & \text{if } 0.00243 \le U < 0.03078 \\ 2, & \text{if } 0.03078 \le U < 0.16310 \\ 3, & \text{if } 0.16310 \le U < 0.47180 \\ 4, & \text{if } 0.47180 \le U < 0.83190 \\ 5, & \text{if } 0.83190 \le U < 1.00000. \end{cases}$$

For example, if we happen to generate a uniform random number $U = 0.45134$, then this would correspond to a simulated value of $X = 3$ tankers arriving.

This method is known as the *discrete inverse transform* method since, after generating a U, we find the value of $X = k$ by finding the interval $[P(k-1), P(k))$ in which U lies—that is, we find the *inverse* of $P(U)$.[4]

Fortunately, Maple users do not have to be concerned about the mechanics of this method since it has already been implemented in Maple. A binomial random variate for the number of tankers arriving, a, can be generated easily using the random[binomiald] command that implements essentially same the procedure we have just described.

```
>   a:=random[ binomiald[ n,p]] (1);
```

$$a := 3.0$$

We assume that at the start of day 1, there are $x_1 = 2$ tankers waiting to be unloaded.

```
>   x[ 1] :=2;
```

$$x_1 := 2$$

```
>   for t from 1 to T do
    a[ t] :=random[ binomiald[ n,p]] (1);
    x[ t+1] :=x[ t] +a[ t] -min (m,x[ t] +a[ t] );

>   od:

>   seq ([ t,[ x[ t] ,a[ t]] ,[ x[ t+1]]] ,t=1..T);
```

---

[4]There are more efficient methods for generating binomial random variates; see, for example, Ross [160, pp. 50–51].

[1, [2, 3.0], [1.0]], [2, [1.0, 3.0], [0]], [3, [0, 4.0], [0]], [4, [0, 4.0], [0]],
[5, [0, 4.0], [0]], [6, [0, 2.0], [0]], [7, [0, 4.0], [0]], [8, [0, 4.0], [0]],
[9, [0, 4.0], [0]], [10, [0, 3.0], [0]], [11, [0, 3.0], [0]],
[12, [0, 4.0], [0]], [13, [0, 2.0], [0]], [14, [0, 2.0], [0]],
[15, [0, 5.0], [1.0]], [16, [1.0, 4.0], [1.0]], [17, [1.0, 3.0], [0]],
[18, [0, 4.0], [0]], [19, [0, 5.0], [1.0]], [20, [1.0, 5.0], [2.0]],
[21, [2.0, 2.0], [0]], [22, [0, 2.0], [0]], [23, [0, 3.0], [0]],
[24, [0, 4.0], [0]], [25, [0, 3.0], [0]]

The simulated number of delayed tankers for T days is as follows.

```
>   Delayed:=[ seq(x[ t] ,t=1..T)] ;
```

Delayed := [2, 1.0, 0, 0, 0, 0, 0, 0, 0, 0, 0, 0, 0, 0, 0, 0, 1.0, 1.0, 0, 0, 1.0, 2.0, 0, 0, 0, 0]

The average of the simulated values over 25 days is found as 0.32 tankers.

```
>   describe[ mean] (Delayed);
```

$$.3200000000$$

Finally, we plot the histogram of the number of delayed tankers in Figure 10.1.

```
>   Ranges:=[ 0..1,1..2,2..3,3..4,4..5,5..6,6..7] ;
```

Ranges := [0..1, 1..2, 2..3, 3..4, 4..5, 5..6, 6..7]

```
>   with(transform) :
>   List:=tallyinto(Delayed,Ranges);
```

List := [Weight(0..1, 19), Weight(1..2, 4), Weight(2..3, 2),
Weight(3..4, 0), Weight(4..5, 0), Weight(5..6, 0),
Weight(6..7, 0)]

```
>   with(statplots) :
>   histogram(List);
```

Naturally, running the simulation for a "run length" of only $T = 25$ days and only *once*—"number of runs"—would not provide a very accurate estimate of the average number of delayed tankers. It is important to emphasize that the proper choice of the run length and the number of runs is a crucial factor in understanding the system behavior.

Since the result of each run is a statistical sample, it is advisable to have several runs of sufficient length to obtain accurate results from the simulation study. In fact, there are procedures for determining the correct number of runs, R, in order to obtain a specified precision to estimate the performance measures of the system under study. A *rough* estimate for determining R is as follows:

A $(1 - \alpha)$% confidence interval for the true mean μ of a performance measure is given by

$$\Pr(\bar{X} - h \leq \mu \leq \bar{X} + h) = 1 - \alpha.$$

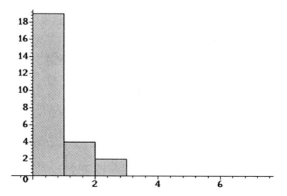

FIGURE 10.1. Distribution of delayed tankers obtained by simulating the process for a run length of $T = 25$ days.

Here, $\bar{X}$ is the sample mean obtained from an initial sample of, say, R_0 runs and h is the half-width of the desired confidence interval

$$h = \frac{z_{\alpha/2}s}{\sqrt{R}}. \tag{10.1}$$

In equation (10.1), s is the sample standard deviation obtained from the initial runs and (for a large R) $z_{\alpha/2}$ is the Z-score, which bounds a right tail equal to $\alpha/2$. If we want to choose R such that $h \leq \delta$, then

$$R \geq \left(\frac{z_{\alpha/2}s}{\delta}\right)^2 \tag{10.2}$$

is the minimum number of runs that should be conducted.

For this example, we increase the value of the run length T to 1000 days and simulate the system for $R_0 = 5$ times. This gave 0.518, 0.383, 0.445, 0.513 and 0.458 as the average number of delayed tankers. Taking the average of these five samples gives $\bar{X} = 0.463$.

```
>    restart: # NOR.mws
>    with(stats,describe,statevalf);
```

$$[\textit{describe, statevalf}]$$

```
>    Means:=[ .518,.383,.445,.513,.458] ;
     R[ 0] :=nops(Means);
```

$$\textit{Means} := [.518, .383, .445, .513, .458]$$

$$R_0 := 5$$

```
>    XBar:=describe[ mean] (Means);
```

$$\textit{XBar} := .4634000000$$

We choose $\delta = 0.01$, $\alpha = 0.05$ and find that $z = 1.9599$. The sample standard deviation is $s = 0.0495$. Using (10.2) gives $R \geq 94$ runs.

```
>    delta:=.01;
```
$$\delta := .01$$
```
>    s:=describe[ standarddeviation] (Means);
```
$$s := .04954432359$$
```
>    alpha:=.05;
```
$$\alpha := .05$$
```
>    z:=statevalf[ icdf,normald] (1-alpha/2);
```
$$z := 1.959963985$$
```
>    R:=(s*z/delta)^2;
```
$$R := 94.29398485$$

For a careful discussion of determining the number of runs, see Banks, Carson and Nelson [12, Sec. 12.4] and Law and Kelton [118, Chapter 8].

We continue this chapter with a discussion of the techniques for generating (pseudo-) random numbers and the "quality" of Maple's uniform random number generator. This is followed by some examples of *static* Monte Carlo simulation where time plays no role. The chapter ends with examples of *dynamic* simulation models that represent systems evolving over time (such as the port operations in Example 97).

10.2   Generating (Pseudo-) Random Numbers

As we indicated, the most basic computational ingredient in a simulation study is the generation of uniform random variables U distributed between 0 and 1 with the probability density function

$$f(u) = \begin{cases} 1, & 0 \leq u \leq 1 \\ 0, & \text{otherwise.} \end{cases}$$

It is easy to show that this r.v. has mean $E(U) = \frac{1}{2}$, second moment $E(U^2) = \frac{1}{3}$, third moment $E(U^3) = \frac{1}{4}$ and variance $\text{Var}(U) = \frac{1}{12}$.

```
>    restart: # Uniform.mws
>    f:=u->1;
```
$$f := 1$$
```
>    EU:=int (u* f (u),u=0..1);
```
$$EU := \frac{1}{2}$$

```
>   EU2:=int(u^2*f(u),u=0..1);
```

$$EU2 := \frac{1}{3}$$

```
>   EU3:=int(u^3*f(u),u=0..1);
```

$$EU3 := \frac{1}{4}$$

```
>   VarU:=int(u^2*f(u),u=0..1)-EU^2;
```

$$VarU := \frac{1}{12}$$

We now discuss a widely used method for generating uniform random numbers.

10.2.1   Mixed-Congruential Method

A sequence of random numbers generated by a computer is not truly random since the computer uses a deterministic equation $r_{n+1} = f(r_n)$ to generate the $(n + 1)$st random number r_{n+1} from the nth random number r_n. For this reason, the computer-generated random numbers are "pseudo-random" numbers that only give the "appearance" of randomness.

There are several methods to generate random numbers and the most widely used among these is the *mixed-congruential method*. This method starts with an initial seed value z_0 and then recursively computes a sequence of integers z_{n+1}, $n \geq 0$ using

$$z_{n+1} = (az_n + b) \quad \text{mod } m, \quad n = 0, 1, 2, \ldots$$

where a (the multiplier), b (the increment) and m (the modulus) are given non-negative integers. Note here that to find z_{n+1}, we divide $(az_n + b)$ by m and the remainder is taken as the value of z_{n+1}. For example, if $a = 5$, $b = 1$, $m = 16$ and $z_0 = 4$, then

$$\begin{aligned}
z_1 &= (5 \times 4 + 1) \quad \text{mod } 16 \\
&= 21 \quad \text{mod } 16 \\
&= 5
\end{aligned}$$

and

$$\begin{aligned}
z_2 &= (5 \times 5 + 1) \quad \text{mod } 16 \\
&= 26 \quad \text{mod } 16 \\
&= 10,
\end{aligned}$$

etc. Since each z_n value is an element of the set $\{0, 1, \ldots, m - 1\}$, the $(n + 1)$st random number r_{n+1} that is in the interval $[0, 1)$ is computed as $r_{n+1} = z_{n+1}/m$, $n = 0, 1, \ldots$ For example, using the above data, we would find $r_1 = z_1/m = 5/16 = .313$ and $r_2 = z_2/m = 10/16 = .625$.

Such methods produce random numbers that always start cycling after some finite number p (called the period of the generator) not exceeding m. Since the selection of the parameters z_0, a, b and m affects the cycle length, it is important to choose these parameters carefully in designing random number generators.

Example 98 *Random numbers and cycle length.* As a simple example of early cycling (after generating the 16th random number), consider the case with $a = 5$, $b = 1$, $m = 16$ and $z_0 = 4$.

```
>   restart: # Congruential.mws
```

```
>   Digits:=3;
```

$$Digits := 3$$

```
>   a:=5; b:=1; m:=16; z[ 0] :=4;
```

$$a := 5$$
$$b := 1$$
$$m := 16$$
$$z_0 := 4$$

Generating the z_n values for $n = 1, \ldots, 25$ suggests that the random numbers r_n will start cycling after the 16th one.

```
>   for n from 0 to 25 do z[ n+1] :=(a* z[ n] +b)  mod m;
    r[ n+1] :=z[ n+1] /m od:
```

```
>   seq(z[ n] ,n=1..25);
```

5, 10, 3, 0, 1, 6, 15, 12, 13, 2, 11, 8, 9, 14, 7, 4, 5, 10, 3, 0, 1, 6, 15, 12, 13

Indeed, we see that the 17th random number is the same as the 1st, the 18th is the same as the 2nd, etc.

```
>   seq([ n,evalf(r[ n] )] ,n=1..25);
```

[1, .313], [2, .625], [3, .188], [4, 0], [5, .0625], [6, .375], [7, .938], [8, .750], [9, .813], [10, .125], [11, .688], [12, .500], [13, .563], [14, .875], [15, .438], [16, .250], [17, .313], [18, .625], [19, .188], [20, 0], [21, .0625], [22, .375], [23, .938], [24, .750], [25, .813]

It has been shown by Fishman [66] that when $c \neq 0$, for a B-bit word computer the maximal (full) period of $p_{max} = m = 2^B$ can be achieved provided that b is relatively prime to m (in other words, 1 is the only positive integer that exactly divides both b and m) and $a = 1 + 4k$ where k is an integer. When $c = 0$, for a 32-bit word computer a satisfactory choice for the modulus and the multiplier are $m = 2^{31} - 1$ and $a = 7^5$. These values have been tested and used extensively; see, Banks, Carson and Nelson [12, p. 295].

10.2.2   Maple's Uniform Random Number Generator

It is easy to generate uniformly distributed random variables using Maple. After loading the `random` subpackage of the `stats` package, entering `random-[uniform](k)` generates k uniform random variables.

```
>    restart: # MapleUniformRN.mws
>    with(stats,random);
```

$$[random]$$

```
>    random[uniform](10);
```

.4274196691, .3211106933, .3436330737, .4742561436, .5584587190,
.7467538305, .03206222209, .7229741218, .6043056139,
.7455800374

10.2.3   Kolmogorov-Smirnov Test for Uniformity

A sequence of uniform random variables $U_1, U_2, \ldots, U_N$ generated for a simulation study must be—naturally—uniformly distributed and independent. There are several tests available for testing these properties. For example, frequency tests are used for testing uniformity, and run tests and autocorrelation tests are used for testing independence; see Banks, Carson and Nelson [12, Section 8.4].

In this section we describe the Kolmogorov-Smirnov test for testing uniformity and use it to test Maple's random number generator. This test compares the theoretical distribution $F(u) = \int_0^u 1\, dx = u$, $0 < u < 1$, of the uniform random variable U, and the empirical distribution $G_e(u)$ of the sample of N observations obtained by generating uniform random numbers. The test involves a few simple steps that can be easily implemented with Maple.

Suppose we generate N sample observations $U_1, U_2, \ldots, U_N$ for which the empirical distribution is defined as

$$G_e(u) = \frac{\text{number of } i\text{'s such that } U_i \leq u}{N}.$$

We start by ranking the observations $U_1, U_2, \ldots, U_N$ in increasing order, i.e., $U_{[1]}, U_{[2]}, \ldots, U_{[N]}$ where $U_{[i]} \leq U_{[i+1]}$. The test statistic d_{max} measures the deviation of the random numbers from the theoretical distribution and it is defined as

$$d_{max} = \max(d^+, d^-)$$

where

$$d^+ = \max_{1 \leq i \leq N} \left(\frac{i}{N} - U_{[i]} \right) \quad \text{and} \quad d^- = \max_{1 \leq i \leq N} \left(U_{[i]} - \frac{i-1}{N} \right).$$

Note that $i/N - U_{[i]}$ is the deviation of $G_e(u)$ above $F(i/N) = i/N$ and $U_{[i]} - (i-1)/N$ is the deviation of $G_e(u)$ below $F((i-1)/N) = (i-1)/N$. Thus, d^+ and d^- correspond to the largest deviation of $G_e(u)$ above and below $F(u)$, respectively. Intuitively, if the maximum d_{max} of the largest deviations is greater than a critical value, then we reject the hypothesis that the samples are drawn from a uniform distribution.

More specifically, the test statistic d_{max} is compared to a critical value d_α for a given significance level of α. If $d_{max} > d_\alpha$, this indicates that the deviations from the theoretical distribution are large, thus the hypothesis that the sample data are from a uniform distribution is rejected. If $d_{max} \leq d_\alpha$, the hypothesis is not rejected. For $N > 35$, the approximate critical values are given by $d_{0.01} = 1.63/\sqrt{N}$, $d_{0.05} = 1.36/\sqrt{N}$ and $d_{0.10} = 1.22/\sqrt{N}$. For a detailed discussion of the Kolmogorov-Smirnov test; see Banks, Carson and Nelson [12, Sec. 8.4].

Example 99 *Kolmogorov-Smirnov uniformity test of Maple's random number generator.* We generate 40 uniform random numbers with Maple's random[uniform] command and apply the Kolmogorov-Smirnov test to see if these samples come from the true uniform distribution. Using a significance level of $\alpha = 0.05$, we would reject the hypothesis that the sample is from the true uniform if the test statistic $d_{max} > d_\alpha = 1.36/\sqrt{40} = 0.215$.

```
>   restart: # KSTest.mws
>   with(stats): #readlib(randomize): randomize():
>   N:=40; d[ .05] :=evalf(1.36/sqrt(N));
```

$$N := 40$$

$$d_{.05} := .2150348809$$

```
>   U:=[ random[ uniform] (N)] :
>   USort:=sort(U);
```

$USort := [.005862664913, .03206222209, .03916959416, .07481365622,$
$.08843057167, .1319057546, .1464863072, .1555907635,$
$.2197600994, .2598119527, .2726006090, .2813387792,$
$.3100754872, .3137460865, .3211106933, .3436330737,$
$.4274196691, .4293926737, .4537470195, .4742561436,$
$.5254285110, .5584587190, .6043056139, .6283634430,$
$.6438424438, .6440313953, .6720753584, .6759829338,$
$.6764707883, .7229741218, .7455800374, .7467538305,$
$.7512095393, .7924959004, .7971794905, .8129204579,$
$.8454735095, .9206249473, .9510535301, .9604988341]$

```
>   d[ a] :=max(seq(i/N-USort[ i] ,i=1..N));
```

$$d_a := .0870795421$$

```
>   d[ b] :=max(seq(USort[ i] -(i-1)/N,i=1..N));
```

$$d_b := .0543056139$$

But we find $d_{max} = 0.087 < d_a$; thus we do not reject the hypothesis and conclude that no difference has been detected between the theoretical uniform distribution and the distribution of the sample $\{U_1, U_2, \ldots, U_N\}$.

```
>    dmax:=max(d[ a] ,d[ b] );
```

$$dmax := .0870795421$$

```
>    if dmax>d[ .05]  then "Reject" else "Don' t Reject"
     fi;
```

                    "Don't Reject"

The next few Maple statements generate and plot the empirical distribution $G_e(u)$ and the theoretical uniform distribution $F(u) = u, 0 < u < 1$. Both graphs are displayed in Figure 10.2.

```
>    G:=seq([ i/N,USort[ i]] ,i=1..N):
>    with(plots):
>    GPlot:=pointplot([ G] ,style=line,
     connect=true,linestyle=1,color=black,thickness=2):
>    FPlot:=plot(u,u=0..1,linestyle=2,
     color=black,thickness=2):
>    display({ GPlot,FPlot} );
```

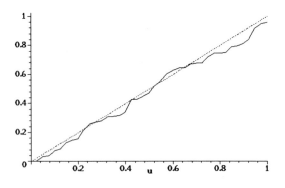

FIGURE 10.2. Comparison of the empirical distribution $G_e(u)$ drawn as a solid curve and the theoretical uniform distribution $F(u)$ drawn as a dotted line.

10.3   Generating Random Variates from Other Distributions

Assuming that a uniform random number has been generated using the method(s) described, it can then be used to generate another random variable with a given distribution. (This is known as generating a *random variate*.) In this section, we discuss the inverse transform method and also explain the use of Maple commands that can generate random variates from 18 continuous and discrete distributions.

Consider a random variable X with density $f(x)$ and distribution $F(x)$. If the inverse F^{-1} of the distribution function can be easily computed, then the inverse transform method can be used to generate a random variate from the distribution of X. This is done by setting $F(X) = U$ where U is a uniform number and solving for $X = F^{-1}(U)$.

10.3.1 Exponential Random Variates

As an example, we apply this method to generate random variates from the exponential distribution with parameter λ. Recall that an exponential random variable X with parameter λ has the density function $f(x) = \lambda e^{-\lambda x}$ and the distribution function $F(x) = \int_0^x \lambda e^{-\lambda t}\, dt = 1 - e^{-\lambda x}$, $x \geq 0$. Since the inverse F^{-1} of the distribution function of exponential is available, the inverse transform method can be easily used to generate exponential random variates.

Example 100 *Generation of exponential random variates.* We start by defining the density and the distribution of the exponential.

```
>   restart: # ITMExponential.mws
>   f:=x->lambda* exp(-lambda* x);
```

$$f := x \rightarrow \lambda\, e^{(-\lambda x)}$$

```
>   F:=x->int(f(t),t=0..x);
```

$$F := x \rightarrow \int_0^x f(t)\, dt$$

After inverting the distribution (i.e., solving $F(X) = U$ for X), we find that $X = -\log(1 - U)/\lambda$. Letting $\lambda = 3$ and generating uniform random numbers, we can now sample from the exponential distribution.

```
>   X:=unapply(solve(F(X)=U,X),U);
```

$$X := U \rightarrow -\frac{\ln(1 - U)}{\lambda}$$

```
>   lambda:=3;
```

$$\lambda := 3$$

```
>   X(.3562);
```

$$.1467890534$$

```
>   with(stats);
```

[*anova, describe, fit, importdata, random, statevalf, statplots, transform*]

Maple's `random[ uniform] (10)` command generates 10 uniform random numbers. After mapping X to them, we obtain 10 samples from the exponential distribution.

```
>   U:=[ random[ uniform] (10)] ;
```

$U := [.3211106933, .3436330737, .4742561436, .5584587190,$
$.7467538305, .03206222209, .7229741218, .6043056139,$
$.7455800374, .2598119527]$

> map(X,U);

$[.1290990629, .1403451025, .2143137166, .2724945872, .4577977537,$
$.01086249093, .4278814513, .3090377060, .4562563270,$
$.1002836690]$

The inverse transform method is useful in generating random variates from the exponential, uniform defined over the interval (a, b), Weibull, triangular and empirical continuous distributions since the inverse F^{-1} of these distributions can be found with relative ease. But this method fails for a number of important continuous distributions including the normal, gamma and beta. There are, of course, other methods that can be used to sample from these distributions and these are discussed in textbooks on simulation, e.g., Banks, Carson and Nelson [12], Ross [160], and Pritsker, O'Reilly and LaVal [153]. We should also mention that as we demonstrated in Example 97 on port operations, the inverse transform method is also applicable to sampling from a wide variety of discrete distributions including the binomial and Poisson.

10.3.2   Maple's Random Variate Generators

Maple can generate random variates from a total of 18 distributions (13 continuous, 4 discrete and 1 empirical discrete) using the command random[distribution_name] . Additional information on these distributions can be obtained by entering ? stats, distributions. The list of 13 continuous and 5 discrete distributions and the corresponding Maple syntax is provided in Tables 10.1 and 10.2, respectively.

We now present an example where Maple generates 5 random variates from some of the distributions that arise frequently in operations research applications.

Example 101 *Random variates from some distributions.* The next set of commands generates random variates from beta, exponential, gamma, normal, uniform and Weibull continuous distributions.

> restart: # MapleRandomVariates.mws

> with(stats); #with(random); nops(with(random));

[*anova, describe, fit, importdata, random, statevalf, statplots, transform*]

> random[beta[1,2]] (5);

$.1586913674, .4696943744, .01327905798, .4781283152, .04138715231$

> random[exponential[2]] (5);

$.9336948690, .5642329350, .1651826067, .7863020910, .6955721300$

> random[gamma[2,4]] (5);

Distribution	Maple Command	Parameters
Beta	beta[n_1, n_2]	$n_1, n_2 \in \mathbb{N}^+$
Cauchy	cauchy[a, b]	$a \in \mathbb{R}, b \in \mathbb{R}^+$
Chi-square (χ^2)	chisquare[n]	$n \in \mathbb{N}^+$
Exponential	exponential[λ]	$\lambda \in \mathbb{R}^+$
Fisher's F	fratio[n_1, n_2]	$n_1, n_2 \in \mathbb{N}^+$
Gamma	gamma[a, b]	$a, b \in \mathbb{R}^+$
Laplace	laplaced[a, b]	$a \in \mathbb{R}, b \in \mathbb{R}^+$
Logistic	logistic[a, b]	$a \in \mathbb{R}, b \in \mathbb{R}^+$
Lognormal	lognormal[μ, σ]	$\mu \in \mathbb{R}, \sigma \in \mathbb{R}^+$
Normal	normald[μ, σ]	$\mu \in \mathbb{R}, \sigma \in \mathbb{R}^+$
Student's t	studentst[n]	$n \in \mathbb{N}^+$
Uniform	uniform[a, b]	$a, b \in \mathbb{R}, a \leq b$
Weibull	weibull[a, b]	$a, b \in \mathbb{R}^+$

TABLE 10.1. Maple can generate random variates from this list of 13 continuous distributions.

Distribution	Maple Command	Parameters
Binomial	binomiald[n, p]	$n \in \mathbb{N}^+, p \in (0, 1)$
Discrete Uniform	discreteuniform[a, b]	$a, b \in \mathbb{N}, a \leq b$
Negative Binomial	negativebinomial[n, p]	$n \in \mathbb{N}^+, p \in (0, 1)$
Poisson	poisson[λ]	$\lambda \in \mathbb{R}^+$
Empirical	empirical[Probs]	$\sum \text{Probs} = 1$

TABLE 10.2. Maple can generate random variates from this list of 5 discrete distributions.

10.83376082, .4130222104, 11.26025842, 12.10647440, 14.76067790

```
>    random[ normald[ 1,2]] (5);
```

−2.199385336, 2.636270022, 1.170950535, 1.189344489, −1.815978260

```
>    random[ uniform[ 2,7]] (5);
```

5.147399353, 5.682254920, 4.007973530, 2.891574531, 6.574125685

```
>    random[ weibull[ 1,5]] (5);
```

1.651659671, 3.026604177, 7.327432705, .4207307627, 4.645956875

In the next group, we generate random variates from binomial, discrete uniform and Poisson discrete distributions.

```
>    random[ binomiald[ 5,.5]] (5);
```
3.0, 4.0, 3.0, 4.0, 3.0

```
>    random[ discreteuniform[ 1,5]] (5);
```
3.0, 5.0, 3.0, 5.0, 1.0

```
>    random[ poisson[ 5]] (5);
```
8.0, 4.0, 2.0, 2.0, 5.0

Maple can also generate random variates from discrete empirical probability distributions defined over the integers $1, \ldots, N$. Once these distributions are defined, their random variates can be generated using the `random` command. In the next example, we define an empirical distribution $p(k)$ that can take 6 values $k = 1, \ldots, 6$:

$$p(k) = \begin{cases} 0.30, & \text{for } k = 1 \\ 0.20, & \text{for } k = 2 \\ 0.10, & \text{for } k = 3 \\ 0.25, & \text{for } k = 4 \\ 0.05, & \text{for } k = 5 \\ 0.10, & \text{for } k = 6. \end{cases}$$

```
>   Probs:=.3,.2,.1,.25,.05,.1;
    sum([ Probs] [ i] ,i=1..nops ([ Probs] ));
```

$$Probs := .3, .2, .1, .25, .05, .1$$

$$1.00$$

```
>   p:=empirical[ Probs] ;
```

$$p := empirical_{.3, .2, .1, .25, .05, .1}$$

The next command generates 10 random variates from this empirical distribution.

```
>   random[ p] (10);
```

$$1.0, 3.0, 1.0, 2.0, 4.0, 4.0, 1.0, 4.0, 1.0, 1.0$$

10.4   Monte Carlo Simulation

A static simulation where time does not play a role is known as *Monte Carlo simulation*. This method has been found useful in the numerical evaluation of definite integrals. It can also be used in the simulation of some one-period probabilistic problems.

10.4.1   Numerical Evaluation of Definite Integrals

Suppose we wish to evaluate

$$\mu = \int_0^1 g(x)\,dx$$

where $g(x)$ is a given function—such as $g(x) = \log(x^2 + e^{-x^2})$—whose integral cannot be computed directly. To evaluate this integral by simulation, we proceed as follows.

First note that if U is uniform over $(0, 1)$—i.e., if U's density is $f_U(u) = 1$, $0 < u < 1$—then we can write

$$\mu = E[g(U)] = \int_0^1 g(u) f_U(u)\, du = \int_0^1 g(u)\, du.$$

Thus, if we can generate independent and uniform random variables $U_1, \ldots, U_k$ defined over $(0, 1)$, then we can approximate the integral $\int_0^1 g(x)\, dx$ using

$$\lim_{k \to \infty} \frac{1}{k} \sum_{i=1}^k g(U_i) = E[g(U)] = \mu$$

where the $g(U_1), \ldots, g(U_k)$ are also independent and identically distributed random variables. The result follows by using the strong law of large numbers.

Generating a sufficiently large number of uniform random numbers $u_1, \ldots, u_k$ and taking the average of $g(u_1), \ldots, g(u_k)$ gives the value of the integral.

Example 102 *Evaluation of a definite integral.* Consider the problem of evaluating the definite integral $\int_0^1 g(x)\, dx$ where $g(x) = \log(x^2 + e^{-x^2})$. We generate 1000 uniform random numbers and evaluate the integral using Monte Carlo.

```
>   restart: # MonteCarloIntegral.mws
>   with(stats,random); with(stats,describe);
```

$$[random]$$

$$[describe]$$

```
>   u:=[ random[ uniform] (1000)] :
>   g:=u->log(u^2+exp(-u^2));
```

$$g := u \to \log(u^2 + e^{(-u^2)})$$

```
>   #plot(g(x),x=0..1);
>   mu:=map(g,u):
```

We find 0.07401 as the simulated value of the integral.

```
>   describe[ mean] (mu);
```

$$.07401925167$$

Maple is unable to evaluate the integral using the `int()` command. But numerical integration of the function—using `evalf(int())`—gives a result that is close to what we found using Monte Carlo simulation.

```
>   int(g(x),x=0..1);
```

$$\int_0^1 \ln(x^2 + e^{(-x^2)})\, dx$$

```
>   evalf(int(g(x),x=0..1));
```

$$.07296304693$$

10.4.2   Simulation of a Static (Single-Period) Problem

In our next example, we use simulation to solve the (in)famous "car and goats" problem.

Example 103 *The car and goats problem—The Monty Hall Dilemma.* Suppose you are on Monty Hall's TV game show "Let's Make a Deal," and Monty gives you a choice of three doors. Behind one door is a shiny new car, and behind each of the other two doors is a smelly goat. You choose a door, say, number 1, and Monty, who knows what is behind the doors, opens another door, say number 3, which has a goat. He says to you, "Do you want to pick door number 2?" Is it to your advantage to switch your choice of doors? That is, do you have a better chance of winning the car if you switch?

This problem created a lot of an excitement in 1991 when it appeared in Marilyn vos Savant's column "Ask Marilyn" in *Parade* magazine and received many wrong answers from readers—many people, including some professional mathematicians, thought that after Monty reveals the goat, the probability of winning is 1/2; see vos Savant's book *The Power of Logical Thinking* [165] for transcripts of the readers' letters. Vos Savant answered the question correctly by stating that if the contestant switches doors then the probability of winning is 2/3, not 1/2.[5]

To solve this problem using simulation we will assume, without loss of generality, that you pick door 1. Monty knows where the car is and will open an empty door. Here are the outcomes if you always switch.

- If the car is really behind door 1, Monty is equally likely to open door 2 or 3.

    - If he opens door 2 and if you switch, you lose.

    - If he opens door 3 and if you switch, you again lose.

- If the car is really behind door 3, Monty will open door 2.

    - In this case, if you switch, you win.

- If the car is really behind door 2, Monty will open door 3.

    - In this case, if you switch, you win.

Thus, using the policy of always switching, you would win the car if it is behind door 2 or 3. With the policy of never switching, you would win the car if it is behind door 1.[6]

---

[5]She argued as follows: Suppose there are 1 million doors and the contestant picks door 1. Monty, who knows what is behind all the doors and always avoids the one with the car, opens all *except* door number 777,777. In this case, the contestant would surely pick door number 777,777.

[6]It is perhaps now obvious (even without using simulation) that, if you switch, the probability of winning the car is 2/3.

At the start of the game, the car is equally likely to be behind any of the three doors. We generate uniform random numbers using discreteuniform[1, 3] to simulate the actual location of the car.

```
>    restart: # MontyHall.mws
>    with(stats); #readlib(randomize): randomize():
```

[*anova, describe, fit, importdata, random, statevalf, statplots, transform*]

```
>    p:=discreteuniform[1,3] ;
```

$$p := discreteuniform_{1,3}$$

We simulate the game 1000 times and find that 67.5% of the time the contestant wins the car under the policy of always switching and 34% of the time the contestant wins under the policy of never switching. Marilyn was right!

```
>    N:=1000;
```

$$N := 1000$$

```
>    # Policy 1: Always Switch
>    Win:=0:
>    for n from 1 to N do
     CarBehindDoor:=random[p](1);
     if CarBehindDoor=1.0 then Win:=Win+0;
     elif (CarBehindDoor=2.0 or CarBehindDoor=3.0)
     then Win:=Win+1;
     fi od:
>    evalf(Win/N);
```

$$.6750000000$$

```
>    # Policy 2: Never Switch
>    Win:=0:
>    for n from 1 to N do
     CarBehindDoor:=random[p](1);
     if CarBehindDoor=1.0
     then Win:=Win+1;
     elif (CarBehindDoor=2.0 or CarBehindDoor=3.0)
     then Win:=Win+0;
     fi od:
>    evalf(Win/N);
```

$$.3400000000$$

10.5 Dynamic Simulation Models

Many real-life systems encountered in operations research applications exhibit probabilistic behavior as they evolve over time. For example, inventory level in a periodic-review inventory control problem is a stochastic process that is influenced by the randomness of demand and the choice of the inventory ordering policy. In a queueing system, the waiting time of consecutive customers is also a stochastic process that is influenced by the randomness in arrival and service patterns.

For problems of this type, it may be important to determine the long-run behavior of the system (e.g., average inventory/backorders or the average time spent in the queue) in order to choose policies that optimize the performance of the system. Naturally, if the problem is simple enough, the system may be analyzed using one of the available analytic models. But in many cases this may not possible and simulation becomes an important tool in understanding the behavior of the system.

In this section we present two such examples—one from inventory and one from queueing—and determine the long-run average behavior of the systems using simulation.

10.5.1 Simulation of an Inventory System with Random Yield

Example 104 *The periodic-review (s, S) inventory policy with random yield.* We consider a single product for which demand Y_n during successive weeks $n = 1, 2, \ldots$ are i.i.d. random variables with density $\alpha_k = \Pr(Y_n = k)$, $k = 0, 1, \ldots$. The inventory level X_n at the end of week n is observed and an order decision is made according to the following (s, S) policy: If $X_n < s$, then we order $Q_n = S - X_n$ units; otherwise (i.e., if $s \leq X_n \leq S$), no order takes place. We assume that deliveries are instantaneous and that any shortages are backordered.

To complicate matters, we will assume that the amount actually received, R_n, after ordering Q_n units is a random fraction of the order quantity given by $R_n = \lfloor \beta Q_n \rfloor$ where β is a continuous random variable with density $g(p)$ defined over $(0, 1)$ and $\lfloor z \rfloor$ is the greatest integer less than or equal to z.[7] With these assumptions, the inventory level X_n can be written as

$$X_{n+1} = \begin{cases} \lfloor \beta(S - Y_{n+1}) \rfloor & \text{if } X_n < s \\ X_n - Y_{n+1} & \text{if } s \leq X_n \leq S. \end{cases} \tag{10.3}$$

The fact that the yield is random makes this a nontrivial problem. If the yield were deterministic (i.e., if $\beta \equiv 1$), then the problem could conceivably be solved as a Markov chain after specifying the transition probability matrix of the inventory level process. But since β and Y_n (for given n) are random variables, simulation appears to be a convenient way to analyze this problem.

```
>    restart: # SsRandomYield.mws
>    with(stats): #readlib(randomize); randomize();
```

We start by assuming that $(s, S) = (5, 15)$ and $X_0 = 15$. The weekly demands are assumed Poisson with rate $\lambda = 10$, i.e., $\alpha_k = \Pr(Y_n = k) = e^{-10} 10^k / k!$,

---

[7]Modeling the randomness of yield has recently become an important research area in operations management; see Gerchak and Parlar [69] and Yano and Lee [197].

and the random yield fraction β is taken as beta with parameters $(a, b) = (6, 2)$, i.e., $g(p) = 42p^5(1 - p)$, $0 < p < 1$. In the simulation of the inventory process, random variates from both the Poisson and beta distributions are generated using Maple's `random` function.

```
>    S:=15; s:=5; lambda:=10; x[ 0] :=15;
```

$$S := 15$$

$$s := 5$$

$$\lambda := 10$$

$$x_0 := 15$$

We perform the simulation for a total of 100 weeks and generate the values of X_n in equation (10.3) using the following Maple commands.

```
>    N:=100;
```

$$N := 100$$

```
>    #plot(statevalf[ pdf,beta[ 6,2]] (x),x=0..1);
>    for n from 0 to N do
     y[ n+1] :=random[ poisson[ lambda]] (1);
     if x[ n] <s then
     b:=random[ beta[ 6,2]] (1);
     x[ n+1] :=floor(b* (S-y[ n+1] ))
     elif x[ n] >=s then
     x[ n+1] :=x[ n] -y[ n+1]
     fi od:
```

The simulated inventory levels (and demand) for the first 10 weeks are found as follows.

```
>    seq([ n,[ x[ n] ,y[ n+1]] ,x[ n+1]] ,n=1..10);
```

> [1, [6.0, 8.0], −2.0], [2, [−2.0, 9.0], 4], [3, [4, 12.0], 2],
> [4, [2, 13.0], 1], [5, [1, 15.0], 0], [6, [0, 7.0], 6],
> [7, [6, 13.0], −7.0], [8, [−7.0, 12.0], 2], [9, [2, 11.0], 2],
> [10, [2, 10.0], 3]

```
>    IL:={ seq(x[ n] ,n=1..N)] : nops (IL):
```

In order to compute the average number of units in inventory and the average number backordered, we need to isolate the nonnegative and negative inventory levels. This is done easily with the `select ()` command.

```
>    ILNonNegative:=select(type,IL,nonneg);
```

ILNonNegative := [6.0, 4, 2, 1, 0, 6, 2, 2, 3, 3, 6, 0, 6, 3, 6, 2.0, 7, 4, 4, 0, 4, 2, 2, 4, 6, 2.0, 6, 5, 3, 5, 1, 1, 6, 2, 0, 2, 0, 1, 4, 2, 2, 2, 7, 0, 4, 5, 1, 4, 3, 1, 2, 1, 5, 0, 4, 3, 3, 6, 3, 9, 1.0, 5, 4, 5, 4, 1, 4, 6, 3, 1, 2, 1, 3, 3, 10, 5, 6]

```
>    ILNegative:=select(type,IL,negative);
```

ILNegative := [−2.0, −7.0, −4.0, −2, −3.0, −5.0, −5.0, −3.0, −4.0, −5.0, −2.0, −2, −2.0, −1.0, −4.0, −6.0, −5.0, −3, −8.0, −4, −1.0, −11.0, −1.0]

The average inventory level and the average number backordered are found as 3.29 and 3.91, respectively.

```
>  with(describe):
>  describe[ mean] (ILNonNegative);
   describe[ mean] (ILNegative);
```

$$3.298701299$$

$$-3.913043478$$

The distribution of the inventory level is found as follows.

```
>  for n from min(op(IL)) to max(op(IL)) do
   k[ n] :=(n..(n+1)) od:
>  Intervals:=[ seq(k[ n] ,n=min(op(IL))..max(op(IL)))] ;
```

Intervals := [−11.0.. − 10.0, −10.0.. − 9.0, −9.0.. − 8.0, −8.0.. − 7.0, −7.0.. − 6.0, −6.0.. − 5.0, −5.0.. − 4.0, −4.0.. − 3.0, −3.0.. − 2.0, −2.0.. − 1.0, −1.0..0, 0..1., 1...2., 2...3., 3...4., 4...5., 5...6., 6...7., 7...8., 8...9., 9...10., 10...11.]

```
>  with(transform):
>  Distribution:=statsort(tallyinto(IL,Intervals));
```

Distribution := [−11.0.. − 10.0, Weight(−10.0.. − 9.0, 0), Weight(−9.0.. − 8.0, 0), −8.0.. − 7.0, −7.0.. − 6.0, −6.0.. − 5.0, Weight(−5.0.. − 4.0, 4), Weight(−4.0.. − 3.0, 4), Weight(−3.0.. − 2.0, 3), Weight(−2.0.. − 1.0, 5), Weight(−1.0..0, 3), Weight(0..1., 7), Weight(1...2., 11), Weight(2...3., 14), Weight(3...4., 11), Weight(4...5., 12), Weight(5...6., 7), Weight(6...7., 11), Weight(7...8., 2), Weight(8...9., 0), 9...10., 10...11.]

Thus, there is one observation in the interval [−11, −10), zero observations in the interval [−10, −9), etc.

The complete distribution of the inventory level X_n is given in Figure 10.3.

```
>  with(statplots):
>  histogram(Distribution);
```

10.5.2   Simulation of a Non-Markovian Queue

Example 105 *Distribution of the waiting time in the queue for a $G/G/1$ queue.* As we saw in Chapter 9, Queueing Systems, the average waiting time in the queue, $\bar{W}$, for the simple Markovian $M/M/1$ model is found analytically as

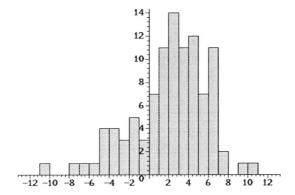

FIGURE 10.3. Distribution of the inventory level X_n in the periodic-review inventory problem random yield for a simulation length of $N = 100$ weeks. Weekly demands are assumed Poisson with parameter $\lambda = 10$ and the random yield fraction β is beta sitributed with parameters $(a, b) = (6, 2)$.

$\bar{W} = \lambda/[\mu(\mu - \lambda)]$ where λ is the rate of the Poisson arrival process and μ is the rate of the exponentially distributed service times.[8] When Markovian assumptions are no longer valid, i.e., when the arrival process is no longer Poisson and/or service time is not exponential, such simple formulas no longer apply.

However, there is a very general set of recursive equations that relate the random queue waiting time W_n of the nth customer to the interarrival time A_n between the $(n-1)$st and nth customer and the service time S_{n-1} of the $(n-1)$st customer.[9] As shown originally by Lindley [122] (see also, Kleinrock [110, p. 277] and Gaver and Thompson [68, p. 572]), we can write

$$W_{n+1} = \max(W_n - A_{n+1} + S_n, 0), \quad n = 0, 1, \ldots.$$

The waiting time in the queue process can now be easily simulated for customers $n = 1, 2, \ldots$, and the distribution of the time in the queue can be developed.

We assume that the interarrival times are distributed as uniform with parameters $(6,8)$ having density $f_A(x) = 1/2$, $6 \leq x \leq 8$ and the service time as Weibull with parameters $(2,6)$ having density $f_S(x) = xe^{-(x/6)^2}/18$, $x \geq 0$.

The Maple commands used to simulate the W_n process are almost identical to the ones we described in Example 104.

```
>   restart: # GG1.mws
>   with(stats):
```

---

[8]Generally, waiting time in the queue is denoted by W_q. In this example we will use the notation W in order not to complicate the notation.

[9]Note that we do not make any probability assumptions about the distributions of A_n and S_n.

```
>   N:=100;  w[ 0] :=0;
```

$$N := 100$$

$$w_0 := 0$$

```
>   #plot(statevalf[ pdf,weibull[ 2,6]] (x),x=0..50);
>   for n from 0 to N do
    a[ n+1] :=random[ uniform[ 6,8]] (1);
    s[ n] :=random[ weibull[ 2,6]] (1);
    w[ n+1] :=max(w[ n] -a[ n+1] +s[ n] ,0)
    od:
>   seq([ n,[ a[ n+1] ,s[ n]] ,w[ n+1]] ,n=0..5);
```

[0, [6.854839338, 3.733992340], 0], [1, [6.687266147, 4.811016670], 0],
[2, [7.116917438, 7.031511744], 0],
[3, [6.064124444, 6.797881782], .733757338],
[4, [7.208611228, 7.019664048], .544810158],
[5, [6.519623905, 3.655438579], 0]

```
>   wList:=[ seq(w[ n] ,n=0..N)] :
>   for n from min(op(wList)) to max(op(wList))  do
    k[ n] :=(n..(n+1))  od:
>   Intervals:=[ seq(k[ n] ,n=min(op(wList)))..
    max(op(wList)))] :
>   with(describe):
```

We see that the mean waiting time in queue is obtained as 2.004 with a standard deviation of 3.101.

```
>   describe[ mean] (wList);
    describe[ standarddeviation] (wList);
```

$$2.004485394$$

$$3.101872397$$

```
>   with(transform):
```

Following is the complete distribution of the waiting times where we find that out of 100 observations, 64 fall in the interval [0,1), 7 fall into interval [1,2) and so on.

```
>   Distribution:=statsort(tallyinto(wList,
    Intervals));
```

Distribution := [Weight(0..1, 64), Weight(1..2, 7), Weight(2..3, 3),
Weight(3..4, 3), Weight(4..5, 6), Weight(5..6, 5),
Weight(6..7, 2), Weight(7..8, 3), Weight(8..9, 5), 9..10,
Weight(10..11, 0), 11..12, Weight(12..13, 0),
Weight(13..14, 0), 14..15]

```
>   with(statplots):
```

A histogram of the distribution of the waiting times is presented in Figure 10.4.

```
>   histogram(Distribution);
```

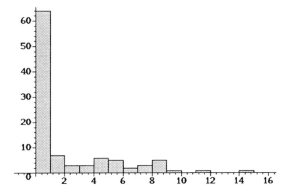

FIGURE 10.4. Distribution of the waiting time in queue, W_n, for the $G/G/1$ queue with uniform interarrival times and Weibull service times for a simulation length of $N = 100$ customers. The uniform distribution has parameters $(6, 8)$ and the Weibull has parameters $(2, 6)$.

It is also possible to obtain a running average of the simulated waiting times, $\frac{1}{n} \sum_{i=1}^{n} w_i$, and plot the result using `pointplot()`. The Maple commands to compute these averages and plot them follow. They can be executed after removing the "#" symbol before each command. Note that as the simulated number of customers (n) approaches the run length N, the graph of the running average should converge to the simulated final average of 2.004.

```
>    #for n from 1 to N do wRun[ n] :=sum(w[ i] ,i=1..n)/n
od:
>    #L:=seq([ n,wRun[ n]] ,n=1..N):
>    #with(plots):
>    #pointplot({ L} ,symbol=point);
```

10.6   Optimization by Random Search

The optimization techniques presented in Chapter 5, Nonlinear Programming, required the differentiability of the function that was to be optimized. In certain cases the function is not differentiable or it may be discontinuous or even specified as a table of values, or worse, it may have a large number of local optima. In those cases, the classical optimization techniques may have to be supplemented by alternative methods in order to optimize a function of several variables.

One of the most useful of such techniques relies upon generation of random numbers to search for the (global) optimum. *Adaptive random search (ARS)* is a simpler version of the global optimization method known as simulated annealing;

see, e.g., Gottfried and Weisman [75, Section 3.4] for a description of ARS and
Laarhoven and Aarts [115] for a detailed account of simulated annealing.

For a problem with, say, two decision variables x and y, starting with a feasible
point in the two-dimensional region that is known to contain the optimal solution
(x^*, y^*), ARS randomly generates the next feasible point and compares it to the
previous point. If the new point is better than the previous one and it is feasible,
it is kept; otherwise, the new point is discarded.[10] For example, if x^* is known to
be in the interval $[x_{min}, x_{max}]$ and if x_{old} is the previous point, then the new point
is generated using the formula $x_{new} = x_{old} + (x_{max} - x_{min})(2r - 1)^v$ where r is
a random number between 0 and 1 and v is an odd integer. As recommended by
Gall [67], initially, v is chosen as 3 or 5 and later, when the improvements in the
objective function become smaller, it is increased to 7 or 9. Note that the larger the
interval of uncertainty $(x_{max} - x_{min})$, the larger the step size $(x_{max} - x_{min})(2r - 1)^v$.
But increased values of v have the effect of shortening the step size since $-1 <
2r - 1 < 1$.

Example 106 *Minimization of a discontinuous and nondifferentiable function
with two local minima.* Consider a function $z(x, y)$ defined as

$$z(x, y) = \begin{cases} 5\sqrt{(x - 2)^2 + (y - 2)^2}, & \text{for } (x - 2)^2 + (y - 2)^2 \leq 4 \\ 3|x - 5| + |y - 5| + 3, & \text{otherwise.} \end{cases}$$

As the three-dimensional graph in Figure 10.5 indicates, $z(x, y)$ is a discontinuous and nondifferentiable function that has two local minima, one at $(2,2)$ and
the other at $(5,5)$.

```
>    restart: # TwoMinima.mws
>    z:=proc(x,y) if (x-2)^2+(y-2)^2<=4 then
     5*sqrt((x-2)^2+(y-2)^2) else
     3*abs(x-5)+abs(y-5)+3 fi end;
```

$$z := \mathbf{proc}(x, y)$$
$$\quad \mathbf{if}\,(x - 2)^2 + (y - 2)^2 \leq 4 \,\mathbf{then}\, 5 \times \text{sqrt}((x - 2)^2 + (y - 2)^2)$$
$$\quad \mathbf{else}\, 3 \times \text{abs}(x - 5) + \text{abs}(y - 5) + 3$$
$$\quad \mathbf{fi}$$
$$\mathbf{end}$$

```
>    z(2,2); z(5,5);
```

$$0$$

$$3$$

```
>    plot3d(z,0..10,0..10,axes=boxed,
     orientation=[ 140,68] ,shading=none);
```

While a visual inspection in this case reveals that the global minimum of $z(x, y)$
is at $(x, y) = (2, 2)$ with $z(0, 0) = 0$, the complicated nature of this type of

---

[10]In simulated annealing, the new point that is worse than the old one is not immediately discarded,
but it is kept with a specific probability.

function normally makes it very difficult to find the global optimum with the standard optimization methods.

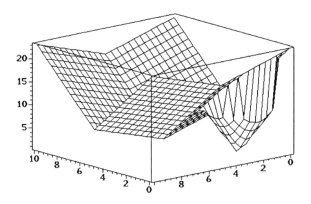

FIGURE 10.5. Graph of the $z(x, y)$ function.

To minimize $z(x, y)$, we now use ARS.

```
>   restart: # ARSTwoMinima.mws
>   with(stats): with(student):
    #readlib(randomize); randomize();
```

We let $N = 10000$ be the maximum number of random points that will be generated. Clearly, many of these points will be quickly discarded because they could be either infeasible or inferior to the last "good" point. The convergence criterion is $|z_n - z_{n-1}| < \varepsilon$ where z_n is the value of the objective function at the nth good point that improves the objective and $\varepsilon = 10^{-6}$.

```
>   N:=10000; epsilon:=10.^(-6);
```

$$N := 10000$$

$$\varepsilon := .1000000000\,10^{-5}$$

The search is performed over the set $\Xi = [0, 5] \times [0, 5]$ which is (presumably) known to contain the optimal solution. The initial point is generated randomly in the set Ξ and the initial value of the objective is set at $z = 999$.

```
>   xMin:=0: xMax:=5: yMin:=0: yMax:=5:
>   r:=random[ uniform] ();
```

$$r := .4274196691$$

```
>   xOpt:=xMin+r*(xMax-xMin);
```

$$xOpt := 2.137098346$$

```
>   r:=random[ uniform] ();
```

$$r := .3211106933$$

```
>    yOpt:=yMin+r* (yMax-yMin);
```

$$yOpt := 1.605553467$$

```
>    zOpt:=999;
```

$$zOpt := 999$$

The list $L_n = [x_{\text{Opt}}, y_{\text{Opt}}, z_{\text{Opt}}]$ is used to keep track of the points that correspond to progressively better values of the objective function.

```
>    n:=0;  L[ 0] :=[ xOpt,yOpt,zOpt] ;
```

$$n := 0$$

$$L_0 := [2.137098346, 1.605553467, 999]$$

The function $z(x, y)$ is defined as a procedure and we generate the new points randomly using $v = 3$. The search stops either (i) when the differences in the improved objective function values become smaller than ε or (ii) when N is reached.

```
>    z:=proc(x,y) if (x-2)^2+(y-2)^2<=4 then
     5* sqrt((x-2)^2+(y-2)^2) else
     3* abs(x-5)+abs(y-5)+3 fi end;
```

$z :=$ **proc**(x, y)

if $(x - 2)^2 + (y - 2)^2 \leq 4$ **then** $5 \times$ sqrt$((x - 2)^2 + (y - 2)^2)$

else $3 \times$ abs$(x - 5) +$ abs$(y - 5) + 3$

fi

end

```
>    for i from 1 to N do:
     r:=random[ uniform] ():
>    x:=xOpt+(xMax-xMin)* (2* r-1)^3:
     r:=random[ uniform] ():
>    y:=yOpt+(yMax-yMin)* (2* r-1)^3:
>    if (x<xMin or x>xMax or y<yMin or y>yMax) then
     next fi:
>    if z(x,y)>zOpt then next fi:
>    xOpt:=x; yOpt:=y; zOpt:=z(x,y);
     n:=n+1:
     L[ n] :=[ xOpt,yOpt,zOpt] ;
     Delta[ n] :=L[ n] -L[ n-1] ;
     if abs(Delta[ n][ 3] )<epsilon then break fi;
>    od:
```

Once the iterations end, we check the final values of n and i. The value of $n = 23$ indicates that out of 10000 iterations only 23 "good" points were found that improved the objective function. However, the value of $i = 10001$ signals that the convergence criterion was not yet satisfied. But, examining the values of L_n and $L_n - L_{n-1}$, we note that the solution found by ARS, i.e.,

$$x = 2.000044975, \quad y = 2.000078133, \quad \text{and} \quad z = .00045076369,$$

is reasonably close to the global optimum $(2, 2, 0)$.

```
>    n; i;
```

$$23$$

$$10001$$

```
>    L[ n] ;
```

$$[2.000044975, 2.000078133, .0004507636940]$$

```
>    L[ n] -L[ n-1] ;
```

$$[.000122649, -.000112829, -.0005800095080]$$

Following are some values of $[k, x_{Opt}, y_{Opt}, z_{Opt}]$ for $k = \lfloor 3n/4 \rfloor$ to n where we observe the quick convergence of the points generated by ARS.

```
>    ListxOptyOptzOpt:=seq([ k,[ L[ k][ 1] ,L[ k][ 2]] ,
     L[ k][ 3]] ,k=floor(3*n/4)..n);
```

ListxOptyOptzOpt := [17, [1.999697073, 2.000495347], .002903159562],
[18, [1.999688219, 2.000251433], .002002660390],
[19, [1.999688006, 2.000219307], .001906802404],
[20, [1.999848550, 2.000217551], .001325382399],
[21, [1.999892000, 2.000207601], .001170065973],
[22, [1.999922326, 2.000190962], .001030773202],
[23, [2.000044975, 2.000078133], .0004507636940]

```
>    x:=' x' ; y:=' y' ;
```

$$x := x$$

$$y := y$$

Improving values of the objective function can also be plotted using `point-plot()`. [The graph can be displayed after removing the # sign preceding the `pointplot()` command.]

```
>    with(plots):
>    ListzOpt:=seq([ k,L[ k][ 3]] ,k=floor(n/2)..n);
```

ListzOpt := [11, .006973344915], [12, .004455756688],
[13, .004023786549], [14, .003955266465], [15, .003893983199],
[16, .003338421731], [17, .002903159562], [18, .002002660390],
[19, .001906802404], [20, .001325382399], [21, .001170065973],
[22, .001030773202], [23, .0004507636940]

```
>    #pointplot([ ListzOpt] ,view=[ 1..n,-1..1] );
```

10.7   Summary

Many interesting and important problems in operations research cannot (yet) be solved analytically. For example, a large number of problems in queueing networks are not amenable to solution since the available mathematical theory is not sophisticated enough to find a closed-form solution for the operating characteristics of such systems. Simulation is a very flexible (albeit time-consuming) solution technique that has found wide applicability. We started this chapter by describing Maple's uniform random number generator. We also gave examples of using Maple's nearly 20 continuous and discrete random variate generators including the exponential, normal, binomial and Poisson. Next, Monte Carlo simulation was used to estimate the definite integral of a complicated function and to simulate the optimal policy in the "car and goats" TV game. Maple's random variate generators were then used to simulate an inventory system with random yield and a non-Markovian queue. The chapter ended with an optimization application where the global optimum of a discontinuous and nondifferentiable function was located using the method of adaptive random search.

10.8   Exercises

1. Two friends who have unpredictable lunch hours agree to meet for lunch at their favorite restaurant whenever possible. Neither wishes to eat at that particular restaurant alone and each dislikes waiting for the other, so they agree that:

   - Each will arrive at a random time between noon and 1:00 p.m.
   - Each will wait for the other either for 15 minutes or until 1:00 p.m. at the latest. For example, if one arrives at 12:50, he will stay until 1:00 p.m.

   (a) Use simulation to find the probability that the friends will meet.
   (b) Use your knowledge of probability to find the exact solution for this problem.

2. If $z_0 = 5$ and $z_{n+1} = 7z_n + 3 \mod 100$, find $z_1, z_2, \ldots, z_{20}$.

3. Consider two independent standard normal random variables $Z_1 \sim N(0, 1)$ and $Z_2 \sim N(0, 1)$ and let R and Θ denote the polar coordinates of the vector (Z_1, Z_2), i.e.,

$$R^2 = Z_1^2 + Z_2^2$$
$$\tan \Theta = \frac{Z_2}{Z_1}.$$

   (a) Show that the joint density of Z_1 and Z_2 is given by

$$f_{Z_1, Z_2}(z_1, z_2) = \frac{1}{2\pi} e^{-\frac{1}{2}(z_1^2 + z_2^2)}.$$

(b) To determine the joint density of (R^2, Θ) make the change of variables

$$s = z_1^2 + z_2^2$$
$$\theta = \tan^{-1}\left(\frac{z_2}{z_1}\right)$$

and show that

$$f_{R^2,\Theta}(s,\theta) = \left(\frac{1}{2}e^{-\frac{1}{2}s}\right) \times \frac{1}{2\pi}, \quad 0 < s < \infty, \quad 0 < \theta < 2\pi.$$

(c) Since R^2 is exponential with mean 2, and Θ is uniform over $(0, 2\pi)$, argue that the vector (Z_1, Z_2) can be generated using the formula

$$Z_1 = R\cos(\Theta) = \sqrt{-2\log U_1}\cos(2\pi U_2)$$
$$Z_2 = R\sin(\Theta) = \sqrt{-2\log U_1}\sin(2\pi U_2)$$

where U_1 and U_2 are random numbers between 0 and 1. This method is due to Box and Muller [36].

4. Consider the function $f(x) = 20x^2(1-x)^3$.

   (a) Find the area under $f(x)$ between 0 and 1.
   (b) Use Maple's int() function to find the exact solution for this problem.

5. Daily demand for the *National Post* newspaper at a particular newsstand in Toronto is randomly distributed as follows:

Number Sold	Probability
150	.05
175	.15
200	.35
225	.25
250	.20

The owner of the newsstand makes a profit of 40 cents on every paper sold, but loses 15 cents on every paper unsold. Use 100 days to simulate the owner's expected profit if he orders 150, 175, 200, 225 or 250 newspapers. What is the maximum order quantity and the maximum expected profit?

6. The standard Brownian motion (BM) $\{Z(t); t \geq 0\}$ is a continuous-time, continuous state-space stochastic process with the property that, (i) $Z(0) = 0$, (ii) $Z(t)$ has stationary and independent increments, and (iii) for every $t \geq 0$, $Z(t) \sim N(0, t)$, i.e., $Z(t)$ is normal with zero mean and variance t. This implies that the increment $\Delta Z(t)$ of the standard BM can

be written as $\Delta Z(t) = Y\sqrt{\Delta t}$ where $Y \sim N(0, 1)$, $E[\Delta Z(t)] = 0$ and $\text{Var}[\Delta Z(t)] = \Delta t$. Simulate the standard Brownian motion over the unit time interval $[0, 1]$ by taking a step size of $\Delta t = 0.001$ and plot the resulting sample path.

7. The function $f(x, y) = 100(y - x^2)^2 + (1 - x)^2$ is known as Rosenbrock's banana function since its contours near the minimum point resemble a banana.

   (a) Plot the surface of this function and its contours.

   (b) Use the adaptive random search method to find the minimum.

8. Consider a dam that generates electricity and supplies water for irrigation. Define X_n as the water level at the start of period n and let r_n be the amount of water released in period n. Let ξ_n be the random amount of rainfall during period n distributed as Poisson with parameter λ, i.e., $\Pr(\xi_n = k) = e^{-\lambda}\lambda^k / k!$. Assume that there is a dam capacity of M units.

   (a) Write a difference equation that relates X_{n+1} to X_n, r_n, ξ_n and M.

   (b) Let $\lambda = 5$, $M = 10$ and use the release policy as

   $$r_n = \begin{cases} X_n - m, & \text{if } m \leq X_n < M \\ 0, & \text{if } 0 \leq X_n < m \end{cases}$$

   with $m = 4$. [What is the relationship between this (m, M) policy and the (s, S) policy used in inventory modeling?] Assuming $X_0 = 10$, simulate the water level process for the next 100 days. What is the average water level in the dam using this policy?

References

[1] J. Abate and W. Whitt. Numerical inversion of probability generating functions. *Operations Research Letters*, 12:245–251, 1992.

[2] J. Abate and W. Whitt. Numerical inversion of Laplace transforms of probability distributions. *ORSA Journal on Computing*, 7:36–43, 1995.

[3] R. L. Ackoff. The meaning, scope and methods of operations research. In R. L. Ackoff, editor, *Progress in Operations Research*, volume I. John Wiley, New York, 1961.

[4] R. L. Ackoff and M. W. Sasieni. *Fundamentals of Operations Research*. John Wiley, New York, 1968.

[5] R. M. Adelson. Compound Poisson distributions. *Operational Research Quarterly*, 17:73–75, 1966.

[6] S. C. Aggarwal. A review of current inventory theory and its applications. *International Journal of Production Research*, 12:443–482, 1974.

[7] H. Anton. *Elementary Linear Algebra*. John Wiley, New York, 6th edition, 1991.

[8] T. M. Apostol. *Mathematical Analysis*. Addison-Wesley, Reading, Mass., 1969.

[9] K. J. Arrow. Historical background. In K. J. Arrow, S. Karlin, and H. Scarf, editors, *Studies in the Mathematical Theory of Inventory and Production*, pages 3–15. Stanford University Press, Stanford, California, 1958.

[10] K. J. Arrow. *Essays in the Theory of Risk-Bearing*. Markham, Chicago, 1971.

[11] K. J. Arrow, T. Harris, and J. Marschak. Optimal inventory policy. *Econometrica*, 19:250–272, 1951.

[12] J. Banks, J. S. Carson, and B. L. Nelson. *Discrete-Event System Simulation*. Prentice Hall, Upper Saddle River, N.J., 2nd edition, 1996.

[13] J. Banks and W. J. Fabrycky. *Procurement and Inventory Systems Analysis*. Prentice-Hall, Englewood Cliffs, N.J., 1987.

[14] S. K. Bar-Lev, M. Parlar, and D. Perry. Optimal sequential decisions for incomplete identification of group-testable items. *Sequential Analysis*, 14(1):41–57, 1995.

[15] E. Barancsi, G. Banki, R. Borloi, A. Chikan, P. Kelle, T. Kulcsar, and G. Meszena. *Inventory Models*. Kluwer Academic Publishers, Dordrecht, 1990.

[16] D. J. Bartholomew. *Stochastic Models for Social Processes*. John Wiley, Chichester, 1982.

[17] M. Baxter and A. Rennie. *Financial Calculus: An Introduction to Derivative Pricing*. Cambridge University Press, Cambridge, 1996.

[18] M. S. Bazaraa and C. M. Shetty. *Nonlinear Programming: Theory and Algorithms*. John Wiley, New York, 1979.

[19] M. J. Beckmann. *Dynamic Programming of Economic Decisions*. Springer-Verlag, Berlin, 1968.

[20] S. Beer. *Decision and Control*. John Wiley, New York, 1966.

[21] R. Bellman. On the theory of dynamic programming. *Proceedings of National Academy of Sciences*, 38:716–719, 1952.

[22] R. Bellman. Dynamic progamming and a new formalism in the calculus of variations. *Proceedings of the National Academy of Sciences*, 39:1077–1082, 1953.

[23] R. Bellman. *Dynamic Programming*. Princeton University Press, Princeton, N.J., 1957.

[24] R. Bellman. *Adaptive Control Processes: A Guided Tour*. Princeton University Press, Princeton, N.J., 1961.

[25] R. Bellman and K. L. Cooke. *Modern Elementary Differential Equations*. Dover, New York, 2nd edition, 1971.

[26] R. Bellman and S. Dreyfus. Dynamic programming and the reliability of multicomponent devices. *Operations Research*, 6:200–206, 1958.

[27] R. Bellman and R. E. Kalaba. Dynamic programming and statistical communication theory. *Proceedings of the National Academy of Sciences, USA*, 43:749–751, 1957.

[28] R. Bellman and R. E. Kalaba. On the role of dynamic programming in statistical communication theory. *IRE Transactions on Information Theory*, IT-3:197–203, 1957.

[29] A. Bensoussan, E. G. Hurst, and B. Näslund. *Management Applications of Modern Control Theory*. North-Holland, Amsterdam, 1974.

[30] D. P. Bertsekas. *Dynamic Programming: Deterministic and Stochastic Models*. Prentice-Hall, Englewood Cliffs, 1987.

[31] D. P. Bertsekas. *Dynamic Programming and Optimal Control*, volume I. Athena Scientific, Belmont, Massachusetts, 1995.

[32] U. N. Bhat. A controlled transportation queueing process. *Management Science*, 16(7):446–452, 1970.

[33] U. N. Bhat. *Elements of Applied Stochastic Processes*. John Wiley, New York, 2nd edition, 1984.

[34] F. Black and M. Scholes. The pricing of options and corporate liabilities. *Journal of Political Economy*, 81:637–659, May–June 1973.

[35] J. H. Bookbinder and D. L. Martell. Time-dependent queueing approach to helicopter allocation for forest fire initial attack. *INFOR*, 17:58–70, 1979.

[36] G. E. P. Box and M. F. Muller. A note on the generation of random normal deviates. *Annals of Mathematical Statistics*, 29:610–611, 1958.

[37] L. Breiman. Stopping-rule problems. In E. F. Beckenbach, editor, *Applied Combinatorial Mathematics*, pages 284–319. John Wiley, New York, 1964.

[38] R. S. Brooks and J. Y. Lu. On the convexity of the backorder function for an EOQ policy. *Management Science*, 15(7):453–454, 1969.

[39] A. E. Bryson and Y.-C. Ho. *Applied Optimal Control*. Halstead, New York, 1975.

[40] G. Carter, J. M. Chaiken, and E. J. Ignall. Response areas for two emergency units. *Operations Research*, 20:571–594, 1972.

[41] S. Çetinkaya and M. Parlar. Optimal nonmyopic gambling strategy for the generalized Kelly criterion. *Naval Research Logistics*, 44(4):639–654, 1997.

[42] S. Cetinkaya and M. Parlar. Nonlinear programming analysis to estimate implicit inventory backorder costs. *Journal of Optimization Theory and Applications*, 97(1):71–92, 1998.

[43] Y. S. Chow, H. Robbins, and D. Siegmund. *The Theory of Optimal Stopping*. Houghton Mifflin, Boston, 1971.

[44] C. W. Churchman. *The Systems Approach*. Delta, New York, 1968.

[45] C. W. Churchman, R. L. Ackoff, and E. L. Arnoff. *Introduction to Operations Research*. John Wiley, New York, 1958.

[46] V. Chvátal. *Linear Programming*. W. H. Freeman and Company, New York, 1983.

[47] E. Çınlar. *Introduction to Stochastic Processes*. Prentice-Hall, Englewood Cliffs, 1975.

[48] C. W. Cobb and P. H. Douglas. A theory of production. *American Economic Review*, 18(Supp. No. 2):139–165, 1928.

[49] L. Cooper and M. W. Cooper. *Introduction to Dynamic Programming*. Pergamon Press, New York, 1981.

[50] R. B. Cooper. *Introduction to Queueing Theory*. CEE Press, Washington, D.C., 1990.

[51] R. M. Corless. *Essential Maple*. Springer-Verlag, New York, 1995.

[52] D. R. Cox. The analysis of non-Markovian stochastic processes by the inclusion of supplementary variables. *Proceedings of Cambridge Philosophical Society*, 51:33–41, 1955.

[53] D. R. Cox and H. D. Miller. *The Theory of Stochastic Processes*. Chapman and Hall, London, 1965.

[54] D. R. Cox and W. L. Smith. *Queues*. Chapman and Hall, London, 1961.

[55] T. B. Crabill, D. Gross, and M. J. Magazine. A classified bibliography of research on optimal design and control of queues. *Operations Research*, 25:219–232, 1977.

[56] M. A. Crane and A. J. Lemoine. *An Introduction to the Regenerative Method for Simulation Analysis*. Springer-Verlag, New York, 1977.

[57] S. Danø. *Nonlinear and Dynamic Programming*. Springer-Verlag, New York, 1975.

[58] G. B. Dantzig. Maximization of a linear function of variables subject to linear inequalities. In T. C. Koopmans, editor, *Activity Analysis of Production and Allocation*, pages 339–347. John Wiley, New York, 1951.

[59] G. B. Dantzig. *Linear Programming and Extensions*. Princeton University Press, Princeton, N.J., 1963.

[60] M. D. Davis. *Game Theory: A Nontechnical Introduction*. Dover, Mineola, N.Y., 1983.

[61] N. Derzko, S. P. Sethi, and G. L. Thompson. Dirtributed parameter systems approach to the optimal cattle ranching problem. *Optimal Control Applications and Methods*, 1:3–10, 1980.

[62] R. L. Disney and P. C. Kiessler. *Traffic Processes in Queueing Networks: A Markov Renewal Approach*. John Hopkins University Press, Baltimore, MD, 1987.

[63] W. Edwards. Dynamic decision theory and probabilistic information processing. *Human Factors*, 4:59–73, 1962.

[64] R. M. Feldman and C. Valdez-Flores. *Applied Probability and Stochastic Processes*. PWS Publishing Company, Boston, 1996.

[65] W. Feller. *An Introduction to Probability Theory and its Applications*, volume II. John Wiley, New York, 2nd edition, 1971.

[66] G. S. Fishman. *Principles of Discrete Event Simulation*. Wiley, New York, 1978.

[67] D. A. Gall. A practical multifactor optimization criterion. In A. Lavi and T. P. Vogel, editors, *Recent Advances in Optimization Techniques*. John Wiley, New York, 1966.

[68] D. P. Gaver and G. L. Thompson. *Programming and Probability Models in Operations Research*. Brooks/Cole, Monterey, Calif., 1973.

[69] Y. Gerchak and M. Parlar. Yield randomness, cost tradeoffs and diversification in the EOQ model. *Naval Research Logistics*, 37:341–354, 1990.

[70] Y. Gerchak and M. Parlar. Investing in reducing lead-time randomness in continuous-review inventory models. *Engineering Costs and Production Economics*, 21:191–197, 1991.

[71] Y. Gerchak, M. Parlar, and S. S. Sengupta. On manpower planning in the presence of learning. *Engineering Costs and Production Economics*, 20:295–303, 1990.

[72] A. Ghosal, S. G. Loo, and N. Singh. *Examples and Exercises in Operations Research*. Gordon and Breach, London, 1975.

[73] P. E. Gill, W. Murray, and M. H. Wright. *Practical Optimization*. Academic Press, New York, 1981.

[74] R. Goodman. *Introduction to Stochastic Models*. Benjamin/Cummings, Menlo Park, 1988.

[75] B. S. Gottfried and J. Weisman. *Introduction to Optimization Theory*. Prentice-Hall, Englewood Cliffs, New Jersey, 1973.

[76] D. Gross and C. M. Harris. *Fundamentals of Queueing Theory*. John Wiley, New York, 2nd edition, 1985.

[77] R. L. Gue and M. E. Thomas. *Mathematical Methods in Operations Research*. Macmillan, London, 1968.

[78] Ü. Gürler and M. Parlar. An inventory problem with two randomly available suppliers. *Operations Research*, 45(6):1–15, 1997.

[79] G. Hadley. *Linear Programming*. Addison-Wesley, Reading, Mass., 1962.

[80] G. Hadley. *Nonlinear and Dynamic Programming*. Addison-Wesley, Reading, Mass., 1964.

[81] G. Hadley and T. M. Whitin. *Analysis of Inventory Systems*. Prentice-Hall, Englewood Cliffs, N.J., 1963.

[82] B. Harris. *Theory of Probability*. Addison-Wesley, Reading, Mass., 1966.

[83] F. W. Harris. How many parts to make at once. *Factory, The Magazine of Management*, 10:135–136, 1913.

[84] A. C. Hax and D. Candea. *Production and Inventory Management*. Prentice Hall, Englewood Cliffs, N.J., 1984.

[85] K. M. Heal, M. L. Hansen, and K. M. Rickard. *Maple V Learning Guide*. Springer-Verlag, New York, 1998.

[86] A. Heck. *Introduction to Maple*. Springer-Verlag, New York, 2nd edition, 1996.

[87] J. M. Henderson and R. E. Quandt. *Microeconomic Theory: A Mathematical Approach*. McGraw-Hill, New York, 1958.

[88] D. P. Heyman and M. J. Sobel. *Stochastic Models in Operations Research, Volume I: Stochastic Processes and Operating Characteristics*. McGraw-Hill, New York, 1982.

[89] D. P. Heyman and M. J. Sobel. *Stochastic Models in Operations Research, Volume II: Stochastic Optimization*. McGraw-Hill, New York, 1984.

[90] D. P. Heyman and M. J. Sobel, editors. *Stochastic Models*. Handbooks in Operations Research and Management Science. North-Holland, Amsterdam, 1990.

[91] F. B. Hildebrand. *Methods of Applied Mathematics*. Prentice-Hall, Englewood Cliffs, N.J., 2nd edition, 1965.

[92] F. S. Hillier and G. J. Lieberman. *Introduction to Operations Research*. Holden-Day, Oakland, Calif., 4th edition, 1986.

[93] C. C. Holt, F. Modigliani, J. F. Muth, and H. A. Simon. *Planning Production, Inventories, and Work Force*. Prentice-Hall, Englewood Cliffs, N.J., 1960.

[94] I. Horowitz. *Decision Making and the Theory of the Firm*. Holt, Rinehart and Winston, New York, 1970.

[95] R. A. Howard. Dynamic programming. *Management Science*, 12:317–348, 1966.

[96] J. P. Ignizio. *Linear Progamming in Single- and Multiple Objective Systems*. Prentice-Hall, Englewood Cliffs, N.J., 1982.

[97] R. B. Israel. *Calculus the Maple Way*. Addison-Wesley, Don Mills, Ontario, 1996.

[98] J. R. Jackson. Networks of waiting lines. *Operations Research*, 5:518–521, 1957.

[99] M. I. Kamien and N. L. Schwartz. *Dynamic Optimization: The Calculus of Variations and Optimal Control in Economics and Management*. North-Holland, New York, 1981.

[100] L. V. Kantorovich. *Mathematical Models of Organizing and Planning Production*. Leningrad State University, Leningrad, 1939. In Russian—English translation appeared in *Management Science*, 6, pp. 366-422, (1959–60).

[101] E. P. C. Kao. *An Introduction to Stochastic Processes*. Duxbury, Belmont, Calif., 1997.

[102] S. Karlin and H. M. Taylor. *A First Course in Stochastic Processes*. Academic Press, San Diego, 2nd edition, 1975.

[103] J. L. Kelly. A new interpretation of information rate. *Bell System Technical Journal*, 35:917–926, 1956.

[104] W. D. Kelton, R. P. Sadowski, and D. A. Sadowski. *Simulation with Arena*. McGraw-Hill, Boston, 1998.

[105] D. G. Kendall. Some problems in the theory of queues. *Journal of the Royal Statistical Society, Series B*, 13:151–185, 1951.

[106] D. G. Kendall. Stochastic processes occurring in the theory of queues and their analysis by the method of imbedded Markov chains. *Annals of Mathematical Statistics*, 24:338–354, 1953.

[107] M. Khouja. The single-period (news-vendor) problem: Literature review and suggestions for future research. *Omega*, 27:537–553, 1999.

[108] D. E. Kirk. *Optimal Control Theory*. Prentice-Hall, Englewood Cliffs, New Jersey, 1970.

[109] C. F. Klein and W. A. Gruver. Dynamic optimization in Markovian queueing systems. In E. Roxin and P. T. Liu, editors, *Kingston Conference on Differential Games and Control Theory*, pages 95–118. Marcel Dekker, New York, 1978.

[110] L. Kleinrock. *Queueing Systems, Volume 1: Theory*. John Wiley, New York, 1975.

[111] G. Klimek and M. Klimek. *Discovering Curves and Surfaces with Maple*. Springer-Verlag, New York, 1997.

[112] P. J. Kolesar, K. L. Rider, T. B. Crabill, and W. E. Walker. A queueing-linear programming approach to scheduling police patrol cars. *Operations Research*, 23:1045–1062, 1975.

[113] B. O. Koopman. Air-terminal queues under time-dependent conditions. *Operations Research*, 6:1089–1114, 1972.

[114] H. W. Kuhn and A. W. Tucker. Nonlinear programming. In J. Neyman, editor, *Proceedings of 2nd Berkeley Symposium on Mathematical Statistics and Probability*, Berkeley, 1951. University of California Press.

[115] P. J. M. Laarhoven and E. H. L. Aarts. *Simulated Annealing: Theory and Applications*. Kluwer Academic, Boston, 1987.

[116] M. S. Lane, A. H. Mansour, and J. L. Harpell. Operations research techniques: A longitudinal update 1973–1988. *Interfaces*, 23(2):63–68, 1993.

[117] R. E. Larson and J. L. Casti. *Principles of Dynamic Programming. Part I: Basic Analytic and Computational Methods*. Marcel Dekker, New York, 1978.

[118] A. M. Law and W. D. Kelton. *Simulation Modeling and Analysis*. McGraw-Hill, New York, 1982.

[119] H. Lee and C. A. Yano. Production control for multi-stage systems with variable yield losses. *Operations Research*, 36:269–278, 1988.

[120] H. L. Lee and S. Nahmias. Single-product, single-location models. In S. C. Graves, A. H. G. Rinnooy Kan, and P. H. Zipkin, editors, *Logistics of Production and Inventory*, pages 3–55. North-Holland, Amsterdam, 1993.

[121] B. Leonardz. *To Stop or Not to Stop: Some Elementary Optimal Stopping Problems with Economic Interpretations*. Almqvist and Wiksell, Stockholm, 1974.

[122] D. V. Lindley. The theory of queues with a single server. *Proceedings of Cambridge Philosophical Society*, 48:277–289, 1952.

[123] S. A. Lippman and J. J. McCall. The economics of job search: A survey. Part I. *Economic Inquiry*, 14:155–189, 1976.

[124] S. A. Lippman and J. J. McCall. The economics of job search: A survey. Part II. *Economic Inquiry*, 14:347–368, 1976.

[125] J. D. C. Little. A proof for the queueing formula $L = \lambda W$. *Operations Research*, 9:383–387, 1961.

[126] R. F. Love, J. G. Morris, and G. O. Wesolowsky. *Facilities Location: Models and Methods*. North-Holland, New York, 1988.

[127] D. G. Luenberger. *Linear and Nonlinear Programming*. Addison-Wesley, Reading, Mass., 1984.

[128] D. G. Luenberger. *Investment Science*. Oxford University Press, New York, 1998.

[129] J. Medhi. *Recent Developments in Bulk Queueing Models*. Wiley Eastern Limited, New Delhi, 1984.

[130] J. Medhi. *Stochastic Models in Queuing Theory*. Academic Press, New York, 1991.

[131] J. Medhi. *Stochastic Processes*. John Wiley, New York, 1994.

[132] M. B. Monagan, K. O. Geddes, G. Labahn, and S. Vorkoetter. *Maple V Programming Guide*. Springer-Verlag, New York, 1996.

[133] J. Mossin. Optimal multi-period portfolio policies. *Journal of Business*, 41:215–229, 1968.

[134] S. Nahmias. Inventory models. In A. Holzman J. Belzer and A. Kent, editors, *The Encyclopedia of Computer Sciences and Technology*, volume 9, pages 447–483. Marcel Dekker, New York, 1978.

[135] S. Nahmias. *Production and Operations Analysis*. Irwin, Homewood, Ill., 2nd edition, 1993.

[136] S. N. Neftçi. *An Introduction to Financial Derivatives.* Academic Press, San Diego, 1996.

[137] R. Nelson. *Probability, Stochastic Processes, and Queueing Theory: The Mathematics of Computer Performance Modeling.* Springer-Verlag, New York, 1995.

[138] G. L. Nemhauser. *Introduction to Dynamic Programming.* John Wiley, New York, 1966.

[139] Operations Research Center, MIT. *Notes on Operations Research 1959.* The Technology Press, Cambridge, Massachusetts, 1959.

[140] M. Parlar. A decomposition technique for an optimal control problem with "PQDZ" cost and bounded controls. *IEEE Transactions on Automatic Control,* AC-27(4):947–951, 1982.

[141] M. Parlar. Optimal dynamic service rate control in time dependent M/M/S/N queues. *International Journal of Systems Science,* 15(1):107–118, 1984.

[142] M. Parlar. EXPIM: A knowledge-based expert system for production/inventory modelling. *International Journal of Production Research,* 27(1):101–118, 1989.

[143] M. Parlar. Probabilistic analysis of renewal cycles: An application to a non-Markovian inventory problem with multiple objectives. *Operations Research,* 48(2), 2000.

[144] M. Parlar and Y. Gerchak. Control of a production system with variable yield and random demand. *Computers and Operations Research,* 16(4):315–324, 1989.

[145] M. Parlar and R. Rempała. Stochastic inventory problem with piecewise quadratic holding cost function containing a cost-free interval. *Journal of Optimization Theory and Applications,* 75(1):133–153, 1992.

[146] M. Parlar and R. G. Vickson. An optimal control problem with piecewise-quadratic cost functional containing a "dead-zone". *Optimal Control Applications and Methods,* 1:361–372, 1980.

[147] M. Parlar and Z. K. Weng. Designing a firm's coordinated manufacturing and supply decisions with short product life-cycles. *Management Science,* 43(10):1329–1344, 1997.

[148] A. L. Peressini, F. E. Sullivan, and J. J. Uhl. *The Mathematics of Nonlinear Programming.* Springer-Verlag, New York, 1988.

[149] S. R. Pliska. Management and optimization of queueing systems. In S. Özekici, editor, *Queueing Theory and Applications*, pages 168–187. Hemisphere Publishing Corporation, New York, 1990.

[150] E. L. Porteus. Numerical comparisons of inventory policies for periodic review systems. *Operations Research*, 33(1):134–152, January/February 1985.

[151] E. L. Porteus. Stochastic inventory theory. In Heyman and Sobel [90], pages 605–652.

[152] N. U. Prabhu. *Stochastic Processes*. Macmillan, New York, 1965.

[153] A. A. B. Pritsker, J. J. O'Reilly, and D. K. LaVal. *Simulation with Visual SLAM and AweSim*. John Wiley, New York, 1997.

[154] D. Redfern and E. Chandler. *Maple O.D.E. Book*. Springer-Verlag, New York, 1996.

[155] S. M. Roberts and J. S. Shipman. *Two-Point Boundary Value Problems: Shooting Methods*. American Elsevier, New York, 1972.

[156] S. Ross. *Applied Probability Models with Optimization Applications*. Holden Day, San Francisco, Calif., 1970.

[157] S. Ross. *Stochastic Processes*. John Wiley, New York, 1983.

[158] S. Ross. *Introduction to Probability Models*. Academic Press, Orlando, Fla., 3rd edition, 1985.

[159] S. M. Ross. *Introduction to Stochastic Dynamic Programming*. Academic Press, New York, 1983.

[160] S. M. Ross. *A Course in Simulation*. Macmillan, New York, 1991.

[161] T. L. Saaty. *Mathematical Methods of Operations Research*. McGraw-Hill, New York, 1959.

[162] A. P. Sage and C. C. White. *Optimum Systems Control*. Prentice Hall, Englewood Cliffs, N.J., 2nd edition, 1977.

[163] İ. Şahin. *Regenerative Inventory Systems: Operating Characteristics and Optimization*. Springer-Verlag, New York, 1990.

[164] M. Sasieni, A. Yaspan, and L. Friedman. *Operations Research: Methods and Problems*. John Wiley, New York, 1959.

[165] M. Vos Savant. *The Power of Logical Thinking*. St. Martin's Griffin, New York, 1996.

[166] H. Scarf. The optimality of (S, s) policies in the dynamic inventory problem. In K. J. Arrow, S. Karlin, and P. Suppes, editors, *Mathematical Methods in the Social Sciences*. Stanford University Press, Stanford, Calif., 1960.

[167] J. W. Schmidt and R. P. Davis. *Foundations of Analysis in Operations Research*. Academic Press, New York, 1981.

[168] L. B. Schwarz, editor. *Multi-Level Production/Inventory Control Systems: Theory and Practice*. Studies in the Management Sciences, Volume 16. North-Holland, Amsterdam, 1981.

[169] S. P. Sethi and G. L. Thompson. *Optimal Control Theory: Applications to Management Science*. Martinus Nijhoff Publishing, Boston, 1981.

[170] M. Shaked and J. G. Shantikumar. Reliability and maintainability. In Heyman and Sobel [90], pages 653–713.

[171] C. Shannon. A mathematical theory of communication. *Bell System Technical Journal*, 27:379–423, 1948.

[172] W. Shih. Optimal inventory policies when stockout results from defective products. *International Journal of Production Research*, 18:677–686, 1980.

[173] E. A. Silver. Establishing the reorder quantity when the amount received is uncertain. *INFOR*, 14:32–29, 1976.

[174] E. A. Silver. Operations research in inventory management: A review and critique. *Operations Research*, 29:628–645, 1981.

[175] E. A. Silver and R. Peterson. *Decision Systems for Inventory Management and Production Planning*. John Wiley, New York, 2nd edition, 1985.

[176] J. Singh. *Operations Research*. Penguin, Harmondsworth, England, 1968.

[177] B. D. Sivazlian and L. E. Stanfel. *Analysis of Systems in Operations Research*. Prentice-Hall, Englewood Cliffs, 1975.

[178] J. S. Smart. *The Sonnets of Milton*. Maclehose, Jackson, Glasgow, 1921.

[179] M. J. Sobel. Optimal average cost policy for a queue with start-up and shut-down costs. *Operations Research*, 17:145–162, 1969.

[180] S. Stidham. $L = \lambda W$: A discounted analogue and a new proof. *Operations Research*, 20:1115–1126, 1972.

[181] D. Stirzaker. *Elementary Probability*. Cambridge University Press, Cambridge, 1994.

[182] H. A. Taha. *Operations Research: An Introduction*. Macmillan, New York, 1987.

[183] L. Takács. *Stochastic Processes*. Methuen, London, 1966.

[184] H. M. Taylor. Optimal replacement under additive damage and other failure models. *Naval Research Logistics Quarterly*, 22:1–18, 1975.

[185] R. J. Tersine. *Principles of Inventory and Materials Management*. North Holland, New York, 3rd edition, 1988.

[186] H. C. Tijms. *Stochastic Modeling and Analysis*. John Wiley, Chichester, 1986.

[187] H. C. Tijms. *Stochastic Models: An Algorithmic Approach*. John Wiley, Chichester, 1994.

[188] A. F. Veinott. Optimal policy for a multi-product, dynamic nonstationary inventory problem. *Management Science*, 12:206–222, 1965.

[189] R. G. Vickson. Lecture notes in MS 635, 1993. Department of Management Sciences, University of Waterloo.

[190] H. M. Wagner. *Principles of Operations Research with Applications to Managerial Decisions*. Prentice-Hall, Englewood Cliffs, N.J., 1969.

[191] H. M. Wagner and T. M. Whitin. Dynamic version of the economic lot size model. *Management Science*, 5:89–96, 1958.

[192] A. Wald. *Sequential Analysis*. John Wiley, New York, 1947. Republished by Dover in 1973.

[193] Z. K. Weng and M. Parlar. Integrating early sales with production decisions: Analysis and insights. *IIE Transactions on Scheduling and Logistics*, 31(11):1051–1060, 1999.

[194] P. Wilmott. *Derivatives: The Theory and Practice of Financial Engineering*. John Wiley, Chichester, 1998.

[195] R. H. Wilson. A scientific routine for stock control. *Harvard Business Review*, 13:116–128, 1934.

[196] R. Wolff. *Stochastic Modeling and the Theory of Queues*. Prentice-Hall, Englewood Cliffs, N.J., 1989.

[197] C. A. Yano and H. L. Lee. Lot sizing with random yields: A review. *Operations Research*, 43:311–334, 1995.

[198] S. Zionts. *Linear and Integer Programming*. Prentice-Hall, Englewood Cliffs, N.J., 1974.

[199] P. H. Zipkin. *Foundations of Inventory Management.* McGraw-Hill, Boston, 2000.

Index

468   Index